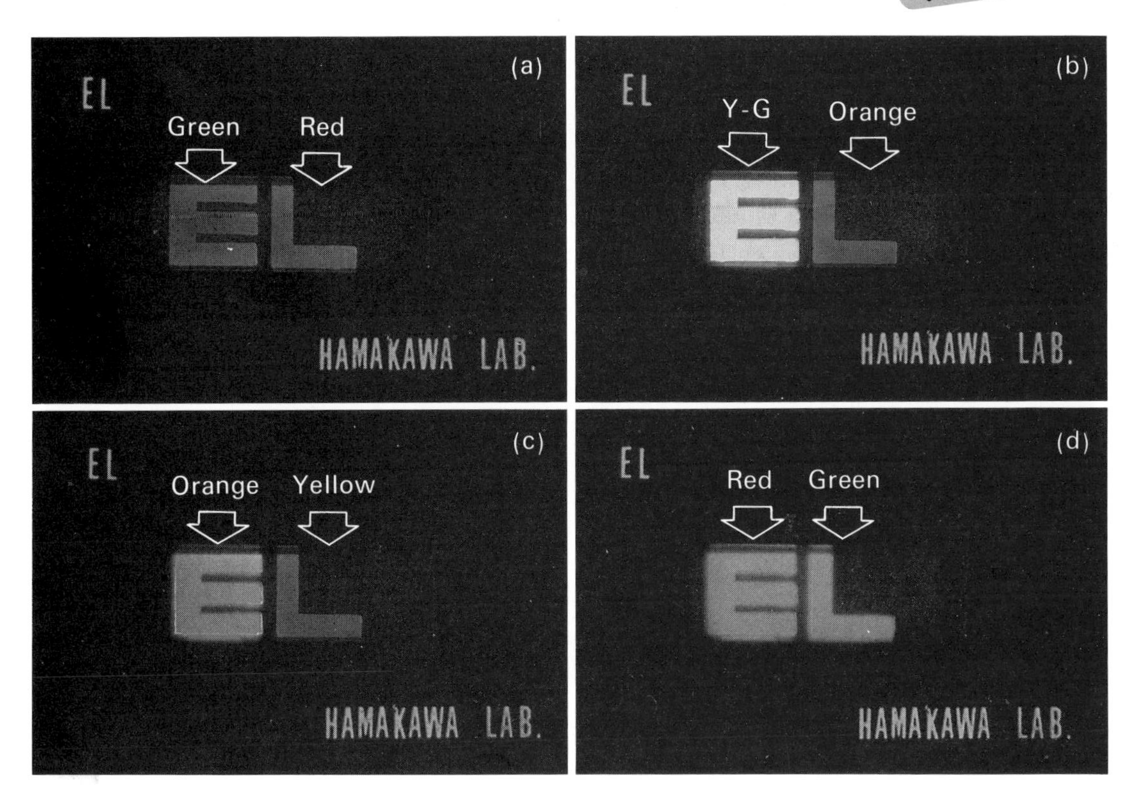

Photograph of the tunable-color electroluminescent (EL) device. Emission color spectra can be changed continuously from red to green by controlling color regulation voltage (0–20 V_{rms}) superposed with sustaining voltage (100 V_{rms}) as shown (see Chapter 18)

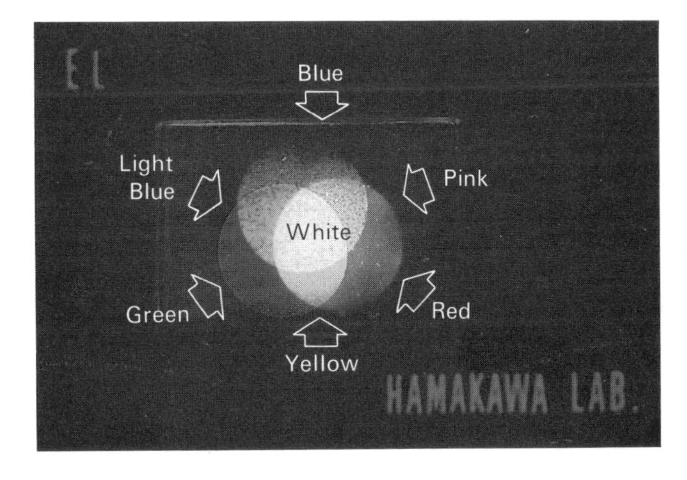

Desk pattern of color mixing for the full-color electroluminescent (EL) device. With this device any color emission can be displayed by controlling the color regulation voltage (see Chapter 18)

JAPAN ANNUAL REVIEWS IN ELECTRONICS
COMPUTERS & TELECOMMUNICATIONS
Vol. 19

SEMICONDUCTOR TECHNOLOGIES

Editor

Jun-ichi Nishizawa	*Professor of Research, Institute of Electrical Communication, Tohoku University, and Director of the Semiconductor Research Institute*

Advisory Committee

Masaharu Aoki	*Science University of Tokyo*
Yoshihiro Hamakawa	*Osaka University*
Akio Hiraki	*Osaka University*
Seijiro Furukawa	*Tokyo Institute of Technology*
Susumu Namba	*Osaka University*
Hisao Oka	*Mitsubishi Electric Corp.*
Takuo Sugano	*The University of Tokyo*
Toshinori Takagi	*Kyoto University*
Yoshiyuki Takeishi	*Toshiba Corp.*
Shoji Tanaka	*The University of Tokyo*
Yasuo Tarui	*Tokyo University of Agriculture and Technology*
Takashi Tokuyama	*Hitachi, Ltd.*
Michiyuki Uenohara	*Nippon Electric Co., Ltd.*
Makoto Watanabe	*Nippon Denshi-Gijutsu Co., Ltd.*
Masatami Yasufuku	*Fujitsu, Ltd.*

#15672117

JAPAN ANNUAL REVIEWS IN ELECTRONICS, COMPUTERS & TELECOMMUNICATIONS
Vol. 19

SEMICONDUCTOR TECHNOLOGIES

1986

Editor J. NISHIZAWA

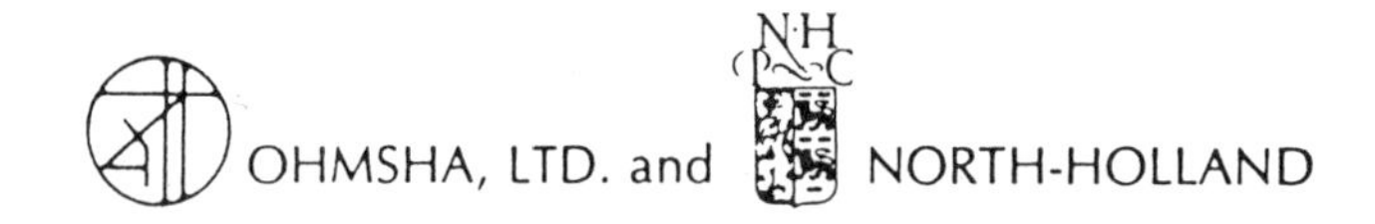

ISBN: 4-274-03112-8 (OHMSHA)
ISBN: 0-444-87934 X (NORTH-HOLLAND)
ISSN: 0167-5036

Co-published by
OHMSHA, LTD.
1-3 Kanda Nishiki-cho, Chiyoda-ku, Tokyo 101, Japan
Sole distributors for Japan
and
ELSEVIER SCIENCE PUBLISHERS B.V. (NORTH-HOLLAND)
P.O. Box 1991, 1000 BZ Amsterdam, the Netherlands
Sole distributors for outside Japan

Printed in The Netherlands

PREFACE

Before World War II, researchers in Japan and other countries studied semiconductor compounds, e.g., CdSe, ZnS, and ZnSe, mainly as photo-luminescent or electroluminescent materials.

After 1960, several research studies began on GaAs as a material for the Esaki diode, the variable capacitance diode and later for the Gunn diode. It is interesting to note that efforts to produce such diodes were effective in improving crystal quality, though the early diodes were not so much used. However, these studies failed to produce a stable material, and efforts are still being made today to improve the quality of the GaAs crystals.

Even after 1970, the science and technology of GaAs crystals lagged behind those of silicon. Moreover, even at that time many discussions centered on the question of whether dislocation-free silicon crystals were better than those with dislocations.

In 1969, a group at Tohoku University published an experimental result of stoichiometric characteristics as a function of arsenic vapor pressure in terms of heat treatment (Jpn. J. Appl. Phys. 8 (1969) 632). Nearly the same results were published soon thereafter by J. L. Moll, E. Munoz and W. L. Snydov. The findings were later extended by the Tohoku group (Jpn. J. Appl. Phys. 13 (1974) 64). These findings suggest that the equilibrium arsenic vapor pressure is higher than was usually mentioned till 1969.

The same effect was also found in the case of solution growth by the same Tohoku University group (Rept. Tech. Group on SSD, IECE of Japan, No. SSD-71-10 (1971) and J. Appl. Phys. 44 (1973) 1638) who called it the temperature difference method (TDM) under controlled vapor pressure (CVP). This report was quickly followed by one from K. Kaneko et al. (Proc. IEEE 61 (1973) 884) for polycrystalline growth named SSD (Synthesized Solute Diffusion Growth). By 1975, the Tohoku University group had successfully grown nearly-dislocation-free crystals by LPE.

The experimental results of the work by S. Akai et al., on GaAs melt growth based on the same idea, are introduced in this volume. The techniques described therein are responsible for the production of high quality GaAs crystals supplied around the world. The results of these discoveries were also adopted and applied to the Czochralski method in GaAs, as also included in this volume.

All observed phenomena went completely against the classical understanding that saturation solubility is independent from vapor pressure, and dependent only on temperature. However, this understanding needs to be amended such that the classical saturation solubility is applicable only at the start of crystal segregation. Subsequently, it is required to set the chemical potentials in the vapor phase described as a function of vapor pressure and in the liquid phase described as a function of solubility concentration, equal with the chemical potential in the solid phase described as a function of the shift of stoichiometry and the solubility concentration of impurities. This idea published by the Tohoku University group, and theoretically analyzed by Dr. A. I. Ivaschenko in Mordavia, USSR, is also introduced in this volume.

This volume contains several papers, as mentioned above, which introduce the

Japanese contribution to high quality III-V compound crystals followed by other papers introducing devices and processing works.

December, 1985 Jun-ichi Nishizawa

CONTENTS

1 HORIZONTAL BRIDGMAN GROWTH OF GaAs BY THREE-TEMPERATURE-ZONE TECHNIQUE

Shin-ichi AKAI and Takashi SUZUKI†

Abstract

The most up-to-date three-temperature-zone horizontal Bridgman technique (3T-HB)[1] for growing large-size substrate-grade GaAs single crystals is reviewed. The basic principle of the 3T-HB technique, the 3T-HB furnace system suitable for growth of large size GaAs single crystal, and the effect of residual oxygen on the concentration of Si and O (n_{Si} and n_O) are described. Semiconductor-laser-grade, low-dislocation substrates with ≤2-inch diameter, and solar-cell- and IC-grade substrates with ≤3-inch diameter are discussed in terms of residual impurities, dopants, EPD, and semi-insulating properties, all of which depend on the thermal growth conditions and doping conditions. Large-size HB-GaAs single crystals having both semiconducting and semi-insulating properties have been grown successfully with a mass of 8.8 kg, a width of 85 mm and a length of 650 mm. Two-inch diameter low-dislocation ($\leq 2000\ cm^{-2}$) GaAs substrates are available by heavily doping with Si or Zn. The best result obtained for 2- and 3-inch diameter semi-insulating GaAs is an average EPD of around $5000\ cm^{-2}$, and that obtained for a D-shaped wafer area of $18\ cm^2$ is less than $2000\ cm^{-2}$.

Keywords: Gallium Arsenide, Crystal Growth, Dislocation Dopants, Semi-insulator

1.1. Introduction

Since the early work by Weisberg et al.,[2] the HB technique including the gradient freezing technique (GF)[3] has been traditionally utilized as the most economical technique for growing substrate-grade GaAs single crystals, while the liquid encapsulated Czochralski (LEC) technique has been enthusiastically investigated because it provides an easy way to obtain so-called "undoped" semi-insulating GaAs, since the two-step LEC/pBN method[4] and the one-step LEC/pBN method[5] have been successfully applied to grow "undoped" semi-insulating GaAs following the LEC/Al_2O_3 method first reported by Weiner et al.[6]

Optoelectronic devices such as lasers and avalanche photodiodes require substrates with low defect density, e.g. etch pit density (EPD) ≤ 1000–$2000\ cm^{-2}$ in highly Si-doped n^+-GaAs substrates or, more recently, highly Zn-doped p^+-GaAs substrates.[7,8] High-performance FETs, and OE-ICs which are integrated from lasers and/or avalanche photodiodes plus electrical devices, and more recently high-electron-mobility transistor (HEMT) ICs[9,10] also require semi-insulating substrates with low defect density (EPD

† Sumitomo Electric Industries, Ltd., 1-1-3 Shimaya, Konohana-Ku, Osaka 554.

JARECT Vol. 19, Semiconductor Technologies (1986), *J. Nishizawa (ed.)*
OHMSHA, LTD. and North-Holland

$\leqslant 10^4$ cm^{-2} or preferably $\leqslant$1000–2000 cm^{-2}). Generally speaking, currently available LEC GaAs[11] samples with diameters larger than 2 inches have higher EPD than 10,000 cm^{-2}. Therefore, the HB technique has great practical importance for obtaining low-defect-density substrates at present.

The purpose of this chapter is to review the most up-to-date 3T-HB technique suitable for growing high-quality substrate-grade GaAs single crystals. First, the principles and basic procedure of this technique are summarized (Section 1.2). Then, the effect of residual oxygen on the concentration of Si and O (n_{Si} and n_O) is analyzed and compared with the experimental results (Section 1.3). The application of the 3T-HB technique to grow large-size GaAs single crystals is described in Section 1.4: high-quality semiconducting GaAs is discussed in Section 1.4.1 and semi-insulating GaAs in Section 1.4.2.

1.2. Principles and procedure of the 3T-HB technique

There are three kinds of boat growth techniques: (a) two-temperature-zone HB technique (2T-HB), (b) three-temperature-zone HB technique (3T-HB), and (c) gradient freezing technique (GF). A diffusion barrier provided between the high-temperature zone (T_1) and the lower-temperature zone is essential[12] to prevent gaseous products (SiO, Ga_2O) generated in the T_1 zone from diffusing out, which results in the well-known phenomena of wetting and Si contamination from the quartz boat. Single crystals grow from the seed end to the tail end by movement of heaters or fused quartz reaction tube (HB techniques) or by decrease of temperature in the T_1 zone (GF technique). In the case of the 2T-HB technique, the temperature (T_2) of the diffusion barrier decreases as the single-crystal growth proceeds, which results in condensation of gaseous products (SiO, Ga_2O)[12] at the diffusion barrier; therefore, the 2T-HB technique is not applicable for growing long GaAs single crystals.

According to the 3T-HB technique,[13,14,15] the temperatures T_1, T_2, and T_3 are kept in the ranges 1245–1270°C, 1080–1220°C, and 605–620°C, respectively.

In equilibrium conditions, the following reactions take place:[12]

$$4\mathrm{Ga(\ in\ Ga\text{–}As\ melt)} + \mathrm{SiO_2(s)} = 2\mathrm{Ga_2O(g)} + \mathrm{Si(in\ Ga\text{–}As\ melt)}, \tag{1.1}$$

$$\mathrm{Si(in\ Ga\text{–}As\ melt)} + \mathrm{SiO_2(s)} = 2\mathrm{SiO(g)}, \tag{1.2}$$

and

$$\mathrm{As_4(g)} = 2\mathrm{As_2(g)}, \tag{1.3}$$

in the T_1 zone,

$$3\mathrm{Ga_2O(g)} + \mathrm{As_4(g)} = \mathrm{Ga_2O_3(s)} + 4\mathrm{GaAs(s)}, \tag{1.4}$$

and

$$\mathrm{SiO(g)} = \mathrm{SiO(s)}, \tag{1.5}$$

in the T_2 zone, and

$$4\mathrm{As(s)} = \mathrm{As_4(g)} \tag{1.6}$$

in the T_3 zone. According to the phase rule, the system is bivariant in the T_1 zone and univariant in the T_3 zone. In the T_2 zone, the condensing phase consists of Ga_2O_3 and GaAs under the condition of very low Si activity in the GaAs melt (highly oxygen doping condition): that is, the reaction (1.4) is dominant. On the contrary, SiO condenses under the condition of very high Si activity like the case of heavily doping of Si: that is, the reaction (1.5) is dominant. In both cases, the system is bivariant in the T_2 zone. If the temperature T_3 is kept constant, the vapor pressure of As_4 is controlled to be constant. The state of the T_2 zone is determined by the temperature T_2 and the arsenic vapor pressure under conditions of high oxygen or Si doping, that is the vapor pressure of Ga_2O or SiO is determined. Therefore, the state of the whole system is determined by keeping both the temperatures T_2 and T_3 constant in these cases. This is the basic principle of the 3T-HB technique. However, Si activity in the GaAs melt is usually selected between the above two cases. The equilibrium Si activity is, therefore, controlled by the initial doping condition.

The effect of the T_2 zone on the Si activity and Si concentration (n_{Si}) is illustrated in Fig. 1.1.[14,15] If we choose the initial condition of Si activity $(A_{Si}) < A_L$ or $A_{Si} > A_U$ then the reactions (1.4) or (1.5) proceed until A_{Si} reaches A_L or A_U,which results in the wetting phenomena of the melt with the quartz boat. As the temperature T_2 is lowered, A_L becomes higher while A_U becomes lower; therefore, if T_2 is lower than T_2^c (1045°C), there is no region of $A_L < A_{Si} < A_U$. In the practical 3T-HB technique, the temperature T_2 is controlled at a value between 1080 and 1220°C.

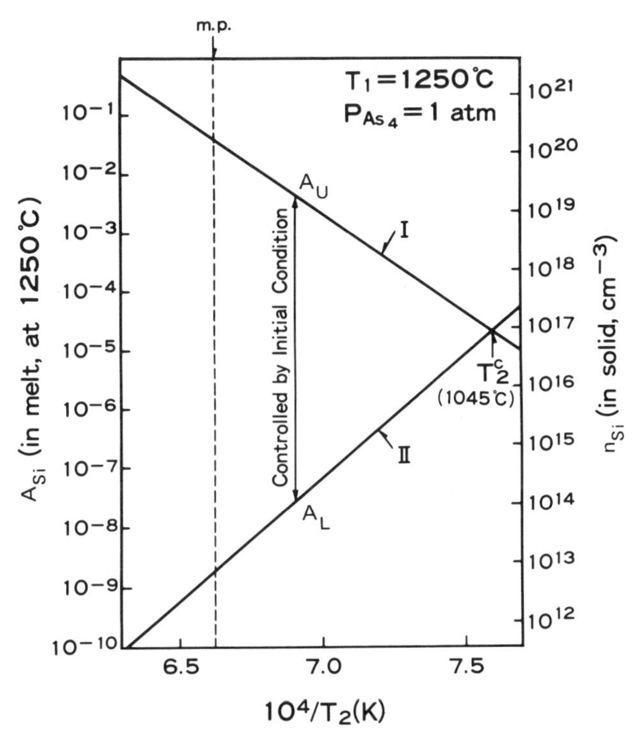

Fig. 1.1 Effect of the intermediate temperature (T_2) on the Si concentration in the 3T-HB growth system. A_{Si} is the Si activity in the GaAs melt, and n_{Si} is the Si concentration in the grown crystals. T_2^c is the minimum intermediate temperature for the 3T-HB technique.

The most up-to-date 3T-HB furnace is designed to make it easy to control the temperature profiles from top to bottom as well as along the longitudinal axis and to balance the temperatures at the right and left sides in the furnace, and also make it possible to program the variation of the temperature with time.

Fig. 1.2 shows a schematic representation of such a 3T-HB furnace and a cross-sectional diagram of the furnace in the vicinity of the growth interface. It is essential to control the longitudinal temperature gradient between 1 and 5°C/cm in order to grow 2-inch or larger diameter single crystals of GaAs with good reproducibility. A multi-section heater and after heater make this possible. The multi-section heater is useful to keep the liquid–solid interface flat. Heaters H_3 and H_4 change the growth interface from

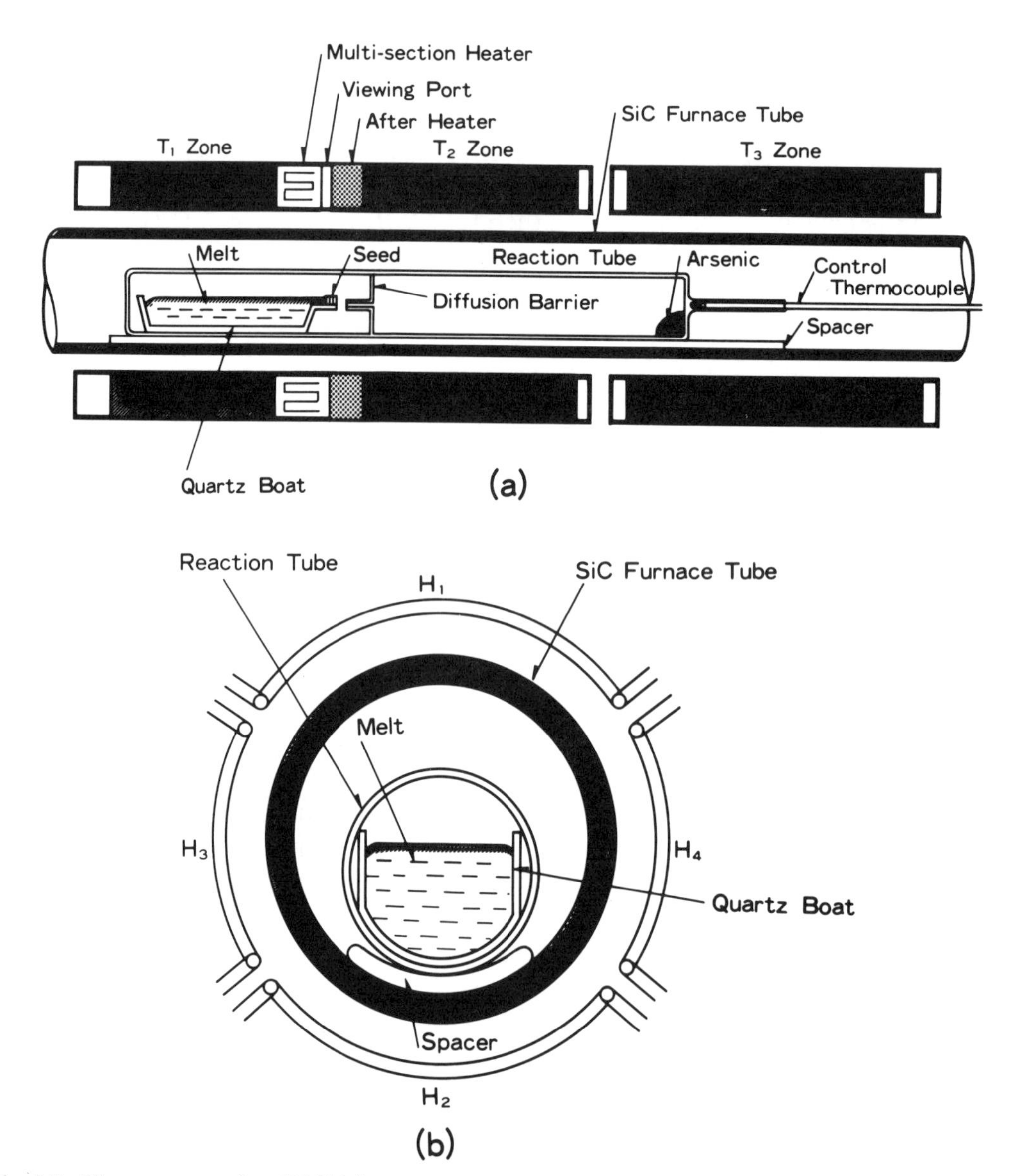

Fig. 1.2 The most up-to-date 3T-HB furnace system (a) and cross-section of four-section heater in the vicinity of growth interface (b).

Table 1.1 Basic growth parameters of the 3T-HB technique.

Growth Parameters	
Method	The 3T-HB Technique
T_1	1245−1270℃
T_2	1080−1220℃
T_3	605− 620℃
$\Delta T/\Delta Z$	1 − 5℃/cm
$\Delta T/\Delta H$	1 − 5℃/cm
Growth Axis	〈111〉
Growth Rate	2 − 5 mm/hr

concave to convex. The top portion is controlled at a lower temperature than the bottom portion by adjusting the heat output of the H_1 and H_2 sections. The vertical temperature gradient is selected at a value between 1 and 5°C/cm at the growth interface. Table 1.1 summarises the basic growth parameters of the 3T-HB technique.

Regarding the vapor pressure P_{As_4} in the T_3 zone which determines the P_{As_4} and P_{As_2} in the T_1 zone, we must choose the optimum vapor pressure carefully. If we keep $T_3 = 614.5°C$, then $P_T(=P_{As_4}+P_{As_2})$ in the T_1 zone is controlled at 0.976 atm which results in the stoichiometric melt[16] at the liquid–solid interface. Usually, values of P_T slightly higher than 1.0 atm give us practically optimum conditions by maintaining $T_3 \geqslant 616°C$.[14,15] Nishizawa et al.[17] have determined the optimum P_T, namely

$$P_T(\text{optimum}) \fallingdotseq 2.6\times 10^6 \exp(1.05/kT)(\text{Torr}), \tag{1.7}$$

by the temperature-difference LPE method under controlled vapor pressure. If we extrapolate this equation to the liquidus curve,[16] we obtain P_T (optimum) = 1.067 atm, N_{As} (mole fraction of As in the melt) = 0.514, and $T_3 = 618°C$. Parsey et al.[18] have also determined the P_T (optimum) corresponding to $T_3 = 617°C$ using a precision Bridgman-

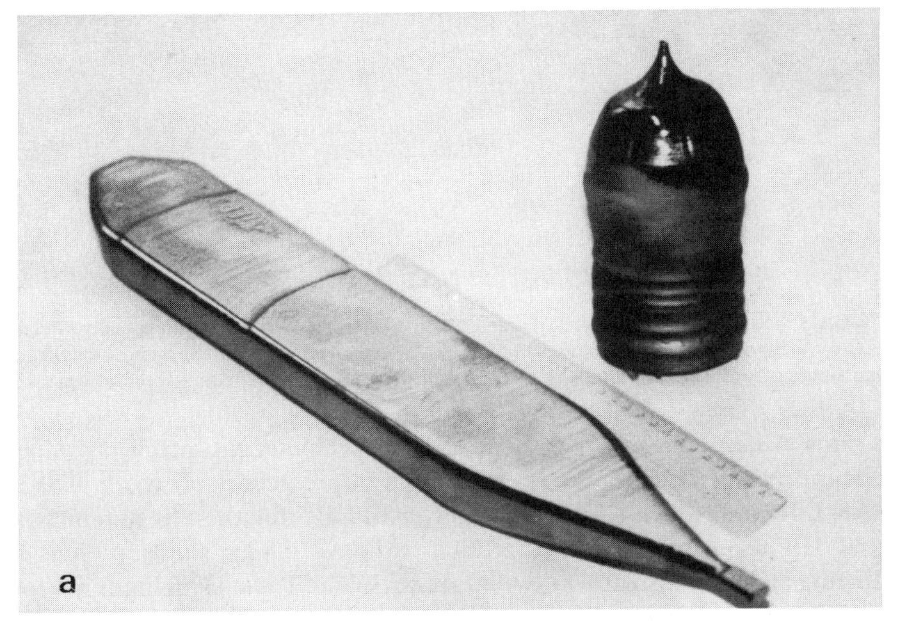

Fig. 1.3(a).

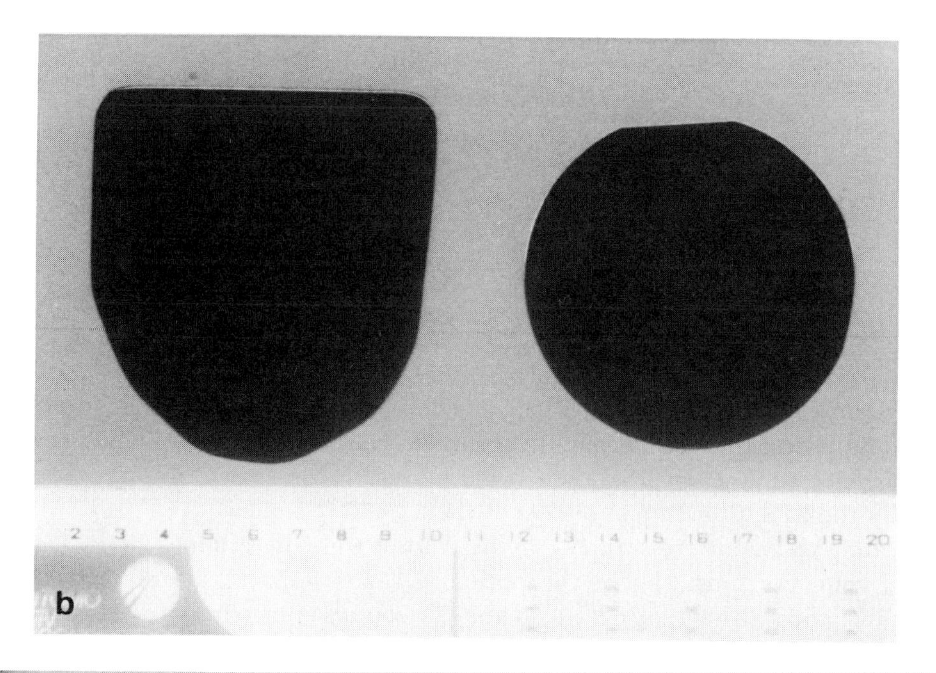

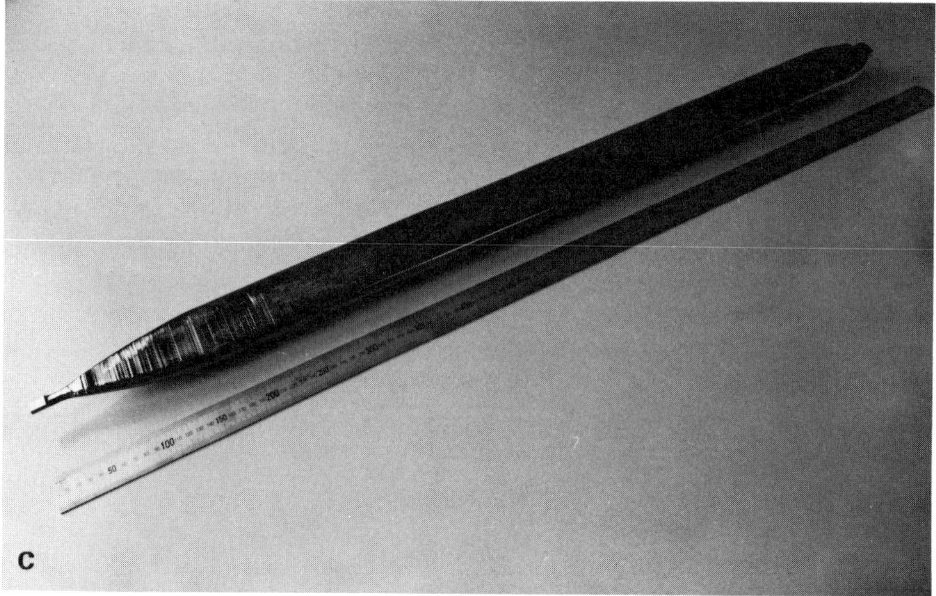

Fig. 1.3 HB-GaAs ingot designed for 3-inch diameter round wafers shown with commercially available 3-inch diameter LEC-GaAs ingot (a), U-shaped wafer and 3-inch diameter round wafer cut from the 8.8 kg ingot (b), and very long single crystal 1 m long and 8 kg in weight designed for 2-inch diameter round wafers (c).

type apparatus. A more recent experiment on the stochiometry-controlled compensation in liquid encapsulated Czochralski GaAs[19] gives a prediction of $N_{As} \fallingdotseq 0.51$ and P_T (optimum) $\fallingdotseq 1.043$ atm which correspond to $T_3 = 617.6°C$ for the HB system.

The 3T-HB technique makes it possible to grow 8.8 kg single crystals of both semiconducting and semi-insulating GaAs as shown in Fig. 1.3(a). This ingot has a width of 85 mm and a length of 650 mm. We can cut 3-inch diameter round wafers from this ingot as shown in Fig. 1.3(b). Moreover, this technique can grow very long single crystals ($\leqslant 1$ m) as shown in Fig. 1.3 (c).

1.3. Control of Si contamination from quartz boat

GaAs crystals grown in quartz boats under the ideally undoped condition are generally contaminated with Si of (5-10) $\times 10^{16}$ cm^{-3}.[2,20] The effect of oxygen doping on the residual Si concentration has been investigated for the HB growth system.[20,21] Even semi-insulating (SI) GaAs can be obtained by oxygen doping, but it is generally difficult to grow SI GaAs with good crystalline quality because of sticking.[21] Usually, the quartz boat is sand blasted with about 50 μm Al_2O_3 or SiC powder to prevent the melt from wetting the boat.[18,20,22] The 3T-HB technique along with the addition of As_2O_3 is especially useful to reduce Si contamination and also to prevent wetting.[15]

Oxygen is the other main residual impurity. However, little investigation has been performed on incorporation of oxygen into HB-GaAs.[21] Borisova et al. have measured the solubility limit of oxygen in GaAs crystal and the distribution coefficient of oxygen at 700–1200°C.[23] Akai et al. have shown a thermodynamic model to calculate the concentrations of Si (n_{Si}) and oxygen (n_O) as a function of added quantity of Si or O_2 taking as a parameter the residual oxygen concentration in the 3T-HB growth system[24] using extrapolated values of the solubility limit ($n_O^{sat} \simeq 1 \times 10^{18}$ cm^{-3}) and distribution coefficient ($k_O \simeq 0.13$) by Borisova et al.

The theoretical concentrations, n_{Si} and n_O, have been calculated as a function of N_{Si}^0 or $N_{O_2}^0$ in ref. 24. Here, we show the result for n_{Si} in Fig. 1.4. Akai et al.[24] have selected the ratio $v^{-1} = V_g/V_m$ (free volume over the melt/melt volume) of about 4.5 from experimental conditions. N_{Si}^0 and $N_{O_2}^0$ are defined by

$$N_{Si}^0 = x_{Si}^0/(V_m\rho/M) \tag{1.8}$$

and

$$N_{O_2}^0 (= \tfrac{3}{2} N_{As_2O_3}^0) = x_{O_2}^0/(V_m\rho/M), \tag{1.9}$$

where x_A^0 (mole) is the quantity of added substance A (Si, As_2O_3) or residual oxygen (oxide film on raw materials and absorbed oxygen or oxide gas on wall of reaction tube), ρ is the density of the melt, and M is the average atomic weight of Ga and As. Fig. 1.4 shows the measured carrier concentration (n_{Hall}) and Si concentration (n_{Si}) of HB GaAs weighing 1.5–3.0 kg, designed for 1.6–2.0 inch diameter round wafers and doped with Si or As_2O_3 (oxygen dopant).

The theoretical background level of n_{Si}, i.e. n_{Si}^0, is 1.24×10^{16} cm^{-3} under the ideally undoped condition, and that of n_O, i.e. n_O^0 is 1.70×10^{16} cm^{-3} for v^{-1} of 4.5. The value of n_{Si}^0 varies with $v^{-1} = V_g/V_m$ from 4.5×10^{16} cm^{-3} ($v^{-1} = 100$) to 1×10^{16} cm^{-3} ($v^{-1} \leqslant 1$). Fig. 1.4 also shows that SI GaAs can be grown by oxygen doping[20,21,25] under the condition of high diffusion barrier temperature. The value n_{Si}^L (T_2) in Fig. 1.4 is the theoretical lower limit of n_{Si} when the temperature of the diffusion barrier is T_2.[15]

As_2O_3-doped GaAs grown in the 3T-HB system has lower values of n_{Si} and n_{Hall} at the tail end than at the seed end, though the segregation coefficient (k_{Si}) is less than unity (0.14). It is believed that this is caused by segregation of oxygen at the tail end and the resultant decrease of Si concentration in the melt by the reaction of Si and oxygen to form

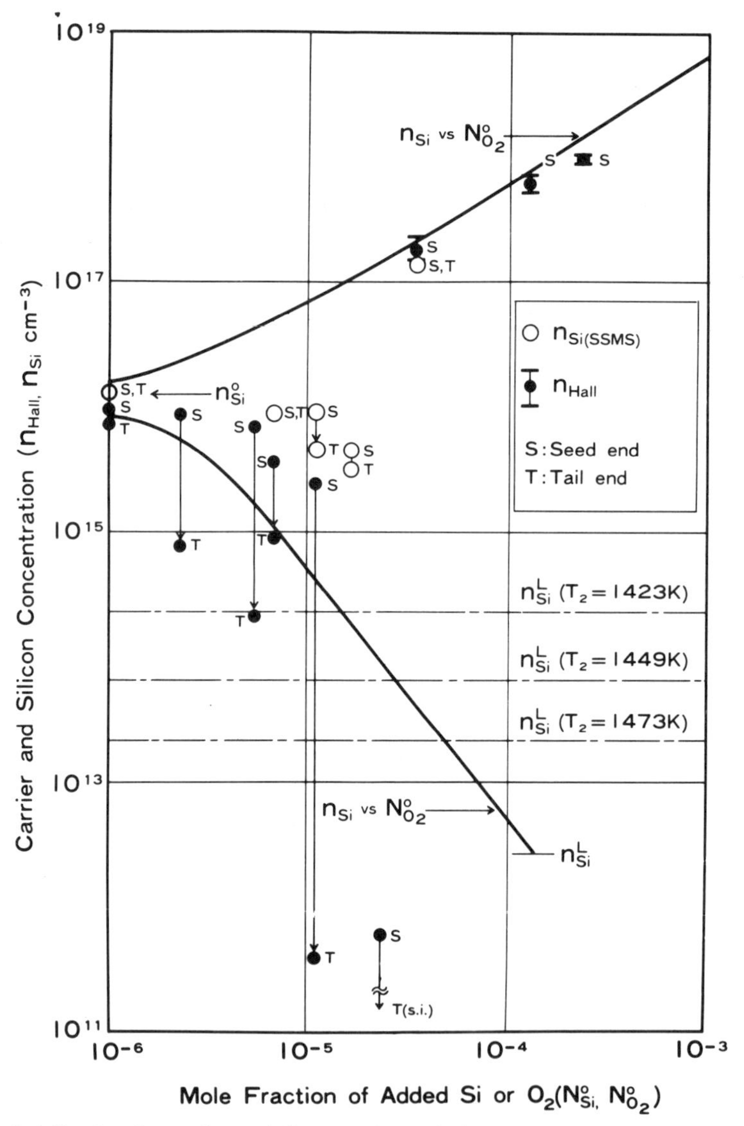

Fig. 1.4 Theoretical (lines) and experimental (dots) carrier and Si concentration as function of mole fraction of added Si or O_2 for the 3T-HB growth system. Oxygen is added in the form of As_2O_3.

SiO_2 or SiO. The increased partial pressure of Ga_2O ($P_{Ga_2O} \propto N_O$ in the melt) is maintained because of the high diffusion-barrier temperature (T_2).

Typical spark source mass spectrometry analysis (SSMS), carrier concentrations, and electron Hall mobilities of undoped and oxygen-doped GaAs ingots (8.8 kg) designed for 3-inch diameter round wafers are shown in Table 1.2. As_2O_3-doped GaAs crystals contain residual Si of $<1\times10^{16}\ cm^{-3}$, which was also confirmed from graphite furnace atomic absorption spectrometry (GFAA), while undoped GaAs has residual Si of $(1\text{–}2)\times10^{16}\ cm^{-3}$, which agrees with the theoretical value. The oxygen concentration of GaAs doped with As_2O_3 ($n^0_{O_2}=6.7\times10^{-6}$) is measured as $(2\text{–}3)\times10^{17}\ cm^{-3}$ by secondary ion mass spectrometry (SIMS), which is higher than the theoretical value of $6\times10^{16}\ cm^{-3}$. The SIMS data, however, show an upper limit of n_O because of the high oxygen background

Table 1.2 SSMS data and electrical properties of undoped and oxygen-doped GaAs.

Crystal		Undoped	Oxygen Doped
Designed Value	N_{Cr} (wt-ppm)	0	0
	$N_{O_2}^{\circ}$	0	2.2×10^{-5}
SSMS analysis (ppma)	B	ND	0.03
	Si	0.49	0.1
	Cr	0.05	0.01
	Cu	0.01	ND
Carrier Conc.	(cm^{-3})	2.4×10^{16}	2.0×10^{12}
Hall Mobility	($cm^2/v \cdot s$)	4.6×10^3	5.0×10^3
Resistivity	(Ω-cm)	5.7×10^{-2}	6.3×10^2

ND: non-detected

[(2–3) × 10^{17} cm^{-3}] during the analysis.[24] The carrier concentration decreases remarkably with the oxygen doping and the electron mobility increases with oxygen doping.[21]

Growth of single crystals by the above 3T-HB method has been carried out using pre-synthesized high-purity polycrystals, which have a carrier concentration of $\leq 5 \times 10^{15}$ cm^{-3} at the front end and $\leq 10^{12}$–10^{15} cm^{-3} at the tail end, and an electron Hall mobility of (5–6) × 10^3 cm^2/Vs, as the starting material.

1.4. Growth and characterization of low-dislocation GaAs

1.4.1. *Semiconducting GaAs*

A growth technique of dislocation-free-grade (≤ 100 cm^{-2}) GaAs with practical size (≥ 10 cm^2) has been developed[26] by combining optimization of the thermal growth conditions and impurity hardening phenomenon.[27] From a practical point of view, large and usable (i.e. without microprecipitates) low-defect GaAs ingots are only available in Si-doped HB GaAs.[28,29]

Roughly speaking, the longitudinal temperature gradient (dT/dz) should be smaller than $(dT/dz)_c$, namely

$$(dT/dz)_c = K\tau_c/D_c \qquad (1.10)$$

in order to obtain dislocation-free GaAs, where τ_c is the critical shear stress, D_c is the crystal diameter, and K is nearly constant.

Swaminathan et al.[30] have shown that Si-doped GaAs has a greater yield stress than undoped GaAs and that Cr doping does not significantly affect the yield stress at temperatures ranging from 250–550°C.

Fig. 1.5 shows the EPD of Si-doped HB-GaAs with (100) wafer area of about 10 cm^2 as a function of carrier concentration for two different thermal conditions. The liquid–solid interface in case (b) is considered more planar than in case (a). Case (c) shows the result recently obtained for smaller crystals with (100) wafer area of ≤ 3 cm^2.[31] Parsey et al. have also succeeded in growing small undoped GaAs crystals (10 × 8 mm^2) with dislocation density less than 500 cm^{-2} by precise control of the arsenic vapor pressure.[18,32]

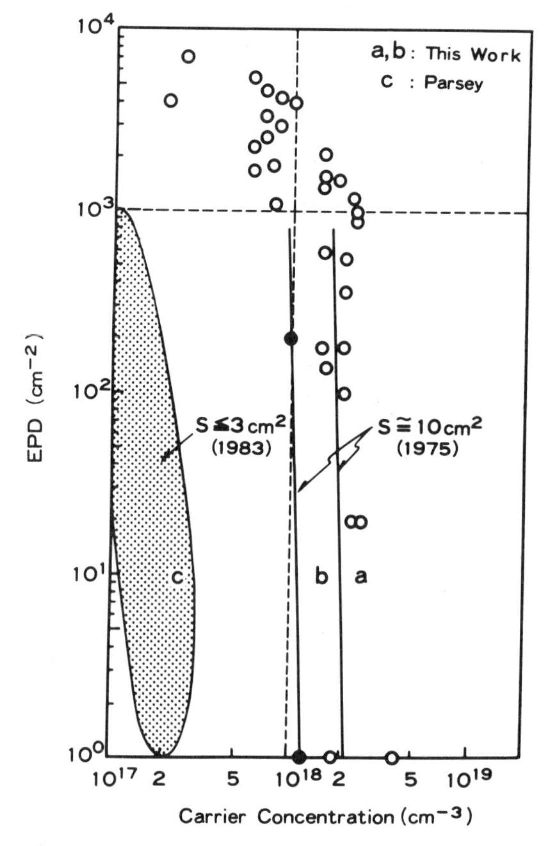

Fig. 1.5 EPD of Si-doped HB-GaAs with (100) wafer area of about $10\,cm^2$ as a function of carrier concentration for two different thermal conditions. The liquid–solid interface in case (b) is considered more planar than in case (a). Case (c) shows the result recently obtained for smaller crystals with (100) wafer area of less than $3\,cm^2$.[31]

Fig. 1.6 Etch pit pattern (molten KOH etch) and EPD distribution of 2-inch diameter low-EPD Si-doped GaAs wafer.

Fig. 1.6 shows an etch pit pattern (molten KOH etch) and EPD distribution of 2-inch diameter low-EPD Si-doped GaAs. The cut-off regions at the upper corners have a dislocation density of about $5 \times 10^3\,cm^{-2}$. At present, it is difficult to grow low-EPD Si-doped GaAs with a diameter of 3 inches. The average EPD of 3-inch diameter Si-doped GaAs is about $3 \times 10^3\,cm^{-2}$ at the seed end and $(5\text{–}10) \times 10^3\,cm^{-2}$ at the tail end [see also Fig. 1.8(a)]. Recently, low-EPD GaAs doped with Si, and with $45 \times 45\,mm^2$ wafer area has also been grown by the GF technique,[33] resulting in a crystal having an average EPD of $500\,cm^{-2}$ except for the peripheral region in which the EPD is about $5 \times 10^3\,cm^{-2}$.

1.4.2. *Semi-insulating GaAs*

At present, it is difficult to grow low-EPD semi-insulating GaAs of practical size.[11] For example, a typical Cr-doped GaAs with (100) area of $18\,cm^2$ and designed for 40 mm diameter round wafers has a dislocation density of $(1\text{–}5) \times 10^3\,cm^{-2}$ at the seed end and $(5\text{–}15) \times 10^3\,cm^{-2}$ at the tail end. The EPD of Cr-doped GaAs is independent of the Cr concentrations ranging from 0.1 to about 5 wt ppm.

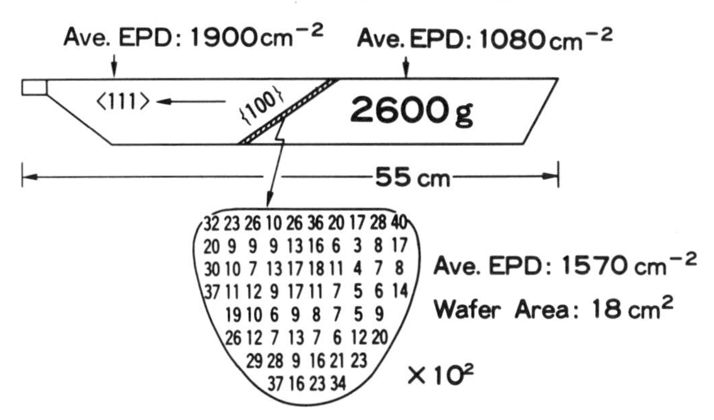

Fig. 1.7 EPD distribution obtained for semi-insulating GaAs doped with Cr and oxygen and with (100) wafer area of 18 cm^2. Average EPDs are less than 2×10^3 cm^{-2} throughout the ingot.

Fig. 1.7 shows one of the best result of EPD obtained for SI GaAs doped with Cr and oxygen, and with (100) area of 18 cm^2. Average EPDs are less than 2×10^3 cm^{-2} throughout the ingot. The best result obtained for 2-inch diameter SI GaAs doped with Cr and/or oxygen has been an average EPD of around 3×10^3 cm^{-2}.

Fig. 1.8 shows the EPD distribution of a (Cr, O)-doped 3-inch diameter round wafer (45 cm^2) along with that of a Si-doped 3-inch diameter round wafer.

It is essential to dope oxygen with mole fraction of $n^0_{O_2} \geqslant 10^{-5}$ (Fig. 1.4) and to reduce the residual Si concentration n_{Si} to $\leqslant$0.1 ppma in order to obtain SI GaAs with n_{Cr} of $\leqslant$0.1 wt ppm by the 3T-HB technique. It is noteworthy that the residual oxygen (from water in B_2O_3) is as high as 6×10^{-5} in mole fraction in the LEC growth system.[24]

Nishine et al. have developed 3-inch diameter low Cr-doped HB-GaAs for IC substrate applications.[34] They have doped with Cr in the range 0.05–0.39 wt ppm and

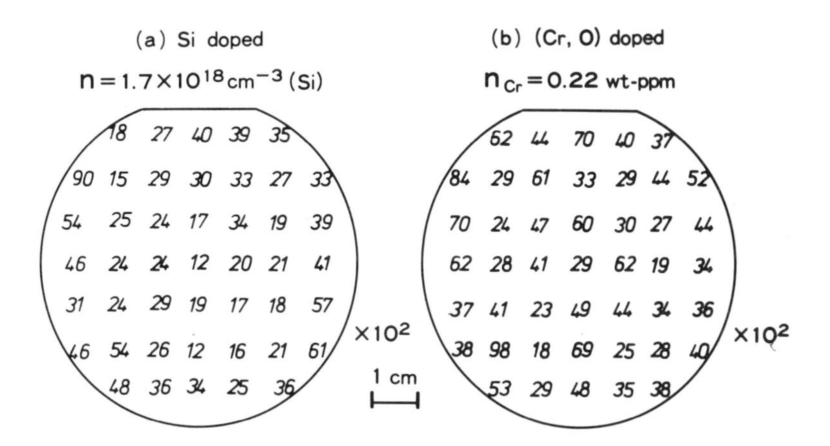

Fig. 1.8 EPD distribution of a (Cr, O)-doped 3-inch diameter round wafer (45 cm^2) along with that of a Si-doped 3-inch diameter round wafer.

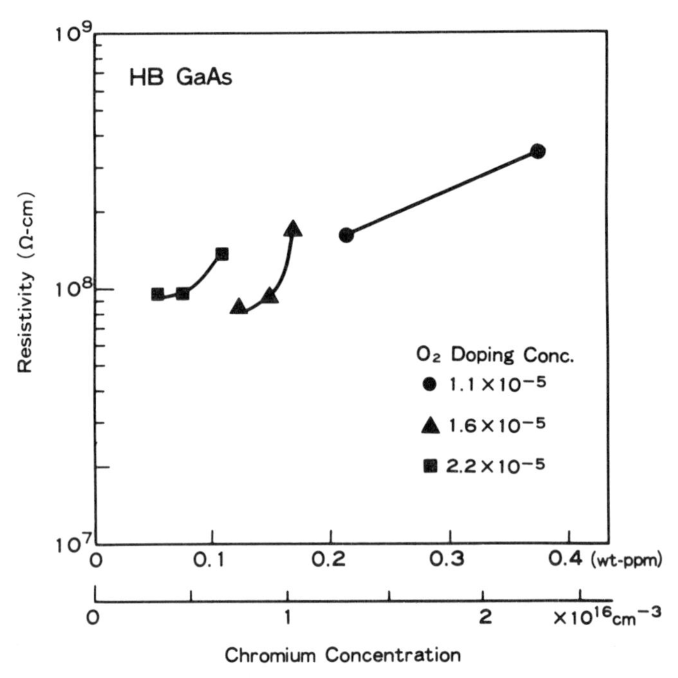

Fig. 1.9 Measured resistivity of (Cr, O)-doped GaAs as a function of n_{Cr} using as a parameter the mole fraction of added oxygen ($N^0_{O_2}$) in the form of As_2O_3. All ingots are designed for 3-inch diameter round wafers.

oxygen ($n^0_{O_2}$) higher than 1.1×10^{-5} mole fraction. The residual Si concentration (n_{Si}) is reduced to less than 0.1 ppma for the higher oxygen doping as shown in Table 1.3. Fig. 1.9 shows the measured resistivity of (Cr, O)-doped GaAs as a function of n_{Cr} with a parameter of mole fraction of added oxygen ($N^0_{O_2}$) in the form of As_2O_3. Direct ion implantation of SI GaAs was carried out using a Si-ion beam at 180 keV without capping at various doses, and followed by annealing without capping (face-to-face configuration) at 820°C for 20 min in nitrogen gas. Hall measurements were carried out by the Van der Pauw method. Fig. 1.10 shows the electron Hall mobilities of implanted layers in 3-inch diameter SI HB-GaAs at room temperature. The lower Cr-doped SI HB-GaAs crystals show fairly high electron mobility in implanted layers similar to undoped SI LEC GaAs crystals.[35] The

Table 1.3 SSMS data of (Cr, O)-doped GaAs.

Crystal		303W(Seed)	303W(Tail)	305W(Seed)	305W(Tail)
Designed Value	N_{Cr} (wt-ppm)	0.21	0.74	0.14	0.27
	$N^0_{O_2}$	1.65×10^{-5}	1.65×10^{-5}	1.65×10^{-5}	1.65×10^{-5}
SSMS analysis (ppma)	B	0.01	0.01	ND	ND
	Si	0.1	0.05	0.1	0.05
	S	0.03	0.02	0.02	0.02
	Cr	0.21	0.53	0.15	0.27
	Fe	0.01	ND	0.01	ND
	Cu	ND	0.01	ND	ND

ND: non-detected

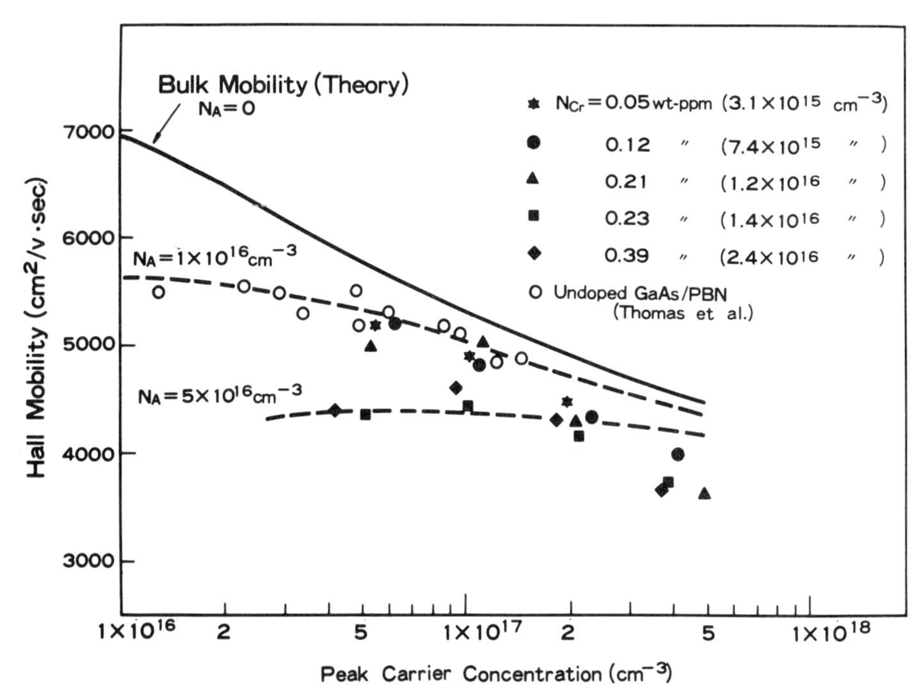

Fig. 1.10 Electron Hall mobilities of Si-ion implanted layers in 3-inch diameter semi-insulating HB-GaAs crystals at room temperature. The activation efficiencies, corrected by taking into consideration the surface depletion layer contribution, range from 84% for Cr doping level of 0.39 wt ppm to about 100% for Cr doping levels less than about 0.2 wt ppm.

Hall mobilities were highly uniform over most of the wafer area of SI GaAs doped with Cr in the range 0.05–0.39 wt ppm.

1.5. Conclusion

The most up-to-date three-temperature-zone horizontal Bridgman technique (3T-HB) for growing large-size substrate-grade GaAs single crystals has been reviewed.

Table 1.4 summarizes the results obtained for HB-GaAs grown by the 3T-HB technique. The background carrier concentration (n_b) can be reduced to less than

Table 1.4 The results obtained for HB-GaAs grown by the 3T-HB technique.

3T-HB	Multi-Section Heater	3D Temperature Profile Program Control
Size	Weight Length	8.8 kg → 20 kg 60 cm → 1 m
Dislocation	2-in ϕ (Si) 2-in ϕ (Zn) 3-in ϕ (Si), (Cr, O) 40 ϕ (Cr, O)	$\leq 1000\ cm^{-2}$ $\leq 2000\ cm^{-2}$ $\leq 5000\ cm^{-2}$ $\leq 2000\ cm^{-2}$
Semi-insulator	n_b Minimum Cr $\mu_{77K}(l^2)$	$\leq 10^{12} cm^{-3}$ 0.05 wt-ppm 10,000 cm^2/v.s.

10^{12} cm^{-3}, and electron Hall mobility of Si-ion implanted layers reaches up to 10,000 cm^2/V s at 77 K. However, undoped SI HB-GaAs is not likely to be obtainable with good reproducibility so long as quartz boats are utilized, though small SI GaAs crystals have been grown using the 3T-HB technique[36] or under controlled pressure of Ga_2O during HB growth.[25] A future development may be anticipated in the technical approach combining the use of a pBN boat, doping with isoelectronic impurities,[37,38] and precise stoichiometric control[18,31,32] as well as optimization of thermal growth conditions. The use of a magnetic field[39] may eliminate micro-inhomogeneities such as micro-precipitates and micro-striations in heavily doped HB-GaAs.

Acknowledgements

The authors wish to thank Mr. Kei-ichiro Fujita, Dr. Shiro Nishine, and Nobuhiro Kito for co-investigation of growth and characterization of HB-GaAs, Dr. Shigeo Murai and Mr. Masahiro Shibata for impurity analysis. They are indebted to Mr. Masaaki Sekinobu for his helpful discussion.

References

1) The major part of this paper was presented at the Electrochemical Society Symposium (May 8–13, 1983, San Francisco) with the title "Horizontal Bridgman growth of optoelectronics-grade GaAs".
2) L. R. Weisberg, F. D. Rosi, and P. G. Herkart: in Properties of Elemental and Compound Semiconductors Vol. 5, Ed. H. C. Gatos (Interscience, New York, 1960) pp. 25–67.
3) J. M. Woodall: Electrochem. Technol., 2 (1964) 167.
4) R. L. Henry and E. M. Swiggard: Gallium Arsenide and Related Compounds (St Louis) 1976 (Inst. Phys. Conf. Ser. No. 33b, 1977) p. 28.
5) T. R. AuCoin, R. L. Ross, M. J. Wade, and R. O. Savage: Solid State Technol., 22-1 (1979) 59.
6) M. E. Weiner, D. T. Lassota, and B. Schwartz: J. Electrochem. Soc., 118 (1971) 301.
7) S. Yamamoto, H. Hayashi, S. Yano, T. Sakurai, and T. Hijikata: Appl. Phys. Lett., 40 (1982) 372.
8) S. Yamamoto, H. Hayashi, T. Hayakawa, N. Miyauchi, S. Yano, and T. Hijikata: Appl. Phys. Lett., 41 (1982) 796.
9) T. Mimura, K. Johin, S. Hiyamizu, K. Hirosaka, and M. Abe: Japan J. Appl. Phys., 20 (1981) L598.
10) K. Nishiuchi, N. Kobayashi, S. Kuroda, S. Notomi, T. Mimura, M. Abe, and M. Kobayashi: 1984 IEEE Int. Solid State Circuits Conf., Digest of Technical Papers, (1984) 48.
11) Very recently, the authors' group has succeeded in growing low EPD (0–200 cm^{-2}) semi-insulating GaAs with a diameter of 65 mm by an improved LEC technique. However, detailed information is not available yet: Nikkei Electronics (1984) 3–12, 127. [in Japanese] See also Ref. 40).
12) C. N. Cochran and L. M. Foster: J. Electrochem. Soc., 109 (1962) 149.
13) T. Shimoda and S. Akai: Japan. J. Appl. Phys., 8 (1969) 1352.
14) T. Suzuki and S. Akai: Bussei, 12-3 (1971) 144. [in Japanese]
15) S. Akai: On the growth of low-defect-density GaAs single crystals by the 3T-HB technique; Doctoral Thesis (1975). [in Japanese]
16) C. D. Thurmond: J. Phys. Chem. Solids, 26 (1965) 785.
17) J. Nishizawa, Y. Okuno, and H. Tadano: J. Crystal Growth, 31 (1975) 215.
18) T. M. Parsey, Jr., Y. Nanishi, J. Lagowski, and H. C. Gatos: J. Electrochem. Soc., 129 (1982) 388.
19) D. E. Holmes, R. T. Chen, K. R. Elliot, and C. G. Kirkpatrick: Appl. Phys. Lett., 40 (1982) 46.
20) J. M. Woodall: Trans. Metall. Soc. AIME., 239 (1967) 378.
21) J. F. Woods and N. G. Ainslie: J. Appl. Phys., 34 (1963) 1469.
22) T. S. Plaskett, J. M. Woodall, and A. Segmüller: J. Electrochem. Soc., 118 (1971) 115.
23) L. A. Borisova, Z. L. Akkerman, and A. N. Dorokov: Neorg. Mater., 13 (1977) 908.
24) S. Akai, K. Fujita, M. Sasaki, and K. Tada: Gallium Arsenide and Related Compounds 1981 (Inst. Phys. Conf. Ser. No. 63, 1982) p. 13.
25) J. M. Woodall, H. Rupprecht, R. J. Chicotka, and G. Wicks: Appl. Phys. Lett., 38 (1981) 639.
26) T. Suzuki, S. Akai, K. Koe, Y. Nishida, K. Fujita, and N. Kito: Sumitomo Electric Tech. Rev., No. 18 (1978) 105.

27) Y. Seki, H. Watanabe, and J. Matsui: J. Appl. Phys., 42 (1978) 822.
28) G. Jacob, J. P. Farges, C. Schémali, M. Duseaux, J. Hallais, W. J. Bartels, and P. J. Roksnoer: J. Crystal Growth, 57 (1982) 245.
29) Recently, dislocation-free-grade Si-doped GaAs was grown by the LEC technique; see R. Fornari, C. Paorici, L. Zanotti and G. Zuccalli: J. Crystal Growth, 63 (1983) 415, and S. Akai: Present status of GaAs single crystal substrate technology, presented at the Kinki Society of Chemical Industry (Nov. 1983) [in Japanese]; see also ref. 11.
30) V. Swaminathan and S. M. Copley: J. Am. Ceram. Soc., 58 (1975) 482.
31) J. M. Parsey, Jr., J. Lagowski, and H. C. Gatos: The effect of melt stoichiometry and impurities on the formation of dislocations in bulk GaAs; Electrochem. Soc. Symp. (May 8–13, 1983, San Francisco).
32) J. M. Parsey, Jr., Y. Nanishi, J. Lagowski, and H. C. Gatos: J. Electrochem. Soc., 128 (1981) 936.
33) T. Toyoshima, S. Mizuniwa, J. Nakagawa, and S. Okubo: Hitachi Cable No. 1 (1981) 35. [in Japanese]
34) S. Nishine, N. Kito, K. Fujita, M. Sekinobu, O. Shikatani, and S. Akai: GaAs IC Symposium, Technical Digest (1982) 58.
35) R. N. Thomas, H. M. Hobgood, G. W. Eldridge, D. L. Barret, and T. T. Braggins: Solid State Electron., 24 (1981) 387.
36) T. Lifang, L. Liying, T. Weizu, L. Qitung, and Z. Yuaxi: in Semi-Insulating III–V Materials, Evian 1982 (Shiva, Nantwich, UK, 1982) p. 248.
37) G. Jacob: in Semi-Insulating III–V Materials, Evian 1982 (Shiva, Nantwich, UK, 1982) p. 2.
38) K. Tada, A. Kawaski, T. Kotani, R. Nakai, T. Takebe, S. Akai, and T. Yamaguchi: in Semi-Insulating III–V Materials, Evian 1982 (Shiva, Nantwich, UK, 1982) p. 36.
39) H. P. Utech and M. C. Flemings: J. Appl. Phys., 37 (1966) 2021.
40) K. Tada, S. Murai, S. Akai, and T. Suzuki: GaAs IC Symp., Technical Digest (Boston, MA, Oct. 23–25, 1984).

2 NEARLY PERFECT CRYSTAL GROWTH IN III–V AND II–VI COMPOUNDS

Jun-ichi NISHIZAWA†, Yasuo OKUNO* and Ken SUTO†

Abstract

The most important point for compound semiconductor such as III–V and II–VI is the deviation from the stoichiometric composition, because the deviation corresponds to imperfections in the crystals. The temperature difference methods under controlled vapor pressure (TDM-CVP) is the most suitable growth method which can control the deviation from the stoichiometric composition by applied vapor pressure upon the solution.

First, the heat-treatment experiment under controlled vapor pressure clarified that the vapor pressure seriously influenced the electrical, optical and crystallographic properties, and the specific vapor pressure at which the crystal becomes stoichiometric was determined. In the case of crystal growth by TDM-CVP, the optimum vapor pressure to produce nearly perfect crystal with very low dislocation density was found and it was in good agreement with that found by heat treatment experiment. TDM-CVP was extended to II–VI compound semiconductors and p-type ZnSe was obtained using vapor pressure control. It was also extended to melt growth of bulk crystals, and the vapor pressure control enabled GaAs crystals to be grown with very low dislocation density compared with the conventional growth methods without controlled vapor pressure. The most important point defect for GaAs at high arsenic vapor pressure was found to be As interstitial atoms. The model for the mechanism of vapor pressure control is described by the concept that the chemical potentials of As in the three phases, vapor, liquid and solid, are equal to each other.

Keywords: TDM-CVP, Perfect Crystal, As Interstitial, Stoichiometry

2.1. Introduction

Nearly perfect crystals of III–V compounds can be grown by the temperature difference method under controlled vapor pressure (TDM-CVP) proposed by J. Nishizawa.[1)] In this method, the vapor pressure of the group-V element is applied on the top of the molten phase during growth in order to control the deviation from stoichiometry of the segregating crystals. Stoichiometry-controlled crystals with lowest defect and dislocation concentrations were obtained by this technique, which was applied to the production of very efficient light emitting diodes in the red to green band.

On the other hand, no suitable growth method has been exploited for growing II–VI compound semiconductors like ZnSe of sufficient high quality: there was no method for

---

† Research Institute of Electrical Communication, Tohoku University Sendai, 980.
* Semiconductor Research Institute, Kawauchi, Sendai, 980.

JARECT Vol. 19, *Semiconductor Technologies* (1986), *J. Nishizawa (ed.)*
OHMSHA, LTD. and North-Holland

obtaining stable p–n junctions of ZnSe because it was difficult to grow p-type ZnSe crystals. The TDM-CVP has been applied to ZnSe, and p–n junctions which emit blue light have been obtained.[2] This is the subject of this chapter.

To decrease the content of point defects due to non-stoichiometry in compound semiconductors, the growth temperature should be as low as possible, and the stoichiometry should be exactly maintained. In 1971, J. Nishizawa and S. Shinozaki published the experimental result on the growth of GaAs, which showed the influence of the applied vapor pressure of As on the top of the molten Sn on the stoichiometry of the crystals segregated on the substrate settled at the lower part of the crucible under the solution. The temperature difference between the molten phase and the substrate accelerates diffusing dissolved GaAs towards the substrate where it segregates. Ga solution was then used instead of Sn.

The influence of the vapor pressure should not be expected to have any influence on the stoichiometry of the segregated crystals if we simply apply the Gibbs phase rule. However, we have found that the stoichiometry is controlled by the vapor pressure.

This method was then applied for the growth of poly-crystals by N. Watanabe et al.[3] who named it SSD.

The author's group extensively continued these experiments using GaAs and GaP and concluded that there seems to be a control of stoichiometry, and the optimum vapor pressure to grow stoichiometric crystals can be represented by

$$P_{\mathrm{GaAs,\,opt}} = 2.6 \times 10^{6} \exp(-1.05\ \mathrm{eV}/kT_{g})\mathrm{Torr} \tag{2.1}$$

$$P_{\mathrm{GaP,\,opt}} = 4.67 \times 10^{6} \exp(-1.01\ \mathrm{eV}/kT_{g})\mathrm{Torr} \tag{2.2}$$

where T_g is the growth temperature. It should be noted that the optimum pressure is just the same as the stoichiometric points of the vapor pressure of heat-treatment.

T. Suzuki and S. Akai extended the experiment towards, the melt growth.[4] The maximum melting temperature of GaAs is 1238°C. Then, the optimum vapor pressure given by the above equation is 813 Torr, which can be related to the temperature of the arsenic deposit chamber (T_{As}) connected by a fine tube of the growth chamber at T_g by

$$P_{g} = P_{As}(T_{g}/T_{As})^{1/2}. \tag{2.3}$$

They confirmed experimentally that the optimum temperature T_{As} was 617°C, which was in very good agreement with the value given by the above equations.

They have been applying this method for production under the name of the three-temperature-zone method. H. C. Gatos and Y. Nanishi et al. repeated nearly the same experiment and obtained a very similar result.[5]

We have recently developed a new growth technique for the melt growth of GaAs by the Czochralski method under controlled As vapor pressure. The dislocation density was as low as $2 \times 10^{3}\ \mathrm{cm}^{-2}$ at the As temperature of 616°C. This result is described in Section 2.3.4.

From all these considerations, the approximation that the saturation solubility is a constant should be corrected to give higher accuracy, and we should say that the saturation solubility depends on the applied vapor pressure and influences the stoichiometry of the segregating crystals.

First, we discuss the heat-treatment experiment of GaAs under the vapor pressure of arsenic, from which the specific vapor pressure under which the exact stoichiometry is realized is determined. Also it is suggested from this experiment that As interstitial atoms are the dominant point defects in the higher vapor pressure region.

2.2. Heat-treatment of III–V compounds

2.2.1. *Effect of heat-treatment on GaAs crystal under arsenic overpressure*[6,7]

The effect of heat-treatment on GaAs crystal under arsenic overpressure was investigated. Specimens were boat-grown (001) oriented GaAs crystals doped with Te. The electron concentration before heat-treatment varied between $8.9 \times 10^{16}\,\mathrm{cm}^{-3}$ and $2.8 \times 10^{18}\,\mathrm{cm}^{-3}$. A specimen and As metal were set in a two-zone quartz ampoule, then evacuated up to 3×10^{-6} Torr and sealed off. Fig. 2.1 shows a schematic diagram of the heat-treatment experiment system. The applied arsenic pressure is controlled by the temperature of the As zone furnace, T_{As}, independently from the heat-treatment temperature in the crystal zone, T_{GaAs}.

The arsenic pressure in the GaAs crystal zone, P_{GaAs} is given by

$$P_{\mathrm{GaAs}} = P_{\mathrm{As}}(T_{\mathrm{GaAs}}/T_{\mathrm{As}})^{1/2} \tag{2.4}$$

if the two zones are connected by a fine tube. The heat-treatment temperatures, T_{GaAs}, were 900°C and 1100°C in this experiment.

The deviation from stoichiometry after heat-treatment was estimated from the change in acceptor concentration, lattice constant and intensity of deep-level photoluminescence. When specimens were heat-treated for over 32 h, the conductivity profile was homogeneous along the thickness direction except for a 60 μm-thickness surface layer. Furthermore, defects in the crystals were in equilibrium with ambient arsenic pressure when the heat-treatment time was over 64 h (Fig. 2.2). This behavior suggests that there are fast and slow processes for stoichiometry control. Our heat-treatment experiments were performed for 67 h and 60 μm surface inhomogeneous layers were etched off before measurements.

The conductivity type of crystals changed from n-type to p-type after heat-treatment for all the As pressures. The acceptor density has a minimum value at a specific arsenic pressure ($P_{\mathrm{As,\,min}}$) which depends on the heat-treatment temperature.

We will show later that these acceptors are due to non-stoichiometric defects and

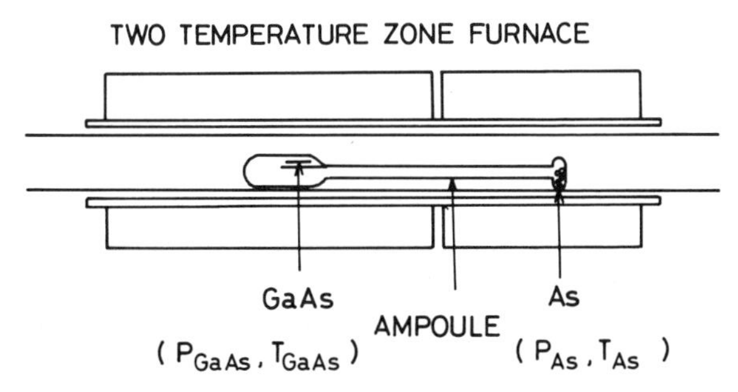

Fig. 2.1 Diagram of experimental set-up of annealing experiment.

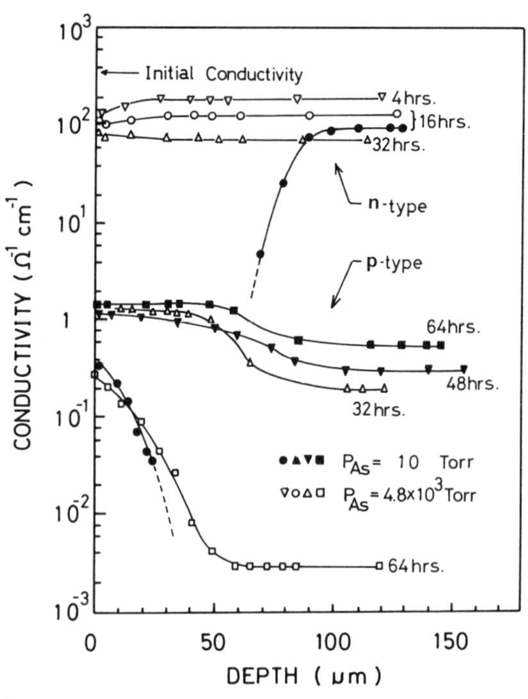

Fig. 2.2 Conductivity profile of annealed samples as a function of annealing time. $T_H = 1100°C$, Te-doped GaAs.

$P_{As, min}$ corresponds to the exact stoichiometry, from comparison with the lattice constant measurements as well as photoluminescence measurements.

Fig. 2.3 shows that the densities of acceptors N_A after heat-treatment are related to the initial electron densities n_i by $N_A \propto n_i^{3/4}$. Even after the initial acceptor densities N_{Ai} were subtracted from N_A, the same tendency was seen. Therefore, these acceptors are likely to be complexes of non-stoichiometric defects and Te impurities, or non-stoichiometric defects which are induced under the presence of Te donors. The activation energies of acceptor levels are also shown in Fig. 2.3. The symbol * represents the sample with 0.15 eV acceptor level, the symbol ** the sample with the 0.18 eV acceptor level and the others, the samples which were n-type even after heat-treatment.

The density of defect-like acceptors after heat-treatment was obtained by the following way. When p-type conversion took place and an acceptor level was found from the temperature dependence of the Hall coefficient, the donor density N_D, the acceptor density N_A and the acceptor energy E_A were calculated by fitting the experimental curve of the temperature dependence of the Hall coefficient to the following equation[8)]

$$\frac{p(p+N_D)}{N_A-N_D-p}=\frac{N_V}{g}\exp\left(-\frac{E_A}{kT}\right). \tag{2.5}$$

Here, N_V is the effective density of states of the valence band and g is the total degeneracy of each state which is assumed to be 2 in this case.

Fig. 2.4 shows the relationship between acceptor density and applied arsenic pressure. The activation energies of acceptors measured by the Hall effect are 0.18 eV except in specimens with low initial electron densities heat-treated at very low and high arsenic

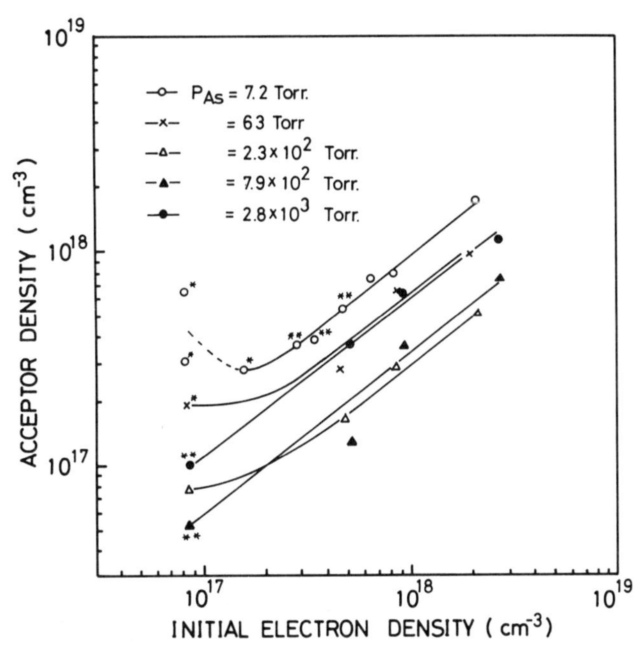

Fig. 2.3 Initial electron density dependence of acceptor density in the annealed GaAs crystals for $T_H = 1100°C$ under various arsenic pressure. The symbols * and ** denote the samples with 0.15 eV and 0.18 eV acceptor level respectively; the other symbols denote the n-type samples after heat treatment.

pressure regions as shown in Fig. 2.5 and Table 2.1. It is shown the acceptor density becomes minimum at the arsenic pressure of about 360 Torr at 1100°C and about 80 Torr at 900°C. We shall denote this arsenic pressure at which the acceptor density becomes minimum as $P_{As, min}$. It is clear that in this range $P_{As, min}$ does not depend significantly on the initial electron density.

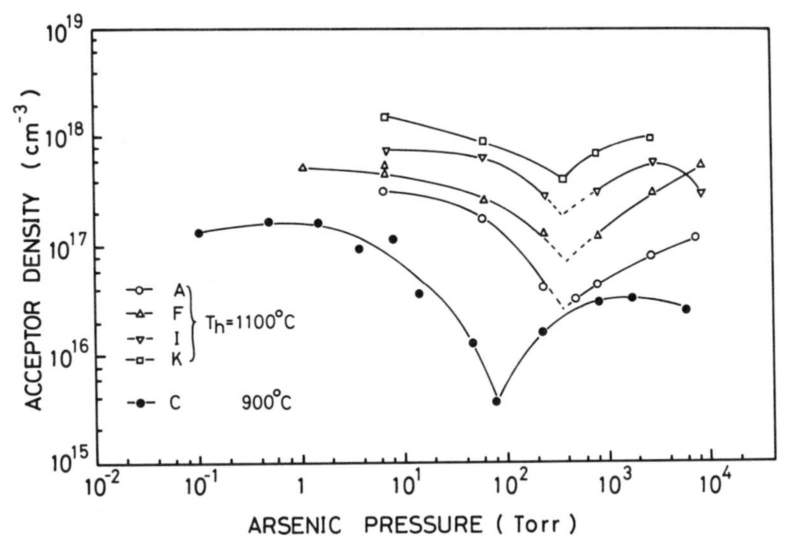

Fig. 2.4 Dependence of acceptor density on arsenic pressure. The acceptor density becomes minimum at a specified arsenic pressure $P_{As,min}$. $t_H = 67$ h. Te-doped GaAs.

Table 2.1 Change of the hole concentration and 1.30 eV band intensity by the isothermal re-annealing at 400°C.

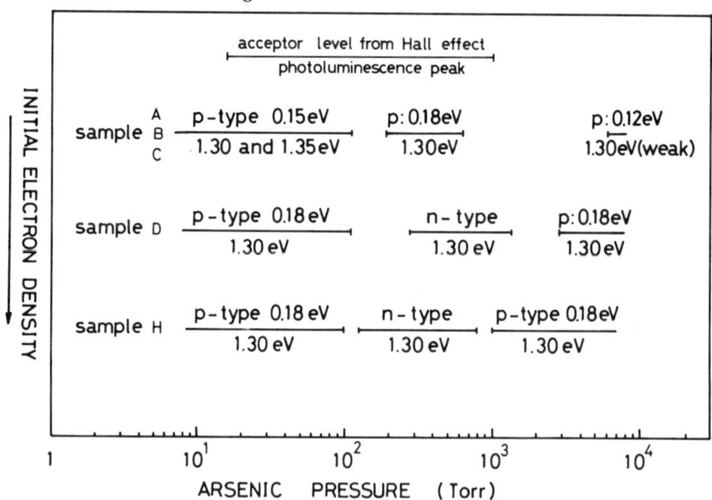

The defect centers should be different types in the arsenic pressure regions, $P_{As} < P_{As,\,min}$ and $P_{As} > P_{As,\,min}$, although the activation energies are similar. Also the photoluminescence spectra show a 1.30 eV peak in most of the region of the $P-T$ diagram in Fig. 2.5 and the emission intensity of the 1.30 eV peak increases with increase of electron density before heat-treatment. Fig. 2.6 shows the relationship between the intensity of the 1.30 eV photoluminescence peak and the applied arsenic pressure. This

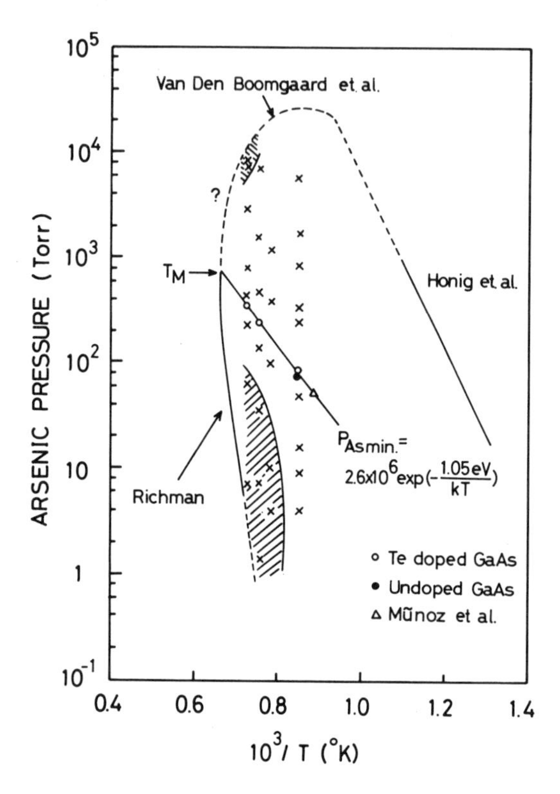

Fig. 2.5 $P_{As,min}$ as a function of heat treatment temperature T_H. The dissociation pressure of GaAs after Richman and the vapor pressure of As after Honig et al. are also shown. Therefore, the central encircled area is thought to correspond to the single phase of gallium arsenide. The symbol × shows the experimental point in this experiment; however, the removed surface layer actually corresponds to the equilibrium phase and the measured inner part does not correspond to the equilibrium state in this diagram. In the shaded area the 0.15 eV acceptor level and the 1.35 eV emission band were observed, and in the dotted area the 0.12 eV acceptor level and the weak emission of the 1.30 eV band were observed. In the other area, the 1.30 eV band emission in all the samples and the 0.18 eV acceptor level in the specimens which were converted to p-type were obtained. The upper side of the line corresponds to Eq. (2.4) where there is expected to be relatively excess arsenic and at the lower side there is relatively excess gallium atoms.

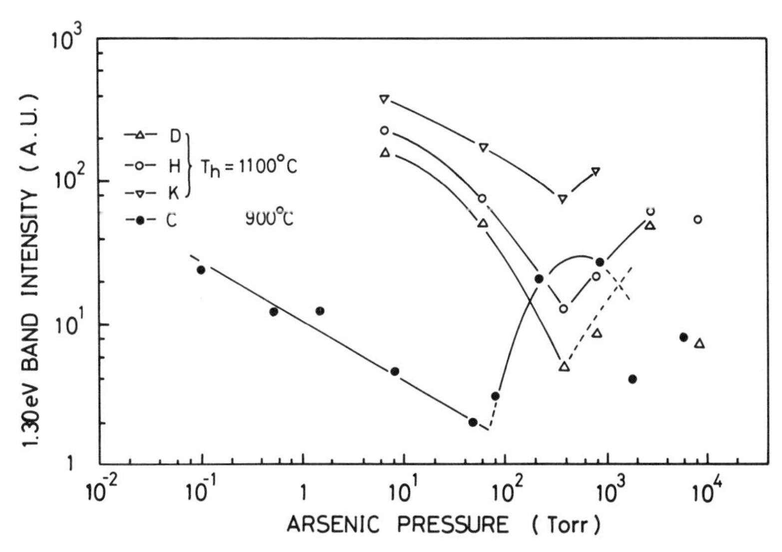

Fig. 2.6 Arsenic pressure dependence of 1.30 eV photoluminescence peak intensity. $T_H = 1100°C$, $t_H = 67$ h, Te-doped GaAs.

value becomes a minimum at the same applied arsenic pressure as $P_{As,\,min}$. This 1.30 eV PL peak is considered to be due to 0.18 eV acceptor levels and associated with non-stoichiometric defects. The emission intensity of the 1.30 eV band and the initial electron density are in accordance with the relation $I_{1.30} \propto n_i^{3/4}$.

Lattice constants of the heat-treated samples were measured by X-ray double crystal method. Cu Kα radiation and GaAs (004)–(004) symmetric configuration were used. It should be noted that the lattice constant also has a minimum value at the same specific arsenic pressure $P_{As,\,min}$ as shown in Fig. 2.7. For all arsenic pressures except $P_{As,\,min}$, the lattice constant becomes larger with increasing annealing temperature, whereas the lattice

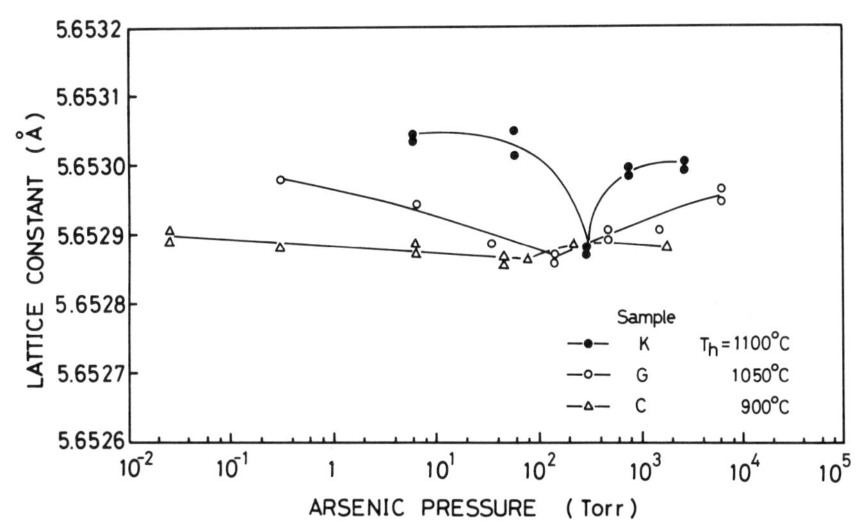

Fig. 2.7 Arsenic pressure dependence of lattice constant. $T_H = 1100°C$, $t_H = 67$ h, Te-doped GaAs.

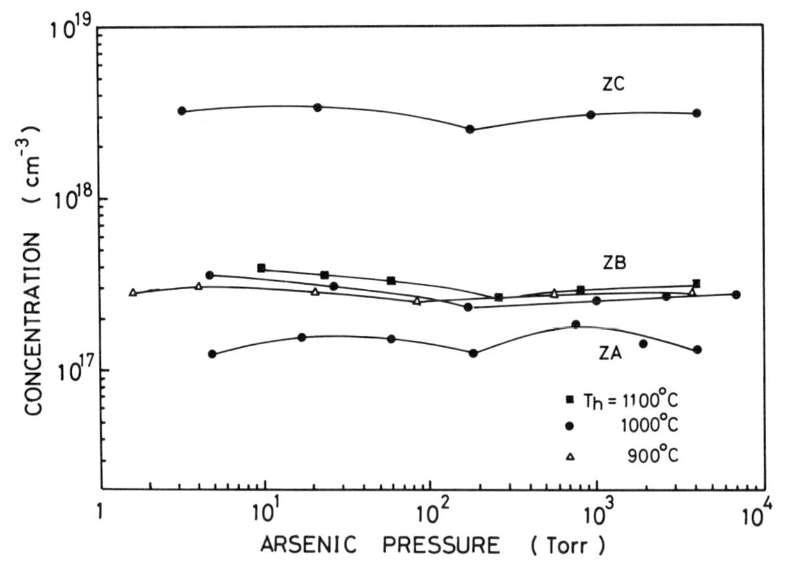

Fig. 2.8 Arsenic pressure dependence of acceptor density after annealing of Zn-doped GaAs.

constant at $P_{\text{As, min}}$ is almost the same and does not depend much on the annealing temperature. From the above data, the temperature dependence of $P_{\text{As, min}}$ can be expressed as follows:

$$P_{\text{GaAs, min}} \approx 2.6 \times 10^6 \exp(-1.05\,\text{eV}/kT)\text{Torr} \tag{2.6}$$

A similar heat-treatment experiment was carried out for Zn-doped p-type GaAs. The results are shown in Figs. 2.8 and 2.9. The carrier concentration depends on the As vapor pressure and becomes a minimum in the neighborhood of $P_{\text{GaAs, min}}$ although the degree of change is not so large. On the other hand, the arsenic pressure dependence of the

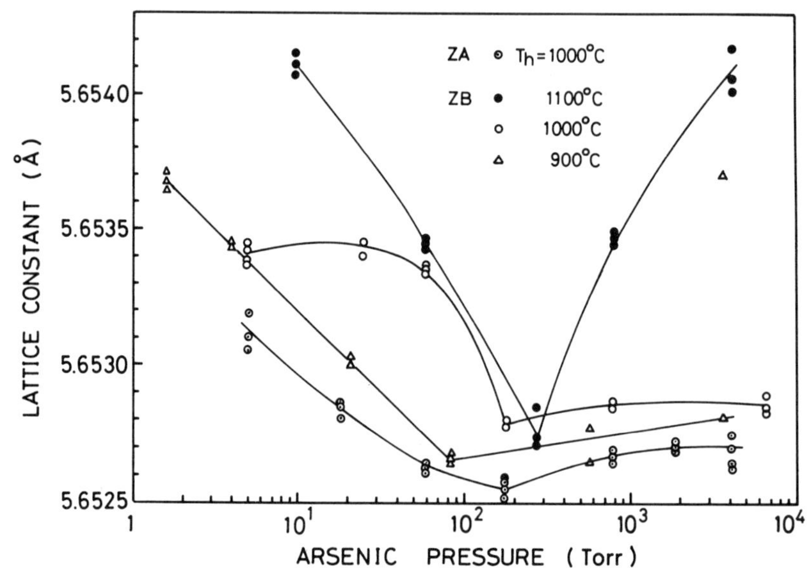

Fig. 2.9 Arsenic pressure dependence of lattice constant after annealed samples of Zn doped GaAs.

lattice constant is clear and the As pressure whose value becomes a minimum is nearly the same as that of n-type GaAs. Therefore, $P_{\mathrm{GaAs,\,min}}$ in Zn-doped p-type GaAs is consistent with the value given by expression (2.6).

The effects of re-annealing on heat-treated crystals have been investigated. When the specimens which have 0.12 eV and 0.15 eV acceptor levels are re-annealed at 300–400°C, their hole densities decrease and their Hall mobilities at 77 K increase. Also the 0.12 eV acceptor levels disappear on re-annealing and the 0.18 eV acceptor level is observed instead. The emission intensity of the 1.30 eV band increases with increasing re-annealing time. The results which show how the hole density and the emission intensity of the 1.30 eV band are changed by the re-annealing time are shown in Fig. 2.10. The 0.12 eV acceptor level is due to the defects which are very unstable and act as non-radiative recombination centers.

The specimens with the 0.15 eV acceptor level always follow the 1.35 eV emission band. On re-annealing, the hole density decreases and the 0.18 eV acceptor level appears instead of 0.15 eV. On the other hand, in the specimens with the 0.18 eV acceptor level, the hole density, acceptor level and the photoluminescence spectra are not changed on re-annealing. Therefore, the defects responsible for the 0.12 eV and 0.15 eV acceptor levels are unstable and changes to the stable 0.18 eV level by the formation of the complexes.

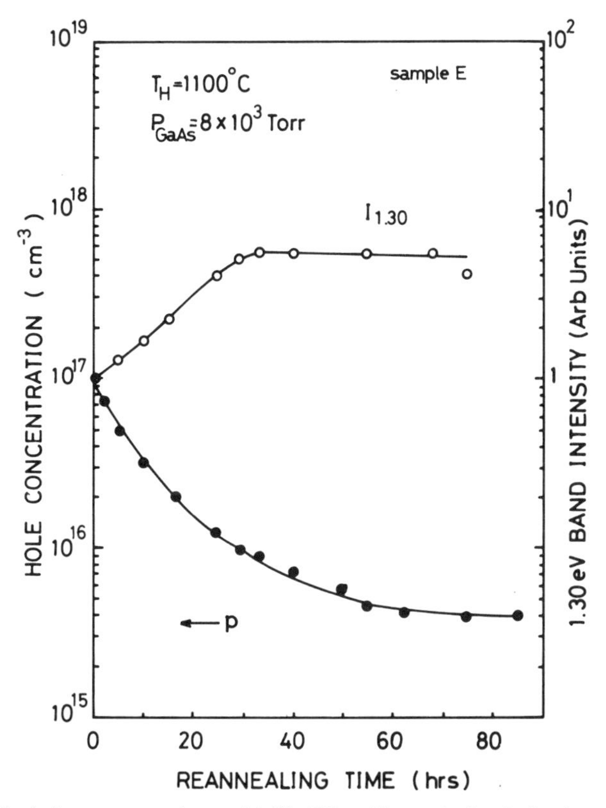

Fig. 2.10 Change of the hole concentration and 1.30 eV band intensity by the isothermal re-annealing at 400°C.

2.2.2. *Interstitial As atoms*

There is, however, some evidence that the dominant defects are arsenic interstitial atoms or their complexes with impurities in the higher vapor pressure region. Fig. 2.11 shows the measured mass density of heat-treated Zn-doped GaAs crystals. The density does not show a minimum at $P_{As, min}$, but increases monotonically with increasing vapor pressure. The observed incremental increase may be due to arsenic interstitial atoms. They

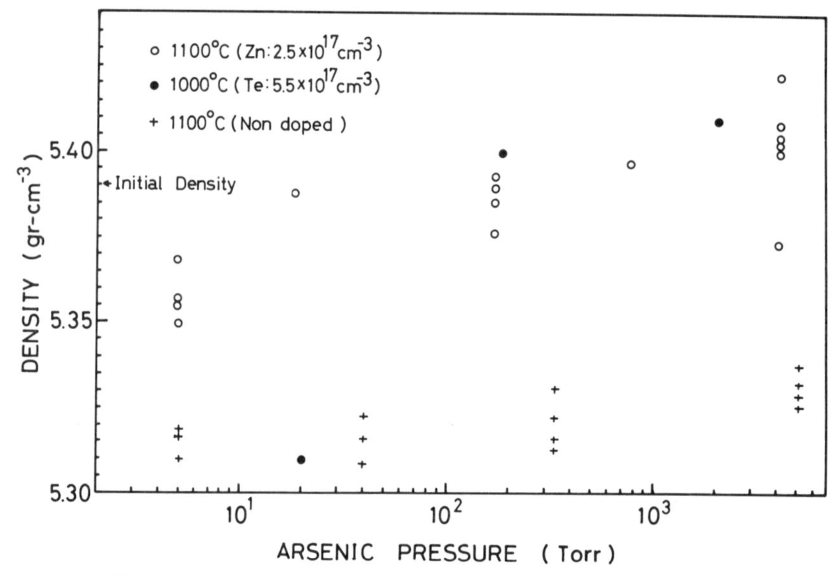

Fig. 2.11 Arsenic pressure dependence of the density of crystals.

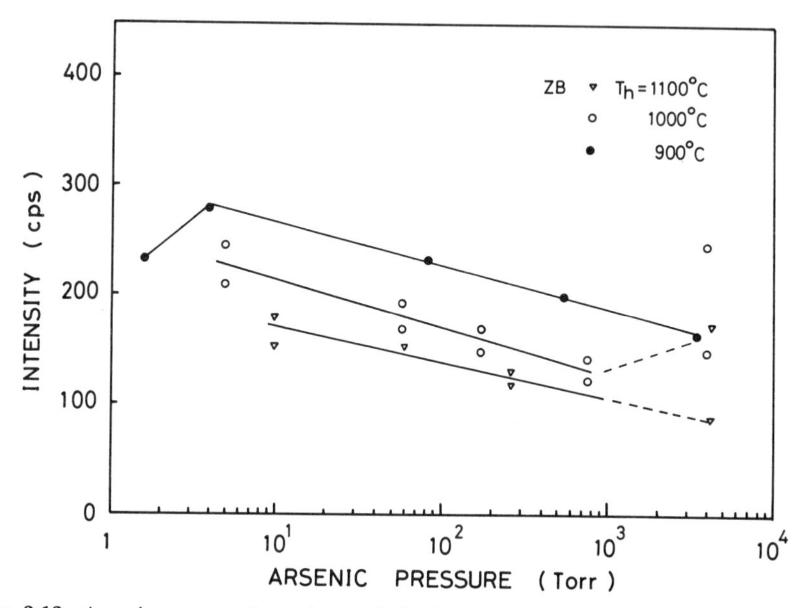

Fig. 2.12 Arsenic pressure dependence of the intensity of anomalous transmission X-ray.

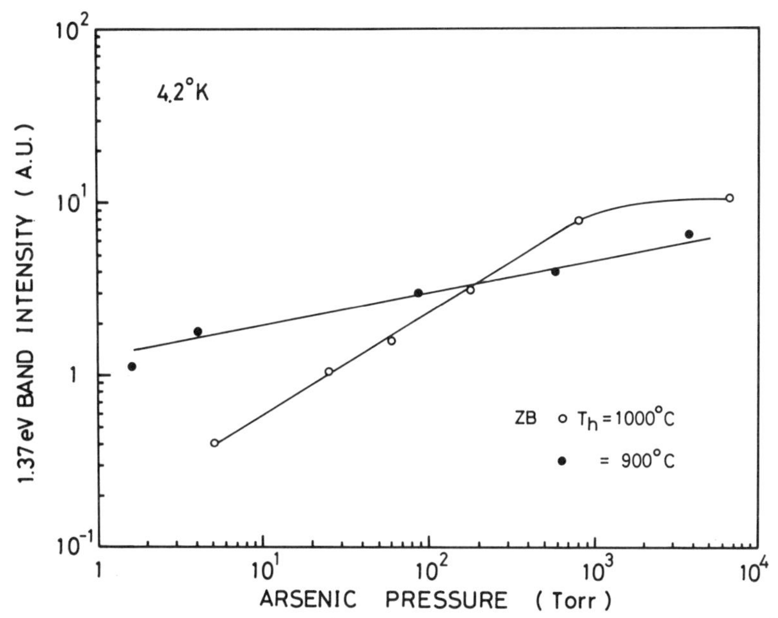

Fig. 2.13 Arsenic pressure dependence of 1.37 eV photoluminescence peak intensity. Samples are Zn-doped GaAs.

may have generated clusters or precipitates of arsenic because the observed increase of mass density is very large. As further evidence, it is seen from Fig. 2.12 that the anomalous X-ray transmission whose intensity is sensitive to the lattice irregularity at the interstitial sites decreases monotonically with increasing applied vapor pressure. Also, we have observed deep-level luminescence at 1.37 eV which may be associated with interstitial arsenic atoms or their complexes in Zn-doped GaAs. Fig. 2.13 shows that the 1.37 eV

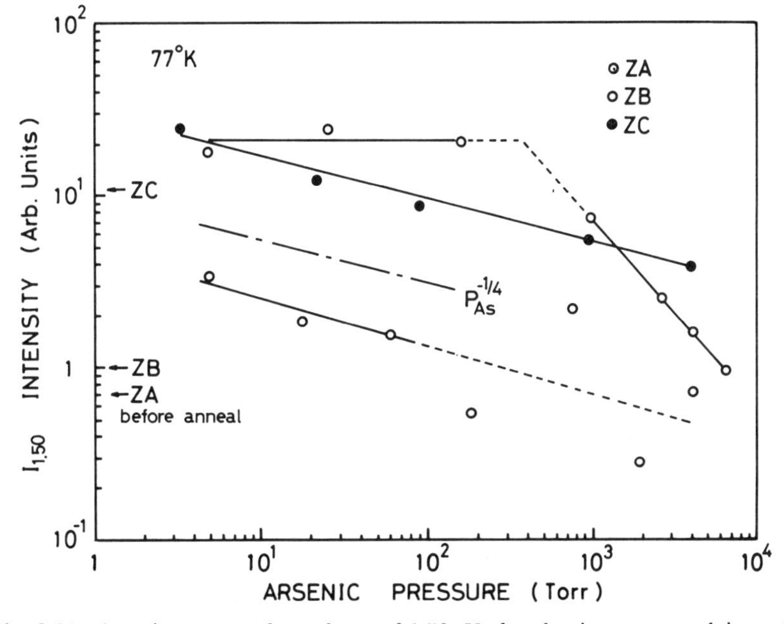

Fig. 2.14 Arsenic pressure dependence of 1.50 eV photoluminescence peak intensity.

luminescence intensity increases with increasing applied vapor pressure, while the intensity of the band-to-band luminescence decreases with increasing vapor pressure in the relation of $P_{As}^{-1/4}$ as shown in Fig. 2.14.

As a more conclusive evidence, we have recently observed Rutherford back-scattering (RBS) of He ions from interstitial arsenic atoms.[9]

2.2.3. *Effect of heat-treatment on GaP crystal under phosphorus overpressure*

A similar heat-treatment experiment was also performed on n-type S-doped GaP under controlled phosphorus vapor pressure.[10] The experimental results on the carrier concentrations and the lattice constants are shown in Figs. 2.15 and 2.16. The electron concentration shows a minimum, and the lattice constant shows a maximum at a specific vapor pressure, in contrast to the case of GaAs. The temperature dependence of the phosphorus pressure where the lattice constant becomes a maximum is shown in Fig. 2.17. This specific vapor pressure, $P_{\mathrm{GaP,min}}$ can be expressed as a function of temperature as follows:

$$P_{\mathrm{GaP,min}} \approx 4.67 \times 10^6 \exp -(1.01\ \mathrm{eV}/kT_g)\mathrm{Torr}. \tag{2.7}$$

We are sure that $P_{\mathrm{GaP,min}}$ in the above experiment gives the stoichiometric condition in GaP, although the reason for the decrease of the lattice constant in non-stoichiometric GaP is not yet clear.

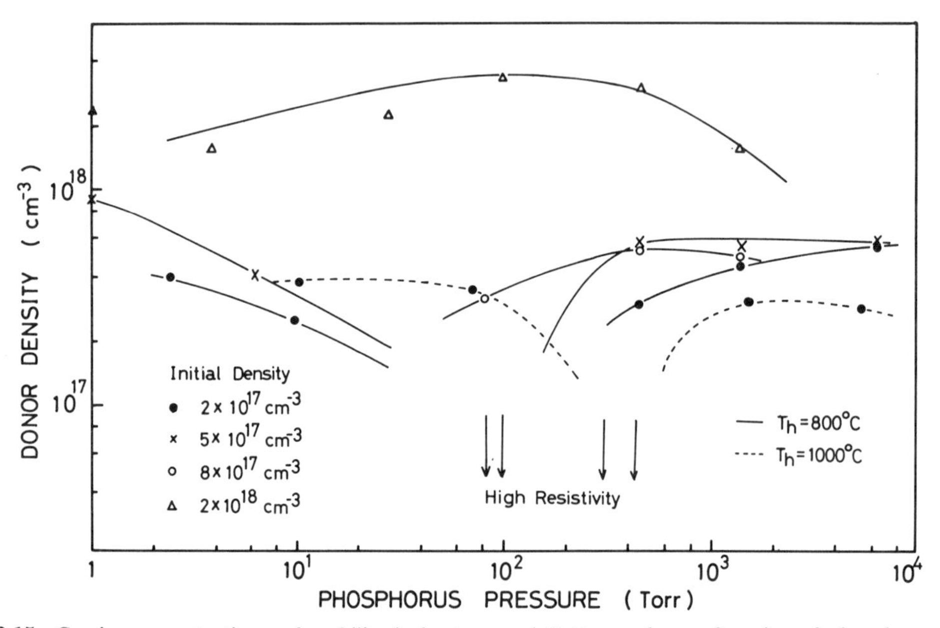

Fig. 2.15 Carrier concentration and mobility in heat-treated GaP crystals as a function of phosphorus vapor pressure.

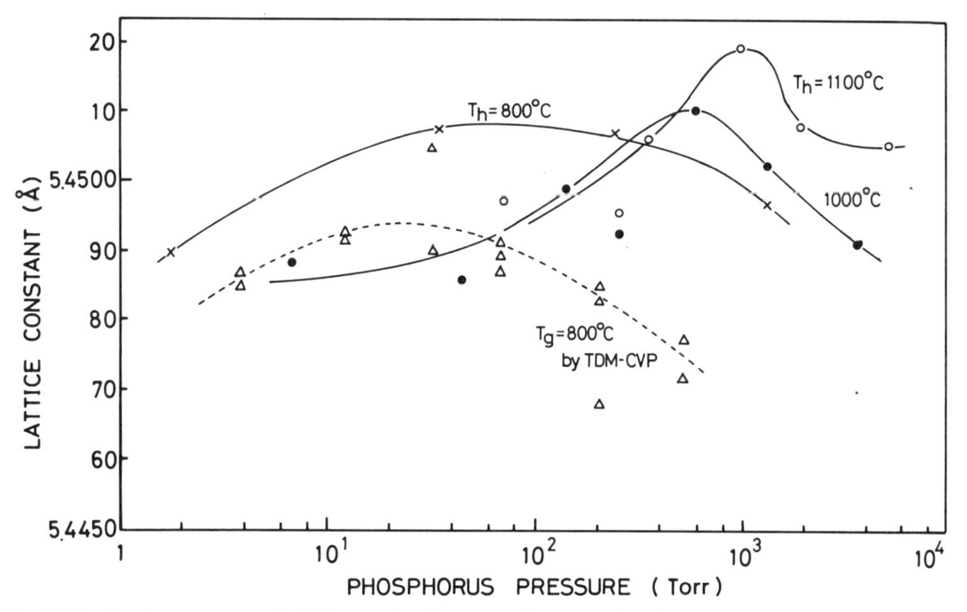

Fig. 2.16 Lattice constant of GaP crystals after annealing as a function of phosphorus vapor pressure.

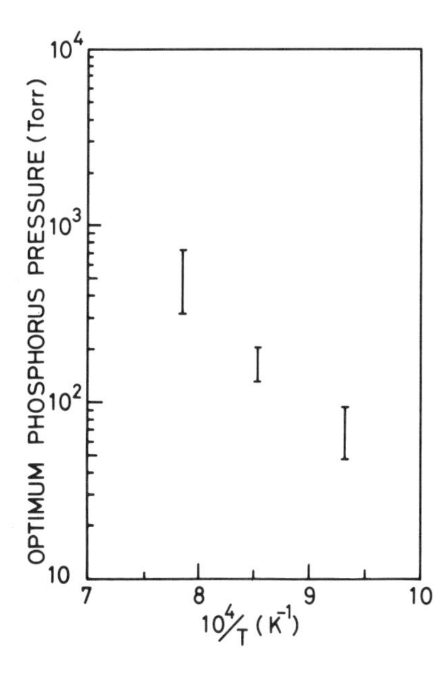

Fig. 2.17 Temperature dependence of the phosphorus pressure where the lattice constant becomes a maximum.

2.3. Crystal growth of III–V compounds under controlled vapor pressure of the group-V element (TDM-CVP)[1,11–16]

2.3.1. *Temperature difference method*[11,12]

In the conventional crystal growth by liquid-phase epitaxy, the crystal segregates due to a decrease of solubility as the temperature is lowered; however, the temperature change in the process of crystal growth usually changes the properties of crystals.

The essentials of TDM-CVP and the apparatus used in this experiment are shown in Fig. 2.18. The vertical portion contains the solution and excess solute (GaAs or GaP). The solution temperature is made higher than the substrate temperature with an additional heater to accelerate the diffusion dissolved GaAs (or GaP), and the arsenic (or phosphorus) vapor pressure is applied from the arsenic (or phosphorus) chamber. Two sets of experiments were done: the first set of experiments was done without adding the vapor pressure of the group-V element.

The temperature T_1 of the upper region of the solution containing the source material is higher by ΔT than the temperature T_2 of the lower region where the epitaxial layer is grown. The saturation solubility at the higher temperature T_1 is larger than that of the T_2 region, and the continuously dissolved material diffuse toward the lower

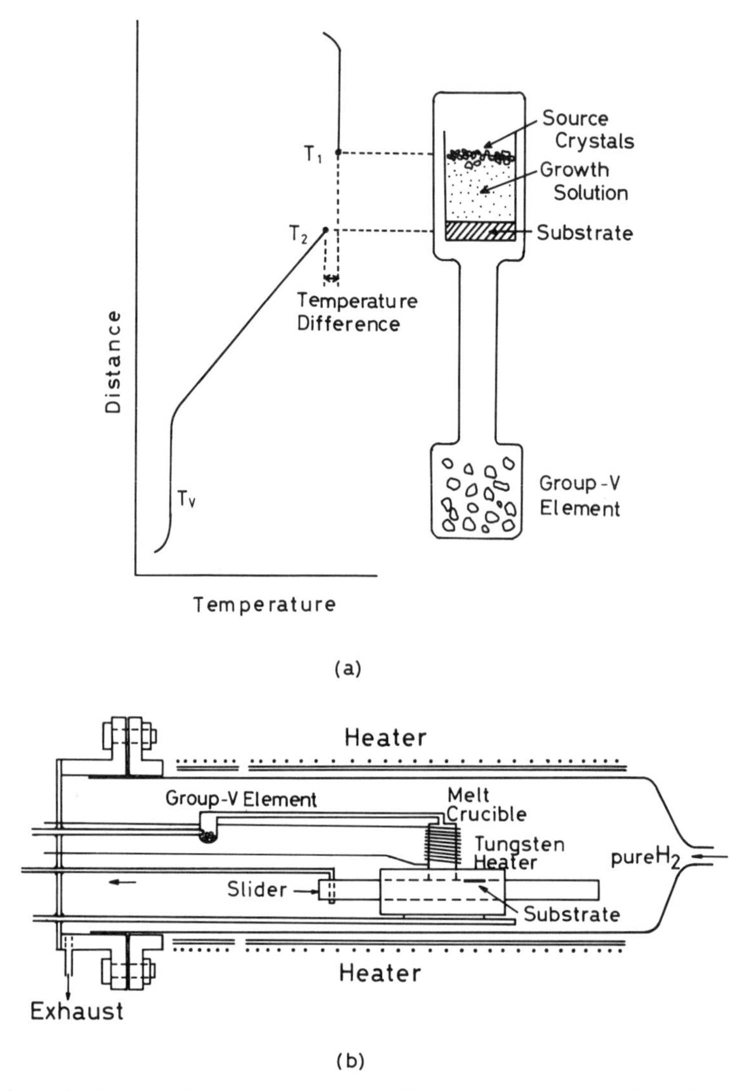

Fig. 2.18 (a) Schematic diagram for the temperature difference method (TDM) under controlled vapor pressure (CVP). (b) Apparatus for LPE by TDM-CVP method. Growth takes place in pure hydrogen which passed through Pd film.

temperature region because of the concentration and kinetic energy difference. The material supersaturates in the T_2 region and recrystallizes on the substrate. The degree of supersaturation depends on the temperature difference ΔT. This growth method is named the temperature difference method (TDM). It is noted that in TDM growth proceeds without changing the furnace temperature. Therefore, the crystal properties (i.e. crystal perfection, impurity concentration, etc.) along the growth direction are uniform, and the growth temperature can be made lower than those of the usual slow-cooling method as the growth thickness is not limited by the saturation solubility at the growth temperature. The growth thickness of the crystal is proportional to the input power of the additional heater and the growth time.

As a typical example, the growth rate of GaP epitaxial layers is about 10 μm/h at the growth temperature of 800°C and at the temperature difference of 20°C produced by the additional heater with power of 24 W. The as-grown surface morphology of GaP shows a step-like structure, and the as-grown surface and cleaved face at the same portion of the crystal are shown in Fig. 2.19 for comparison. The substrates used in this experiment are S-doped GaP[(2–3) $\times 10^{17}/cm^3$] grown by liquid encapsulated Czochralski and have surfaces parallel to the ⟨111⟩ plane. Stain etching to reveal the root-like faults and stacking faults in the cleaved face was performed in a solution of $HF + H_2O_2 + 3H_2O$ at 20°C.[17] It is evident that the large steps on the as-grown surface correspond to the root-like faults, which were first observed by the authors.[11,12] These faults seem to originate from planar defects produced by temperature fluctuation.

Fig. 2.20(a)–(f) shows the dependence of the morphologies of the cleaved faces of the grown layers on temperature and time duration under the growth conditions shown by the accompanying schematic diagrams. Temperature control was carried out by the main heater wound on the reaction tube, while an auxiliary heater was used to provide

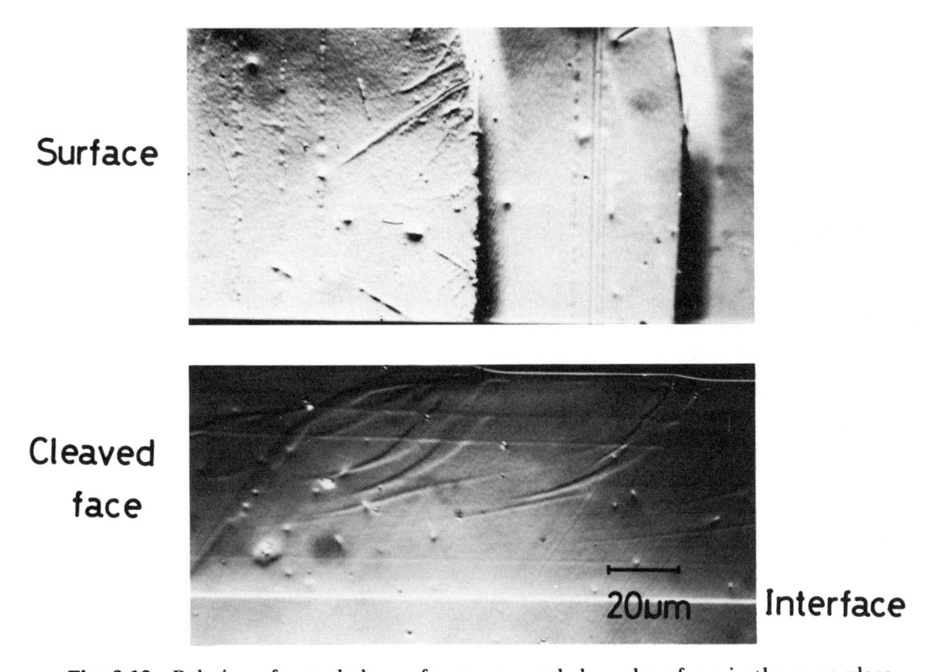

Fig. 2.19 Relation of morphology of as-grown and cleaved surfaces in the same place.

temperature difference uniformity. A number of root fault and lateral stripes due to steep temperature changes during growth have been observed as shown in Fig. 2.20(a)–(e).

It is noted that the growth can be realized even by the temperature procedure as shown in Fig. 2.20(b). This result is evidence that the growth is performed by the temperature difference in the solution. By comparing these figures, it is clear that the growth temperature of TDM is lower by 200 to 300°C than the conventional slow-cooling method. Moreover, the cleaved face morphology is perfect in TDM.

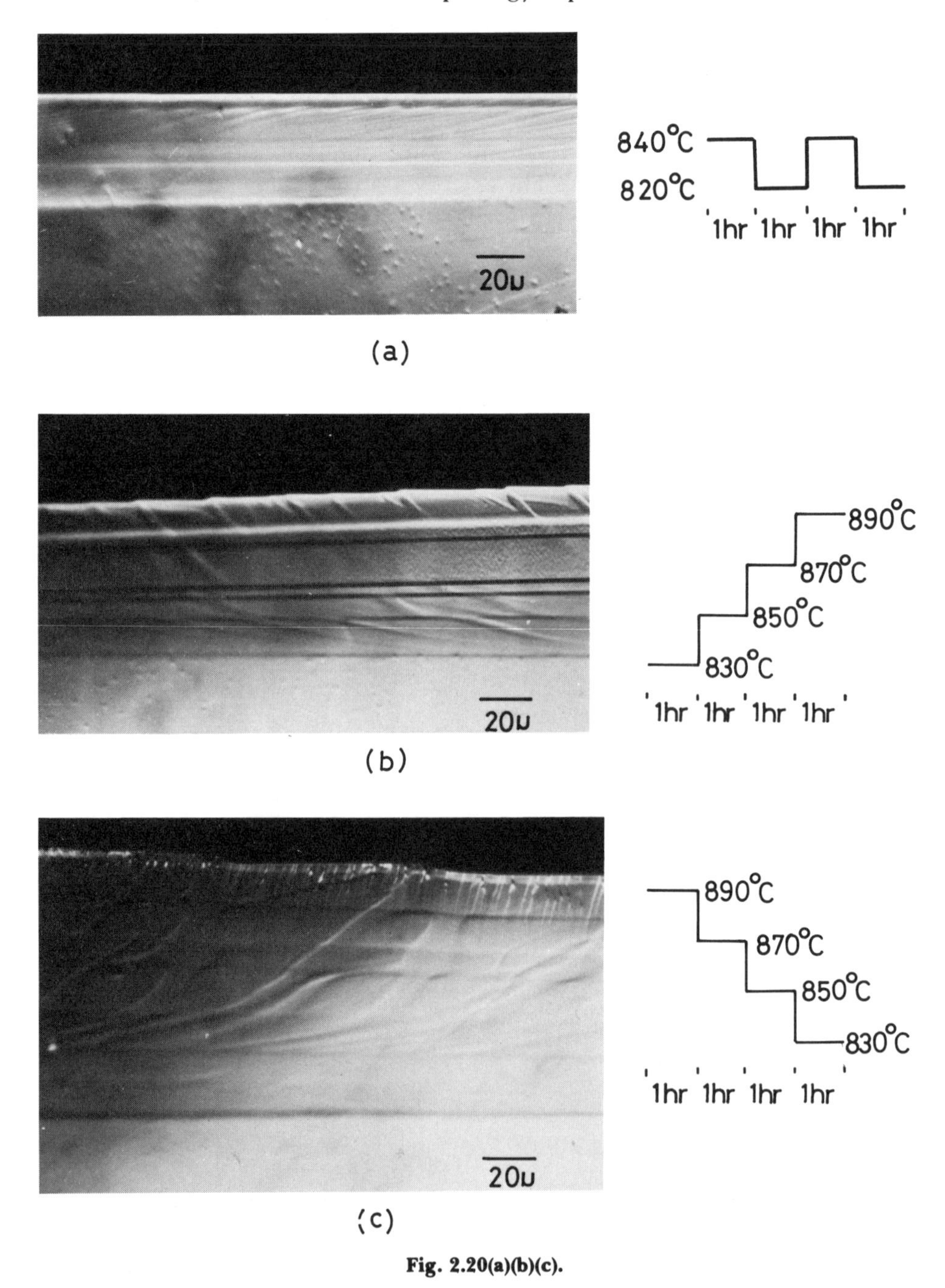

Fig. 2.20(a)(b)(c).

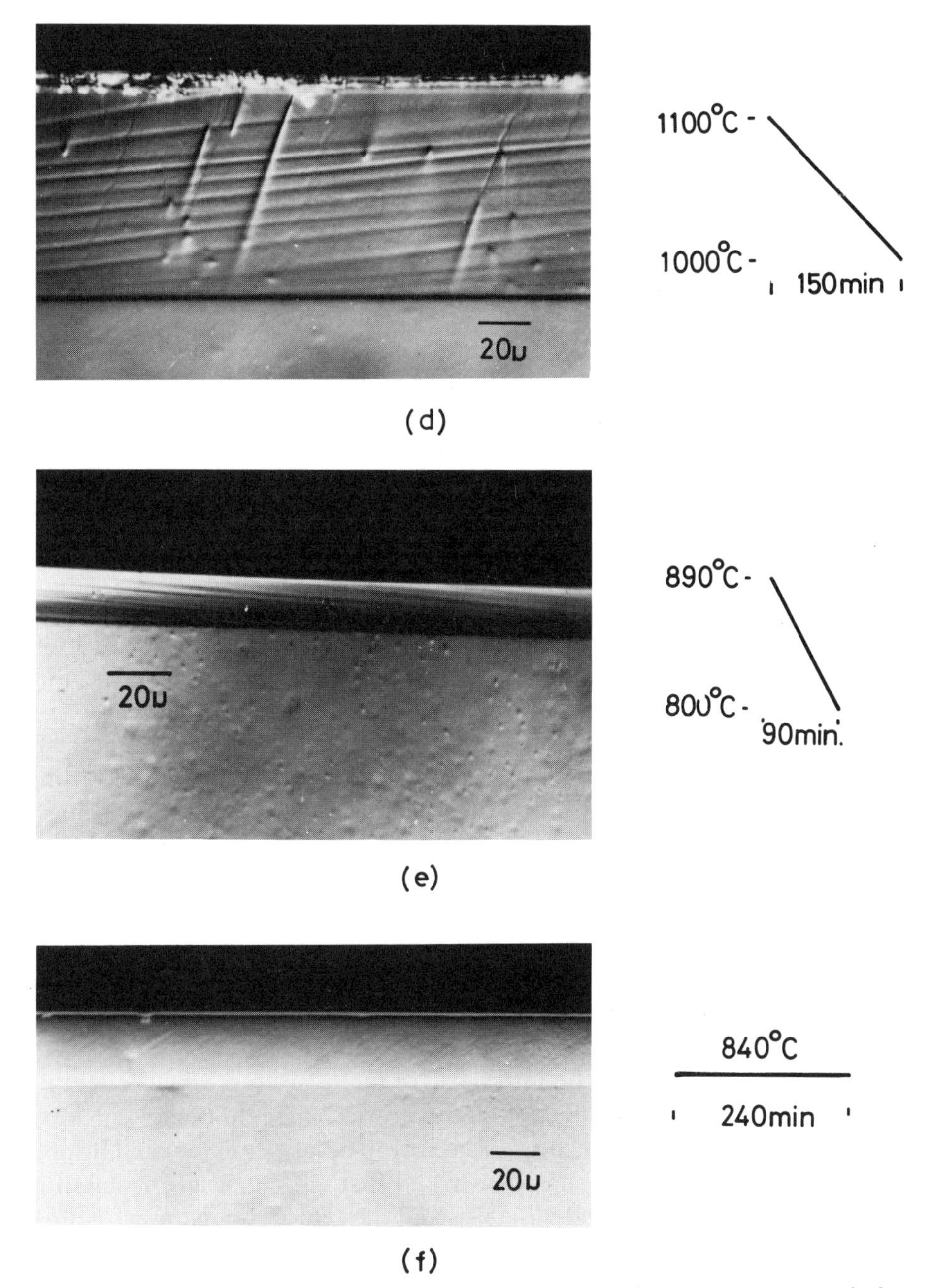

Fig. 2.20 Cleaved surface morphology and schematic diagram of growth temperature process is shown in the additional figures.

Fig. 2.20(f) shows the morphologies of the cleaved face and as-grown surface when the temperature was kept constant during the whole growth time for 2 h and growth rate was about 0.2 μm/min. In this case, neither lateral stripes nor root-like faults were observed. These results show that the temperature change for these growth layers generates stripes parallel to the interface.

Therefore it is clear that the temperature difference method is much more effective to grow defect-free crystals than the conventional slow-cooling method.

In the next section, the effect of the combination of TDM and the controlled vapor pressure (CVP) is described.

2.3.2. *Effect of vapor pressure on the epitaxial growth*

The method for applying the vapor pressure of the group-V element is shown in Fig. 2.18(a). The phosphorus (or arsenic) chamber at a temperature T_P (or T_{As}) provides the vapor pressure on the surface of the molten phase through a quartz tube, the vapor pressure being given by the same equation as Eq. (2.3) given for the heat-treatment experiment. The growth method is called the temperature difference method under controlled vapor pressure (TDM-CVP). The epitaxial growths of GaP and GaAs were carried out by TDM-CVP.

2.3.2.1. *GaP*

The phosphorus pressure added to the melt crucible is constant during each growth. In several runs, the phosphorus pressure P_2 was controlled in the range 1 to 1000 Torr, while the temperature of the red phosphorus ranges from 250 to 425°C. Substrate dimensions were $1.2 \times 1.2 \times 0.03$ cm^3. Growth rate of the epitaxial layers was 10 μm/h.

Fig. 2.21(a)–(d) show the morphology of the cleaved faces of the epitaxial layers grown at several phosphorus pressure for 4 h under the growth temperature cycling at an interval of 1 h at 840 and 810°C shown by the accompanying schematic diagrams.

At the phosphorus pressures of 1, 10 and 1000 Torr, a large number of root-like faults are observed. The lines due to temperature change are shown in the figures, but some are difficult to see and are shown as dashed lines. At the phosphorus pressure of 100 Torr, these faults, even the lateral stripes due to temperature change, are not observed, as shown in Fig. 2.21(c). These photographs are clear examples that the generation of faults is suppressed at an optimum phosphorus pressure. This phosphorus pressure is defined as the optimum phosphorus pressure ($P_{GaP,opt}$).

The lateral stripes parallel to the interface are faults produced by the temperature change and show the detrimental effect of temperature change on the crystal perfection. We found that the root-like faults in epitaxial layers can be largely eliminated by providing an optimum pressure ($P_{GaP,opt}$) of phosphorus, and that $P_{GaP,opt}$ is a function of growth temperature. The relationship between $P_{GaP,opt}$ and the growth temperature is almost the same that between $P_{GaP,min}$ and the annealing temperature of crystals in the heat treatment of GaP.[10)]

Figs. 2.22(a)–(f) show X-ray topographs taken by the Berg–Barrett method. The epitaxial layers were grown under several phosphorus pressures for a constant temperature (800°C), temperature difference (20°C) and growth time (3 h). A large number of defects are observed in the substrate, but the crystal perfection of crystals grown under CVP is improved compared with crystals grown without CVP.

The topographs show the clear phosphorus pressure dependence and no defect is revealed in the epitaxial layer grown at the optimum phosphorus pressure. In the crystals grown with phosphorus pressure lower or higher than $P_{GaP,opt}$, a greater number of

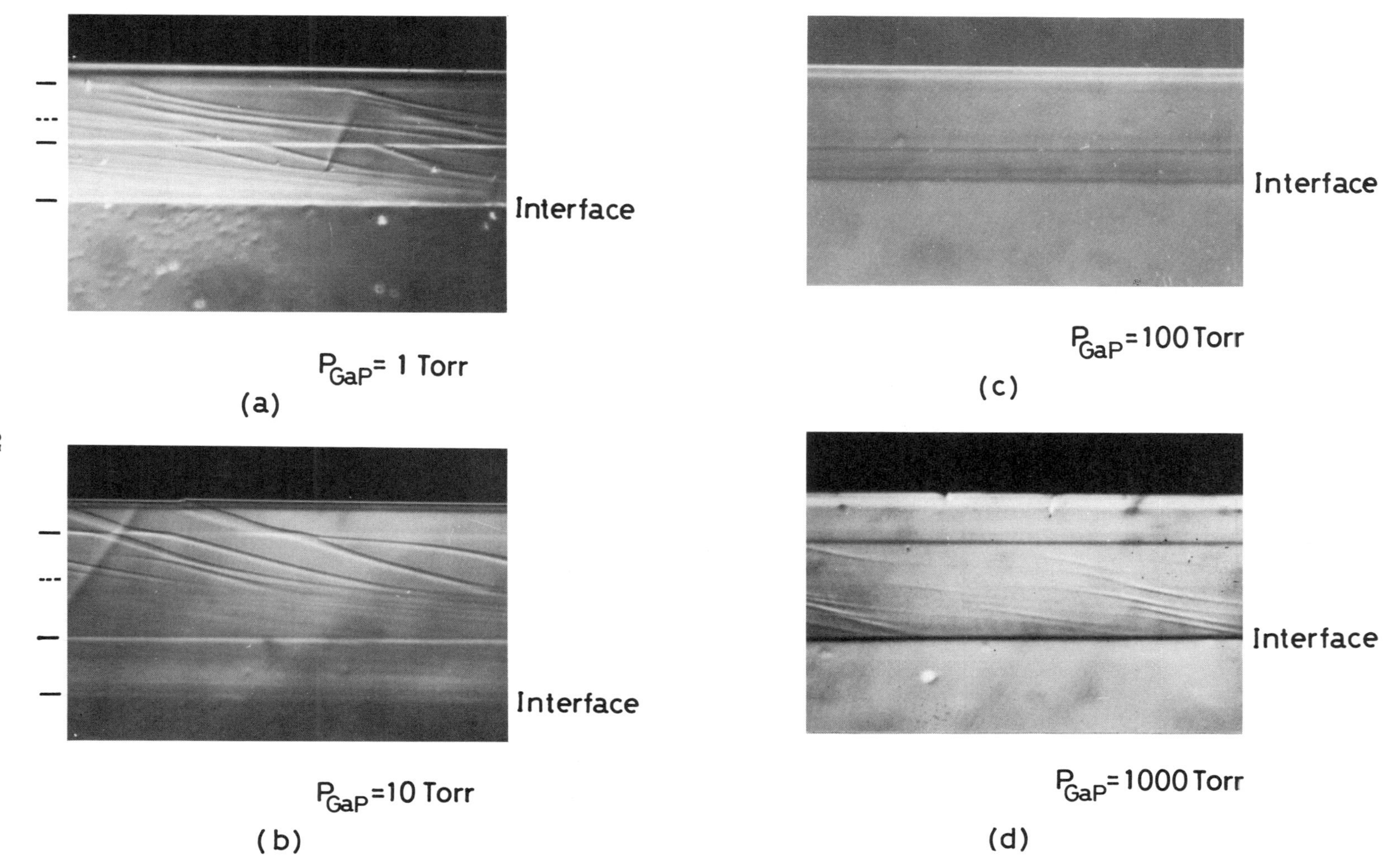

Fig. 2.21 Cleaved surface of layer grown at phosphorus vapor pressure of 1 Torr (a), 10 Torr (b), 100 Torr (c), and 1000 Torr (d).

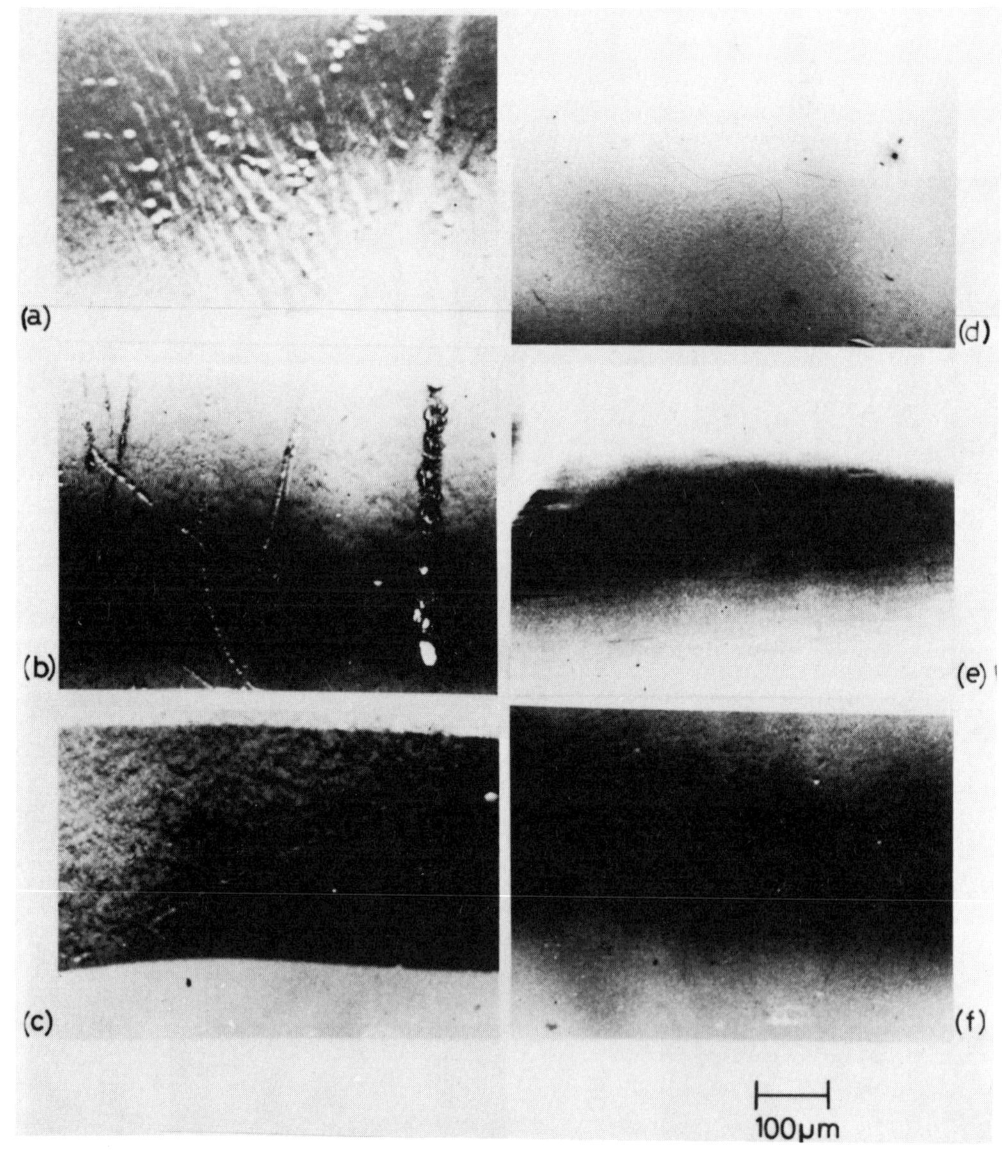

Fig. 2.22 X-ray topographs of grown layers at various phosphorus pressures: (a) without control of phosphorus pressure, (b) 1 Torr, (c) 10 Torr, (d) at optimum phosphorus pressure, (e) 300 Torr, (f) 1000 Torr.

defects are generated and the habits of the crystals such as surface morphology are also affected.

Fig. 2.23 shows the cleaved face of a layer grown at 750°C with the optimum pressure of 42.5 Torr; it exhibits a morphology without any visible defects. The dependence of dislocation density in the epitaxial layer on the phosphorus pressure applied during growth is shown in Fig. 2.24. We have obtained dislocation-free crystals by applying a pressure of $P_{GaP} = 67$ Torr at the growth temperature of 800°C. It is noted that the vapor pressure range to obtain perfect crystal is very narrow.

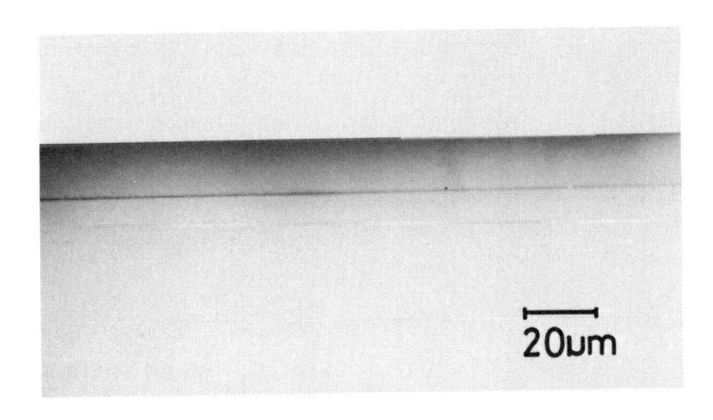

Fig. 2.23 Cleaved surface of layer grown at 750°C at phosphorus pressure of 42.5 Torr.

This optimum phosphorus pressure is in good agreement with the value of $P_{\mathrm{GaP,\,min}}$ obtained from the heat-treatment experiment given by Eq. (2.7) as shown in Fig. 2.17. Therefore, $P_{\mathrm{GaP,\,opt}}$ can be expressed by

$$P_{\mathrm{GaP,\,opt}} = 4.67 \times 10^{6} \exp - (1.01\ \mathrm{eV}/kT_{\mathrm{g}})\ \mathrm{Torr}. \tag{2.8}$$

The electrical properties of the GaP grown layer are examined by Hall measurements and C–V measurements of Schottky diodes as shown in Fig. 2.25. The donor concentration N_{D} was calculated from the temperature dependence of the carrier concentration. The surface carrier concentration was measured over the whole surface. The donor

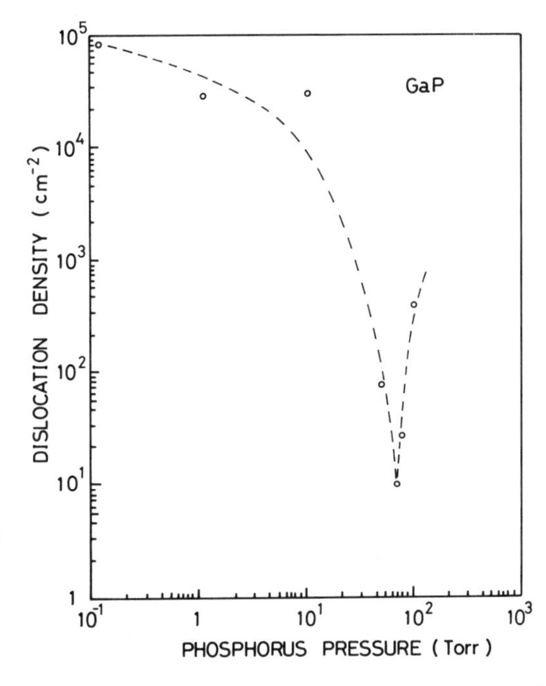

Fig. 2.24 Phosphorus pressure dependence of dislocation density in the epitaxial layer. Nearly dislocation-free crystals were obtained by applying $P_{\mathrm{GaP}} = 67$ Torr. The growth temperature was 800°C.

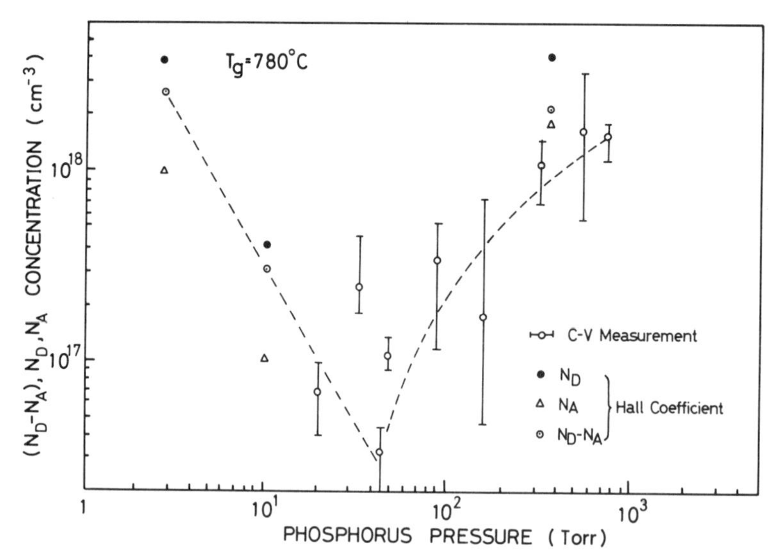

Fig. 2.25 Dependence of electrical properties of GaP epitaxial layers on phosphorus pressure: ○: surface carrier concentration measured by C–V; ●: donor concentration measured by Hall measurement.

concentration N_D becomes a minimum at about the same $P_{GaP, opt}$ obtained in surface morphology and X-ray topography studies. The value of $P_{GaP, opt}$ is slightly lower because of the lower growth temperature (780°C).

The photocapacitance characteristics were measured to study deep-level impurities in the GaP light-emitting diode constituted by the double epitaxial layers grown by TDM-CVP at the growth temperature of 800°C as shown in Fig. 2.26. Two peaks exist in

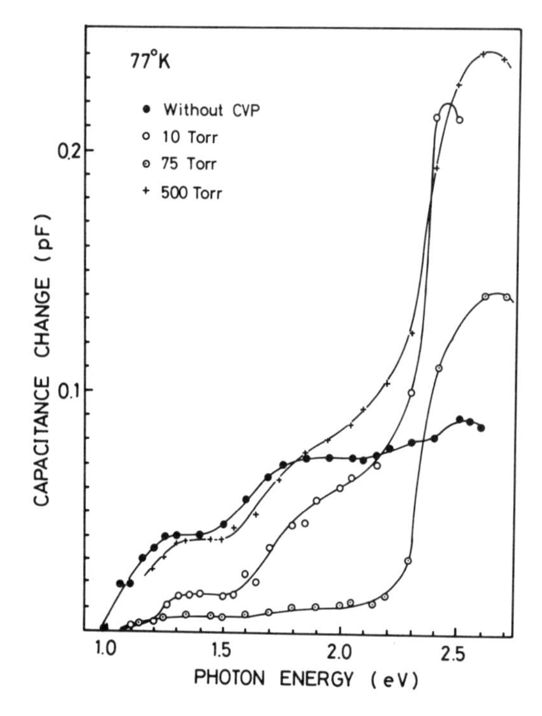

Fig. 2.26 Typical photocapacitance spectra at each phosphorus pressure. The peaks in the lower energy region do not exist at $P_{GaP} = 75$ Torr.

the lower-energy region in the case of $P_{GaP} = 10$ Torr, 500 Torr and without CVP. However, the diode grown under the optimum pressure ($P_{GaP} = 75$ Torr) has no peaks in the lower-energy region. It is clear that the introduction of deep impurities or defects can be suppressed by applying the vapor pressure of $P_{GaP} = 75$ Torr.

2.3.2.2. *GaAs*[1,13,14)]

A first experiment of TDM-CVP was carried out for the epitaxial growth of GaAs from Sn solvent. Fig. 2.27 shows the arsenic pressure dependence of the lattice constant. It is observed that the lattice constant of the grown layer is a little larger than that of the substrate and shows a maximum at about 4 Torr at the growth temperature of 630°C. Following this experiment, the epitaxial growth of the GaAs using Ga solvent was carried out by a similar growth system as that of GaP.

The substrates used in this experiment were Cr-doped GaAs with surface parallel to

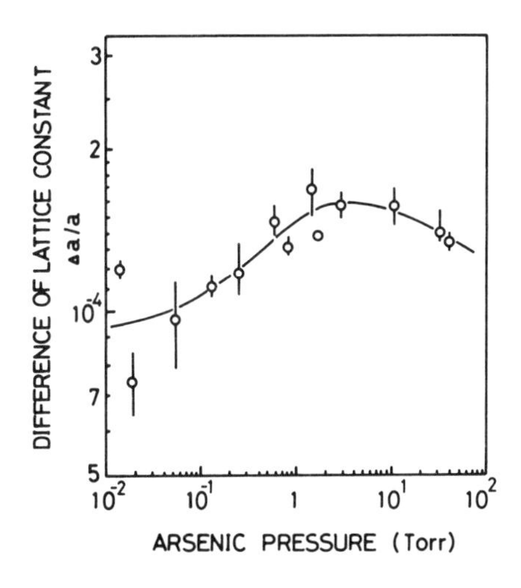

Fig. 2.27 Arsenic pressure dependence of the difference of lattice constant $\Delta a/a$ in the substrate and Sn-doped GaAs layers.

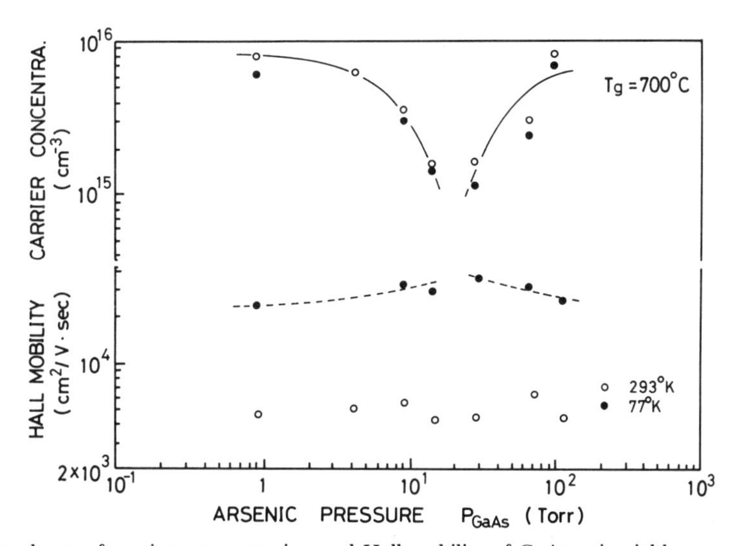

Fig. 2.28 Dependence of carrier concentration and Hall mobility of GaAs epitaxial layers on arsenic pressure.

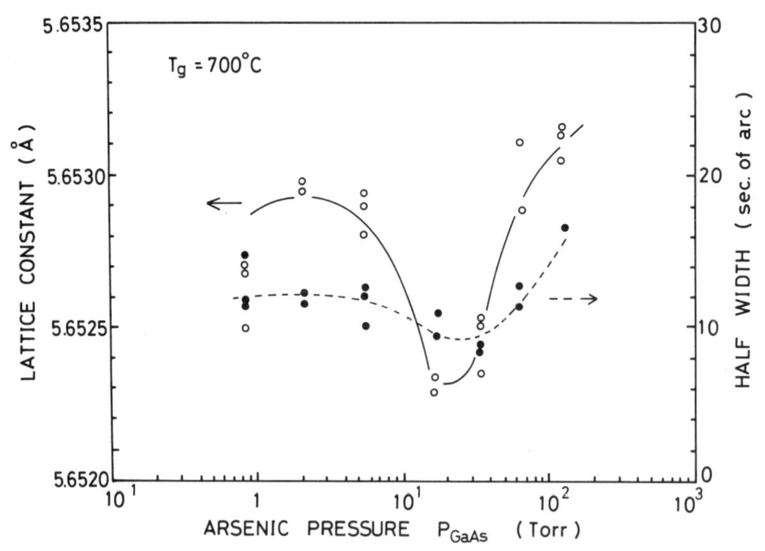

Fig. 2.29 Dependence of lattice constant and half-width of rocking curve on arsenic pressure.

the (100) plane. The electrical properties of epitaxial layers grown under various constant arsenic pressures at growth temperatures from 530 to 800°C, were measured. The arsenic pressure dependence of the electrical properties are nearly the same and then the results of crystal grown at 700°C is shown as a representative example. Fig. 2.28 shows the result for $T_g = 700$°C. Epitaxial layers are n-type and carrier concentrations are reduced to minimum and the Hall mobility becomes a maximum at a specified arsenic pressure. This arsenic pressure is defined as the optimum arsenic pressure ($P_{GaAs, opt}$) and this value covers a narrow pressure range.

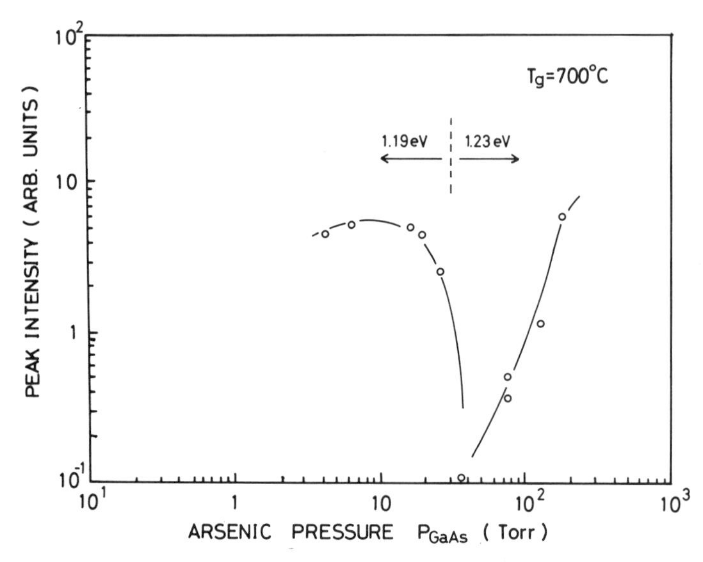

Fig. 2.30 Dependence of photoluminescence intensity on arsenic pressure.

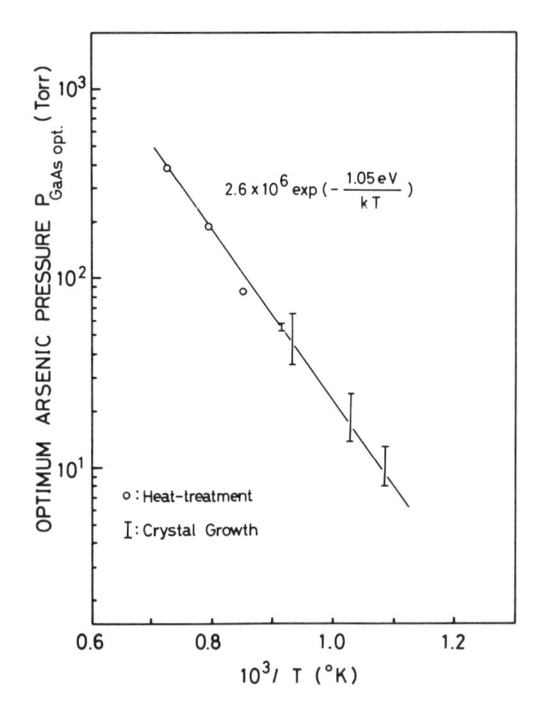

Fig. 2.31 Relation between optimum arsenic pressure $P_{GaAs,opt}$ and growth temperature. The solid circle is the specific arsenic pressure $P_{GaAs,min}$.

There are two possibilities for n-type conductivity; one is that non-stoichiometric defects for both low and high vapor pressure regions are donor-like, another is that incorporation of impurities are controlled by the concentration of non-stoichiometric defects.

Lattice constants and half widths of X-ray rocking curve measured by the double crystal method have a minimum at the same $P_{GaAs, opt}$ as shown in Fig. 2.29. Photoluminescence measurements were also made on the same samples. No arsenic pressure dependence of the peak intensity of the band-to-band transition was observed, but the deep-level luminescence at 1.23 eV and 1.19 eV showed an apparent arsenic pressure dependence as shown in Fig. 2.30. Different types of defect are considered to contribute in the lower and higher arsenic pressure regions. It is clear that the concentration of deep centers responsible for the 1.23 eV and 1.19 eV peaks decreases as $P_{GaAs, opt}$.

Fig. 2.31 shows the optimum arsenic pressure as a function of growth temperature, together with the specific arsenic pressure $P_{GaAs, min}$ for the heat-treatment described in Section 2.2. These two kinds of pressure are on the same line described as follows:

$$P_{GaAs, opt} \approx 2.6 \times 10^6 \exp(-1.05\,\mathrm{eV}/kT)(\mathrm{Torr}). \quad (2.9)$$

This result clearly shows that deviation from stoichiometry of GaAs grown by TDM-CVP can be controlled by the control of arsenic pressure on the surface of the molten phase, and $P_{GaAs, opt}$ should give the exact stoichiometry.

Then, we investigated the dislocation behaviors in non-doped GaAs epitaxial layers grown by TDM-CVP. MSO (modified super oxisol) etchant which is a solution of $HF + H_2O_2 + 10H_2O$ reveals clear etch projection corresponding to dislocations, the etching depth being as thin as 3000 Å.[18] The etch projection density (EPD) of epitaxial

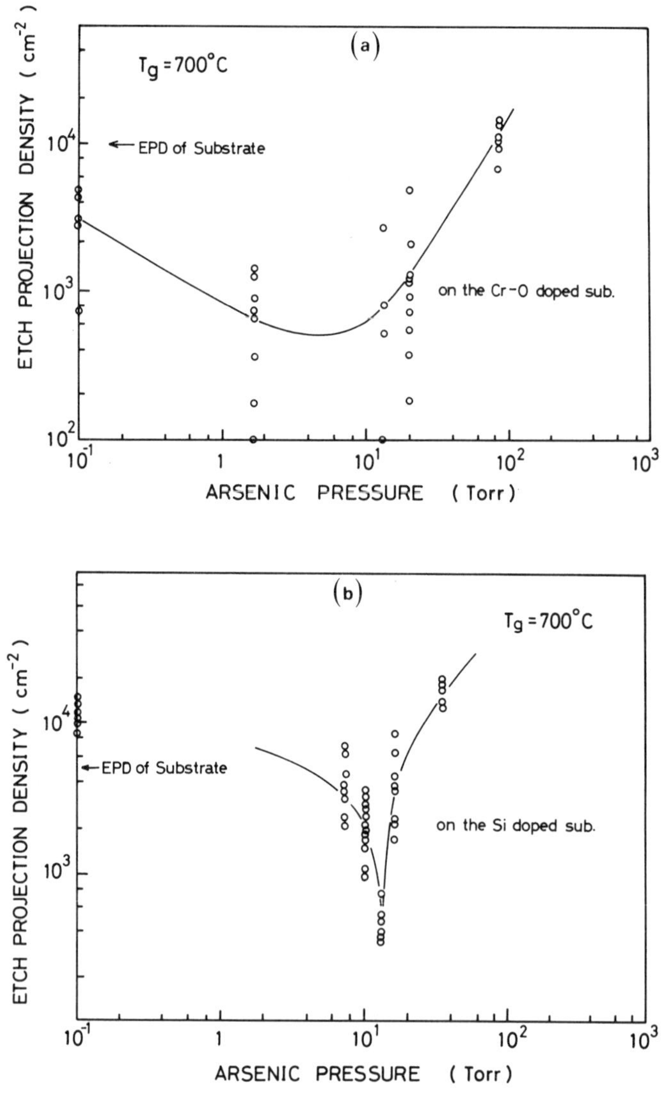

Fig. 2.32 Arsenic pressure dependence of etch projection density (EPD) of epitaxial layer on Cr–O-doped GaAs (a) and Si-doped GaAs (b).

layers grown on Cr–O-doped and Si-doped substrates varies with arsenic overpressure as shown in Fig. 2.32(a), (b).

It should be noted that EPD has its minimum value at $P_{\mathrm{GaAs,\,opt}}$. This fact suggests that the arsenic overpressure during crystal growth affects the propagation of dislocations from the substrate into the epitaxial layer and the introduction of dislocations from the crystal surface.

Although the vapor pressure is likely to control the concentrations of non-stoichiometric point defects like As vacancies and As interstitial atoms as well as Ga vacancies and Ga interstitial atoms, it is noted that the introduction and propagation of dislocations needs the supply of these point defects.

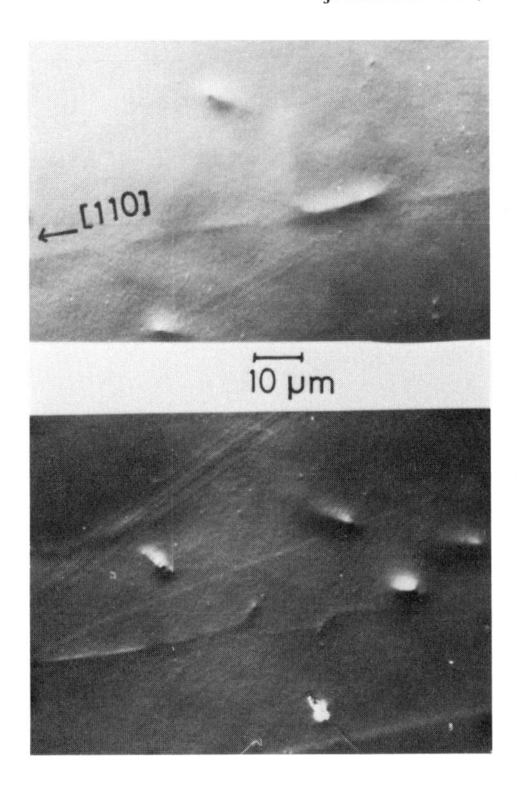

Fig. 2.33 Dislocation lying along ⟨110⟩ direction observed in a interface region of epitaxial layer revealed by MSO etchant.

After removal of substrates, we investigated dislocation images in epitaxial layers by using anomalous transmission X-ray topography. The density of dislocations clearly decreases at $P_{GaAs,opt}$ as shown in Fig. 2.32. One mechanism for the decrease of dislocations in the epitaxial layers is reduction of the propagation of dislocations from the substrate into the epitaxial layers. Fig. 2.33 shows the dislocations lying along the ⟨110⟩ direction observed in the interface region between the substrate and the epitaxial layer. By precise investigation using X-ray topography, it has been found that these dislocations are the so-called 60° type. In addition, transmission X-ray topographs of the cleaved face reveal directly that the dislocations in the substrate seem to interact with these dislocations lying along the ⟨110⟩ direction in the interface region as shown in Fig. 2.34. So some of the dislocations in the substrate bend in the interface region and do not propagate into the epitaxial layers. Arsenic overpressure is thought to affect propagation phenomenon.

In the LPE samples grown by TDM-CVP, the density of stacking faults increases with increase of arsenic overpressure as shown in Fig. 2.35. Anomalous transmission X-ray topographs reveal partial dislocation images surrounding stacking faults as shown in Fig. 2.36. From the precise investigation of dislocation images using various kinds of diffraction planes, it has been found that these stacking faults are of the so-called Frank-type with the Burgers vector of $\frac{1}{3}a\langle 111\rangle$.[18] In general, Frank-type stacking faults are formed by the preferential condensation of vacancies, interstitial atoms and various kind of impurities on the ⟨111⟩ plane. These stacking faults relate to As interstitial atoms.

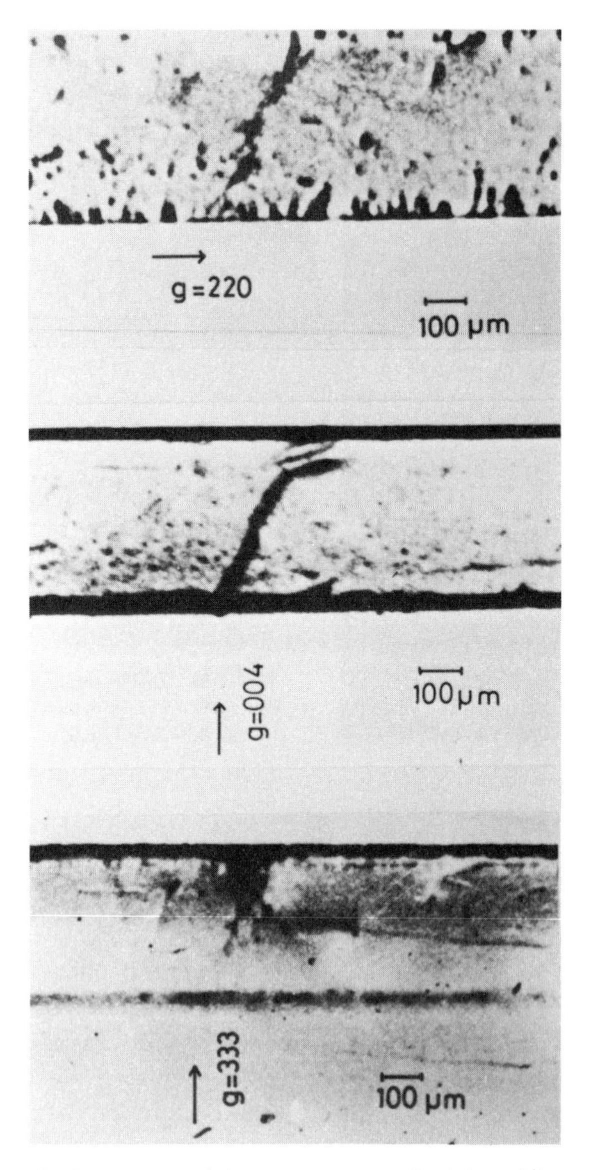

Fig. 2.34 Transmission X-ray topograph of cleaved face.

2.3.3. *Impurity effect*

The effect of impurities on the non-stoichiometry is important. Lattice sites of silicon in the GaAs were found to change as a function of arsenic vapor pressure.

Epitaxial growth of Si-doped GaAs was carried out by TDM-CVP. Contamination of the growth solution by oxygen must be prevented in this experiment, because oxygen combined with Si atoms to form silicon oxide which is inactive as an electrical carrier. Therefore special care was given to avoid the oxidation of the graphite crucible and to remove arsenic oxide on the surface of metallic As. We have thus grown high-purity GaAs layers of lower than $1 \times 10^{13}/cm^3$ carrier density in the case of no impurity doping. The

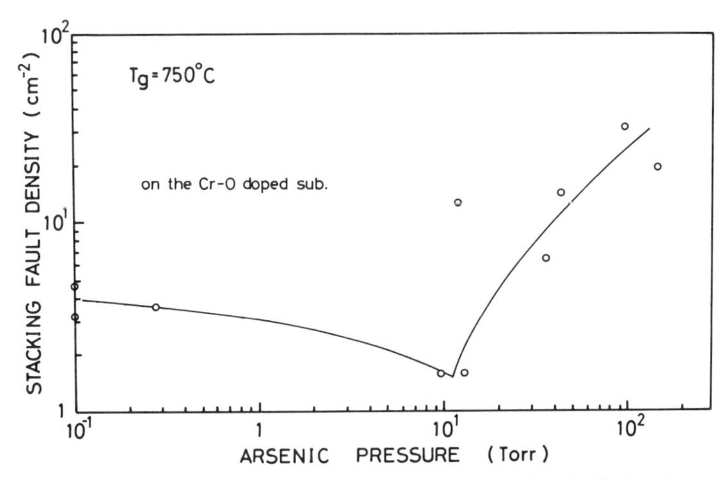

Fig. 2.35 Arsenic pressure dependence of stacking fault density.

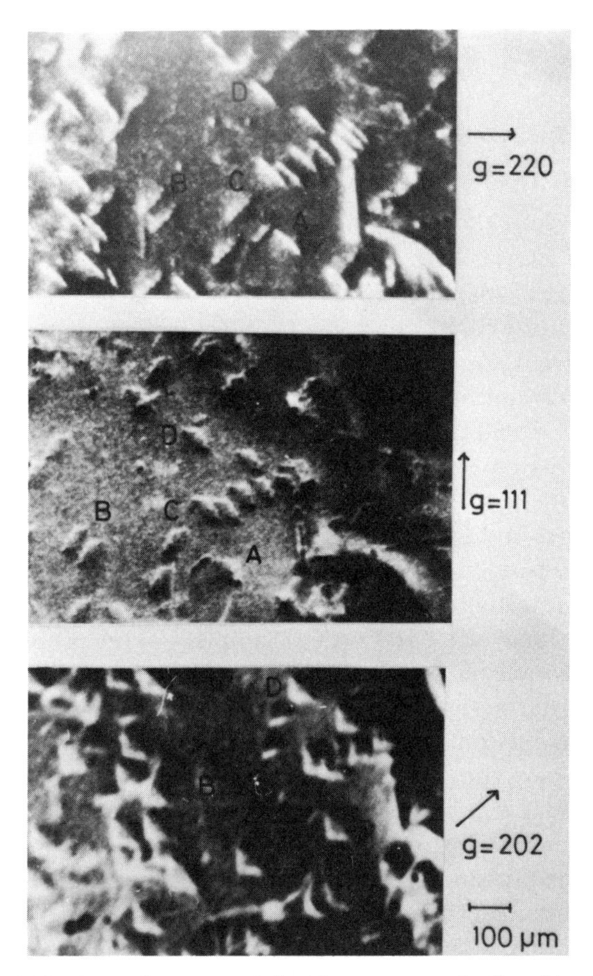

Fig. 2.36 Anomalous transmission X-ray topograph of stacking faults.

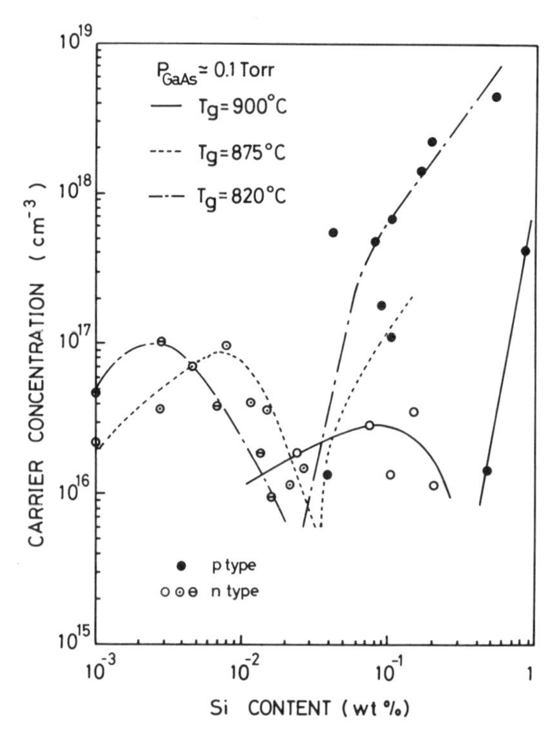

Fig. 2.37 Relation between carrier concentration and Si content in Ga solution.

growth thickness of the layers was about 20 μm irrespective of the applied As pressure.

The dependence of the electrical properties on the doping amounts of Si was investigated and the result was shown in Fig. 2.37. Epitaxial growth was performed at the constant temperatures of 820, 875 and 900°C under a constant As pressure of 0.1 Torr. Crystals segregated from the solution with lower concentration of Si showed n-type conduction. The carrier concentration increased with Si content and showed a maximum at 3×10^{-3} wt% for $T_g = 820$°C and at 5×10^{-3} wt% for $T_g = 875$°C. With increasing Si content, the conductivity type converted to p-type and the hole density increased with Si content. The Si contents where the p–n conversion occurred were different according to the growth temperature and has a temperature dependence.

The As pressure dependence of the electrical properties with the doping quantities of Si kept constant was investigated. The Si contents in the Ga solution were selected as 3.0×10^{-3}, 9.1×10^{-3} and 1.1×10^{-1} wt%. Samples were prepared at constant of 820°C under controlled arsenic pressure (P_{GaAs}) in the range of 0.1 to 500 Torr. As the pressure dependences of the carrier density and Hall mobility when Si/Ga is 3×10^{-3} wt% are shown in Figs. 2.38 and 2.39. Two well-defined types of conversions were observed at certain As pressures where the carrier concentrations become a minimum. Especially, the conversion As pressure at $P_{GaAs}\approx 55$ Torr was very close to the value of the optimum As pressure where stoichiometric composition was achieved was mentioned above. The crystals grown at lower pressures than 55 Torr should have higher concentration of As vacancies and the Si atoms replacing the As lattice sites are expected to act as an acceptor or vice versa. Therefore, the Si-doped GaAs crystals will be converted from p-type to n-type at the boundary of $P_{GaAs} = 55$ Torr while the As pressure dependence of the Hall

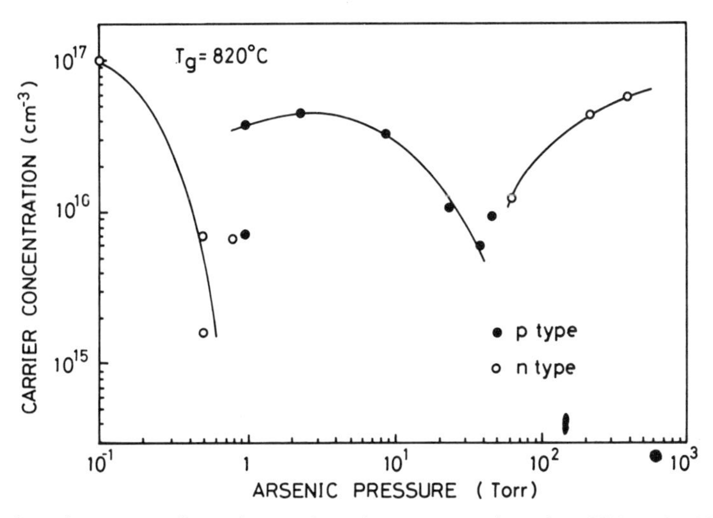

Fig. 2.38 Arsenic pressure dependence of carrier concentration when Si/Ga $\approx 3 \times 10^{-3}$ wt%.

mobility showed that the value became larger towards $P_{GaAs} \approx 55$ Torr. The half-width of the X-ray rocking curves increased with P_{As} higher than the optimum pressure which gave a minimum half-width and the best crystallographic quality as shown in Fig. 2.40. This may be attributed to the increase of the concentration of interstitial As atoms incorporated with Si atoms which replaced some of the Ga lattice sites in the high As pressure region. There were insufficient data on the other conversion at about 1 Torr for any definite conclusion to be drawn.

It appeared that as the content of Si atoms increased the first conversion at about 55 Torr shifted towards lower pressures. Hence Si atoms can readily replace the Ga lattice sites or carry out the replacement of the As sites to the extent attainable. The minimum

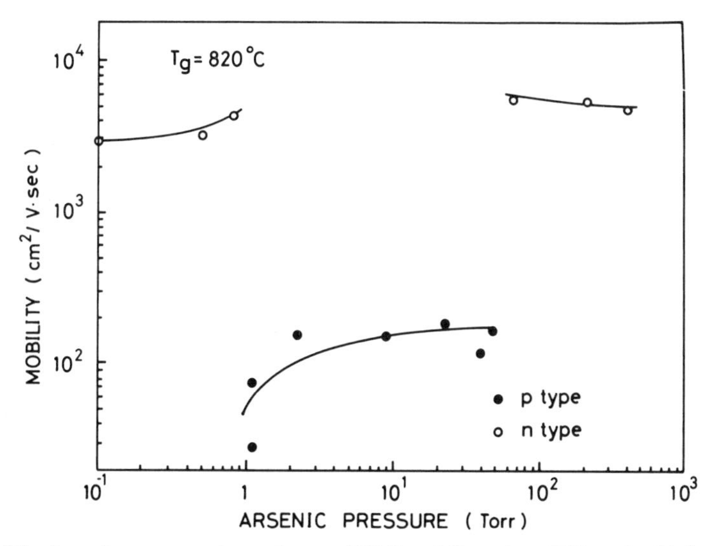

Fig. 2.39 Arsenic pressure dependence of Hall mobility when Si/Ga $\approx 3 \times 10^{-3}$ wt%.

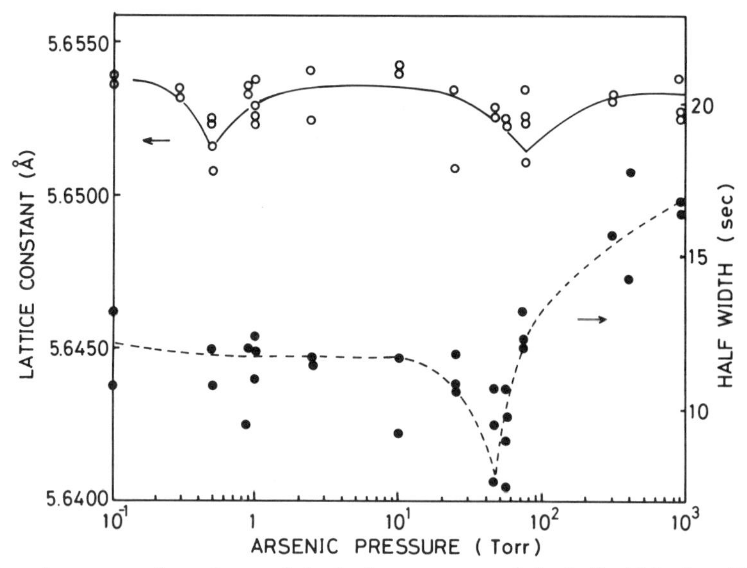

Fig. 2.40 Arsenic pressure dependence of the lattice constant and the half-width of rocking curve.

lattice constants were observed at 20 Torr for Si/Ga $\approx 9.1\times10^{-3}$ wt% and at 2 Torr for Si/Ga $\approx 1.1\times10^{-1}$ wt%.

2.3.4. *Bulk crystal growth of GaAs by As pressure controlled Czochralski method*

As mentioned in Section 2.1, TDM-CVP can be applied to the melt growth. For this purpose the authors developed a new growth technique by the Czochralski method under a controlled As vapor pressure as shown in Fig. 2.41.[19] The arsenic pressure is controlled by the temperature of the As deposition zone, T_{As}.

The quartz chamber prevents the escape of arsenic overpressure. An effective seal between the chamber and the pull rod as well as the crucible supporting rod is made by a sealing material, molten B_2O_3, in Mo lids.

Two crystals of GaAs, No. 1 and No. 2, were pulled under the As-pressure controlling temperature of 616°C and 620°C. In both cases, the temperature gradient just above the melt surface was made about 10°C/cm. This value was much lower than the case of the commercial LEC method (100°C/cm).

The radial etch-pit density distributions across wafers cut from the shoulder of the rod are shown in Fig. 2.42. The EPD distributions are U-shaped and the EPDs around the center of wafers are as low as 2×10^3 cm^{-2}, which is lower than the values in commercial undoped LEC crystals by a factor of 10. Another important result is that the EPD in the crystal 1 is lower than that in the crystal 2. This fact suggests the preciseness of Eq. (2.1) for the optimum vapor pressure because the optimum As temperature for the melt growth is 617°C as was mentioned in Section 2.1. Further investigation based on the precise control of the As pressure is now being made to evaluate the net effect of the stoichiometry on the dislocation density.

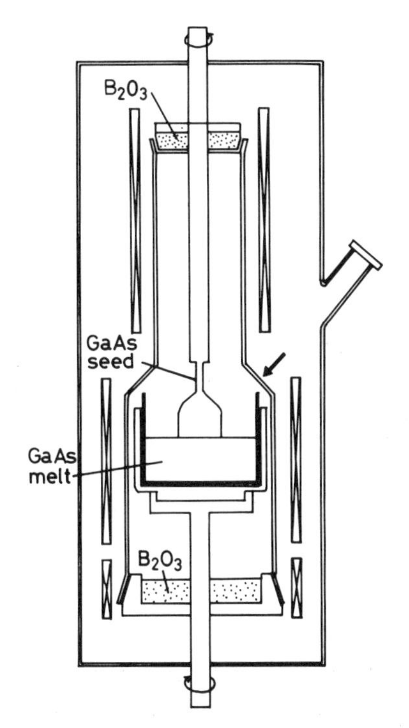

Fig. 2.41 Schematic diagram of the furnace of Czochralski method under controlled As vapor pressure.

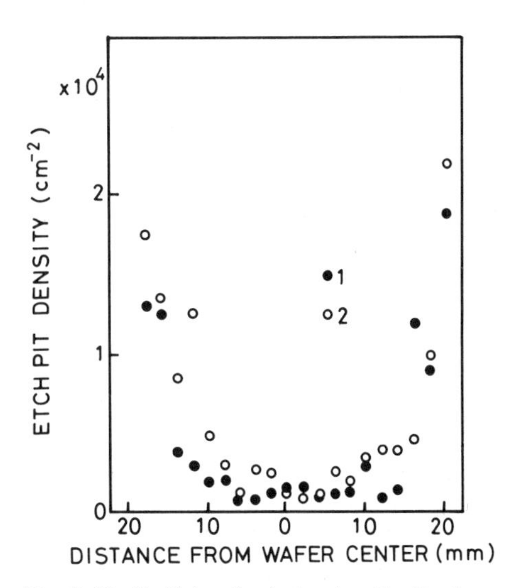

Fig. 2.42 Radial etch-pit density distribution across wafer cut from the shoulder of the rod.

2.4. Photocapacitance spectroscopy of deep levels related to non-stoichiometry

The origins of deep levels in semiconductors are various impurities, lattice defects and complex defects. The segregation of impurities and the formation of lattice defects are closely related to the deviation from the stoichiometric composition of the compound semiconductor.

The "photocapacitometry", which the authors were the first to develop, measures the change of the capacitance of p–n junctions or Schottky diodes as a function of photon energy of illuminating monochromatic light, and is very sensitive for the detection of deep levels.

Fig. 2.43 shows the photocapacitance spectra of nitrogen-free (N-free) and nitrogen-doped (N-doped) GaP LED at each phosphorus pressure applied during crystal growth.[20] The numerous deep levels are detected at lower or higher phosphorus pressure, and at an optimum phosphorus pressure these deep levels are suppressed. Comparing N-doped GaP with N-free GaP, nitrogen atoms acting as emission centers also seem to form various deep levels.

The concentration of deep levels ($E_T = 0.65$ eV, 1.49 eV, 1.90 eV) in N-free GaP depends on the applied phosphorus pressure as shown in Fig. 2.44. These data indicate that some of these deep levels have a close relation with the deviation from the stoichiometric composition.[21] Another characteristic of these deep levels is shown in Fig. 2.45. The deep-level concentration (N_T) of $E_T = 0.65$ eV is related to the shallow donor

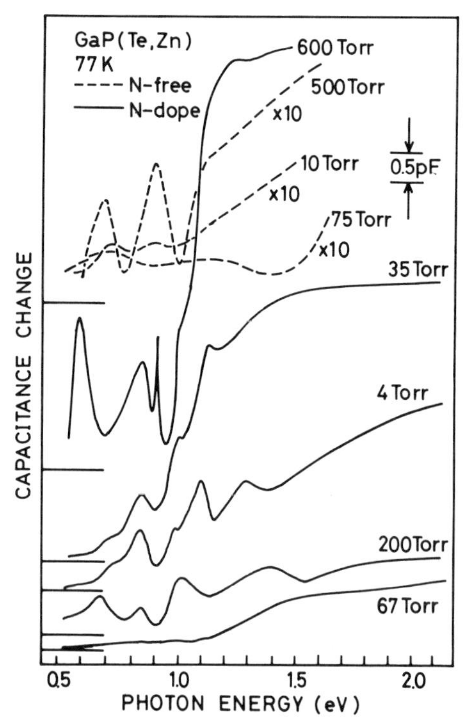

Fig. 2.43 Photocapacitance spectra of N-free and N-doped GaP LED at each phosphorus pressure. The numerous deep levels appear at lower or higher phosphorus pressure.

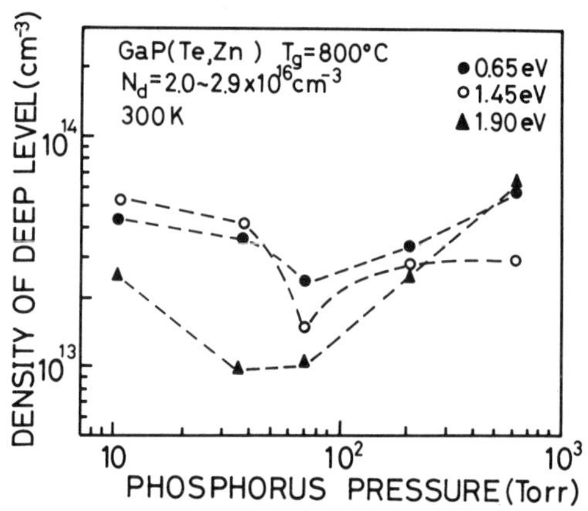

Fig. 2.44 Phosphorus pressure dependence of concentration of deep levels in N-free GaP.

concentration (N_D) by $N_T \propto N_D^{0.8}$. This tendency implies that these 0.65 eV deep levels are complex defects which involve donor impurity atoms and non-stoichiometric defects such as phosphorus vacancies or phosphorus interstitial atoms.

Similar structures of deep levels were detected at 2.10 eV and 2.20 eV in S-doped and Te-doped GaP, respectively, as shown in Fig. 2.46(a), (b). They are excitation probability

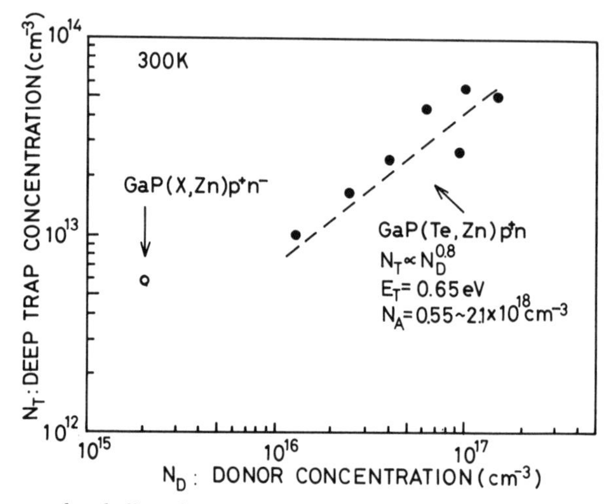

Fig. 2.45 Relation between the shallow donor concentration and the deep trap concentration of $E_C - E_T \fallingdotseq$ 0.65 eV.

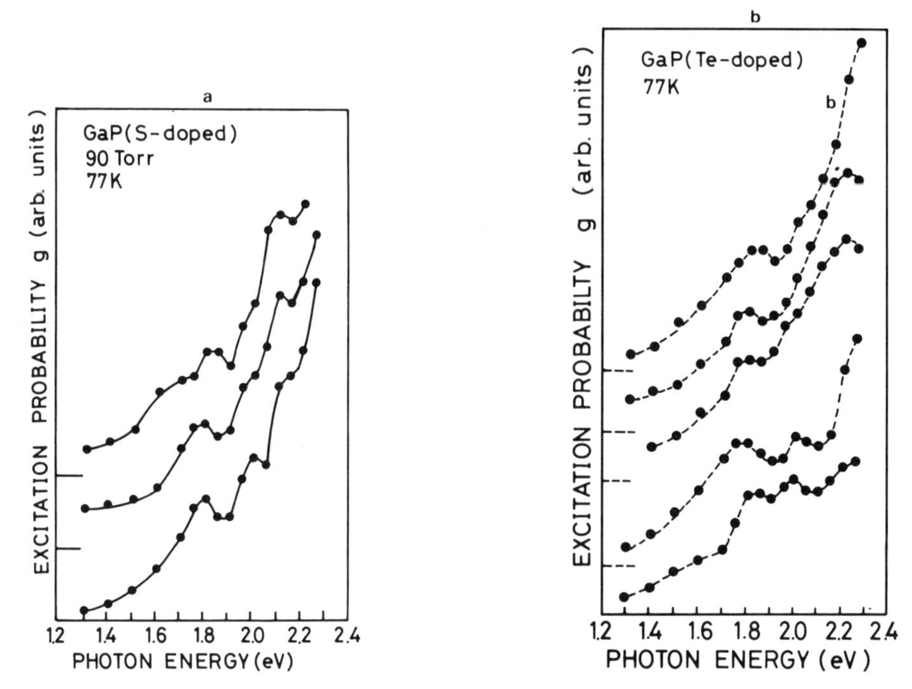

Fig. 2.46 Excitation probability at each photon energy; (a) S-doped GaP, (b) Te-doped GaP.

spectra, which are very effective for measuring deep-level energies in detail, compared with the saturated photocapacitance spectra shown in Fig. 2.47. Their densities are higher at lower phosphorus pressure. Also, in view of the fact that the concentration of the 2.10 eV deep level increases with the sulphur donor concentration as shown in Fig. 2.48, these deep levels are considered to be complex defects of donor atoms and non-stoichiometric defects such as phosphorus vacancies.

Another type of deep-level structure at 2.00 eV is found intensively at higher phosphorus pressure, which can be distinguished by the comparison in Fig. 2.49.

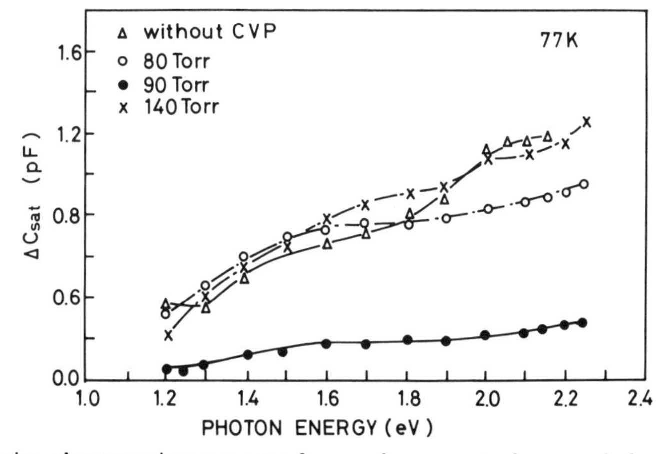

Fig. 2.47 Saturation photocapacitance spectra for samples grown under several phosphorus pressures.

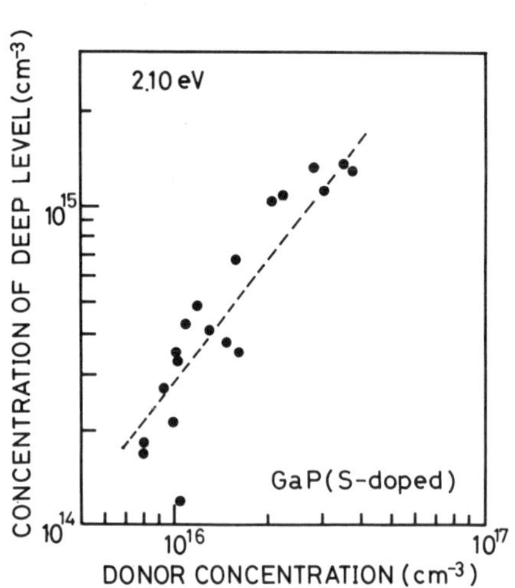

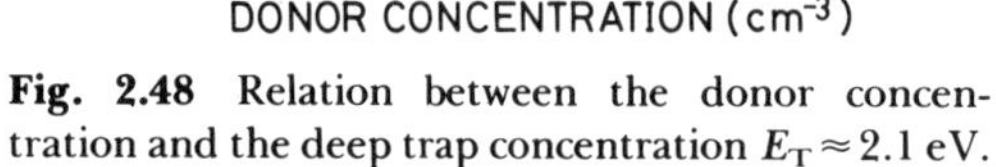

Fig. 2.48 Relation between the donor concentration and the deep trap concentration $E_T \approx 2.1$ eV.

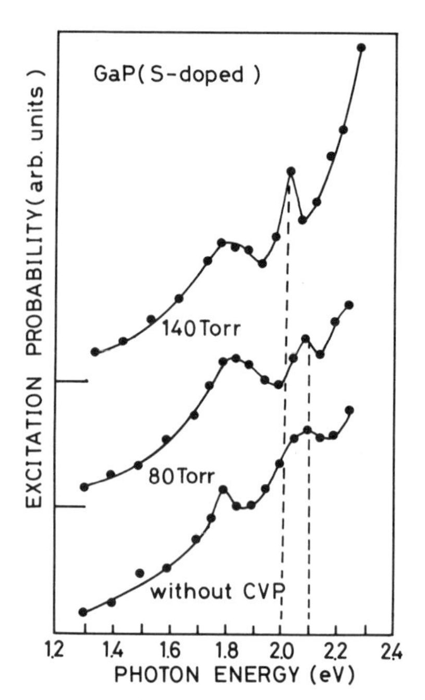

Fig. 2.49 Excitation probability for the samples grown under several phosphorus pressures.

Fig. 2.50 confirms that this 2.00 eV deep level is inclined to be formed at higher phosphorus pressure. The 2.00 eV level is detected independently of the kinds of donor impurity atoms, so one of the models of this level could be considered such as interstitial phosphorus atoms.

These deep levels significantly reduce the emission efficiency of pure green GaP LED. Fig. 2.51 shows the correlation of brightness of LED's with the 0.65 eV deep trap concentration.

As stated above, the deep levels in GaP depend strongly on the applied phosphorus

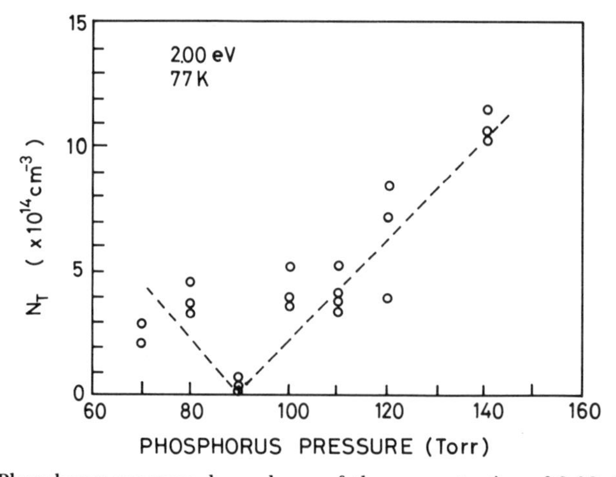

Fig. 2.50 Phosphorus pressure dependence of the concentration of 2.00 eV deep level.

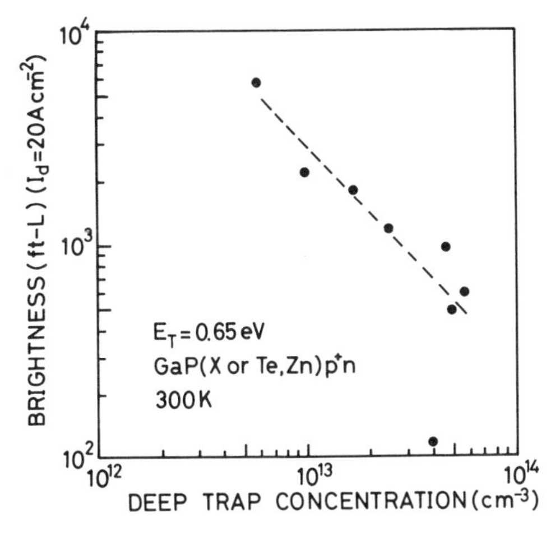

Fig. 2.51 Correlation of brightness of LED with the 0.65 eV deep trap concentration.

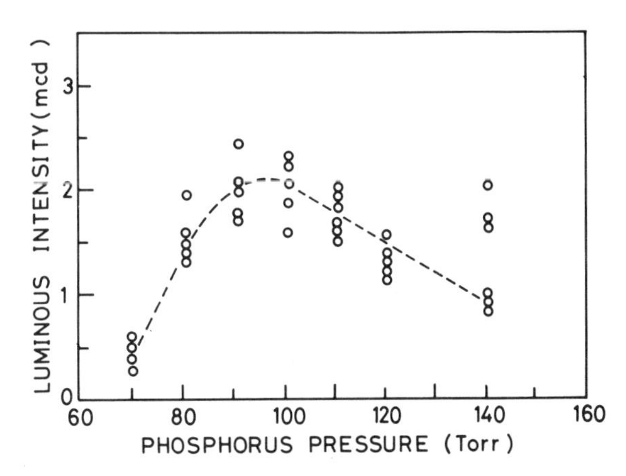

Fig. 2.52 Phosphorus pressure dependence of the luminous intensity of GaP LED.

pressure. The origins of deep levels are the non-stoichiometric defects or the complex defects due to combination of impurity atoms and non-stoichiometric defects. Under the optimum phosphorus pressure, these deep-level defects can be suppressed, and the emission efficiency of GaP LED can be enhanced as shown in Fig. 2.52.[22)]

2.5. Lattice fitting in heterojunctions

Fig. 2.53 shows the misfit dislocation at the GaAs–GaAlAsP heterojunction grown by TDM-CVP.[23)] The interaction between misfit dislocations lying along the [110] and [110] direction where the electron microscope observation was made in the direction vertical to the interface. The lattice constant of the epitaxial layer changed with the vapor pressure applied during the growth as shown in Fig. 2.29. This technique can also be applied to the

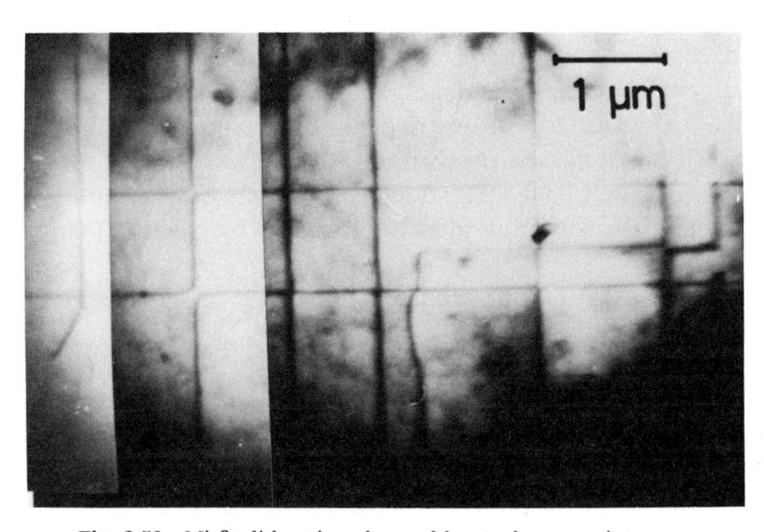

Fig. 2.53 Misfit dislocation observed by an electron microscope.

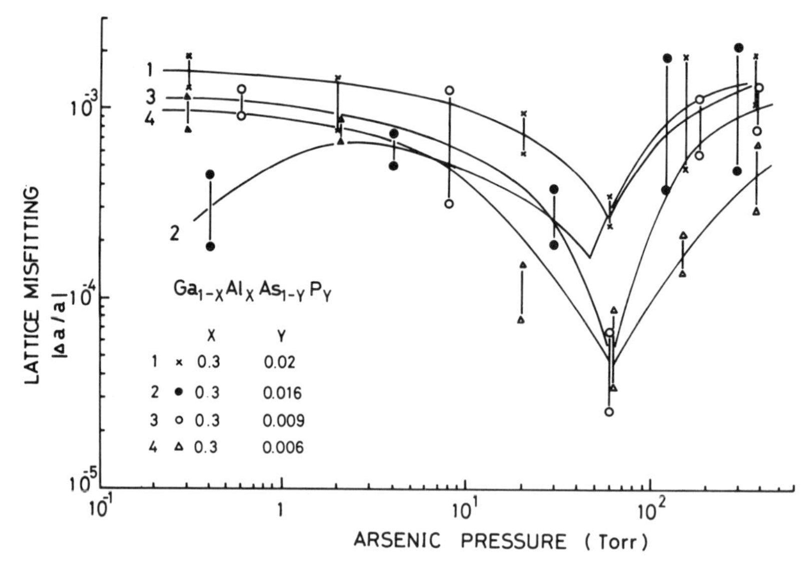

Fig. 2.54 Arsenic pressure dependence of the lattice misfit between the GaAlAsP epitaxial layer and the GaAs substrate.

growth of perfect lattice-fitting heterojunctions. Nishizawa has proposed that the operation lifetime of the semiconductor laser diode with double heterostructure is limited by the lattice misfit at the heterojunction.[24)] The lattice constant of the grown layer cannot be determined by the growth condition in the conventional growth method in which the vapor pressure is not controlled. Therefore the control of vapor pressure plays an important role and the lattice constant of the layer can be controlled through the control of non-stoichiometry of the grown layer. The composition of the mixed crystal did not change when the vapor pressure was applied during growth in TDM-CVP.

At first the lattice fitting of the heterojunction was carried out by the adjustment of the composition of mixed crystal. The epitaxial growth of the $Ga_{1-x}Al_xAs_{1-y}P_y$ layer was performed on a GaAs substrate, then the value of y varied from 0.006 to 0.02 with x nearly equal to 0.3. Constant As pressures of 0.3 to 400 Torr were applied to the solution. The lattice misfitting $\Delta a/a_0$ was measured from the lattice constants of the GaAs substrate and the GaAlAsP epitaxial layer and the dependence of this value of the As vapor pressure is shown in Fig. 2.54. The lattice misfitting depends on the applied As pressure and becomes as small as 2×10^{-5} when the As pressure of 60 Torr was applied for crystals with the composition y of 0.006 and 0.009. The double heterojunction laser without lattice strain can be realized under the condition mentioned above.

2.6. Crystal growth of II–VI compounds by TDM-CVP

2.6.1. *Aspects of II–VI compounds*

The properties of II–VI compounds published in the literature[25)] are shown in Table 2.2. No compound with amphoteric conductivity has been obtained by the conventional growth method. This may be explained from the correlation of the vapor pressure of the constituent element. The relation between the temperature and the vapor pressure of

Table 2.2 Some properties of II–VI compounds.

Item / Crystal	Conductivity Type	Energy Gap (eV)	Lattice Constant (Å)	Melting Point (°C)
ZnS	n	3.66	5.409	1830
ZnSe	n	2.67	5.669	1520
ZnTe	p	2.26	6.104	1295
CdS	n	2.41	5.802	1475
CdSe	n	1.67	6.05	1239

constituent element in each II–VI compound in Table 2.2 is shown in Fig. 2.55.[26] In the case that the vapor pressure of the group-VI element (P_{VI}) is higher than that of group-II element, the conductivity is n-type (this case corresponds to ZnS, ZnSe, CdS and CdSe). On the other hand, the conductivity type of ZnTe, for which P_{II} is higher than P_{VI}, is p-type. Therefore, it can be considered that the conductivity type of II–VI compounds is determined by the non-stoichiometry of the crystal and the vapor pressure control during crystal growth is much more important for II–VI compounds than for III–V compounds.

It is clear that ZnSe is more suitable than ZnS, because the energy gap of ZnSe corresponds to blue emission and a band-to-band transition can be expected, from Table 2.2.

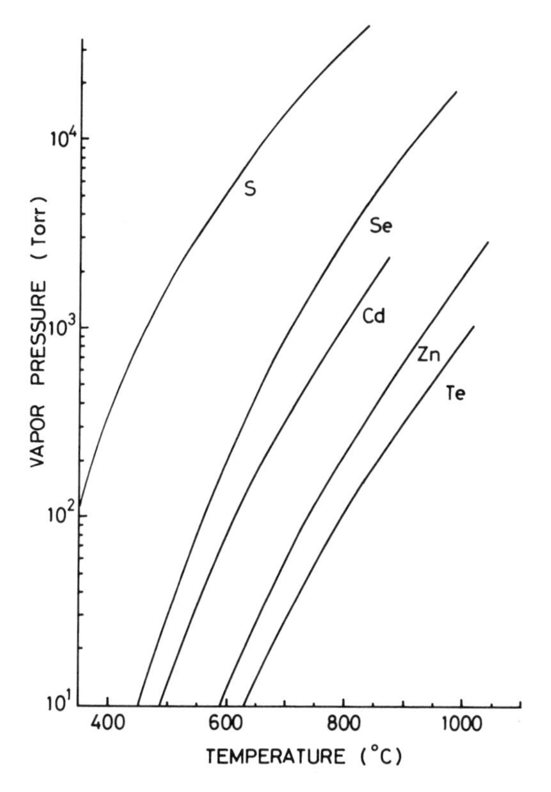

Fig. 2.55 Vapor pressure data for some constituent elements of II–VI compounds.

2.6.2. *Crystal growth of ZnSe by TDM-CVP*

A schematic diagram of the growth system and the temperature distribution of the furnace of TDM-CVP is shown in Fig. 2.56. A temperature difference in the solvent (Zn or Se) is formed and ZnSe source crystals are disposed on the top surface of the solvent at the relatively higher temperature. The vapor pressure of constituent (Se or Zn) is controlled by the temperature of the chamber connected to the growth chamber by the fine quartz tube in order to isolate thermally between the growth and controlled vapor pressure chamber. The temperatures of growth and the controlled vapor pressure chamber are maintained to be constant during growth.

Crystal growth should be carried out in the Zn solution with the lower vapor pressure under controlled Se vapor pressure if the growth is performed using the same idea as for the III–V compounds. However, the solubility of ZnSe in Zn is much lower than that in Se solution at the same temperature. So, as a first trial, ZnSe crystals were grown from Se solvent under controlled Zn pressure at the growth temperature of 1050°C which is considerably lower than the melting point of 1520°C. The purity of Zn, Se and powder ZnSe used in this experiment was 99.9999%. The ZnSe powder was baked out at about 1000°C for 1 h under the vacuum better than 2×10^{-6} Torr in order to remove impurities, and then the quartz tube was sealed off. The solidification of powder ZnSe was performed at the temperature of about 1250°C for 50 h. The quartz ampoule with the shape as schematically shown in Fig. 2.56 was baked under vacuum, and then Se, a ZnSe source crystal, of appropriate quality and metal Zn which are etched by the conventional etchant, were charged in the ampoule. The ampoule was set into the furnace. The crystal growth was performed at a constant temperature difference and also a constant growth temperature where the applied Zn pressure was varied at each growth run. The crystal was grown along the inner wall of the crucible and was conical shaped. The group Ia element

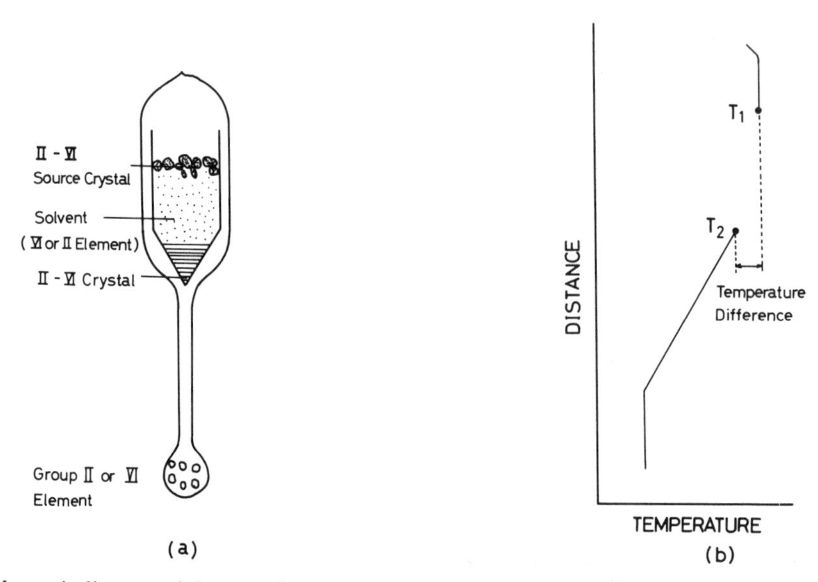

Fig. 2.56 Schematic diagram of the growth system of II–VI compounds by TDM-CVP (a), and the temperature distribution of furnace (b).

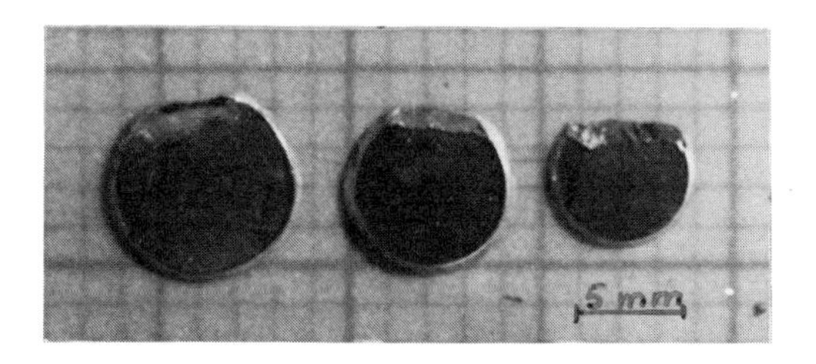

Fig. 2.57 Photograph of wafer cut from conical shaped crystal.

Li, which is expected to act as a p-type impurity, was doped in the range of 5×10^{-3} to 5×10^{-2} mol% for the Se solution.

The grown crystals were almost 10 mm long yellowish transparent single crystals having twins. A photograph of sliced p-type crystal is shown in Fig. 2.57. The quality of the crystal was investigated by X-ray double crystal spectrometry. The best value of the half-width of the rocking curve was 15 s, which was nearly the same value as that of the epitaxial GaAs layer grown by TDM-CVP. This result was the best for any II–VI compound grown up to the present.

2.6.3. *Properties of grown ZnSe crystal*

The electrical properties of the as-grown p-type crystal have been measured. The carrier concentration and Hall mobility are measured by the van der Pauw method.[27] The size of the sample for measurement was $3\times3\times0.5$ mm^3 and Au conteact of 0.5 mm square was evaporated onto the four corners of the surface as an ohmic contact. Typical measurement values at room temperature are shown in Table 2.3. These values depend on the doping concentration of Li in the Se solvent and the value of applied Zn pressure during growth.

The temperature dependence of the carrier concentration and hole Hall mobility of the p-type crystal have been measured in the temperature range of 90 K to 300 K. Fig. 2.58 shows the temperature dependence of the carrier concentration p for one of the p-type crystals. The temperature dependence of p was analyzed using the single-valley model:[9]

$$\frac{p(p+N_D)}{N_A-N_D-p}=\frac{2(2\pi m^*kT)^{3/2}}{gh^3}\exp\left(-\frac{E_A}{kT}\right) \qquad (2.5)$$

where N_D and N_A are the concentrations of donors and acceptors, respectively, E_A is the ionization energy of the acceptors, g is the spin degeneracy and m^* the density of states effective mass. The acceptor density N_A, the donor density N_D and the acceptor energy E_A

Table 2.3 Carrier concentration and Hall mobility of representative samples measured by the van der Pauw method.

Sample No.	Hole conc. (cm^{-3})	Mobility (cm^2/V·s)
GHP-53	6.4×10^{12}	78
GHP 95	3.0×10^{15}	20

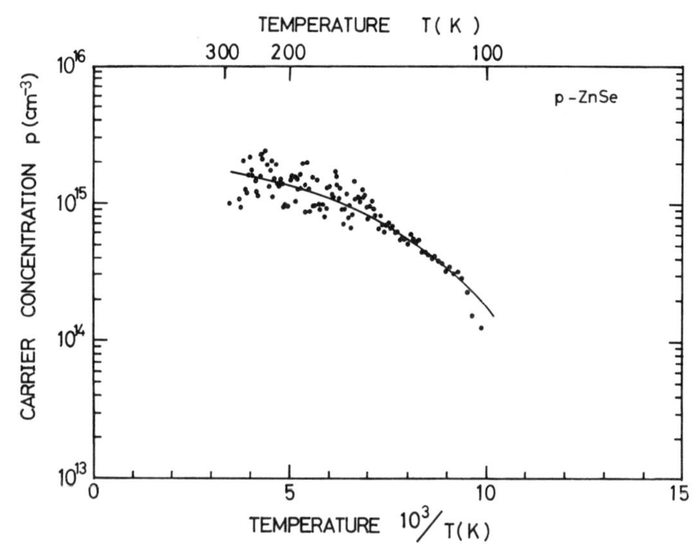

Fig. 2.58 Hole concentration of p-type ZnSe versus inverse temperature. The solid line shows the fitted values with $E_A = 0.085$ eV, $N_A = 5 \times 10^{15}$ cm^{-3}, $N_D = 5 \times 10^{14}$ cm^{-3}.

determined from the experimental curve of carrier concentration were 5×10^{15} cm^{-3}, 5×10^{14} cm^{-3} and 85 meV, respectively.

The temperature dependence of the hole Hall mobility of the same sample is shown in Fig. 2.59. The highest mobility observed was 200 cm^2/V s at low temperature and this value decreased as the temperature increased. At temperatures between 150 K and 300 K, the mobility was independent of temperature and was in the range 30 to 70 cm^2/V s.

Also the resistivity of the grown p-type crystal depends on the applied Zn pressure when the concentration of Li in the solution is nearly the same and shows a maximum at a certain Zn pressure. The values of the resistivity range from 3.0 Ω cm to 4×10^5 Ω cm under these experimental conditions and show a drastic change with a slight deviation

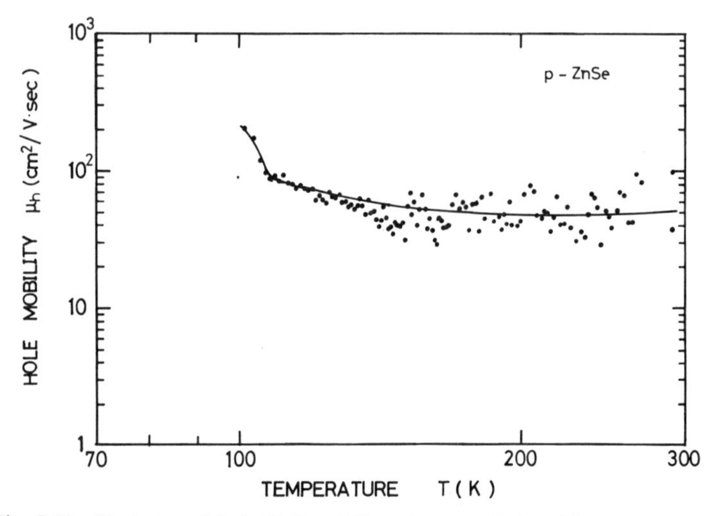

Fig. 2.59 Variation of hole Hall mobility of p-type ZnSe with temperature.

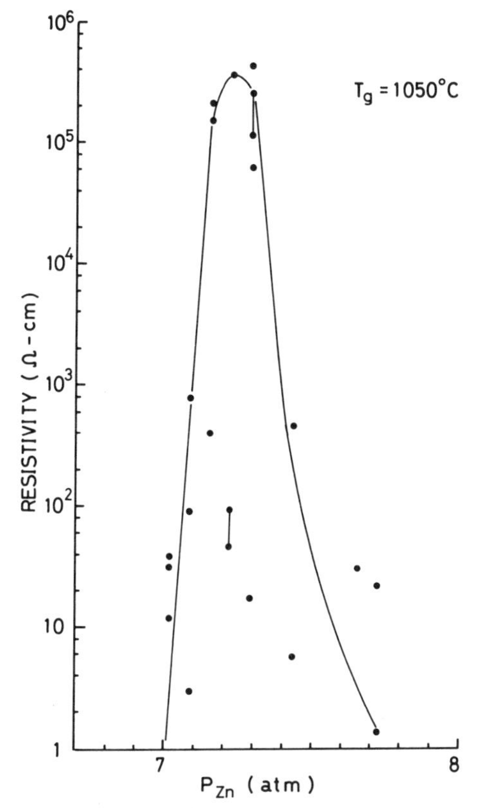

Fig. 2.60 Relation between resistivity of grown p-type ZnSe and applied Zn pressure.

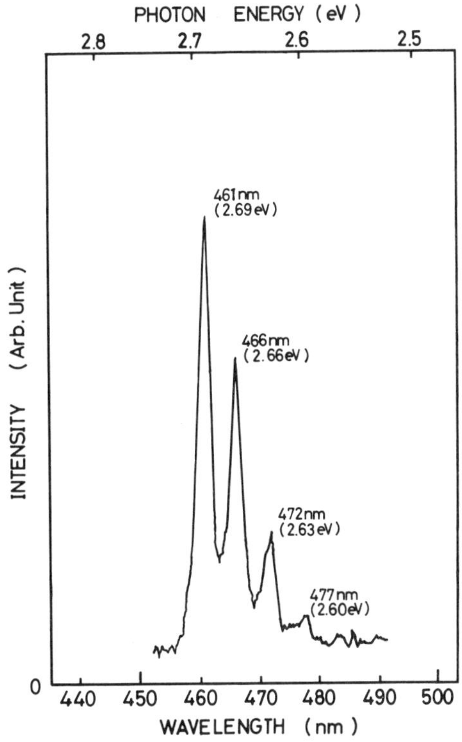

Fig. 2.61 A typical photoluminescence spectrum of p-type ZnSe crystal measured at 4.2 K.

from the certain Zn pressure as shown in Fig. 2.60. It can be said that the saturation solubility depends on the applied Zn pressure and influence on the stoichiometry of the segregated ZnSe crystal. The Zn pressure which shows a maximum resistivity is named the optimum pressure and is about 7.2 atm at the growth temperature of 1050°C.

Photoluminescence of as-grown p-type crystals was also measured at 4.2 K. Crystals are immersed in liquid helium and are excited with a 500 W high-pressure mercury-arc lamp. A typical emission spectrum is shown in Fig. 2.61. The exciton lines in the higher energy band than 2.70 eV are not observed because of the weak excitation level.[28] However, three sharp peaks at 2.69, 2.66 and 2.63 eV in the blue band appear and the latter two peaks seem to be replicas of the 2.69 eV peak. In addition to these peaks, broad peaks due to deep levels are observed in the green (~2.4 eV) and orange (~1.95 eV) bands in crystals which are grown under the Zn pressure deviating from the optimum. The intensity ratio of P_{green} (peak intensity of green band) and P_{orange} (peak intensity of orange band) to P_{blue} (peak intensity of blue band) has been investigated. The 2.69 eV peak of the maximum peak in the blue band is selected as the emission intensity of the blue band. These values depend on the applied Zn pressure during the crystal growth and become minimum at the same Zn pressure as shown in Fig. 2.62. Therefore the blue emission without any deep emission is realized at the Zn pressure of 7.2 atm. This pressure coincides with the Zn pressure at which the resistivity shows a maximum value and gives a stoichiometric composition.

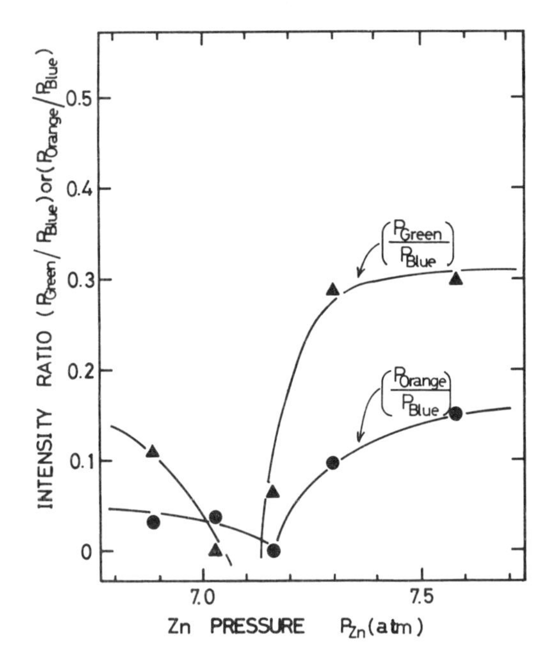

Fig. 2.62 Zn pressure dependence of the intensity ratio of P_{green} and P_{orange} to P_{blue} in the photoluminescence spectrum.

It is clear that stoichiometry-controlled perfect crystal growth of other II–VI compounds and their conductivity control can also be made by the TDM-CVP. It can be said that the TDM-CVP is applicable to most compound crystals which contain volatile elements to obtain stoichiometry-controlled perfect crystals.

2.7. Application to electronic devices

The TDM-CVP growth technology has been applied to the mass production of super brightness LEDs of GaAlAs and GaP.

The superbright GaAlAs red LED can be fabricated by the application of the optimum arsenic pressure at 20 mA in 2400 mcd and the external quantum efficiency is 17% at the wavelength of 660 nm. This value is the highest brightness achieved anywhere.

In GaP LEDs grown under the optimum phosphorus vapor pressure non-radiative transitions are reduced so that doping of nitrogen becomes unnecessary and bound or free exciton recombination becomes dominant. 555 nm very pure green emission without notrogen isoelectronic level becomes available, the quantum efficiency being 0.2% and brightness 150 mcd at 20 mA.[22] Also, GaAs LED with more than 10% quantum efficiency can be obtained by the same method.

The blue light emission from ZnSe p–n junction grown under the optimum Zn pressure has been obtained. The wavelength is 480 nm and the halfwidth is 7 nm.[2] The brightness of the pure blue color is 2 mcd at 2 mA. Details are described in Chapter 21 of this volume.

2.8. Consideration of the mechanism of TDM-CVP

In the temperature difference method under controlled vapor pressure (TDM-CVP) it is well established by many experiments that the stoichiometry of the segregated crystal on the substrate settled at the bottom of the molten phase is controlled by the vapor pressure of the volatile component (arsenic in the case of GaAs) applied on top of the molten phase (Section 2.3). The optimum vapor pressure which gives the stoichiometry is given by Eq. (2.1) (Section 2.1) as a function of temperature.

It is important that the above optimum vapor pressure is the same as that of the heat treatment experiment (Section 2.2) in which the solid phase is directly controlled by the applied vapor pressure.

The temperature difference between the molten phase and the substrate accelerates diffusion of dissolved GaAs towards the substrate and its segregation there. The growth rate is controlled by the temperature difference but not by the applied vapor pressure if source crystals are present on top of the molten phase. We also have confirmed that, without temperature difference, the arsenic content in the liquid tends to become constant during application of the vapor pressure. This means that some kind of three-phase equilibrium is established for an arbitrary value of applied vapor pressure.

This phenomenon is beyond the expectation from the simple application of the Gibbs phase rule, because this rule states that $f = c - p + 2$ (f, c and p are the numbers of the freedoms, components, and phases, respectively); and for three-phase equilibrium with two components we see that $f = 1$. This means that the saturation solubility is only a function of temperature, and therefore there is no possibility of controlling the stoichiometry of the segregating crystal.

On the other hand, it should be noted that our system is an open system where the arsenic atoms can be fed externally as a form of the applied vapor pressure. In the present paper, we show how the usual phase diagram theory on the basis of the Gibbs phase rule should be corrected.

The authors have already pointed out that the saturation solubility can change as a function of the applied vapor pressure, thus chemical potentials of arsenic in the gas phase and liquid phase can be in equilibrium and then the chemical potential of the segregating crystal is in equilibrium with that of the liquid.

The same idea that chemical potentials are in equilibrium has been followed by A. I. Ivaschenko in USSR,[29)] although the details are not yet known to us. We hope that the readers will refer to their presentation as it will be published in the near future.

The theory presented here is based on the idea that the chemical potentials are in equilibrium, but much more thorough discussion is made, and the experimental optimum vapor pressure can be explained by this theory.[30,31,32]

First, we assume the two different states of arsenic atoms in the liquid as one of calculable models which allow vapor pressure dependence of saturation solubility. The structure of the liquid is not yet well understood compared with that of the solid. Therefore, the corrections of the present model should be examined from experiments. It should be pointed out that usual regular solution treatment of liquid which assumes a mixture of arsenic and gallium liquids gives no good result because high vapor pressure gives an arsenic-rich liquid if liquid–gas equilibrium is assumed.

Moreover, we consider individual elementary reaction processes of give-and-take of the molecular components at each different place; the gas–liquid interface, internal of the liquid, and the liquid–solid interface (namely we deal with an inhomogeneous system). The applied vapor pressure fixes the chemical potential of arsenic. All the elementary processes can be equilibrated to the applied chemical potential of As without discrepancy of Ga chemical potentials except for the process at the gas–liquid interface. As a result of this reaction at the gas–liquid interface, a thin layer of polycrystals may cover the free surface of the liquid. However, this reaction tends to cease and further development of this reaction is hard to occur unless some external stirring action, which needs extra energy, is given.

Another possibility of such reaction processes is the segregation processes on the substrate, i.e. the reaction processes at the liquid–solid interface. However, this is hard to consider if there is no temperature difference because our experiments showed that there was little continuous segregation even for 7 days application of the vapor pressure.

The present model will make clear how the phase diagram should be drawn with the applied vapor pressure as a parameter.

2.8.1. *Elementary reaction processes*

The following three assumptions are made

(1) There are two different sites or states of arsenic atoms in the gallium solution. Their concentrations can be dependent on each other as a result of their chemical reaction being in equilibrium, which means that we have not introduced any new variable in order to fit the Gibbs phase rule.

One of the more visual models of the liquid state of the above mentioned assumption is illustrated in Fig. 2.63; gallium atoms make a bonding matrix with As_l states of arsenic atoms, and other than As_l, there are As^* states which can only change to As_l states when they find a vacant position of the matrix, V_l. Therefore we have the following equation of reaction:

$$As^* + V_l = As_l; \ \Delta G_l \tag{2.10}$$

where ΔG_l is the free energy difference and given by $\Delta G_l = \Delta H_l - T\Delta S_l$. The enthalpy difference, ΔH_l, is practically the energy difference, while ΔS_l is the difference in the entropy of vibration.

Fig. 2.63 should not be confused with a rigid matrix like an amorphous solid. Each arm of bonding means only nearest-neighbor interaction between atoms; fluidicity of liquid is an as-yet unresolved problem. Although dislocations and vacancies have been discussed, both are unrealistic. We simply assume that As_l and liquid Ga atoms can easily move keeping their configurational states without the assistance of the assumed V_l for some quantum mechanical reason.

In the present paper, V_l only means a vacant site to which As^* can be transferred

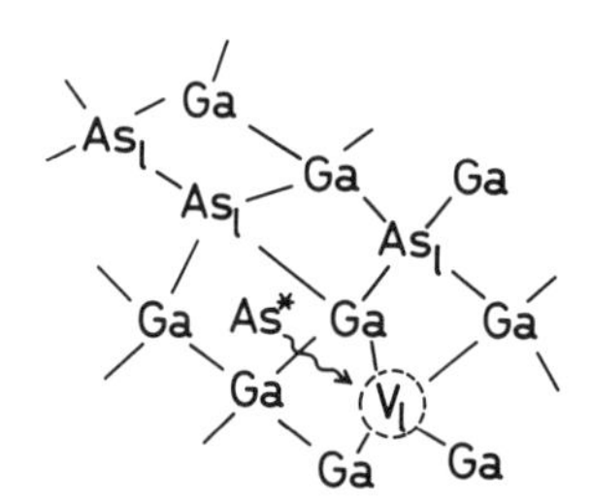

Fig. 2.63 A model of the Ga solution in which two states of As, $GaAs_l$ and As^*, are present.

to become As_l. One example of a liquid which has two states of an element is water with oxygen in H_2O bond and dissolved O_2 particles.

It is not always necessary to assume that the vacant state V_l exists. Otherwise, however, we must assume that the site transfer between As* and As_l is difficult to occur, except at the liquid–gas interface where there should be a number of free arms of bonding matrix of the liquid acting as equivalent to the V_l states, or vaporization of gallium would make a constant concentration of V_l.

(2) The next assumption is less essential than the first. Rather, we assume it for simplicity as will be seen later. Segregation of the solid phase GaAs is mainly due to the sticking of $GaAs_l$ molecular states formed by one of the gallium atoms surrounding the As_l atoms.

This assumption does not mean the neglect of sticking of gallium atoms which are distant from As_l atoms as will be stated later. This assumption will become impractical near the melting point of the GaAs crystal where the solution will include a number of As–As bonds. However, for the usual liquid-phase crystal growth, the arsenic concentration in the liquid is so low that we can make the above assumption.

Making this assumption, the reaction equation (2.10) should be changed as follows:

$$As^* + V_l + Ga = GaAs_l(\text{in liquid});\ \Delta G_l. \tag{2.10$'$}$$

From the above assumption, the following reaction equation for segregation holds:

$$GaAs(\text{lattice}) = GaAs_l(\text{in liquid});\ \Delta G^f_{GaAs}. \tag{2.11}$$

(3) As for the deviation from stoichiometry, we assume the existence of only arsenic vacancies, V_{As}, and arsenic interstitial atoms, I_{As}, because of several experimental results. However, even when we introduce gallium vacancies and gallium interstitials no problem should arise because equilibrium equations between V_{As}, I_{As} V_{Ga}, and I_{Ga} should hold.

The most important reaction equation at the liquid–solid interface is

$$As^*(\text{in liquid}) = I_{As};\ \Delta G_{I_{AS}}. \tag{2.12}$$

On the other hand, V_{As} is formed by the reactions

$$GaAs(\text{lattice}) = Ga(\text{lattice}) + V_{As} + I_{As};\ \Delta G_{FR} \tag{2.13}$$

or GaAs(lattice) = Ga(lattice) + V_{As} + As*(in liquid);

$$\Delta G_V = \Delta G_{FR} - \Delta G_{I_{AS}}. \tag{2.14}$$

The equations (2.13) and (2.14) are dependent on each other because the equation (2.12) holds, so that we need only take into account one of the above two equations for the equilibration of formation of V_{As}.

As well as the above equations we should take into account the $GaAs_l$ molecules in the liquid for the formation of I_{As} and V_{As}, as follows:

$$GaAs_l(\text{in liquid}) = Ga + V_l + I_{As};\ \Delta G_{I_{AS}} - \Delta G_l \tag{2.15}$$

$$GaAs_l(\text{in liquid}) + \{Ga(\text{lattice}) + V_{As}\} = Ga + V_l + GaAs(\text{lattice});\ -\Delta G_V - \Delta G_l. \tag{2.16}$$

Also, the following reaction equation for sticking of Ga atoms should hold:

$$\mathrm{Ga} + \mathrm{V_l} = \{\mathrm{Ga(lattice)} + \mathrm{V_{As}}\}. \tag{2.17}$$

We can again state that it is not necessary to assume V_l, but if we do not assume the presence of V_l the reactions given by the equations

$$\mathrm{GaAs_l(in\ liquid)} = \mathrm{Ga} + \mathrm{I_{As}} \tag{2.15$'$}$$

$$\mathrm{GaAs_l(in\ liquid)} + \{\mathrm{Ga(lattice)} + \mathrm{V_{As}}\} = \mathrm{Ga} + \mathrm{GaAs(lattice)} \tag{2.16$'$}$$

$$\mathrm{Ga} - \{\mathrm{Ga(lattice)} + \mathrm{V_{As}}\} \tag{2.17$'$}$$

should be assumed to be difficult to occur at the liquid–solid interface.

2.8.2. *Solubility of arsenic in the liquid*

In the presence of the solid phase, the reaction process (2.11) is equilibrated in the molten system at a homogeneous temperature. From the coincidence of the chemical potentials, the compositional ratio of $GaAs_l$ in the liquid, X_{GaAs_l}, can be given by

$$\gamma_{\mathrm{GaAs_l}} X_{\mathrm{GaAs_l}} = \exp(-\Delta G^{\mathrm{f}}_{\mathrm{GaAs}}/RT) \tag{2.18}$$

The activity coefficient γ_{GaAs_l} is a result of the difference between the bonding energy of a $GaAs_l$ molecule and one of the surrounding Ga atoms and that of the Ga–Ga bond. Hereafter, we neglect such a difference for simplicity. Then,

$$X_{\mathrm{GaAs_l}} = \exp(-\Delta G^{\mathrm{f}}_{\mathrm{GaAs}}/RT). \tag{2.18$'$}$$

This equation is only a function of temperature, and corresponds to the Gibbs phase rule.

Then, we consider application of the vapor pressure of arsenic on top of the liquid. We assume that gas molecules are in the form of As_4, then the reaction equation at the gas–liquid interface is

$$\tfrac{1}{4}\mathrm{As_4} = \mathrm{As^*(in\ liquid)};\ \Delta G_*. \tag{2.19}$$

From the coincidence of the chemical potentials at the equilibrium, we get

$$\frac{\gamma_{\mathrm{As^*}} X^*_{\mathrm{As}}}{(P_{\mathrm{As_4}})^{1/4}} = \exp\left(-\frac{\Delta G^*}{RT}\right). \tag{2.20}$$

γ_{As^*} is the activity coefficient of As* in the liquid. Within the quasi-chemical approach, which assumes the bonding energies for the nearest neighbor atoms

$$\gamma_{\mathrm{As^*}} = \exp(\Omega/RT)$$

$$\Omega = [H_{\mathrm{As^*-Ga}} - \tfrac{1}{2}(H_{\mathrm{Ga-Ga}} + H_{\mathrm{As^*-As^*}})] > 0$$

as far as $X_{As^*} \ll 1$, where H_{As^*-Ga}, H_{Ga-Ga} and $H_{As^*-As^*}$ are the molar bond energies of As*–Ga, Ga–Ga and As*–As* bonds, respectively.

The equation (2.20) can therefore be rewritten

$$X_{\mathrm{As^*}} = (P_{\mathrm{As_4}})^{1/4} \exp(-\Delta G'_*/RT), \quad \Delta G'_* = \Delta G_* + \Omega. \tag{2.20$'$}$$

From Eqs. (2.18′) and (2.20′) we have got the solubility of arsenic, $X_{As} = X_{GaAs_l} + X_{As^*}$, as a function of temperature and applied vapor pressure. Now, we must take account of the equilibrium between As_l and As^*. From Eq. (2.10′), we must have

$$\frac{X_{GaAs_l}}{\gamma_{As^*}X_{As^*}\gamma_{V_l}X_{V_l}X_{Ga}}=\exp\left(-\frac{\Delta G_l}{RT}\right) \tag{2.21}$$

where X_{Ga} is near to 1, X_{V_l} is given by

$$\gamma_{V_l}X_{V_l}\doteqdot\frac{X_{GaAs_l}}{\gamma_{As^*}X_{As^*}}\exp\left(\frac{\Delta G_l}{RT}\right). \tag{2.21′}$$

This is the necessary condition that X_{GaAs_l} and X_{As^*} are independently given by Eqs. (2.18′) and (2.20′) under the whole equilibrium.

Instead, we need not consider the existence of V_l, so long as the site transfer $As^* \rightarrow As_l$ is difficult to occur inside the liquid. Anyway, it is important that X_{As^*} can be changed independent of the value of X_{GaAs_l}.

2.8.3. *Equilibrium at the liquid–solid interface*

We assume that the concentration of V_l at and near the liquid–solid interface is the same as that in the liquid; this is very different from the liquid–gas interface. It has been found that the growth mechanism is two-dimensional, and a $GaAs_l$ molecule moves on the solid surface until it is caught at a kink or a step. However, we assume that the movement need not require some extra concentration of V_l; it is similar to the fact that the fluidicity of the liquid should be explained by a model other than the "vacancy" model (some quantum mechanical model).

The concentration of the interstitial arsenic atoms, $N_{I_{As}}$ is determined by the reaction process (2.12), then

$$\frac{n_{I_{As}}}{\gamma_{As^*}X_{As^*}}=\exp\left(-\frac{\Delta G_{I_{As}}}{RT}\right) \tag{2.22}$$

is given from the equilibrium condition, where $n_{I_{As}}=N_{I_{As}}/N_I$, and N_I is the density of the interstitial site in the lattice. $\gamma_{As^*}X_{As^*}$ is given by Eq. (2.20), then $n_{I_{As}}$ is given by

$$n_{I_{As}}=(P_{As_4})^{1/4}\exp\left(-\frac{\Delta G^*+\Delta G_{I_{As}}}{RT}\right). \tag{2.22′}$$

This is just the same as the direct vapor pressure application on the crystal as in the heat-treatment experiment because the reaction equations (2.12) and (2.19) give the equation

$$\tfrac{1}{4}As_4=I_{As};\ \Delta G_{I_{As}}+\Delta G^* \tag{2.23}$$

and this reaction equation which holds for the direct vapor pressure control gives the same expression as that of Eq. (2.22′).

The concentration of the arsenic vacancy V_{As} is given by Eq. (2.13),

$$n_{V_{As}}n_{I_{As}}=\exp(-\Delta G_{FR}/RT) \tag{2.24}$$

or as an equivalent expression from Eq. (2.14).

$$n_{V_{As}} = \frac{1}{\gamma_{As^*} X_{As}} \exp\left(-\frac{\Delta G_V}{RT}\right) = (P_{As_4})^{-1/4} \exp\left\{-\frac{(\Delta G_V - \Delta G_*)}{RT}\right\}. \tag{2.24'}$$

Next, we consider the reaction of $GaAs_l$ molecules into I_{As} and V_{As} by Eqs. (2.15) and (2.16). If the concentration of V_l at the liquid–solid interface can be assumed to be the same as that in the liquid, both the reactions are automatically equilibrated. That is, by the equilibrium condition of Eq. (2.15), we have at the interface

$$\frac{X^l_{Ga} \gamma_{V_l} X_{V_l} n_{I_{As}}}{X_{GaAs_l}} = \exp\left(-\frac{\Delta G_{I_{As}} - \Delta G_l}{RT}\right).$$

But this can simply be deduced from Eq. (2.21) and Eq. (2.22). The discussion about Eq. (2.16), and also Eq. (2.17), is quite similar to the above.

For a full discussion of the deviation from stoichiometry of the solid phase, we must take account of the role of electrons and holes, and we will discuss this in a different paper. However, we have shown here that the control of the stoichiometry of the crystal in the presence of the liquid over the crystal gives just the same result as the direct control by the gas phase over the crystal which is the case in the heat-treatment of the crystal under the vapor pressure.

2.8.4. *Reactions at the gas–liquid interface*

At the liquid–gas interface, we must take account of the reaction of As_l other than that of As* given by Eq. (2.19). The reaction may be written as

$$GaAs_l(\text{liquid surface}) = Ga + \tfrac{1}{4}As_4;\ \Delta G^l_{GaAs} = \Delta G^{sub}_{GaAs} - \Delta G^f_{GaAs} \tag{2.25}$$

where ΔG^{sub}_{GaAs} is given by the equation

$$GaAs(\text{lattice}) = Ga + \tfrac{1}{4}As_4;\ \Delta G^{sub}_{GaAs}. \tag{2.26}$$

Equation (2.25) means that surface atoms of gallium can make a molecule $GaAs_l$ without considering about the vacancy state V_l. On the contrary if the hypothetical equation

$$GaAs_l(\text{liquid surface}) = Ga + V_l + \tfrac{1}{4}As_4 \tag{2.27}$$

should hold, then there would occur no problem about equilibrium because this equation is automatically fulfilled. However, it should be unrealistic to consider about the vacant state V_l at the gas–liquid interface. Rather, the surface may be equivalent to be full of vacant states to allow the reaction (2.25) because of the vaporization of gallium atoms and the existence of free arms.

In the two-phase equilibrium without solid phase, the equilibrium condition of Eq. (2.25) is given by

$$X_{GaAs_l} \doteqdot (P_{As_4})^{1/4} \exp\left(\frac{\Delta G^l_{GaAs}}{RT}\right) \tag{2.28}$$

so long as $X_{Ga} \approx 1$.

Three-phase equilibrium at a gas–liquid–solid interface is given by equating Eq. (2.28) to Eq. (2.18′), which uniquely gives the equilibrium vapor pressure $P^{eq}_{As_4}$ in the usual meaning.

Once the applied vapor pressure exceeds $P^{eq}_{As_4}$ the reaction processes (2.25) and (2.11) can no longer be equilibrated and the reaction of the formation of the solid phase proceeds. However, it should be emphasized that such reactions take place only at the gas–liquid interface. Therefore, the free liquid surface where gas molecules directly attack should be covered with thin polycrystals. This gives no problem about the equilibrium conditions inside the liquid and the liquid–solid interface, because the mole ratio X_{As^*} of As* should be given by Eq. (2.20′) irrespective of whether it is above or below $P^{eq}_{As_4}$ so long as the covering on the free liquid surface is not atomically perfect.

We can assume that the thickness of the covering layer is very thin and hard to grow thick, because once the stoichiometry of this layer has been controlled by the direct application of the As_4 vapor pressure by Eq. (2.23), the liquid–solid interface of this layer is then equilibrated by Eq. (2.12), just the repeating of the discussion of Section 2.8.3.

2.8.5. *Comparison with experiment*

In the two-phase equilibrium without the solid phase the solubility X_{As} is given by

$$X_{As} = X_{GaAs_l} + X_{As^*}$$

$$X_{GaAs_l} \doteqdot (P_{As_4})^{1/4} \exp(\Delta G^{l}_{GaAs}/RT) \qquad (2.28)$$

$$X_{As^*} = (P_{As_4})^{1/4} \exp(-\Delta G^{l}_{*}/RT). \qquad (2.20')$$

In the three-phase equilibrium over the vapor pressure $P^{eq}_{As_4}$ at which both Eqs. (2.28) and (2.18′) hold, the saturation solubility $X_{As,sat}$ is given by

$$X_{As,sat} = X_{GaAs_l} + X_{As^*}$$

$$X_{GaAs_l} = \exp(-\Delta G^{f}_{GaAs}/RT) \qquad (2.18')$$

$$X_{As^*} = (P_{As_4})^{1/4} \exp(-\Delta G^{l}_{*}/RT) \qquad (2.20')$$

The temperature and applied vapor pressure dependences of X_{GaAs_l} and X_{As^*} are schematically illustrated in Fig. 2.64. The phase diagram as a function of applied vapor pressure is also schematically shown in Fig. 2.65 with exaggeration of the vapor pressure dependence of the saturation solubility, and the solidus range.

Fig. 2.66 shows the experimental result of the change in the atomic ratio of As in the liquid, X^{total}_{As}, which includes both the liquid and solid phases. Details of the experiment are given in refs. 29, 30, 31. When the applied vapor pressure is reduced from P_1 to P_2 (both being much higher than $P^{eq}_{As_4}$), X^{total}_{As} reduces by vaporization. Although some of the dissolved arsenic may change to the solid during vaporization, we neglect this effect to get a rough estimate. Then, within the present approximation,

$$X^{total}_{As}(P_1) - X^{total}_{As}(P_2) = X_{As^*}(P_1) - X_{As^*}(P_2) \qquad P_1 > P_2 > P^{eq}_{As_4}$$

from which we can get the vapor pressure dependence. The result is shown in Fig. 2.67. The vapor pressure dependence is nearly on the $\frac{1}{4}$ power curve for $T = 770°C$. The experimental points are not enough for $T = 685°C$, but we roughly estimate the temperature dependence from the data at the two temperature. The result shown in Fig.

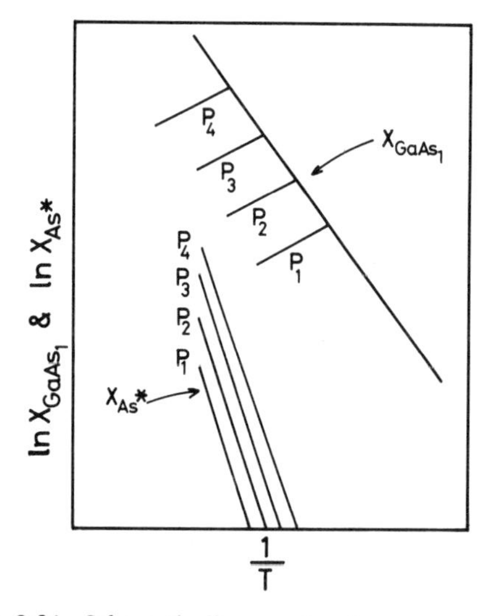

Fig. 2.64 Schematic diagram showing temperature dependence of X_{GaAs_l} and X_{As^*}.

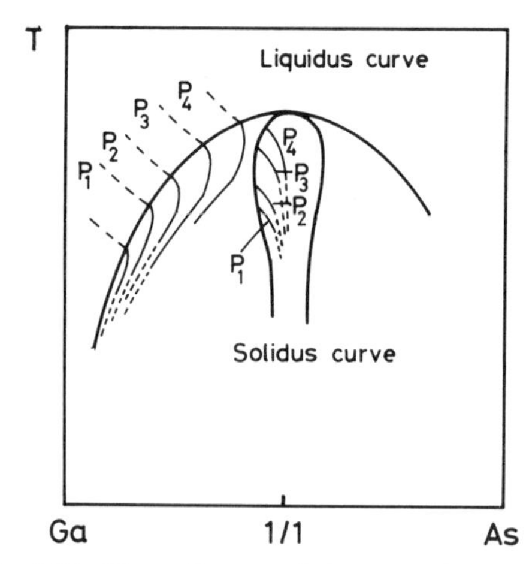

Fig. 2.65 Phase diagram of GaAs as a function of applied vapor pressure.

2.68 gives the activation energy 2.3 eV. Then, the activation energy corresponding to the enthalpy part of $\Delta G'_*$, i.e. $\Delta H'_*$, is nearly 2.3 eV. This value is approximately twice the activation energy of the solubility of GaAs in Ga solution which is known to be 1.0 eV (i.e. $\Delta H^{f}_{GaAs} \approx 1.0$ eV within the framework of the present approximation).

Because the applied vapor pressure is much higher than $P^{eq}_{As_4}$, the change in the saturation solubility as a function of vapor pressure, $\Delta X_{As,sat}$, is nearly equal to X_{As^*} itself. The high activation energy is consistent with the fact that the change in $X_{As,sat}$ is small.

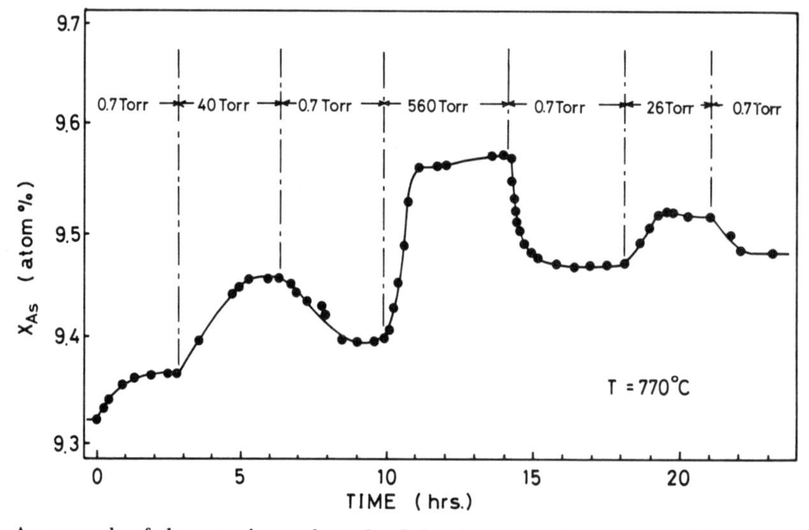

Fig. 2.66 An example of the experimental result of the change in the quantity of As in the Ga solution.

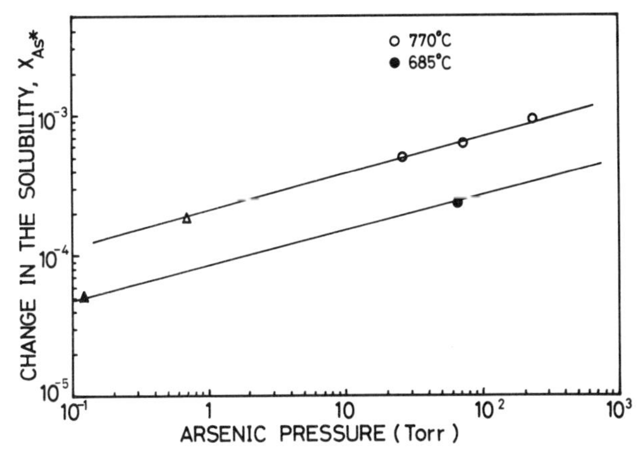

Fig. 2.67 X_{As^*} estimated from the experiment (the data at the highest vapor pressures were omitted because of the large uncertainty).

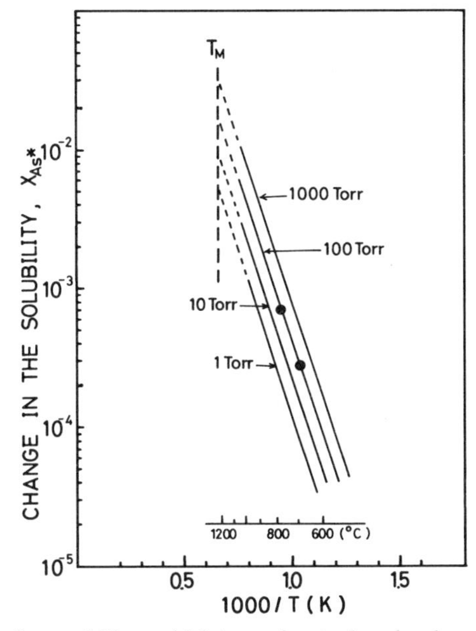

Fig. 2.68 Temperature dependence of X_{As^*}, which is nearly equal to the change in the saturation solubility, as a function of temperature.

2.8.6. *Comparison with the Gibbs phase rule*

For a two-component closed system at three-phase equilibrium, Gibbs' statement can be written as follows if the temperature is given:

$$\mu^{g}_{As}(P, X^{g}_{As}, X^{l}_{As}, X^{s}_{As}) = \mu^{l}_{As}(P, X^{g}_{As}, X^{l}_{As}, X^{s}_{As}) = \mu^{s}_{As}(P, X^{g}_{As}, X^{l}_{As}, X^{s}_{As})$$

$$\mu^{g}_{Ga}(P, X^{g}_{As}, X^{l}_{As}, X^{s}_{As}) = \mu^{l}_{Ga}(P, X^{g}_{As}, X^{l}_{As}, X^{s}_{As}) = \mu^{s}_{Ga}(P, X^{g}_{As}, X^{l}_{As}, X^{s}_{As}) \quad (2.29)$$

where μ^{j}_{i} is the chemical potential of a component i for phase j. P is the total pressure, and X^{j}_{i} is the compositional ratio (of As) for a phase j. The numbers of equations and the independent variables are the same, namely 4, so all the variables are uniquely determined, giving the equilibrium pressure P^{eq}, saturation solubility X^{l}_{As}, and the deviation from stoichiometry $\delta = X^{s}_{As} - 0.5$ at a given temperature.

On the other hand, our system is an open system, i.e. the arsenic atoms can be fed from the arsenic vessel at a different temperature. This was taken into account in Eqs. (2.19), (2.23), (2.25) by simply giving the arsenic chemical potential

$$\mu^{g}_{As}(\text{applied}) = \tfrac{1}{4} RT \ln P_{As_4,\text{applied}} + f(T) \tag{2.30}$$

where P_{As_4} is the applied vapor pressure (assuming only As_4 atoms are dominant) caused by the fed As_4 molecules in equilibrium at an assumed homogeneous temperature of the molten phase. $f(T)$ is only a function of temperature for an ideal gas, and need not be considered here.

In the framework of the Gibbs rule given by Eq. (2.30), the chemical potential of As in the liquid, and therefore the solubility X_{As}, should not change and there should be no change in the deviation from stoichiometry. On the contrary, we have assumed the solubility X_{As} can be changed even when the solid phase is present, so that the chemical potential of arsenic in the liquid should be given by a different expression, $\bar{\mu}^{l}_{As}$. We have assumed the equality of As chemical potentials

$$\mu^{g}_{As}(\text{applied}) = \bar{\mu}^{l}_{As}(P_{\text{applied}}, X^{g}_{As}, X^{l}_{As}, X^{s}_{As}) = \mu^{s}_{As}(P_{\text{applied}}, X^{g}_{As}, X^{l}_{As}, X^{s}_{As}) \tag{2.31}$$

via each elementary process.

However, we must take into account whether the equality of the chemical potentials of Ga

$$\begin{aligned}\mu^{g}_{Ga}(P_{\text{applied}}, X^{g}_{As}, X^{l}_{As}, X^{s}_{As}) &= \bar{\mu}^{l}_{Ga}(P_{\text{applied}}, X^{g}_{As}, X^{l}_{As}, X^{s}_{As}) \\ &= \mu^{s}_{Ga}(P_{\text{applied}}, X^{g}_{As}, X^{l}_{As}, X^{s}_{As})\end{aligned} \tag{2.32}$$

hold or not. We have four independent equations (2.31) and (2.32), and three independent variables X^{g}_{As}, X^{l}_{As}, X^{s}_{As}, so that once we assume the equality of arsenic chemical potentials, at least one of the equations for gallium chemical potentials is no longer equal to the others. This difficulty, which occurs when the discussion of Eq. (2.29) is simply extended to our case as Eq. (2.31) and (2.32), has been avoided in the present paper as follows.

We deal with the inhomogeneous systems in which the gas–liquid interface and liquid–solid interface are at different places, and introduce individual elementary reaction processes for the liquid, solid and the interfaces. Therefore, the number of equations for which chemical potentials should be equated is increased. It should be noted that this does not affect the number of freedoms because some of the reaction equations are dependent on each other, and the numbers of independent equations and independent variables should not change even after this treatment. However, if only one of the whole elementary reaction processes is assumed to be not equated, all the other elementary processes can be equated. In the presence paper, this special process occurs at the gas–liquid interface where the solid covering over the melt may occur.

As for the expected statement that the inhomogeneous system as a whole may have a higher free energy than the homogeneous system in which gas, liquid and solid are making direct interface at any point, we must say that there is no way to make the transition to a homogeneous system unless there is some external disturbance like stirring which needs extra energy.

In the present approximation, the phase diagram of compound semiconductors should be understood as having the parameter of the applied vapor pressure or its chemical potential as illustrated in Fig. 2.65.

2.8.7. *Optimum vapor pressure*

(1) We assume that the concentration of the arsenic interstitial atoms, $N_{I_{As}}$, and the concentration of the arsenic vacancies, $N_{V_{As}}$, are the same at the optimum vapor pressure at which the exact stoichiometry is realized.

The reaction of As*, the component giving the arsenic chemical potential in the liquid, and I_{As}, the arsenic interstitial atom in the solid is given by Eq. (2.12) in Section 2.8.1, which is rewritten here

$$\mathrm{As^*(in\ liquid)} = \mathrm{I_{As}};\ \Delta G_{I_{As}}. \tag{2.12}$$

On the other hand, the reaction of As* and As_4 vapor is given by Eq. (2.19) in Section 2.8.2,

$$\tfrac{1}{4}\mathrm{As_4(gas)} = \mathrm{As^*(in\ liquid)};\ \Delta G^* \tag{2.19}$$

for any applied vapor pressure, P_{As_4}.

From the equilibrium condition of the two reaction equations, the density of the arsenic interstitial atoms, $n_{I_{As}} = N_{I_{As}}/N_L$, is given by Eq. (2.22′)

$$n_{I_{As}} = (P_{As_4})^{1/4} \exp\left(-\frac{\Delta G_{I_{As}} + \Delta G^*}{RT}\right). \tag{2.22′}$$

On the other hand, the reaction equations (2.12) and (2.19) lead to Eq. (2.23):

$$\tfrac{1}{4}\mathrm{As_4(gas)} = \mathrm{I_{As}};\ \Delta G'_{I_{As}} = \Delta G_{I_{As}} + \Delta G^* \tag{2.23}$$

which corresponds to the direct reaction of the solid and vapor phases, i.e. $\Delta G'_{I_{As}}$ is the free energy for the direct reaction between the solid and vapor phase.

The equation (2.22′) is rewritten as

$$n_{I_{As_4}} = (P_{As_4})^{1/4} \exp(-\Delta G'_{I_{As}}/RT). \tag{2.33}$$

(2) The reaction equations for the arsenic vacancies are given by Eq. (2.13) or (2.14) in Section 2.8.1,

$$\mathrm{GaAs(lattice)} = \mathrm{Ga(lattice)} + \mathrm{V_{As}} + \mathrm{I_{As}};\ \Delta G_{FR} \tag{2.13}$$

$$\mathrm{GaAs(lattice)} = \mathrm{Ga(lattice)} + \mathrm{V_{As}} + \mathrm{As^*};\ \Delta G_{V_{As}} = \Delta G_{FR} - \Delta G_{I_{As}}. \tag{2.14}$$

From the equilibrium condition of the two reaction equations (2.19) and (2.14), the density of the arsenic vacancies $n_{V_{As}} = N_{V_{As}}/N_L$ is given by Eq. (2.24′) in Section 2.8.3,

$$n_{V_{As}} = (P_{As_4})^{-1/4} \exp\left(-\frac{\Delta G_{V_{As}} - \Delta G^*}{RT}\right). \tag{2.24′}$$

On the other hand, Eqs. (2.14) and (2.19) give the following reaction equation:

$$\text{GaAs(lattice)} = \text{Ga(lattice)} + \text{V}_{\text{As}} + \tfrac{1}{4}\text{As}_4; \qquad \Delta G'_{\text{V}_{\text{As}}} = \Delta G_{\text{V}_{\text{As}}} - \Delta G^* \tag{2.34}$$

$\Delta G'_{\text{V}_{\text{As}}}$ is the free energy for the direct reaction of the solid and vapor phases and Eq. (2.24′) can be rewritten as

$$n_{\text{V}_{\text{As}}} = (P_{\text{As}_4})^{-1/4} \exp(-\Delta G'_{\text{V}_{\text{As}}}/RT). \tag{2.35}$$

At the optimum pressure, $P^{\text{opt}}_{\text{As}_4}$, at which the solid phase is stoichiometric, $n_{\text{V}_{\text{As}}}$ is equal to $n_{\text{I}_{\text{As}}}$. Therefore, from the equality of Eq. (2.33) and (2.35), we have

$$P^{\text{opt}}_{\text{As}_4} = \exp\left(-\frac{2(\Delta G'_{\text{V}_{\text{As}}} - \Delta G'_{\text{I}_{\text{As}}})}{RT}\right), \tag{2.36}$$

(3) Now, we must estimate the value of $\Delta G'_{\text{V}_{\text{As}}}$ and $\Delta G'_{\text{I}_{\text{As}}}$. The free energy of formation of the arsenic vacancy $\Delta G^{\text{F}}_{\text{V}_{\text{As}}}$ is defined by the following reaction equation:

$$\text{GaAs(lattice)} = \text{Ga(lattice)} + \text{V}_{\text{As}} + \text{As(surface)}; \quad \Delta G^{\text{F}}_{\text{V}_{\text{As}}} \tag{2.37}$$

where As(surface) means the solid As atom brought to the solid GaAs surface. On the other hand, the sublimation of solid arsenic is given by

$$\text{As(surface)} = \tfrac{1}{4}\text{As}_4; \quad \Delta G^{\text{sub}}_{\text{As}} \tag{2.38}$$

where $\Delta G^{\text{sub}}_{\text{As}}$ is the free energy of the sublimation of arsenic.

From Eqs. (2.34), (2.37) and (2.38), we have

$$\Delta G'_{\text{V}_{\text{As}}} = \Delta G^{\text{F}}_{\text{V}_{\text{As}}} + \Delta G^{\text{sub}}_{\text{As}}. \tag{2.39}$$

The free energy of formation of the arsenic interstitial atom, $\Delta G^{\text{F}}_{\text{I}_{\text{As}}}$ is defined by the following equation:

$$\text{As(surface)} = \text{I}_{\text{As}}; \quad \Delta G^{\text{F}}_{\text{I}_{\text{As}}}. \tag{2.40}$$

From Eqs. (2.23), (2.38) and (2.40), we have

$$\Delta G'_{\text{I}_{\text{As}}} = \Delta G^{\text{F}}_{\text{I}_{\text{As}}} - \Delta G^{\text{sub}}_{\text{As}}. \tag{2.41}$$

Therefore, the optimum vapor pressure given by Eq. (2.36) is rewritten as

$$P^{\text{opt}}_{\text{As}_4} = \exp\left(-\frac{4\Delta G^{\text{sub}}_{\text{As}}}{RT}\right) \exp\left(-\frac{2(\Delta G^{\text{F}}_{\text{V}_{\text{As}}} - \Delta G^{\text{F}}_{\text{I}_{\text{As}}})}{RT}\right). \tag{2.36′}$$

The free energy $\Delta G^{\text{sub}}_{\text{As}}$ is given by $\Delta G^{\text{sub}}_{\text{As}} = \Delta H^{\text{sub}}_{\text{As}} - T\Delta S^{\text{sub}}_{\text{As}}$, and the enthalpy part, or the activation energy, $\Delta H^{\text{sub}}_{\text{As}}$ is known to be 0.33 eV. Then, the optimum vapor pressure is expressed as

$$P^{\text{opt}}_{\text{As}_4} = A \exp\left(-\frac{1.32\ \text{eV}}{kT}\right) \exp\left(-\frac{2(\Delta H^{\text{F}}_{\text{V}_{\text{As}}} - \Delta H^{\text{F}}_{\text{I}_{\text{As}}})}{kT}\right). \tag{2.36″}$$

$\Delta H^{\text{F}}_{\text{V}_{\text{As}}}$ and $\Delta H^{\text{F}}_{\text{I}_{\text{As}}}$ are the enthalpy parts of the free energies $\Delta G^{\text{F}}_{\text{V}_{\text{As}}}$ and $\Delta G^{\text{F}}_{\text{I}_{\text{As}}}$, expressed in units of eV.

The fact that the experimental value of the activation energy of the $P^{opt}_{As_4}$ is about 1 eV means that $\Delta H^F_{V_{As}}$ and $\Delta H^F_{I_{As}}$ are not so different.

(4) We can estimate the values of $\Delta H^F_{V_{As}}$ and $\Delta H^F_{I_{As}}$, as follows. The changes in the lattice parameters of heat-treated GaAs crystals should be in proportion to the concentrations of arsenic vacancies in the low arsenic pressure region, and to the concentrations of arsenic interstitial atoms in the high arsenic pressure region. The temperature dependences of the changes in the lattice parameters in heat-treated p-type GaAs crystals are given in Fig. 2.69, for the highest arsenic pressure boundary (solid line) and for the lowest arsenic pressure boundary (dotted line). For the highest arsenic pressure boundary, the corresponding reaction equation is Eq. (2.23), and the arsenic pressure P_{As_4} in Eq. (2.33) is that of pure arsenic liquid given by the following reaction equation:

$$\mathrm{As(liquid)} = \tfrac{1}{4}\,\mathrm{As_4(gas)};\ \Delta G^{vap}_{As} \tag{2.42}$$

from which we have

$$(P^{max}_{As_4})^{1/4} = \exp(-\Delta G^{vap}_{As}/RT). \tag{2.43}$$

Eqs. (2.33) and (2.43) give the change in the lattice parameter at the highest vapor pressure boundary, as follows

$$\Delta a/a \propto n^{max}_{I_{As}} = \exp\left(-\frac{\Delta G'_{I_{As}} + \Delta G^{vap}_{As}}{RT}\right). \tag{2.44}$$

ΔH^{vap}_{As}, which is the enthalpy part of ΔG^{vap}_{As} is known to be 0.064 eV. Fig. 2.69 gives the activation energy of $\Delta a/a$ at the highest vapor pressure boundary as about 1.25 eV, that is

$$\Delta H'_{I_{As}} + \Delta H^{vap}_{As} = 1.25\ \mathrm{eV}$$

from which we get $\Delta H'_{I_{As}} = 1.19$ eV, and $\Delta H^F_{I_{As}} = 1.52$ eV. For the lowest vapor

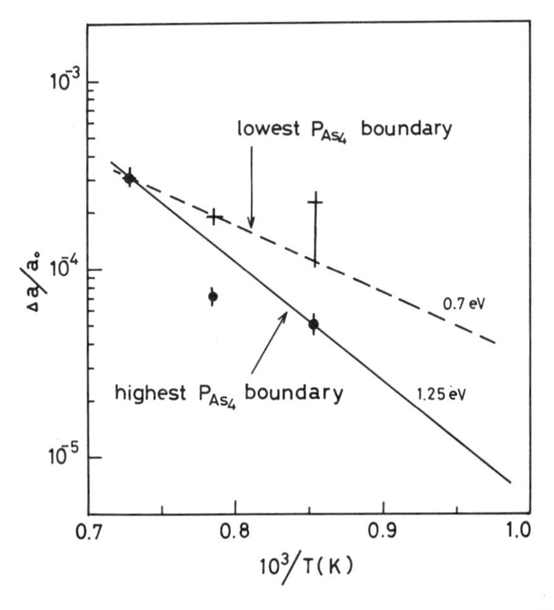

Fig. 2.69 Replot from Fig. 2.9; Δa is the change in the lattice parameter, a_0 the lattice parameter at a stoichiometric point.

pressure boundary, the dominant defects are arsenic vacancies and the corresponding reaction equation is Eq. (2.14).

The vacancy concentration is given by Eq. (2.35), and the arsenic vapor pressure P_{As_4} in this expression is that of the GaAs liquidus line on the Ga-rich side, which is known to be

$$(P_{As_4}^{eq})^{1/4} = \exp(14.0\ \text{cal/mol deg}/R)\exp(-1.01\ \text{eV}/kT)\ \text{(atmosphere)}.$$

Therefore, the change in the lattice parameter $\Delta a/a$ for the lowest vapor pressure boundary is given by

$$\Delta a/a \propto n_{V_{As}}^{max} \propto \exp\left(-\frac{\Delta H'_{V_{As}} - 1.01\ \text{eV}}{kT}\right). \tag{2.45}$$

Fig. 2.69 gives the activation energy for the change in the lattice parameter about 0.7 eV. Therefore, we have $\Delta H'_{V_{As}} = 1.71$ eV, and also $\Delta H_{V_{As}}^F = 1.38$ eV is derived from Eq. (2.39). By using the values $\Delta H_{I_{As}}^F = 1.52$ eV and $\Delta H_{V_{As}}^F = 1.38$ eV, Eq. (2.36″) becomes

$$P_{As_4}^{opt} = A\exp(-1.04\ \text{eV}/kT).$$

In order to compare with the experimental result, we must know the factor A. From Eq. (2.36′), the factor A is given by

$$A = \exp\left(\frac{4\Delta S_{As}^{sub}}{R}\right)\exp\left(\frac{2(\Delta S_{V_{As}}^F - \Delta S_{I_{As}}^F)}{R}\right)\ \text{(atmosphere)}$$

where ΔS_{As}^{sub}, $\Delta S_{V_{As}}^F$, and $\Delta S_{I_{As}}^F$ are the vibrational entropy parts of the free energies ΔG_{As}^{sub}, $\Delta G_{V_{As}}^F$ and $\Delta G_{I_{As}}^F$.

ΔS_{As}^{sub} is known to be 8.49 cal/mol deg (atmosphere). In order to know $\Delta S_{V_{As}}^F$ and $\Delta S_{I_{As}}^F$, we assume that the changes in the lattice parameters of heat-treated crystals are just given by $\Delta a/a = n_{I_{As}}^{max}$ instead of Eq. (2.44) for the highest vapor pressure boundary, and $\Delta a/a = n_{V_{As}}^{max}$ instead of Eq. (2.45) for the lowest vapor pressure boundary, although they should be underestimations for $n_{I_{As}}^{max}$ and $n_{V_{As}}^{max}$. By this assumption, we can estimate that $\Delta S_{V_{As}}^F = +1.18$ cal/mol deg and $\Delta S_{I_{As}}^F = +10.48$ cal/mol deg from Fig. 2.69.

Then, A is given by

$$A = \exp(7.68)\ \text{atmosphere} = 1.64 \times 10^6\ \text{Torr}.$$

or we have

$$P_{As_4}^{opt} = 1.64 \times 10^6 \exp(-1.04\ \text{eV}/kT)\ \text{Torr}. \tag{2.46}$$

This result is in good agreement with the experimental result as compared in Fig. 2.70. Although ΔS_{As}^F and $\Delta S_{I_{As}}^F$ are underestimates, we think that the two underestimations cancel out in the expression for A and we have got good agreement with the experimental result. Even if we put $\Delta a/a = \frac{1}{3}n_{V_{As}}$ or $\frac{1}{3}n_{I_{As}}$, the final result (2.46) does not change. The vibrational entropy $\Delta S_{V_{As}\ \text{or}\ I_{As}}$ is given by

$$\Delta S_{V_{As}\ \text{or}\ I_{As}} = 3R\ln(\nu/\nu')$$

where ν and ν' are the vibrational frequencies of the perfect lattice and around the point defect, respectively, within the oscillator model of the crystal. The experimental obser-

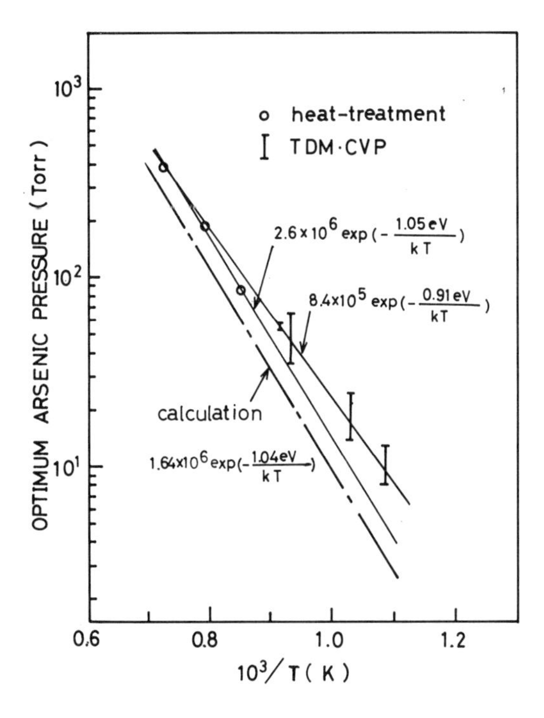

Fig. 2.70 Calculation of the optimum vapor pressure.

vation that lattices containing vacancies or interstitial atoms expand implies that $\nu > \nu'$. This agrees with the estimation that $\Delta S^{F}_{I_{As}}$ and $\Delta S^{F}_{V_{As}}$ are both positive.

In the above treatment we have neglected the electrical activity of V_{As} and I_{As}. We must take into account the electrical activities of these defects in the equilibrium equations if the defect concentration is not much smaller than the intrinsic carrier concentration at the growth temperature. For example, the intrinsic carrier concentration of GaAs at 700°C is about $4 \times 10^{16}\,cm^{-3}$.

2.8.8. *Effect of arsenic vapor pressure on the segregation of impurities*

(1) We consider donor impurities like Te, Se, S and O, which substitute for the As sites in the GaAs lattice. For simplicity, we assume that V_{As} and I_{As} are electrically inactive so that there is no compensation effect, and the density of impurity atoms is much smaller than the intrinsic carrier concentration n_i at the growth temperature (for example, $n_i \approx 4 \times 10^{16}\,cm^{-3}$ at $T = 700°C$) so that there is little effect of free carriers released from impurities on the reaction equations.

On assumption that the impurity, for example Te, in the liquid has the form of a $GaTe_l$ molecule like a $GaAs_l$ molecule, the reaction equation at the liquid–solid interface can be written as follows:

$$GaTe_l(\text{in liquid}) = Ga(\text{lattice}) + Te^{+} + e; \quad \Delta G_{Te} \qquad (2.47)$$

where Te^+ is a donor ion in the GaAs lattice and e is a free electron. Simultaneously, sticking of GaAs molecules and the reactions of As* molecules occur at the liquid–solid interface and the liquid–gas interface, as follows:

$$GaAs(\text{lattice}) = GaAs_l(\text{in liquid});\ \Delta G^{f}_{GaAs} \tag{2.48}$$

$$\tfrac{1}{4}As_4(\text{gas}) = As^*(\text{in liquid});\ \Delta G^* \tag{2.49}$$

$$As^*(\text{in liquid}) = I_{As};\ \Delta G_{I_{As}} \tag{2.50}$$

$$As^* + V_l + Ga_l = GaAs_l(\text{in liquid});\ \Delta G_l \tag{2.51}$$

$$GaAs(\text{lattice}) = Ga(\text{lattice}) + V_{As} + As^*(\text{in liquid});\ \Delta G_{V_{As}}. \tag{2.52}$$

On the other hand, the $GaTe_l$ molecule in the liquid reacts with V_{As} in the lattice as follows:

$$GaTe_l(\text{in liquid}) + V_{As} = Ga_l + Te^+ + e + V_l;\ \Delta G'_{Te}. \tag{2.53}$$

However, the above equation can be derived from Eqs. (2.47) to (2.52). Thus, we have the following relation

$$\Delta G'_{Te} = \Delta G_{Te} + \Delta G_{GaAs} - \Delta G_l - \Delta G_{V_{As}} \tag{2.54}$$

and this reaction is automatically equilibrated if the reactions (2.47) to (2.52) are in equilibrium.

At first, from the equilibrium condition of Eq. (2.47), we have

$$\frac{n_{Te^+}(n/N_c)}{\gamma_{GaTe} X_{GaTe}} = \exp\left(-\frac{\Delta G_{Te}}{RT}\right) \tag{2.55}$$

where $n_{Te^+} = N_{Te^+}/N_L$, n is the free electron concentration in the lattice and N_c is the effective density of states of the conduction band. Assuming that $n \approx n_i$, n_{Te^+} is given by

$$n_{Te^+} = \gamma_{GaTe} X_{GaTe_l} \frac{1}{(n_i/N_c)} \exp\left(-\frac{\Delta G_{Te}}{RT}\right). \tag{2.56}$$

The above equation does not show the effect of the applied vapor pressure. In the above discussion, the effect of the vapor pressure appears as the changes of concentrations of V_{As} and V_l via the reaction equation (2.53) without any effect on the concentration of Te^+.

However, this simplest model is unrealistic because we can assume that a V_{As} in the solid phase can react with a Te atom as illustrated in Fig. 2.71. This Te atom must be considered to be different from a $GaTe_l$ molecule because it can locate a V_{As} site without releasing a Ga_l atom. We denote this atom as Te^*, like As^* in the liquid. The reaction of

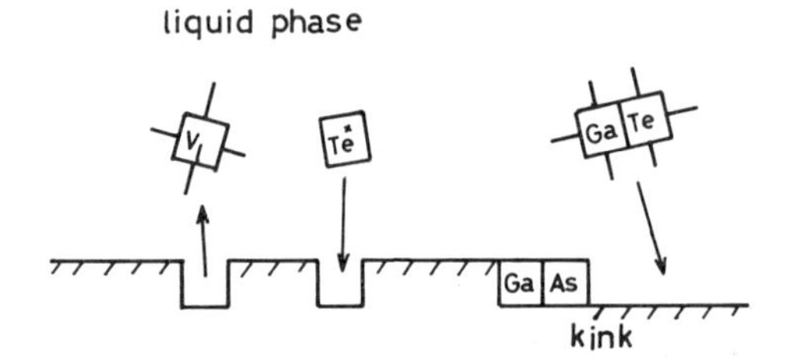

Fig. 2.71 A model of sticking of $GaTe_l$ and Te^*.

Te^* and V_{As} can be written as

$$Te^*(\text{in liquid}) + V_{As} = Te^+(\text{lattice}) + e; \ \Delta G^*_{Te}. \tag{2.57}$$

Te^* and $GaTe_l$ molecules react in the liquid via the presence of V_l as follows:

$$Te^* + V_l + Ga_l = GaTe_l; \ \Delta G^*_{TeV_l}. \tag{2.58}$$

This reaction equation can be derived from Eqs. (2.51), (2.52), (2.57), (2.47) and (2.48). Therefore, the reaction of Te^* and $GaTe_l$ is automatically equilibrated if Eqs. (2.51), (2.52), (2.57), (2.47) and (2.48) are in equilibrium, and we have the following relation:

$$\Delta G^*_{TeV_l} = \Delta G_l + \Delta G_{V_{As}} + \Delta G^*_{Te} - \Delta G_{Te} - \Delta G^f_{GaAs}. \tag{2.59}$$

Therefore, we need not take into account the equilibration of Te^* and $GaTe_l$ molecules in the liquid. From the equilibrium condition of Eq. (2.57) we have

$$\frac{n_{Te^+}(n/N_c)}{\gamma_{Te^*} X_{Te^*} n_{V_{As}}} = \exp\left(-\frac{\Delta G_{Te^*}}{RT}\right). \tag{2.60}$$

We assume that the total concentration of Te atoms in the liquid, X_{Te}, is constant,

$$X_{GaTe_l} + X_{Te^*} = X_{Te} = \text{constant}. \tag{2.61}$$

This should be a good approximation in the case of Te for which we need not consider the external vapor pressure control of the impurity concentration. The situation should be very different in the case of oxygen atoms for which the external oxygen vapor pressure determined by the experimental condition should determine the number of oxygen atoms in the liquid.

From Eqs. (2.56), (2.60) and (2.61), we have

$$n_{Te^+} = \frac{X_{Te}}{(n_i/N_c)\left\{\dfrac{1}{\gamma_{GaTe}\exp(-\Delta G_{Te}/RT)} + \dfrac{1}{\gamma_{Te^*} n_{V_{As}} \exp(-\Delta G_{Te^*}/RT)}\right\}}. \tag{2.62}$$

When $n_{V_{As}}$ is so high that the second term in the denominator can be neglected, the above equation becomes

$$n_{Te^+} \approx \frac{\gamma_{GaTe} X_{Te}}{(n_i/N_c)} \exp\left(-\frac{\Delta G_{Te}}{RT}\right). \tag{2.63}$$

This equation coincides with Eq. (2.55) because we can assume that most of the Te atoms in the liquid are $GaTe_l$ molecules when $n_{V_{As}}$ is high enough. It is important that n_{Te^+} does not depend on $n_{V_{As}}$ in Eq. (2.63).

On the contrary, when $n_{V_{As}}$ is so low that the first term of the denominator of Eq. (2.62) can be neglected, we have

$$n_{Te^+} \approx \frac{\gamma_{Te^*} X_{Te}}{(n_i/N_c)} n_{V_{As}} \exp\left(-\frac{\Delta G_{Te^*}}{RT}\right). \tag{2.64}$$

This equation shows that n_{Te^*} becomes low in proportion to $n_{V_{As}}$ if $n_{V_{As}}$ is low. This behavior of the density of Te donors is illustrated in Fig. 2.72, as a function of applied arsenic vapor pressure.

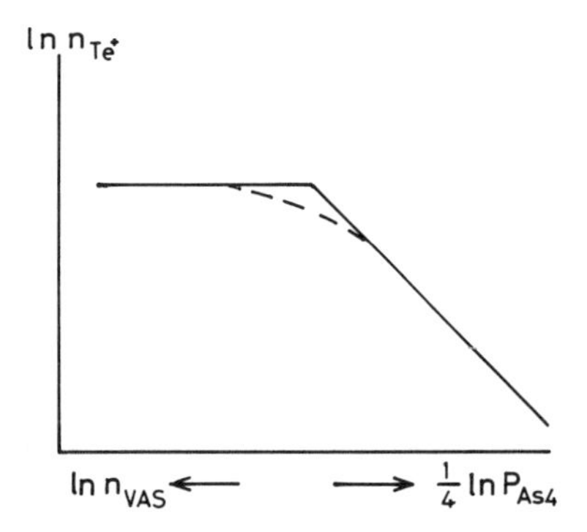

Fig. 2.72 Schematic diagram representing Eq. (2.62).

The vapor pressure above which n_{Te^+} starts to decrease is given from the condition that the first and second terms of the denominator of Eq. (2.62) are equal.

Therefore, $n_{V_{As}}$ and the applied vapor pressure P_{As_4} at this bending point is given by

$$n_{V_{As}} = \frac{\gamma_{GaTe}}{\gamma_{Te^*}} \exp\left(- \frac{(\Delta G_{Te} - \Delta G_{Te^*})}{RT} \right) = (P_{As_4})^{1/4} \exp\left(- \frac{G_{V_{As}}}{RT} \right). \tag{2.65}$$

It is important that the bending point does not depend on the amount of impurities. In the region where n_{Te^+} decreases linearly in proportion to $n_{V_{As}}$, most of the Te atoms in the liquid have the form of Te* instead of $GaTe_l$. Such behavior should be different for different kinds of impurities because the free energy ΔG_{Te^*} should largely be affected by the atomic radius and ionicity of the impurity atoms.

(2) Now, we discuss the experimental fact that the free electron concentration of GaAs shows a minimum at the optimum vapor pressure.

We have three hypotheses:

(a) Both V_{As} and I_{As} behave as donors and the measured free electron concentration corresponds to $n_{V_{As}} + n_{I_{As}}$, which shows a minimum at the stoichiometric point.

(b) The measured free electron concentrations are due to residual donor impurities of different kinds, one which substitutes for the arsenic lattice site, and the other which substitutes for the Ga lattice site like a Si atom. However, this idea is unrealistic because there is no reason why the minimum of the carrier concentration should coincide with the stoichiometric point.

(c) The measured free electron concentrations are due to residual donor impurities; nearly ideal vapor pressure control mechanism works at a vapor pressure near to the optimum vapor pressure which gives the exact stoichiometry. However, if the applied vapor pressure is much smaller or much higher than the optimum vapor pressure, arsenic vacancies or arsenic interstitial atoms aggregate and contribute to the generation and propagation of edge dislocations. Edge dislocations are stoichiometric. However, if arsenic interstitial atoms come to an edge dislocation, they would catch Ga atoms from Ga lattice site, and released V_{Ga} would be going out of the crystal lattice because they would be unstable.

Thus, arsenic interstitial atoms could contribute to the propagation of edge dislocations. A similar discussion is also possible for arsenic vacancies if a gallium vacancy is replaced by a gallium interstitial atom. At the interface of the solid and liquid phases, end points or end regions of dislocations could react with Te* atoms in the liquid just like V_{As} at the interface does. This reaction can be written as

$$\text{Te}^*(\text{in liquid}) = \text{Te}^+(\text{As lattice site at or near a dislocation}) + \text{e};\ \Delta G_D^* \tag{2.66}$$

When the above reaction equilibrates, we have

$$\frac{n_{Te^+}(n/N_c)}{\gamma_{Te^*} X_{Te^*} a n_D} = \exp\left(-\frac{\Delta G_{D^*}}{RT}\right) \tag{2.67}$$

where n_D is the density per square centimeter of dislocations, and a is a constant introduced to reduce n_D to the density per cubic centimeter. The reactions given by Eqs. (2.66) and (2.57) compete with each other.

Therefore, at the vapor pressure regions much higher or lower than the optimum vapor pressure, where many dislocations are formed, the reaction with dislocations prevails over the reactions with point defects like V_{As} and I_{As}. In such a case, the amounts of Te* in the liquid and Te donors in the solid are not controlled by the amount of V_{As}, or the applied vapor pressure. From Eqs. (2.56), (2.61) and (2.67), we have

$$n_{Te^+} = \frac{X_{Te}}{(n_i/N_c)\left\{\dfrac{1}{\gamma_{GaTe}\exp(-\Delta G_{Te}/RT)} + \dfrac{1}{\gamma_{Te^*} a n_D \exp(-\Delta G_D^*/RT)}\right\}}. \tag{2.68}$$

When n_D is high, we have the same equations as Eq. (2.63), corresponding to the flat region of Fig. 2.3. That is, there appears to be no dependence on dislocation density. If we assume that dislocations with $n_D \approx 10^5\ \text{cm}^{-2}$ are produced at a vapor pressure different from the optimum vapor pressure, the corresponding value of $a n_D$ is of the order of $10^{13}\ \text{cm}^{-3}$. Therefore, if $N_{V_{As}}$ at the bending point in Fig. 2.3 is of the order of $10^{13}\ \text{cm}^{-3}$ or smaller, n_{Te^+} is given by Eq. (2.63) and there appears to be no dependence on the density of arsenic vacancies nor dislocations.

This hypothesis can interpret the fact that the free electron concentrations near the optimum vapor pressure are sometimes rather high compared with $N_{I_{As}} + N_{V_{As}}$ expected to be surviving under the stoichiometric condition.

2.9. Conclusion

As the results of the numerous investigations about the characteristics in III–V and II–VI compound crystals, it can be said that the most important point is the control of the deviation from the stoichiometric composition in the crystal. From the heat-treatment experiment of GaAs and GaP the specific vapor pressure for exact stoichiometry was found as a function of temperature which is given by Eqs. (2.1) and (2.2). Then, the authors developed the most ideal crystal growth method called TDM-CVP for III–V and II–VI compounds in which stoichiometry control was possible by the applied vapor pressure onto the top of the molten phase, and found the optimum vapor pressure at which exact stoichiometry was realized and nearly dislocation-free crystals were obtained. This optimum vapor pressure was in good agreement with the vapor pressure for the exact stoichiometry found in the heat-treatment experiment.

This technique was first applied to the III–V compounds and realized super bright LED (red from GaAlAs, pure green from GaP). Then, TDM-CVP was extended to II–VI compounds; in particular we have realized the growth of p-type ZnSe and blue luminescent diodes of ZnSe p–n junctions.

Also TDM-CVP was extended to the melt growth and we obtained GaAs crystals with very few dislocations.

The mechanism of stoichiometry control in TDM-CVP was clarified from the equality of the chemical potentials of arsenic between the gas, liquid and solid phases.

References

1a) S. Shinozaki, K. Ishida, and J. Nishizawa: Rept. Tech. Group on SSD, IECE of Japan, No. SSD-71-10 (1971).
1b) J. Nishizawa, S. Shinozaki, and K. Ishida: J. Appl. Phys., 14 (1973) 1638.
2) J. Nishizawa, K. Ito, Y. Okuno, and F. Sakurai: to be published in J. Appl. Phys.
3) K. Kaneko, M. Ayabe, M. Dosen, K. Morizane, S. Usui, and N. Watanabe: Proc. IEEE, 61 (1973) 884.
4) T. Suzuki and S. Akai: Bussei. 12 (1971) 144. (in Japanese)
5) J. M. Parsey, Jr., Y. Nanishi, J. Lagowski, and H. C. Gatos: J. Electrochem. Soc., 128 (1981) 937.
6) H. Otsuka, K. Ishida, and J. Nishizawa: J. Appl. Phys., 8 (1969) 632.
7) J. Nishizawa, H. Otsuka, S. Yamakoshi, and K. Ishida: Japan. J. Appl. Phys., 13 (1974) 46.
8) J. S. Blakemore: Semiconductor Statistics, International Series of Monographs on Semiconductors, Vol. 3, Ed. Heinz K. Henisch (Pergamon, Oxford, 1962) p. 139.
9) J. Nishizawa et al.: to be published.
10) J. Nishizawa, Y. Okuno, K. Suto, T. Sato, and S. Yamakoshi: Solid State Commun., 14 (1974) 889.
11) J. Nishizawa and Y. Okuno: IEEE Trans. Electron Devices, ED-22 (1975) 716.
12) J. Nishizawa, Y. Okuno, and H. Tadano: J. Crystal Growth, 31 (1975) 215.
13) J. Nishizawa and Y. Okuno: Proc. 2nd Int. School on Semiconductor Optoelectronics (Catniewo, 1978).
14) J. Nishizawa, N. Toyama, Y. Oyama, and K. Inokuchi: Proc. 3rd Int. School on Semiconductor Optoelectronics (Cetniewo, 1981).
15) J. Nishizawa and K. Inokuchi: Proc. Int. Optoelectronics Workshop in Taiwan (1981).
16) J. Nishizawa and M. Koike: Int. School on Defect Complexes in Semiconductor Structures (Hungary, 1982).
17) J. Nishizawa, Y. Oyama, Y. Tadano, K. Inokuchi, and Y. Okuno: J. Crystal Growth, 47 (1979) 434.
18) J. Nishizawa, Y. Oyama, and Y. Okuno: J. Crystal Growth, 52 (1981) 925.
19) K. Tomizawa, K. Sassa, Y. Shimanuki, and J. Nishizawa: to be published.
20) J. Nishizawa, M. Koike, K. Miura, and Y. Okuno: Japan. J. Appl. Phys., 19 (1980) 22.
21) J. Nishizawa, Y. J. Shi, K. Suto, and M. Koike: J. Appl. Phys., 53 (1982) 3878.
22) J. Nishizawa, M. Koike, and C. C. Jin: J. Appl. Phys., 54 (1983) 2807.
23) J. Nishizawa, Y. Okuno, M. Fukase, and H. Tadano: J. Cryst. Mol. Struct. 10 (1980) 123.
24) J. Nishizawa: Ōyō Butsuri (1972). (in Japanese)
25) M. Aven and J. S. Prener, (Eds.): Physics and Chemistry of II–VI compounds (North-Holland, Amsterdam, 1967).
26) R. E. Honig and D. A. Kramer: RCA Rev. (1969) 285.
27) L. J. Van der Pauw: Philips Res. Rep., 13 (1958) 1.
28) J. L. Merz, K. Nassau, and J. W. Shiever: Phys. Rev., B8 (1973) 1444.
29) A. I. Ivaschenko: Conf. Berg und Huttenmanisher tag (Frieberg, 1983).
30) J. Nishizawa, Y. Kobayashi, and Y. Okuno: Japan. J. Appl. Phys., 19 (1980) 345.
31) J. Nishizawa, Y. Okuno, and Y. Kobayashi: Proc. 1978 Int. Conf. on Vapor Growth and Epitaxy, ICVGE-4.
32) J. Nishizawa: J. Japan. Assoc. Crystal Growth, 5 (1978) 211.

3 VAPOR PHASE EPITAXY OF III–V SEMICONDUCTORS

Hisashi SEKI and Akinori KOUKITU†

Abstract

Recent work concerned with VPE growth of III–V compound semiconductors in Japan is reviewed. In particular, this chapter is centered on the current research on the halogen transport VPE of GaAs, GaInAs, InGaAsP and Al alloys. The single flat-temperature zone method for the growth of GaAs is described. The purity of GaAs is discussed in terms of the $AsCl_3$ mole fraction. Studies of the hydride and chloride methods for the growth of ternary and quaternary alloys are mentioned with emphasis on InGaAs and InGaAsP.

Keywords: Hydride and Chloride VPE, SFT Method, Thermodynamics

3.1. Introduction

Vapor phase epitaxy (VPE) is a process of considerable technological importance in the semiconductor industry. For III–V compounds, it provides materials of high quality and purity on a large scale. In addition, alloy semiconductors such as $GaAs_yP_{1-y}$ and $In_{1-x}Ga_xAs_yP_{1-y}$ can be grown easily as epitaxial layers. These advantages have stimulated and facilitated the development of devices such as field-effect transistors, light-emitting diodes and injection lasers.

The halogen transport VPE process is performed by the hydride method[1,2] or by the chloride method.[3] In the hydride method, HCl is used to transport Ga or In metals, and AsH_3 and PH_3 are the source of the As and P. Since the source reactions of Ga or In metals with HCl proceed quite steadily and are easy to control, the hydride method is preferably used for the growth of alloy semiconductors, including the preparation of the superlattice structures. In the chloride method, $AsCl_3$ or PCl_3 is passed over elemental Ga or In to form metal chlorides. To date, the method is used widely to grow high-purity GaAs and InP, and is applied to the fabrication of microwave devices. In contrast to the hydride method, the chloride method has not been extensively investigated yet for the alloy semiconductors, because there is a transient due to the requirement of source saturation during changes in $AsCl_3$ (or PCl_3) flow over the Ga (or In) boat. Recently, however, the difficulty has been overcome by Komeno et al.[4] employing the dual-growth-chamber reactor which was used by Mizutani et al.[5] in the hydride method for the preparation of InGaAsP/InP heterostructures. They have made $In_{0.53}Ga_{0.47}As$/InP superlattices by the chloride method.[4]

In this chapter the current development of the halogen transport VPE in Japan is briefly reviewed. In the growth of GaAs the single flat-temperature zone (SFT) method

---

† Department of Industrial Chemistry, Tokyo University of Agriculture and Technology, Koganei, Tokyo 184.

JARECT Vol. 19, *Semiconductor Technologies* (1986), *J. Nishizawa (ed.)*
OHMSHA, LTD. and North-Holland

will be described. Comparisons of the electrical properties will be made between the SFT and conventional methods in terms of the mole fraction effect. In the growth of ternary and quaternary alloys, special emphasis will be placed on the growth of InGaAs and InGaAsP alloys since these materials are of great importance in the fields of opto-electronic and microwave devices.

3.2. GaAs

The Ga–$AsCl_3$–H_2 vapor transport system[3] is used as the standard method for the growth of high-purity GaAs. Among numerous factors which influence the purity of grown layers, it has been shown that the $AsCl_3$ mole fraction plays a particularly important role.[6–8] In fact, the purity of GaAs grown by this system varies inversely with the $AsCl_3$ mole fraction in the reactor, and high-purity GaAs is obtained only in the high $AsCl_3$ mole fraction range. Recent interesting work in this field is the preparation of high-purity GaAs by the single flat-temperature zone (SFT) method.[9,10] According to this method, the residual impurity concentrations are markedly reduced in the low $AsCl_3$ mole fraction range, and the purity of grown layers is independent of the $AsCl_3$ mole fraction.

Fig. 3.1 shows a schematic view of the apparatus and temperature profiles for the conventional and SFT methods. In the conventional method, the source gallium is kept at 800–900°C, and the substrate is placed at 700–750°C in a temperature gradient zone (profile (a)), while in the SFT system, the source and substrate are maintained at the same temperature as shown in profile (b). The SFT method has the following advantages: (1) easy control of the furnace temperature because only one flat temperature zone is required; (2) Si contamination can be reduced because the source temperature can be made lower than that for the conventional method; (3) uniform grown layers in the thickness and doping are obtained as there is no temperature gradient in the substrate position. The SFT method is possible for four systems: Ga–AsH_3–HCl–H_2, Ga–$AsCl_3$–H_2, GaAs(crust)–$AsCl_3$–H_2, and GaAs(crust)–$AsCl_3$–He–H_2 systems. Of these four, the last one is the most promising, since it provides reproducibly uniform grown layers with high purity.

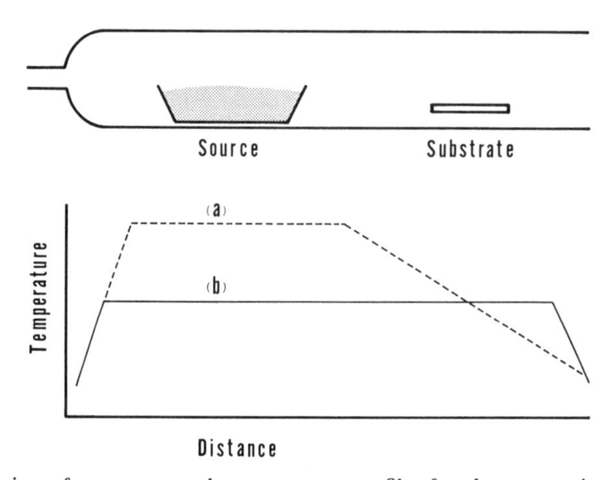

Fig. 3.1 Schematic view of apparatus and temperature profiles for the conventional and SFT methods.

The electrical properties of epitaxial layers obtained both by the SFT and conventional methods are compared in Figs. 3.2 and 3.3. Fig. 3.2 shows the variation of the carrier concentration as a function of the $AsCl_3$ mole fraction for undoped layers. The data of the conventional system are indicated by the filled circles,[11] triangles,[8,12] and squares.[13] The well known inverse dependence of the carrier concentration on the $AsCl_3$ mole fraction is seen for the conventional method, while, although there is scatter in the data, the $AsCl_3$ mole fraction effect is not observed for the SFT methods. The carrier concentrations of the SFT methods are remarkably low at the low $AsCl_3$ mole fractions, compared with those of the conventional system. For example, the difference of the concentrations becomes nearly three orders of magnitude at the lowest region of the $AsCl_3$ mole fraction. The growth at a low $AsCl_3$ mole fraction is suitable for the preparation of pyramid-free surfaces, since the higher the $AsCl_3$ mole fraction, the more pronounced is the pyramid formation during epitaxial growth.[11,12]

In Fig. 3.3 the dependence of Hall mobility at 77 K on the $AsCl_3$ mole fraction is shown for all of the data used in Fig. 3.2. Compared with the conventional method, the SFT method apparently produces higher-mobility epitaxial layers in a low $AsCl_3$ mole

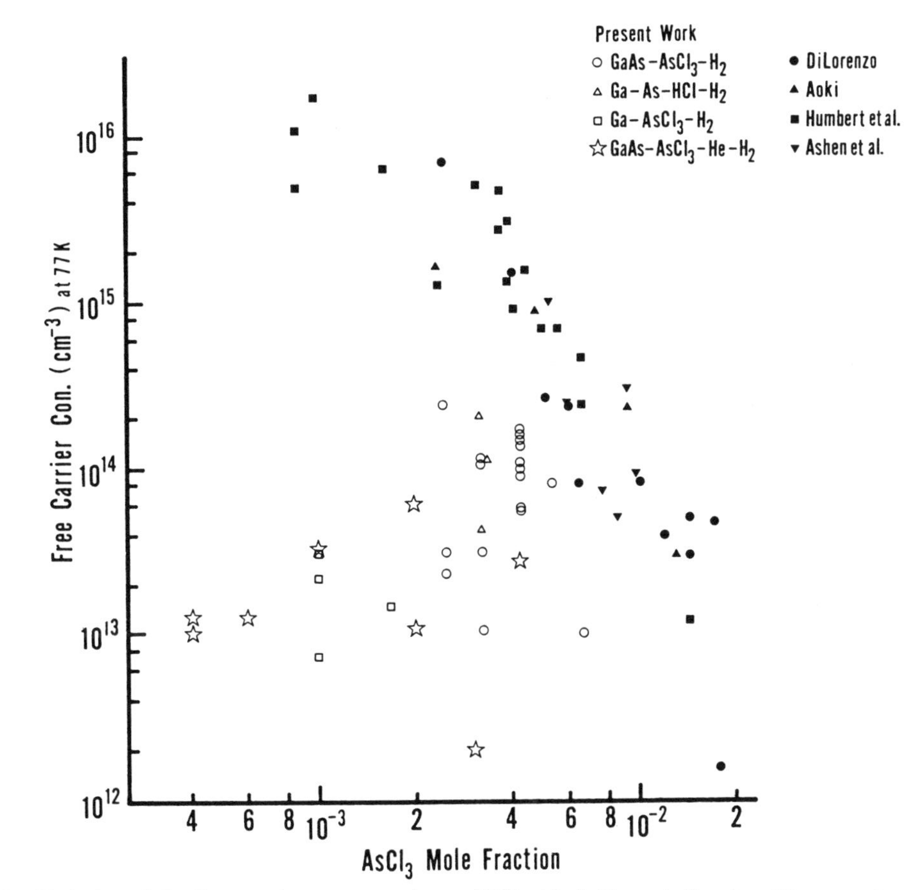

Fig. 3.2 Variation of the free carrier concentration at 77 K with $AsCl_3$ mole fraction. The experimental data were extracted from refs. 11, 12, 13, 8. The respective systems for the SFT method are: (○) GaAs(crust)–$AsCl_3$–H_2; (△) Ga–As–HCl–H_2; (□) Ga–$AsCl_3$–H_2; (☆) GaAs(crust)–$AsCl_3$–He–H_2.[10]

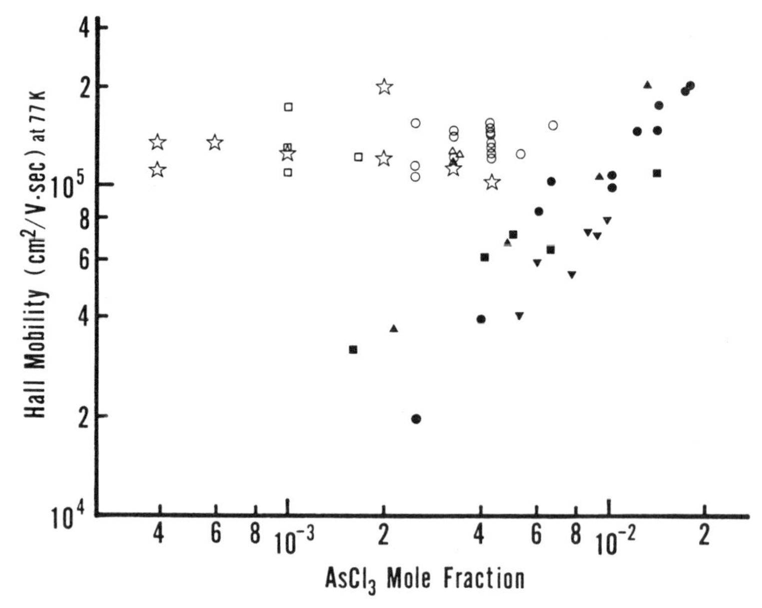

Fig. 3.3 Variation of the Hall mobility at 77 K with the $AsCl_3$ mole fraction.[10]

fraction range, and the values are independent of the $AsCl_3$ mole fraction. It is thought that the decrease of the impurity concentration in the SFT method is ascribed to the decrease of silicon contamination which was brought by the use of the lower source temperature.[14]

The uniformity of thickness and doping is an important factor for the preparation of GaAs devices. An excellent uniformity has been achieved by the SFT method. For example, the thickness and doping are constant to within ±3% on two wafers each of area about 20 cm^2. The kinetically limited growth conditions provide the highly uniform epitaxial layers. The same conclusion has been reported by Komeno et al.[15] in the conventional Ga–$AsCl_3$–H_2 system at low temperatures. They reported that highly uniform layers have been obtained on the substrate with areas of up to about 40 cm^2.

3.3. InGaAs

The ternary alloy $In_{0.53}Ga_{0.47}As$ grown on InP is interesting because of its application to devices such as photodiodes, lasers and microwaves. The VPE growth of this alloy has been performed by three methods: (1) the hydride method, (2) the chloride method, and (3) metal organic chemical vapor deposition (MOCVD). To date, most of the work has been done via the hydride and MOCVD methods. In contrast to these, the chloride method has not been studied extensively because of the necessity of crust formation. It was generally thought that the transient time required for the crust formation would make it difficult to grow multilayered structures with abrupt junctions. However, in the preparation of superlattice structures the method has received considerable attention since the recent successful work of Komeno et al.[4] They have shown that the method can grow high-quality epitaxial layers at a rate as slow as that of molecular-beam epitaxy. Recent studies on the hydride and chloride methods will be described below.

The main chemical reactions that occur in the deposition of InGaAs are:

$$2GaCl + \tfrac{1}{2}As_4 + H_2 = 2GaAs(alloy) + 2HCl \qquad (3.1)$$

$$2InCl + \tfrac{1}{2}As_4 + H_2 = 2InAs(alloy) + 2HCl \qquad (3.2)$$

$$As_4 = 2As_2. \qquad (3.3)$$

Assuming the system to be at chemical equilibrium, the alloy composition and the driving force for the deposition have been calculated for the three (hydride, chloride, and SFT) methods.[16,17] Fig. 3.4 shows a comparison of the deposit composition. The abscissa, *R*, indicates the input mole ratio, which is HCl(Ga)/[HCl(Ga) + HCl(In)] for the hydride system, and is $AsCl_3$(Ga)/[$AsCl_3$(Ga) + $AsCl_3$(In)] for the chloride and SFT systems. Compared with the hydride system, the chloride and SFT systems always give a higher gallium content at a given input mole ratio. However, the differences between the three curves are rather small, since the same reactions dominate in the deposition zone.

In Fig. 3.5, the driving force for deposition, $\Delta P = (P^0_{GaCl} + P^0_{GaCl_3} + P^0_{InCl} + P^0_{InCl_3}) - (P_{GaCl} + P_{GaCl_3} + P_{InCl} + P_{InCl_3}) = P_{HCl} - P^0_{HCl}$, is given as a function of the deposition temperature for the three methods. It is seen that $In_{0.53}Ga_{0.47}As$ tends to deposit more readily in the hydride systems than in the chloride and SFT systems. This results in a slow growth rate in the chloride and SFT methods. The slow growth rate is favourable for the preparation of superlattice structures.

Susa et al. have grown high-quality epitaxial layers by the In–Ga–AsH_3–HCl–H_2 system.[18–20] $AsCl_3$ (7 nines) is used as an HCl source. Monochlorides of indium and gallium formed by reactions of the HCl with source In and Ga metals at 800°C, are mixed with AsH_3 at 825°C and supplied to the growth region at a temperature of 730°C. The ratio of the HCl gas flow rate over the In to that over the Ga is 3. A sliding quartz boat is used to protect InP substrates from thermal decomposition (Fig. 3.6). Substrates are initially covered with a quartz plate which could be slid off to expose the substrate surface for InGaAs deposition, as shown in Fig. 3.6(b). Improved crystalline quality ($n_{77} = 1.0 \times 10^{15}\ cm^{-3}$, $\mu_{77} = 35400\ cm^2/V\ s$) is attributed to this boat.

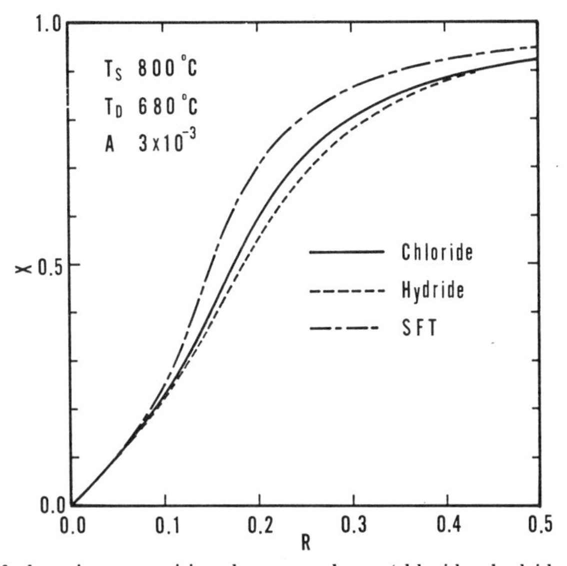

Fig. 3.4 Comparison of deposit composition between three (chloride, hydride (III : V = 1 : 1) and SFT) methods.[17]

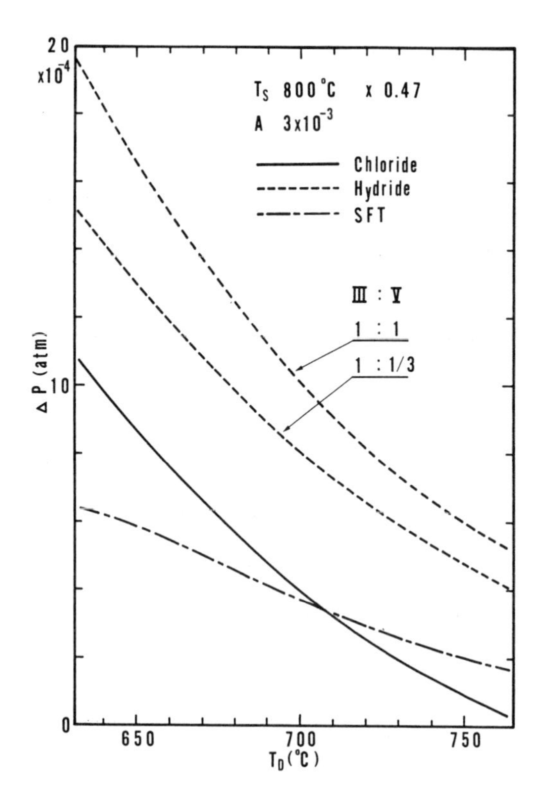

Fig. 3.5 Driving force for deposition as a function of temperature.[17]

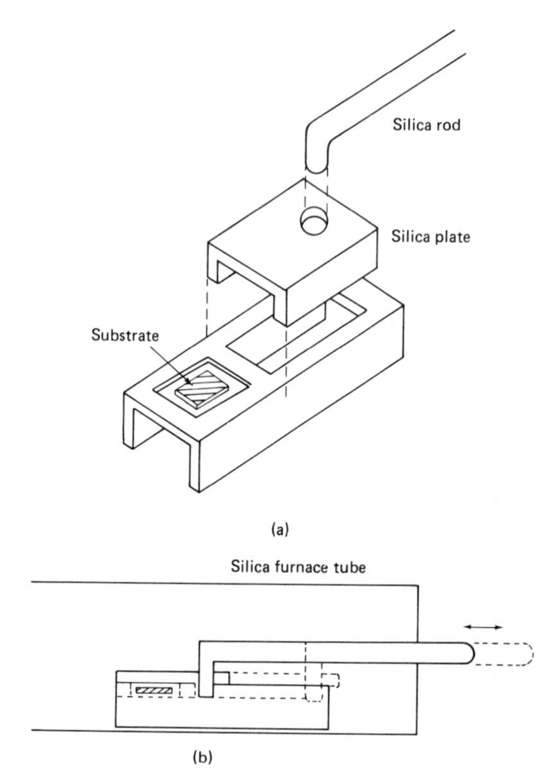

Fig. 3.6 Schematic view of slide boat substrate supporter.[18]

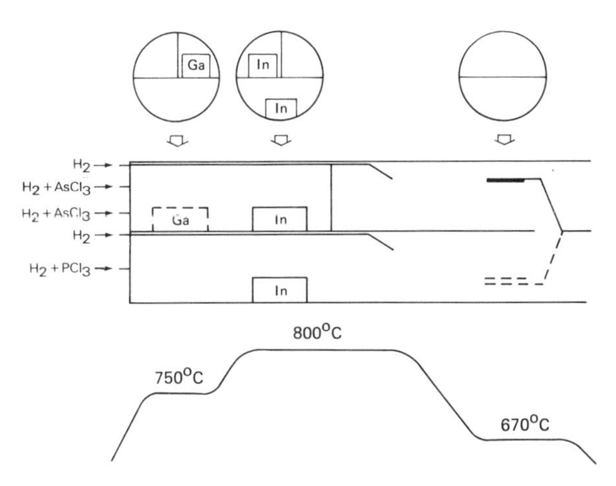

Fig. 3.7 Schematic diagram of the reactor and temperature profile for chloride VPE growth of InGaAs/InP hererostructure.[4)]

The preparation of InGaAs/InP superlattices has been performed by Komeno et al. using chloride method.[4)] The Ga–In–$AsCl_3$–H_2 and In–PCl_3–H_2 systems are used for the growth of InGaAs and InP, respectively. They have claimed that the motivation for employing the chloride method was as follows: (1) the chloride method is safer than MOCVD which uses dangerous AsH_3, PH_3, and metalorganic compounds; (2) this method offers lower cost and higher yield of production compared with MBE; (3) InGaAs and InP layers of very high purity can be grown easily. Fig. 3.7 shows a schematic diagram of their reactor and a temperature profile. The reactor consists of two chambers: the upper one is for InGaAs growth and the lower for InP growth. The transfer of the substrate is effected within a few seconds. An abrupt heterointerface is realized by reduction of the growth rates of InGaAs and InP. For example, the growth rate of InGaAs is about 3 Å/s and that of InP about 12 Å/s. These values are comparable with those used in molecular beam epitaxy.

Fig. 3.8 is a micrograph of the angle-beveled cross-section of an InGaAs/InP superlattice. The twelve layers of InGaAs (~80 Å) appear as the lighter lines while the

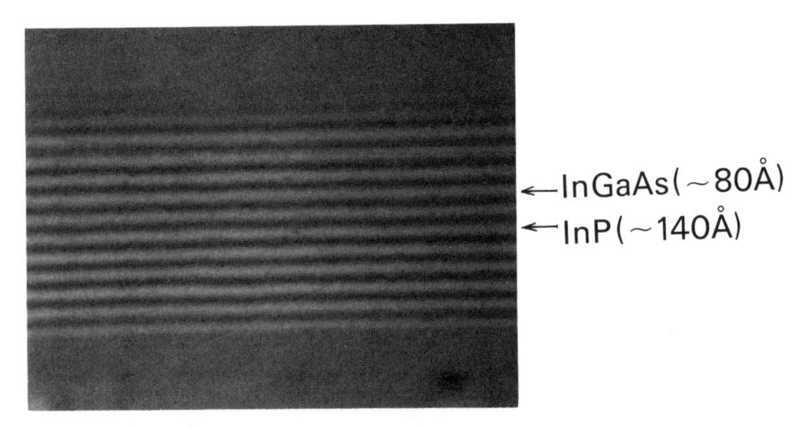

Fig. 3.8 Angle-beveled cross-section of an InGaAs/InP superlattice grown by chloride VPE. The twelve InGaAs layers (80 Å) appear as the lighter lines while the eleven InP layers (140 Å) are the darker material.[4)]

eleven darker lines are InP layers (~140 Å). Heterointerface transition widths less than 30 Å are claimed.

Komeno et al.[21] have also reported the growth of selectively doped InGaAs/InP heterojunctions using the same growth system, and the observation of a high-mobility, two-dimensional electron gas (TDEG) at the interface of the heterostructure. Enhanced electron mobilities as high as 9,400, 71,200 and 10,600 $cm^2/V\,s$ at 300, 77 and 4.2 K, respectively, have been achieved with the sheet electron concentration N_s equal to $2 \times 10^{11}\ cm^{-2}$ at 4.2 K. These values are the highest reported to date for selectively doped heterostructures using an InGaAs layer.

3.4. InGaAsP

InGaAsP quaternary alloys have received much attention as materials for laser and detector diodes in long-distance fiber-optical communication systems. Vapor-phase epitaxy of the materials was attempted at first by Sugiyama et al. using the Ga–In–AsH_3–PH_3–HCl–H_2 system.[22] They investigated the influence of the input mole ratios Ga/(Ga + In) and P/(As + P) on the alloy composition of epitaxial layers and growth conditions for InGaAsP alloys on GaAs substrate with bandgap wavelengths in the range of 0.6–0.9 μm. Enda,[23] Olsen et al.,[24] and Hyder et al.[25] have reported the preparation of alloys lattice-matched to the InP substrate.

A simplified thermodynamic calculation has been made by Nagai to determine the growth aspects of the quaternary alloys.[26] Koukitu and Seki performed a more complete analysis, considering the chemical reactions that occur in the deposition region as follows:[27]

$$2GaCl + \tfrac{1}{2}As_4 + H_2 = 2GaAs(alloy) + 2HCl \tag{3.4}$$

$$2InCl + \tfrac{1}{2}As_4 + H_2 = 2InAs(alloy) + 2HCl \tag{3.5}$$

$$2GaCl + \tfrac{1}{2}P_4 + H_2 = 2GaP(alloy) + 2HCl \tag{3.6}$$

$$2InCl + \tfrac{1}{2}P_4 + H_2 = 2InP(alloy) + 2HCl \tag{3.7}$$

$$As_4 = 2As_2 \tag{3.8}$$

$$P_4 = 2P_2 \tag{3.9}$$

$$AsH_3 = \tfrac{1}{2}As_2 + \tfrac{3}{2}H_2 \tag{3.10}$$

$$PH_3 = \tfrac{1}{2}P_2 + \tfrac{3}{2}H_2 \tag{3.11}$$

Fig. 3.9 is a deposition diagram for $In_{1-x}Ga_xAs_yP_{1-y}$ alloys showing the solid composition versus the input mole ratios. The dashed lines indicate the alloy compositions lattice-matched to InP, GaAs, and $GaAs_{0.6}P_{0.4}$ respectively. From the diagram, one can predict an input mole ratio which is necessary for a given alloy composition. For example, to deposit the lattice-matched alloys to InP, one must choose smaller values for (HCl(Ga)/HCl(Ga) + HCl(In)), less than 0.15. It is worthwhile to note that the curves in the figure are not shifted appreciably by the change of the input values of n_{Cl}/n_H and $n_{(Ga+In)}/n_{(As+P)}$.

Figs. 3.10 and 3.11 show comparisons of the experimental data of Sugiyama et al.[22] with the values derived from such a calculation. The agreement between the experimental

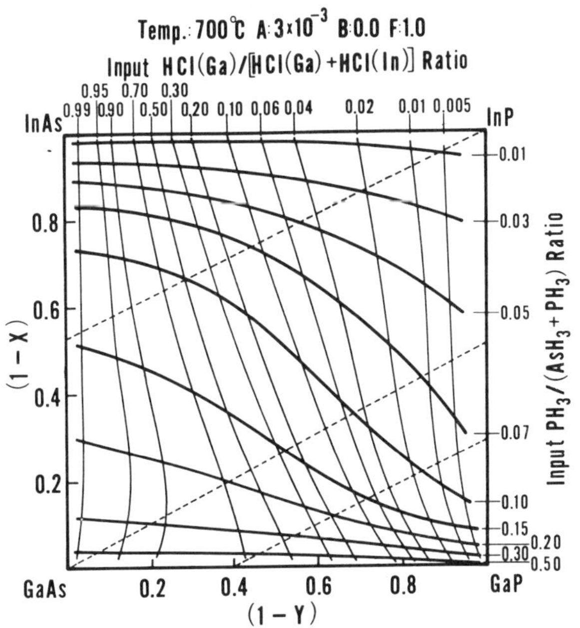

Fig. 3.9 Deposition diagram for $In_{1-x}Ga_xAs_yP_{1-y}$ alloys. The dashed lines indicate the alloy compositions lattice-matched to InP, GaAs and $GaAs_{0.6}P_{0.4}$.[27]

and calculated compositions is remarkably good and shows that the thermodynamic calculation provides a useful tool for predicting the proper conditions for the growth of a desired InGaAsP alloy.

The effects of substrate orientations on both the growth rate and the composition are shown in Table 3.1. Five different substrates (about 5×5 mm) were placed on a quartz plate horizontally, as shown in the upper figure in Table 3.1. The experiment was carried

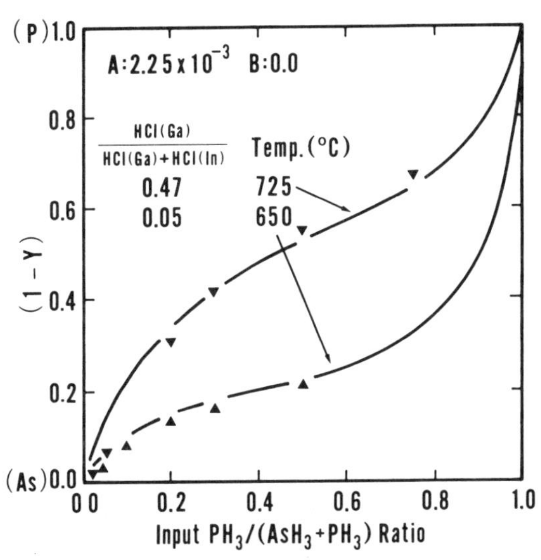

Fig. 3.10 Comparison between the experimental and calculated (solid lines) compositions of quaternary alloys. Dependence of alloy composition on input $PH_3/(AsH_3+PH_3)$ ratio.[27]

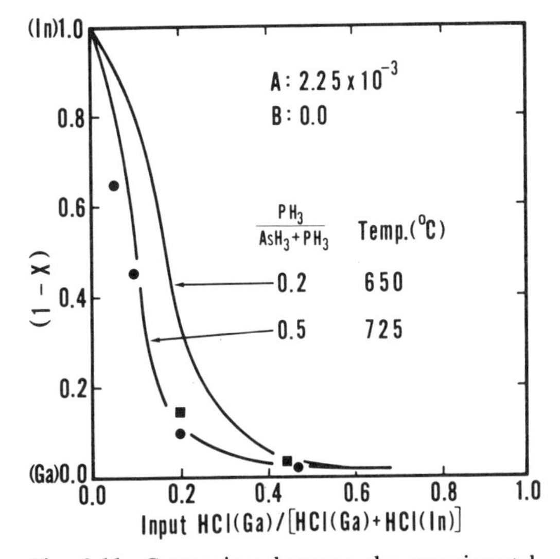

Fig. 3.11 Comparison between the experimental and calculated (solid lines) compositions of quaternary alloys. Dependence of alloy composition on input HCl(Ga)/[HCl(Ga) + HCl(In)] ratio.[27]

Table 3.1 Effects of substrate orientations on both the growth rate and the composition

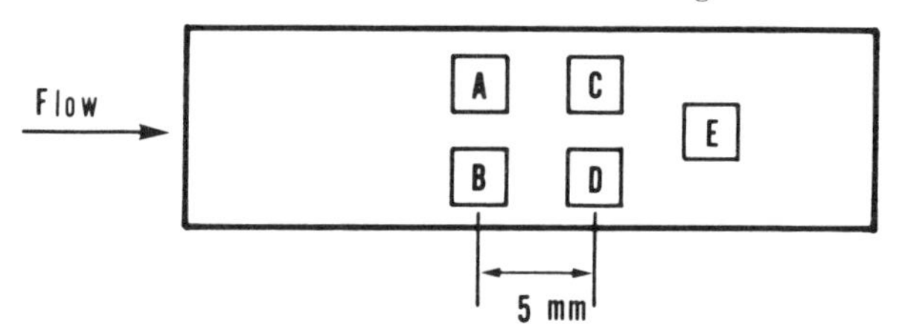

position	Substrate		Growth rate (μm/h)	Composition x	Composition y
A	InP	(100)	5.0	0.35	0.81
E		(100)	5.1	0.35	0.81
B	GaAs	(100)	5.1	0.34	0.82
C		(110)	9.9	0.34	0.82
D		(311)B	10.4	0.36	0.81

out in a single run by the SFT method.[28] The compositions of $In_{1-x}Ga_xAs_yP_{1-y}$ are independent both of the chemical nature of the substrate surfaces (i.e. InP or GaAs) and the substrate orientations. They agree reasonably well with the thermodynamically predicted values ($x = 0.46$, $y = 0.69$). However, the growth rate depends on the substrate orientation. These facts support the conclusion[29] obtained for the ternary alloys that the atom incorporation is dominated by the bulk equilibrium energy but that the growth velocity is dominated by surface phenomena involving a rate transfer mechanism.

The preparation of composition-controlled abrupt heterointerfaces is of considerable importance for practical application. Olsen et al. have used a VPE reactor equipped with a waiting-chamber for the preparation of InGaAsP/InP heterostructure, and succeeded in preparing room-temperature CW operation lasers.[24] In the reactor, the wafer is transferred from the growth position to the waiting-chamber when the gas composition is changed. However, two problems were found in this method.[5] One was the contamination of InP layers by gallium and arsenic, and the other was the difficulty in obtaining epitaxial DH wafers having mirror surfaces with a perfect lattice matching. In order to overcome these problems, Mizutani et al. used the dual-growth chamber reactor for the preparation of InGaAsP/InP heterostructures.[5] The reactor was developed at first by Watanabe et al. to produce abrupt doping profiles in GaAs growth.[30] To date, similar types of reactor have been used widely in the VPE growth of optoelectronic devices, including the preparation of the superlattice structures.[2,31,32,33]

Fig. 3.12 shows the dual-growth-chamber reactor. It consists of one horizontal chamber to grow InP and another parallel chamber to grow InGaAsP, having a common exhaust. The substrate can be transferred between the chambers by withdrawing and sifting the substrate holder. Transfer can be effected within two seconds and heterointerface transition widths less than 50–60 Å are claimed. In the preparation of the heterostructure, source gallium and indium are transported by HCl as metal chlorides at 800°C. Arsine and phosphine are brought in through a separate tube and then mixed with the chlorides in the mixing zone at 850°C. The growth temperature is varied from 660 to

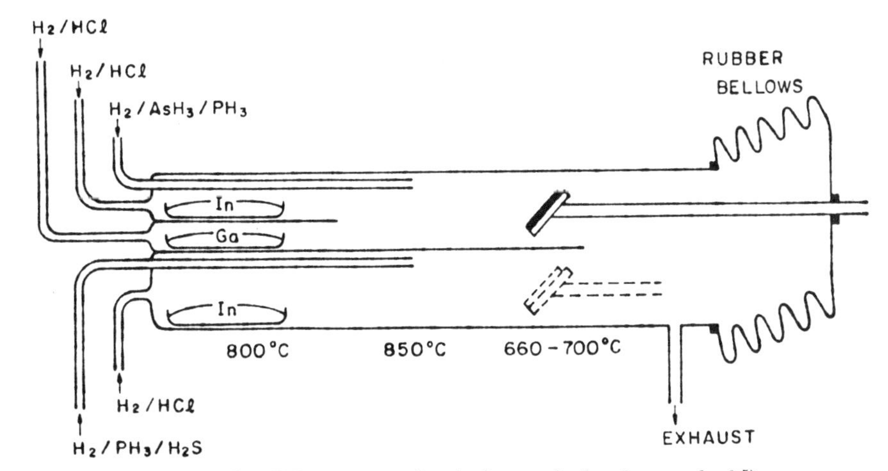

Fig. 3.12 VPE apparatus for dual-growth-chamber method.[5]

700°C, maintaining a flat profile (within ±0.5°C). High-quality InGaAsP/InP DH lasers in CW operation at room temperature have been synthesized in this reactor.

One of the drawbacks of the hydride method is extraneous deposition on the wall of the reactor tube. This is expected from the thermodynamic calculation of InGaAs, since the hydride method has a larger driving force for deposition as shown in Fig. 3.5. In the growth of III–V alloys, this makes harmful effects on the control of both the growth rate and the composition. Recently, Mizutani and Watanabe have shown that the injection of HCl into the growth region is found to be very effective in suppressing the deposition.[34] Injection of HCl is carried out by using a conventional hydride growth reactor and by introducing HCl with the group-V hydride gas to a mixing zone maintained at 850°C through a separate tube which is extended beyond the source region. In the growth of InGaP, Usui et al. have found that the addition of a small amount of oxygen gas (70–100 ppm) is also effective for the suppression of the extraneous deposition.[32] They have reported that the growth of InGaP becomes reproducible and that the X-ray rocking curve half-width is reduced by the addition of oxygen gas.

Very recently, a thermodynamic analysis for the chloride method has been reported.[35] Most VPE work on InGaAsP has been done by the hydride and MOCVD methods, but in view of the recent success in preparing InGaAs layers,[4] the growth by the chloride method with In and Ga crust sources is attractive for the preparation of superlattice structures and high-purity InGaAsP.

3.5. Al alloys

Al-containing alloys have many attractive properties. For example, AlGaInP lattice-matched to GaAs has a maximum direct band gap of ~2.3 eV and suitable properties as materials for a short-wavelength laser and LED. Although crystal growth has been limited by the difficulty of handling Al and Al compounds at high temperature because of their corrosivity, the recent success in preparing AlGaAs films[36] has cast light on halogen transport for the growth of Al-containing materials.

Kitamura et al. have reported the growth of GaAlSb in a closed tube system with iodine as a transport agent.[37,38] For the growth of AlGaInP and AlGaInAs, a thermodynamic analysis has been made.[39,40] It is shown that the deposition of these alloys is feasible by the halogen transport VPE.

3.6. Conclusions

The current development of the halogen transport VPE in Japan has been reviewed. The results are summarized as follows:

(1) In the SFT method, the $AsCl_3$ mole fraction effect is not observed and high-purity GaAs is obtained in the low $AsCl_3$ mole fraction.
(2) The thermodynamic calculation provides fundamental growth aspects of ternary and quaternary alloys.
(3) A dual- or multi-growth-chamber reactor is a useful technique for the preparation of composition-controlled abrupt heterostructures.
(4) The chloride method is promising for the growth of superlattices and high-purity layers of ternary and quaternary alloys.

References

1) J. J. Tietjen and J. A. Amick: J. Electrochem. Soc., 113 (1966) 724.
2) G. H. Olsen and T. J. Zamerowski: Progress in Crystal Growth and Characterization, Vol. II, Ed. B. R. Pamplin (Pergamon, Oxford, 1979) p. 309.
3) J. R. Knight, D. Effer and P. R. Evans: Solid State Electron., 8 (1965) 178.
4) J. Komeno, K. Kodama, M. Takizawa, and M. Oseki: Extended Abst. of 15th Conf. on Solid State Devices and Materials (Tokyo, 1983) p. 57.
5) T. Mizutani, M. Yoshida, A. Usui, H. Watanabe, T. Yuasa, and I. Hayashi: Japan. J. Appl. Phys., 19 (1980) 1113.
6) B. Cairns and R. Fairman: J. Electrochem. Soc., 115 (1968) 327c.
7) J. V. DiLorenzo and G. E. Moore, Jr.: J. Electrochem. Soc., 118 (1971) 1823.
8) D. T. Ashen, P. J. Dean, D. T. Hurle, J. B. Mullin, A. Royle, and A. M. White: Gallium Arsenide and Related Compounds 1974. (Inst. Phys. Conf. Ser. No. 24, 1975) p. 229.
9) A. Koukitu, H. Seki, and M. Fujimoto: Japan. J. Appl. Phys., 18 (1979) 1747.
10) A. Koukitu, S. Kouno, K. Takashima, and H. Seki: Japan. J. Appl. Phys., to be published.
11) J. V. DiLorenzo: J. Cryst. Growth, 17 (1972) 189.
12) T. Aoki and M. Yamaguchi: Japan. J. Appl. Phys., 11 (1972) 1775.
13) A. Humbert, L. Hollan, and D. Bois: J. Appl. Phys., 47 (1976) 4137.
14) H. Seki, A. Koukitu, H. Seki, and M. Fujimoto: J. Cryst. Growth, 45 (1978) 159.
15) J. Komeno, M. Nogami, A. Shibatomi, and S. Ohkawa: Gallium Arsenide and Related Compounds 1980 (Inst. Phys. Conf. Ser. No. 56, 1981) p. 9.
16) H. Nagai: J. Electrochem. Soc., 126 (1979) 1400.
17) A. Koukitu and H. Seki: Japan. J. Appl. Phys., 23 (1984) 74.
18) N. Susa, Y. Yamauchi, H. Ando, and H. Kambe: Japan. J. Appl. Phys., 19 (1980) 117.
19) H. Kanbe, Y. Tamauchi, and N. Susa: Appl. Phys. Lett., 35 (1979) 603.
20) Y. Yamaguchi, N. Susa, and H. Kanbe: J. Cryst. Growth, 56 (1982) 402.
21) J. Komeno, T. Takikawa, and M. Ozeki: Electron. Lett., 19 (1983) 473.
22) K. Sugiyama, H. Kojima, H. Enda, and M. Shibata: Japan. J. Appl. Phys., 16 (1977) 2197.
23) H. Enda: Japan. J. Appl. Phys., 18 (1979) 2167.
24) G. H. Olsen, C. J. Nuese, and M. Ettenberg: Appl. Phys. Lett. 34 (1979) 262.
25) S. H. Hyder, R. R. Saxena and C. C. Hooper: Appl. Phys. Lett., 34 (1979) 584.
26) H. Nagai: J. Cryst. Growth, 48 (1989) 359.
27) A. Koukitu and H. Seki: J. Cryst. Growth, 49 (1980) 325.
28) H. Seki, A. Koukitu, and M. Matsumura: J. Cryst. Growth, 54 (1981) 615.

29) J. B. Mullin and D. T. J. Hurle: J. Lumin., 7 (1973) 176.
30) H. Watanabe, M. Yoshida, and Y. Seki: Electrochem. Soc. Extended Abst., 151st Meeting (Philadelphia, 1977) p. 255.
31) G. Beuchet, M. Bonnet, P. Thebault, and J. P. Duchemin: Gallium Arsenide and Related Compounds 1980 (Inst. Phys. Conf. Ser. No. 56, 1981) p. 9.
32) A. Usui, Y. Matsumoto, T. Inoshita, T. Mizutani, and H. Watanabe: Gallium Arsenide and Related Compounds 1981 (Inst. Phys. Conf. Ser. No. 63, 1982).
33) T. Yanase, Y. Kato, I. Mito, M. Yamaguchi, K. Nishi, K. Kobayashi, and R. Lang: Electron. Lett., 19 (1983) 700.
34) T. Mizutani and H. Watanabe: J. Cryst. Growth, 59 (1982) 507.
35) A. Koukitu and H. Seki: Abstracts 44th Meeting of the Japan Soc. of Appl. Phys. (Tohoku, 1983) p. 563.
36) K. H. Bachem and M. Heyen: Gallium Arsenide and Related Compounds 1980 (Inst. Phys. Conf. Ser. No. 56, 1981).
37) N. Kitamura, M. Kakehi, and T. Wada: J. Cryst. Growth, 45 (1978) 176.
38) J. Shen, N. Kitamura, M. Kakehi, and T. Wada: Japan. J. Appl. Phys., 20 (1981) 1169.
39) A. Koukitu and H. Seki: Japan. J. Appl. Phys., 21 (1982) 1675.
40) H. Seki and A. Koukitu: 2nd Record of III–V Alloy Semiconductor Physics and Electronics Seminar, Japan (March, 1983) p. 19.

4 AlGaAs/GaAs SUPERLATTICE STRUCTURES GROWN BY METALORGANIC CHEMICAL VAPOR DEPOSITION

Hiroji KAWAI, Kazuo KAJIWARA, Ichiro HASE and Kunio KANEKO†

Abstract

Atomically abrupt ultra-thin AlGaAs/GaAs alternate layers have been successfully grown by atmospheric pressure metalorganic chemical vapor deposition. The transition width in composition on the heterointerface was evaluated using both ion-sputtering Auger electron spectroscopy and transmission electron microscopy (TEM). The TEM lattice image of AlAs/GaAs superlattices confirms the heterojunction abruptness to be within one monolayer. The photoluminescence line width from the single quantum wells increases with the decrease in well thickness, but its rate of increase is not so steep as that derived from a model of the formation of the island-like structure on the interface nor so steep as that of MBE-grown structures. The coupling of wavefunctions in a pair of closely spaced quantum wells was established by observation of photoluminescence spectra. Vertical transport diodes consisting of n^+GaAs/undoped-AlGaAs(200 Å thick)/n^+GaAs have been fabricated. The observed current agrees well with the current calculated by taking the depletion and accumulation regions adjacent to the barrier into account.

Keywords: MOCVD, AlGaAs, Superlattice, Quantum Well, AES, TEM, Photoluminescence, Vertical Transport

4.1. Introduction

Since the fabrication of optically pumped quantum–well heterostructure lasers grown by metalorganic chemical vapor deposition (MOCVD) in 1978,[1] the possibility of growing ultra-thin AlGaAs/GaAs alternate layers with sharp heterojunction interfaces has been recognized. The successful fabrication of quantum-well lasers, however, does not mean that abrupt heterointerfaces have been formed. Much effort has been spent on growing ultra-fine heterostructures with abrupt heterojunctions by MOCVD and in evaluating the interface gradings using a variety of analysis methods such as photoluminescence spectroscopy,[2–4] sputtering Auger electron spectroscopy (AES),[5] transmission electron microscopy (TEM)[6–8] and spectroscopic ellipsometry.[9]

Nowadays, it is clear that MOCVD is capable of growing ultra-thin AlGaAs/GaAs alternate layers whose transition width is about one monolayer, the same as that grown by molecular beam epitaxy (MBE).

It is worthwhile investigating whether the MOCVD-grown heterostructures with abrupt hetero-interface are the same as that grown by MBE or not, since the growth

† Sony Corporation Research Center, Fujitsuka-cho 174, Hodogayaku, Yokohama.

JARECT Vol. 19, *Semiconductor Technologies* (1986), *J. Nishizawa (ed.)*
OHMSHA, LTD. and North-Holland

kinetics and the growth temperatures of these two growth techniques are different. One of their differences may be in the formation of the "island"-like structure of the AlGaAs/GaAs interface.[4] The island-like structure, which means the fluctuation in the thickness of the grown layer, is a crucial problem for superlattice devices. Island-like structures one or two atomic layers high and with an area larger than the intrinsic exciton have been observed in superlattices grown by MBE.[10,11] There has been, however, little investigation into the formation of such island-like structures grown by MOCVD.

Little effort[12] has been made to elucidate the electron transport properties perpendicular to the heterojunction plane whereas four works[13–16] were reported on the transport properties parallel to the GaAs/n-AlGaAs heterojunction plane. We have very little data on the structural, optical and electrical properties of quantum structures grown by MOCVD, compared with those of the quantum structures grown by MBE, although practical devices such as quantum well lasers grown by MOCVD have been well reported.

This chapter reports on recent progress in: (1) MOCVD growth of ultra-thin AlGaAs/GaAs superlattices and quantum wells; (2) an evaluation of metallurgical AlGaAs/GaAs heterojunction interfaces by means of the ion-depth profiling of superlattices by Auger electron spectroscopy with Ar ion sputtering and the observation of the lattice image and diffraction spots of superlattices by transmission electron microscopy; (3) a study of the photoluminescence of single and resonantly coupled quantum wells to evaluate their structural and optical qualities and compare them with those of quantum wells grown by MBE; and (4) a discussion of vertical electron transport phenomena through a single AlGaAs thin layer.

4.2. Growth of ultra-thin AlGaAs/GaAs alternate layers

Most common growth systems employ either the vertical reactor originally designed by H. M. Manasevit[17] or the horizontal reactor designed by S. J. Bass,[18] and they are usually operated under atmospheric pressure. Both systems, however, have their problems for growing abrupt hetero-interfaces GaAs/AlGaAs. The conventional vertical reactor has a considerable amount of "dead space" around the susceptor. Because of its considerable volume, the mixing leads to a gradual change in gas composition on the wafer with a time constant about V/v, where V is the dead space and v the gas feed rate. Therefore, the transition width at the hetero-interface is roughly estimated as the product of the time constant and the growth rate.

The conventional Bass-type reactor has an advantage over the aforementioned vertical reactor since convection is suppressed and the flow parallel to the susceptor surface is nearly laminar. However, the gas recirculates between the front edge of the susceptor and the inlet in the reactor tube, because of the sudden change of the cross-sectional area perpendicular to the direction of the flow. This gas recirculation enlarges the transition width between layers with different compositions.

A necessary condition for obtaining an abrupt hetero-interface whose compositional transition width is within one atomic length is to change the reactant gas on the wafer into a new composition within the time required for the growth of one atomic layer. To achieve the above condition, the original quartz reactors should be modified to decrease the dead space and to reduce the gas recirculation in front of the susceptor. There is little detailed information available on the structures of reactors being used because this information tends to be proprietary. M. R. Leys et al.[8] recently reported the novel "chimney" reactor, a vertical reactor specially designed to eliminate the dead volume and to suppress the gas recirculation and also to keep the flow laminar from the inlet to the exhaust in the reactor. In addition, they paid special attention to the pipe work in order to allow fast switching of components and to minimize the stagnant volume between the valve seal and the connecting carrier gas line which leads to the reactor chamber.

Reducing the growth rate is an additional way of growing sharp heterojunctions. Leys et al.[8] have showed that 30 ppm of TMG (trimethyl gallium) fed into the reactor results in a GaAs growth rate of 3.5 Å/s, which is about one fifth of the normal growth rate. It is possible further to reduce the growth rate to that in MBE, but this is not necessary if a growth system which meets the above conditions is employed.

In our growth system care was taken to grow ultra-thin AlGaAs/GaAs superlattices with atomically abrupt heterojunctions by adopting the growth technique described above. We used a vertical reactor operated under atmospheric pressure. The reactor was designed to maintain a laminar flow from the inlet of the reactor to the end of the susceptor. The source materials used were TMG, TMA (trimethyl aluminum), 5% of arsine/H_2. The concentrations of TMG, TMA, and arsine in the reactor were 20 ppm, 8 ppm and 2000 ppm, respectively. The growth rate of GaAs was 4 Å/s at a temperature of 750°C, the same temperature as that adopted by R. D. Dupuis et al.,[5] by P. J. M. Griffiths et al.[6] and by J. M. Brown et al.[7] Epitaxy operation was regulated by a sequence controller in order to control precisely the growth time in each layer.

4.3. Evaluation of the metallurgical AlGaAs/GaAs heterojunction interfaces

4.3.1. *Ion-depth profile of an AlGaAs/GaAs superlattice as observed by Ar^+ ion-sputtering Auger electron spectroscopy*

C. M. Garner et al.[19] first used Auger electron spectroscopy and Ar^+ ion sputter etching to analyze the hetero-interface of AlGaAs/GaAs multilayers grown by MBE, MOCVD and liquid phase epitaxy (LPE). R. D. Dupuis et al.[5] also evaluated the MOCVD-grown AlGaAs/GaAs hetero-interface by the same method and determined that the compositional transition width was smaller than 17 Å. This is as far as we know the best value so far reported.

As has been discussed by J. S. Johannessen et al.,[20] when ion-depth profiling is used with AES, several experimental factors cause broadening of the planar abrupt hetero-interface. One factor is the Auger electron escape depth, which depends upon the electron kinetic energy: the smaller the energy, the smaller the escape depth. The electron escape depth of an Al_{LVV}(~64.5 eV) Auger electron is only 6 Å, whereas that of an Al_{KLL}(~1397 eV) Auger electron is as large as 20 Å. Knock-on mixing is another major factor influencing the depth resolution: the surface atoms are knocked on and pushed into the interior of the crystal lattice by the sputtering Ar^+ ions. It is, therefore, better to decrease the sputtering energy of the Ar^+ ion as low as possible. When samples were sputtered with 250 eV Ar^+, which is the lowest energy ever reported, knock-on broadening was reported to be about 10 Å,[19] while the sputtering rate of AlGaAs was as low as 0.42 Å/min.[5] In addition, sputter-induced surface roughness causes degradation of the depth resolution. It is not, however, clear how the degree of surface roughness is related with the sputtering conditions, e.g. direction of the incident Ar^+ beam, the voltage, the current, the sputtering rate, the sputtering depth, etc.

We have measured the compositional depth profile of a two-period AlAs/GaAs superlattice with AES.[20] The superlattice consists of four layers: (1), AlAs(30 Å thick), (2), GaAs(30 Å thick), (3), AlAs(30 Å thick) and (4), GaAs(55 Å thick) in order from a GaAs : Si substrate just oriented to the (100) direction. Fig. 4.1 shows the Auger depth profile of the superlattice. The AES system used in this study is JEOL JAMP-10S. The Auger profiling conditions were as follows: (1) Auger electrons detected were Al_{LVV}(~64.5 eV), Ga_{MMM}(~55.0 eV) and As_{MNN}(~32.0 eV) for Al, Ga and As, respectively. (2) An Ar^+ beam with 450 V of acceleration voltage sputtered the sample, over which a metal mask was placed to form a crater 500 μm in diameter in order to determine the total sputtering depth and the sputtering rate. The depth was measured as the step height using a Talystep profilometer. The average sputtering rate was 1.1 Å/min. (3) The probing area of the Auger electrons detected in the center of the crater was nominally 5 μm in diameter.

Preliminary experiments were performed to obtain the relationship between the sputtering rate and the Al composition in AlGaAs. The sputtering rate decreased almost linearly with the increase of the Al mole fraction in AlGaAs, as has been shown by C. M. Garner et al.[19]

In Fig. 4.1, the Al_{LVV} peak to peak (p–p) signal changed steeply from zero to the value

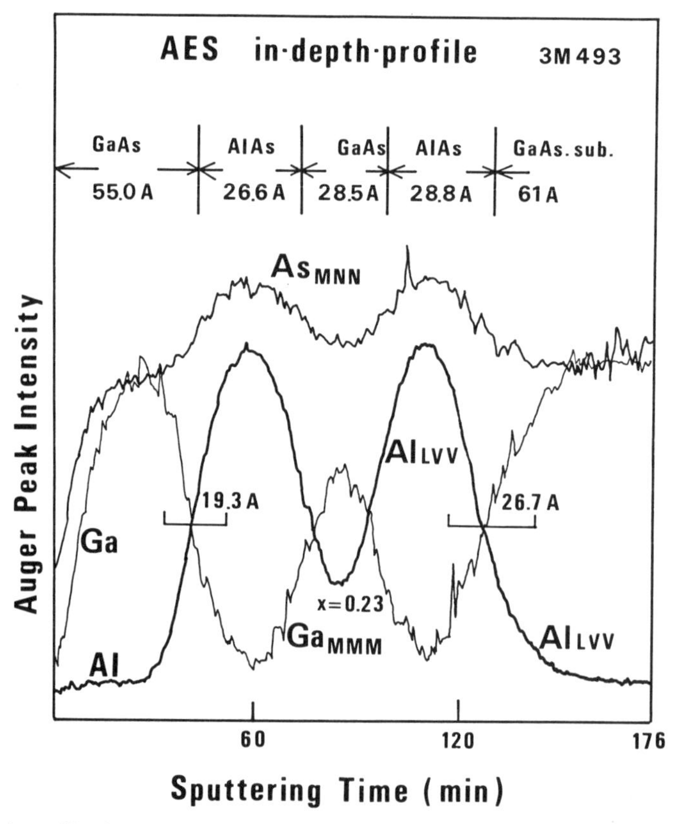

Fig. 4.1 Auger depth profile of a two-period AlAs–GaAs superlattice grown by MOCVD. Auger p–p heights of the Al_{LVV}, Ga_{MMM} and As_{MNN} signals were monitored as a function of sputtering time.

of pure AlAs at the interface between the top GaAs layer and the adjacent AlAs layer. The transition width defined by a 10–90% Al_{LVV} signal change was 19 Å. The escape depth is 6 Å and the knock-on broadening at 450 V is 23 Å, which is a value interpolated from the published data.[19] Our data indicate smaller knock-on broadening than the published data,[19] which may mean the difference in the sputtering conditions in the two cases: the beam angle, beam intensity and so on. One can see narrow plateaux of the Al Auger signal in the two AlAs layers, even though the layer is only 30 Å thick. The transition width of the interface between the GaAs substrate and the first AlAs layer is larger than that of the top hetero-interface. This substrate/epilayer hetero-interface should be metallurgically abrupt since only TMA was introduced into the reactor as the group-III element to grow the AlAs layer on the GaAs substrate. The rather broad transition width obtained in the substrate/epilayer interface and the narrow top transition layer suggest that the microscopic roughness on the GaAs substrate existed prior to the deposition of the first AlAs layer and the roughness was partly cured during deposition. In the middle GaAs layer, the Al_{LVV} signal did not decrease to zero. The observed profile is the result of the superposition of the profiles on both sides of the interface, which has a nominal transition width of about 20 Å. The Ga_{MMM} p–p signal exactly complemented the Al_{LVV} p–p signal over the sputtered region.

4.3.2. *Transmission electron microscopy of the AlGaAs/GaAs ultra-thin multilayers*

The direct method to study the structure of hetero-interface is the observation of the cross-sectional structure image of superlattices by using TEM. P. M. Petroff[21,22] performed the pioneering work in this field, observing the cross-sectional TEM image of a $(GaAs)_2$–$(AlAs)_2$ superlattice grown by MBE. Recently, three works[6–8] have reported TEM studies of the AlGaAs/GaAs superlattices grown by atmospheric-pressure MOCVD. The clear contrast of the TEM lattice images of the AlAs/GaAs heterojunction was observed, indicating that the compositional transition width in the interface is approximately one atomic layer.

We now show examples of TEM analysis of superlattices[23] grown by the present MOCVD system. A thin cross-sectional slice with (110) cleaved surface were prepared in the same manner as that described by J. M. Brown et al.[7] Fig. 4.2 shows the lattice image of the $(GaAs)_{11}$(30 Å)–$(AlAs)_{17}$(50 Å) superlattice. The resolution of the TEM image (obtained by JEOL JEM-200CX with 200 kV acceleration voltage) is estimated to be 1.4 Å. The direct transmission beam, four (111) and two (002) equivalent diffracted beams were used to obtain the lattice image. The clear contrast between the GaAs (the dark regions) and AlAs layers (the bright regions) allows us to see that the transition width at the interface is approximately one atomic layer. The fluctuation in thickness of the GaAs and AlAs layers, which forms the so-called "islands", cannot be seen in this figure, although we cannot deny the existence of the two-dimensional nuclei whose lateral extent is much smaller than that of the island. The nuclei are formed by the thermal activation of the surface.

Fig. 4.3(a) shows the dark-field TEM image of a $(GaAs)_2$–$(AlAs)_5$ superlattice. The beam used to obtain this image was the (002) diffracted beam. Both the GaAs and AlAs layers are highly uniform. Fig. 4.3(b) shows the transmission electron diffraction (TED) pattern of the same sample. The TED pattern exhibited superlattice satellite spots

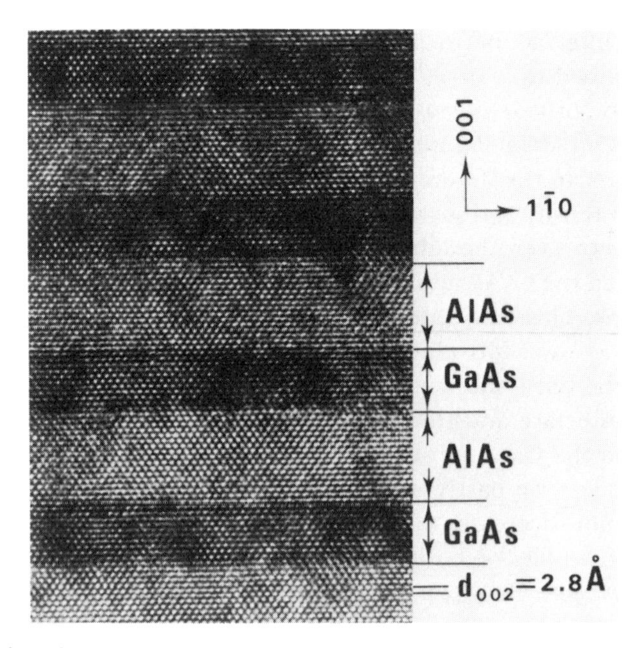

Fig. 4.2 The TEM lattice image of a $(GaAs)_{11}$–$(AlAs)_{17}$ superlattice.

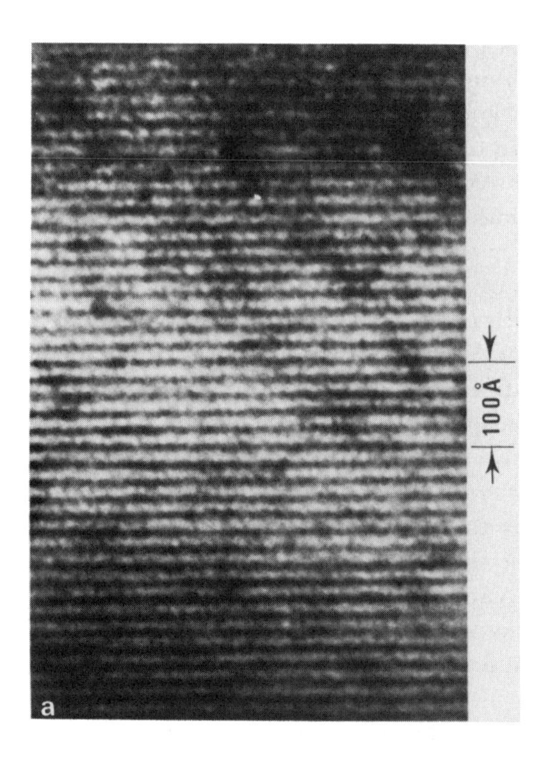

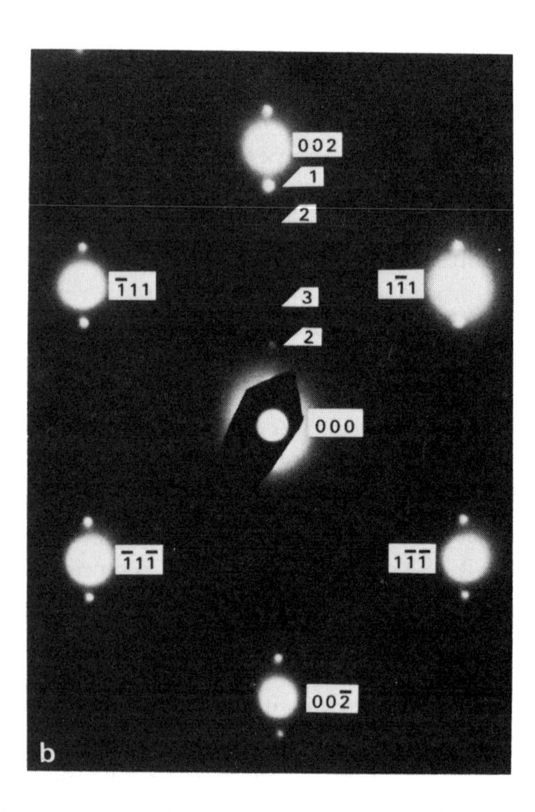

Fig. 4.3 (a) TEM dark-field image and (b) transmission electron diffraction spots of a $(GaAs)_2$–$(AlAs)_5$ superlattice. Note that the layers are highly uniform. TED superlattice satellite spots are seen in third-order, each spaced 1/7 of the interatomic distance in the reciprocal lattice space.

on both sides of the main diffraction spots. Clear satellite spots can be seen even in the third-order with a spacing of $d_s = \frac{1}{7}$ of d_{002}, where d_{002} is the interatomic spacing of 2.83 Å, which indicates that the periodicity of the superlattice is exactly seven atomic layers (20 Å).

We note that the substrate temperature in growing superlattice by MOCVD was around 750°C,[6,7,23] which is much higher than that for growing superlattices by MBE. Petroff et al. reported the existence of the critical temperature T_c above which monolayer superlattices are disordered to form random alloys. For a $(GaAs)_1$–$(AlAs)_1$ superlattice, T_c equals 610°C. Even a $(GaAs)_4$–$(AlAs)_4$ superlattice was partly disordered during deposition at 610°C. The disordering of the superlattice grown above T_c is attributed to the growth of many small islands around the nuclei formed by surface roughening. The surface roughening increases steeply above a critical temperature. It has not yet been well understood why MOCVD only can produce monolayer-scale superlattices at much higher temperatures. Possible reasons are the difference of operating pressure and/or the difference in growth kinetics of the two growth methods. It is well known that both optical and electrical qualities of bulk AlGaAs improves with the increase of growth temperature. MOCVD may have an advantage, therefore, in making high-quality superlattices or quantum wells. Indeed, as will be discussed in the following section, the optical quality of the quantum wells grown by MOCVD is superior to that grown by MBE in terms of the photoluminescence spectrum line width.

4.4. Photoluminescence of single and resonantly coupled quantum wells

4.4.1. *Single quantum well*

There have been few investigations of the optical properties of quantum wells grown by MOCVD compared with those of quantum wells grown by MBE.[24,25] Frijlink and Maluenda[2], in their investigation of quantum wells grown by MOCVD, observed no trace of luminescence in the wavelength region longer than that corresponding to the lowest energy level of the confined particle transitions in single quantum wells. They noted that the photoluminescence peak energies agreed well with the energies calculated without correction for two-dimensional (2D) free exciton formation at 4 K, whereas P. Dawson et al.[26] observed 2D free exciton luminescence from an MBE-grown single quantum well even at room temperature. Vojak et al.[27] also reported the formation of 2D free excitons with binding energies of 20 meV and 13 meV, corresponding to heavy and light hole excitons, respectively, as observed in photo-pumped stimulated emission spectra from a multi-quantum-well double heterostructures at 4.2 K. It is still an unknown problem whether the origin of luminescence is a 2D free exciton or an $n = 1$ confined electron–hole pair at a given temperature.

Fluctuation in the thickness of the grown layer (called the "island" of the AlGaAs/GaAs interface) has been investigated for superlattices grown by MBE.[10,11,28] Little is known, however, about the formation of such island-like structures grown by MOCVD.

Four GaAs layers, 30, 40, 70 and 100 Å thick separated by 500 Å thick $Al_{0.54}Ga_{0.46}As$ layers were grown on a GaAs substrate at 780°C.[4] Fig. 4.4 shows the photoluminescence spectrum of the sample with four quantum wells, measured at 75 K. The arrows indicate the calculated emission wavelength of the transition between the $n = 1$ electron and $n = 1$

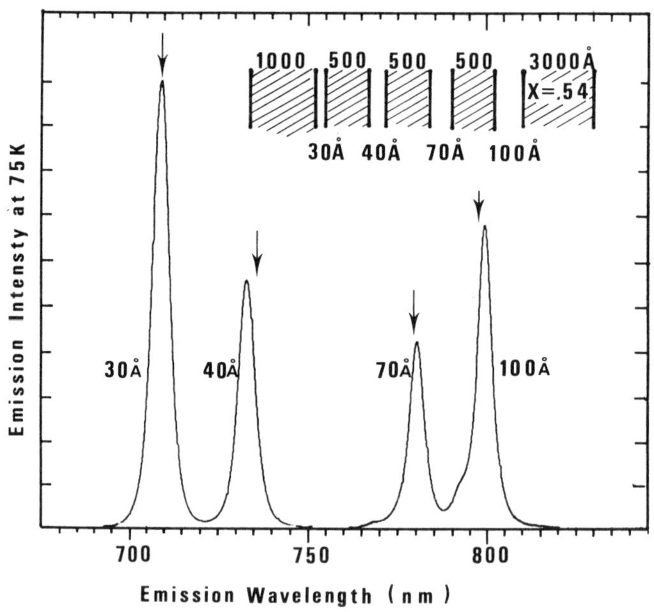

Fig. 4.4 Photoluminescence spectrum at 75 K for a sample with four single quantum wells 100, 70, 40 and 30 Å thick, in order from the GaAs substrate. The arrows correspond to the calculated wavelength.

heavy hole in a finite square potential well. The measured peak wavelength corresponding to each well thickness is close to the calculated emission wavelength. No systematic shift towards shorter wavelength caused by interface grading[2)] is seen. This is a natural consequence of the abrupt heterojunctions as can be seen by referring to the results of the TEM analysis in the previous section. Fig. 4.5 shows the photoluminescence spectrum of

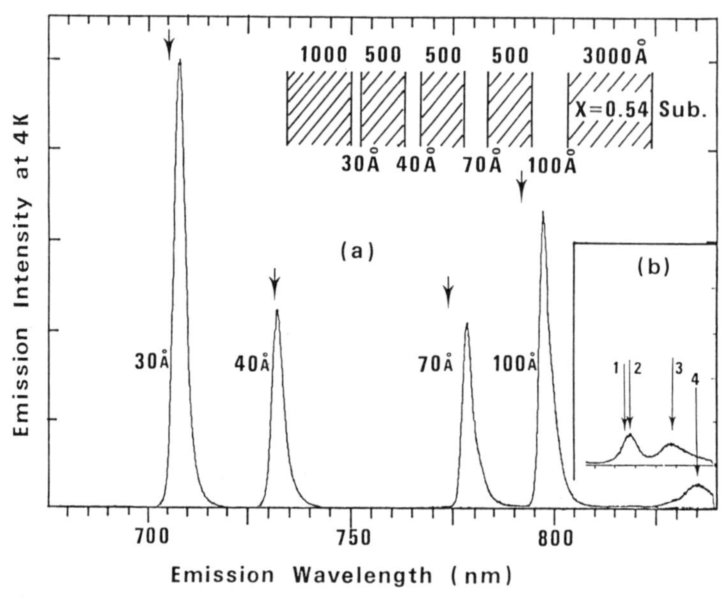

Fig. 4.5 Photoluminescence spectrum at 4.2 K for the same sample as is shown in Fig. 4.4. Calculated emission wavelength are also shown by the arrows. Note that the observed emission wavelength shifts slightly toward the longer-wavelength region, which is the result of the formation of the 2D free exciton at 4.2 K.

the same sample at 4.2 K, in which there is none of the extrinsic luminescence frequently seen in MBE-grown samples.[30] The measured peak energy in each well, however, is lower than the calculated energy at liquid helium temperature. The energy difference between the measured peak energy and the calculated value at 4.2 K should have the same value as that at 75 K, provided the radiative transition takes place in the same transition path as that at 75 K. The down-shift of the emission energy at 4.2 K is significant (9 meV for a 30 Å thick well and 6 meV for 100 Å thick well). This down-shift is the result of the exciton formation which has been observed in both the optical absorption[25] and luminescence excitation[31] spectra at liquid helium temperature.

The well thickness dependence of the photoluminescence line width at both 75 K and 4.2 K is shown in Fig. 4.4 together with some recently published data.[2,10,28] C. Weisbush et al.[11] have suggested that the line width is related to the fluctuation in well thickness within a layer, namely the formation of an island-like structure on the well/barrier interface. If the lateral extent of the island is longer than the exciton diameter or the mean free path length (several hundred Angstroms) for a radiative transition lifetime of an electron–hole pair, the transition energy in the island is different from that in the adjacent area. The spectrum is broadened by the superposition of such transitions with different energies. A more precise analysis of the dependence of the spectrum line width on the fluctuation in well thickness was reported by J. Singh et al.[32] The broken line in Fig.

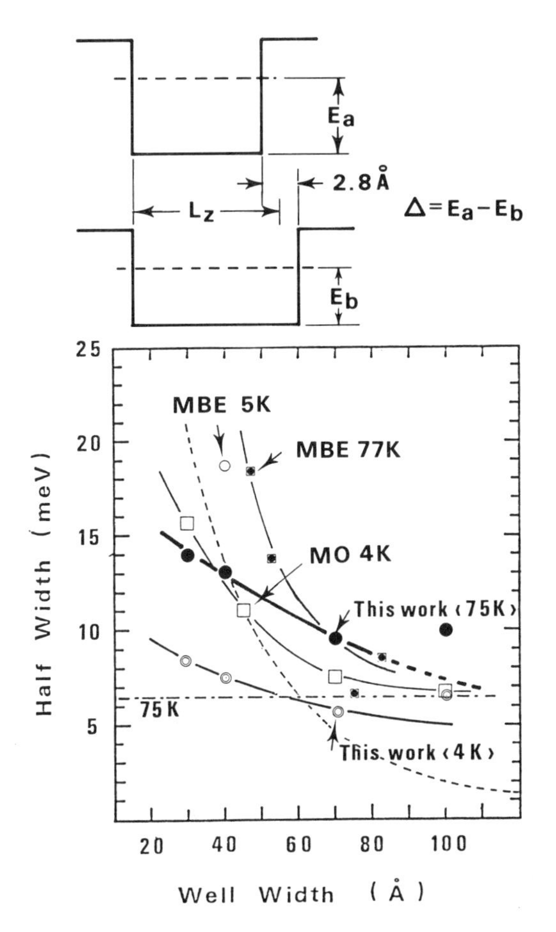

Fig. 4.6 Photoluminescence line width versus quantum well thickness. Some published data are also shown in the figure. The data for the MBE at 77 K, the MBE at 4 K and MO at 5 K are the data of refs. 10, 28 and 2, respectively. The broken line denotes the calculated energy variation against well thickness with fluctuation in thickness of one atomic layer at 0 K. The upper picture shows that situation.

4.6 is a plot of the difference in energy between two quantum wells whose well thickness differs by one monolayer (2.8 Å) at 0 K, as a function of well width. The line width data of the quantum wells grown by MBE suggest that the fluctuation in well thickness is one or two monolayers.[10,11] The present MOCVD data increase with decreasing well layer thickness, but in the region of narrow wells the line width is much smaller than either the MBE data or the calculated curve. It may be concluded, therefore, that the average fluctuation in well thickness over the interface seen by an exciton or an electron–hole pair is smaller than one monolayer, a conclusion supported by the TEM lattice image in which no island or fluctuation in layer thickness could be seen.

4.4.2. *Resonantly coupled double well*

When the width of the barrier between the two wells is very thin, the wavefunctions of both sides of the wells couple and split the degenerate single well states into a doublet state. Multi-quantum wells (from two to ten wells) of barrier width from12 to 18 Å were grown by MBE[29] and the coupling of the single-well states was observed by the optical absorption method at 4.2 K. Luminescence measurements were not, however, made since they require a sample of higher optical quality because the radiative transition is much more sensitive to the crystal qualities than the optical absorption transition.

The author grew at 780°C[12] two GaAs quantum wells 30 Å thick separated by a 20 Å thick $Al_{0.5}Ga_{0.5}As$ barrier. Fig. 4.7 shows the photoluminescence spectrum of the double well both at 75 K and at room temperature. The energy levels of the doublet states can be given by the following equations:

symmetric state:

$$\frac{k\cos(ka)+\alpha\sin(ka)\tanh(\alpha\cdot b)}{k^2\sin(ka)-k\cdot\alpha\cos(ka)\cdot\tanh(\alpha\cdot b)}=\frac{1}{\alpha} \tag{4.1}$$

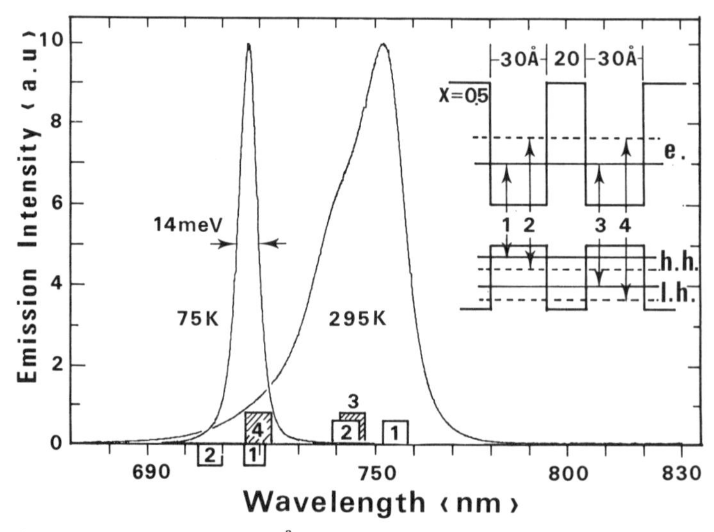

Fig. 4.7 Photoluminescence spectra of a 30 Å thick coupled quantum wells separated by a 20 Å thick $Al_{0.5}Ga_{0.5}As$ barrier at both 75 K and room temperature.

antisymmetric state:

$$\frac{k\cos(ka)+\alpha\sin(ka)\cdot\coth(\alpha\cdot b)}{k^2\sin(ka)-k\alpha\cos(ka)\cdot\coth(\alpha\cdot b)}=\frac{1}{\alpha} \tag{4.2}$$

where

$$k=(2m_w E)^{1/2}/\hbar \tag{4.3}$$

$$\alpha=[2m_b(V-E)]^{1/2}/\hbar \tag{4.4}$$

and v is the barrier height, E the eigenenergy of the doublet, a the well width, b one half the barrier thickness, and m_w and m_b the effective mass of the particle in the well and in the barrier. Optical transitions take place between a symmetrical electron and a symmetrical heavy or light hole and between an antisymmetrical electron and an antisymmetrical heavy or light hole with the same quantum number, because of the overlap integral rule in addition to the parity selection rule. The calculated emission wavelengths are shown as short bars on the horizontal axis in Fig. 4.7. The width of the bars represents the distribution in emission wavelength that would result if the well thickness were distributed from 29 Å to 31 Å. The observed emission peak wavelength agrees well with the calculated emission wavelength. The line width of the luminescence of the coupled wells at 75 K is 14 meV, which is the same value of that of the single well and is the smallest ever achieved in a 30 Å thick single quantum well, so far as we know. The shoulder on the high-energy side of the spectrum corresponds to a superposition of the two kinds of transitions –

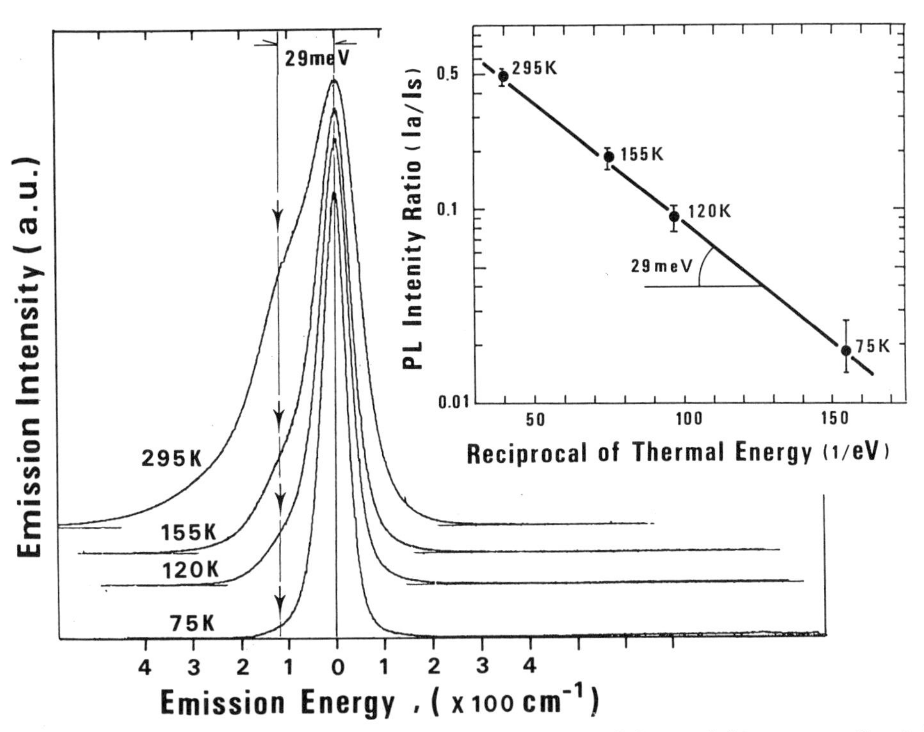

Fig. 4.8 Temperature dependence of the photoluminescence spectra of the coupled quantum wells at both 75 K and room temperature.

antisymmetrical transition to a heavy-hole band and symmetrical transition to a light-hole band – the former being predominant because of the higher density of state of the heavy-hole sub-band.

Fig. 4.8 shows the temperature dependence of the photoluminescence spectra of the coupled wells. The energy separation between the peak and the shoulder is about 29 meV. When the logarithm of the intensity ratio of the shoulder to the main peak was plotted against the reciprocal of the thermal energy, as in the inset to Fig. 4.8, a linear dependence was observed. The slope corresponds to the energy separation between the two transition states when the Boltzmann statistics on the carrier population are assumed. The slope is 29 meV, which is in excellent agreement with the value deduced from the spectra as well as the calculated energy separation of the two transition states: symmetric/symmetric and antisymmetric/antisymmetric, both to a heavy-hole band.

In addition to the successful observation of the coupling well states, it is noted that there is no luminescence caused by alloy clustering around 787 nm at 75 K, which corresponds to a well width of $2L_z + L_b = 80$ Å, as well as no luminescence caused by extrinsic origins such as the recombination with neutral acceptors seen in MBE-grown quantum wells.[30)]

4.5. Vertical electron transport through thin AlGaAs tunnel barriers

The original idea of semiconductor superlattices was born of the transport phenomena of resonant electron tunneling through a series of closely spaced potential barriers.[34)] Vertical transport through a single barrier is the foundation of all multilayer vertical transport phenomena and its understanding provides basic information on how to fabricate transport devices using superlattice structures such as Bloch oscillators, ballistic hot-electron transistors and the other Brillouin-zone folding devices yet to be conceived. In spite of this there are only few data on vertical transport through a single AlGaAs layer grown by MBE,[35,36)] and they are not in agreement with theory. Although there has been a few reports on samples grown by MOCVD,[12,33)] the data agree well with the theoretical predictions.

Vertical transport diodes consisting of $Al_xGa_{1-x}As$ ($x = 0.4$, 0.7) 200 Å thick sandwiched between Se-doped GaAs ($n = 7 \times 10^{17}$) layers on a Si-doped GaAs substrate were grown by MOCVD. We now focus on the results obtained from these diodes whose Al content in the barrier is 0.7. Fig. 4.9 shows the plot of $\log(I/T^2)$ versus $1/kT$ with the applied voltage as a parameter. The linear relation, $\log(I/T^2)$ versus $-(\phi - E_f)/kT$, where I is the current and ϕ the barrier height, holds at constant bias voltage when the thermionic emission current predominates and the tunneling current is ignored. The barrier height determined from the slope of the straight line obtained when the bias is below the 50 mV was 680 meV. The gentle slope and non-linear relation seen when a high voltage is applied is caused by the inclusion of the tunneling emission current through the barrier. The current versus voltage relation at both 77 K and room temperature is shown in Fig. 4.10 together with some calculated $I-V$ curves with barrier height as a parameter. Calculations are performed under the following assumptions: (1) that the electron-accumulated layer in the cathode and the depleted layer in the anode adjacent to the barrier layer are formed under biasing; (2) that the transmission coefficient for tunneling can be derived using the WKB approximation and is set at unity above the potential

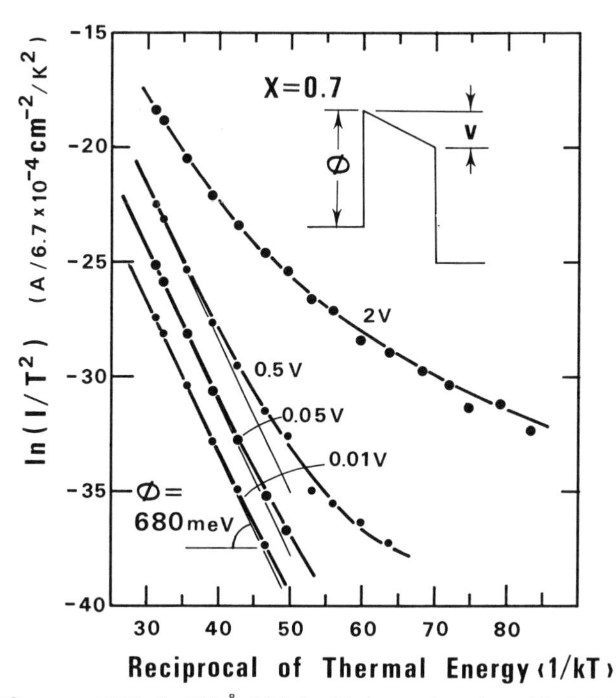

Fig. 4.9 Plot of $\ln(I/T^2)$ versus $1/kT$ of a 200 Å thick double heterojunction $n^+GaAs/Al_{0.7}Ga_{0.3}As/n^+GaAs$ diode at constant bias voltages.

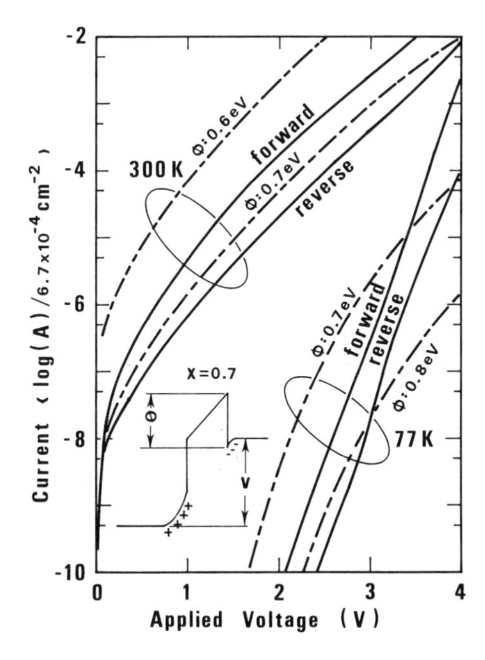

Fig. 4.10 $I-V$ characteristics of the same diode as in Fig. 4.9 at 300 K and 77 K. Broken lines are the calculated curves with barrier height as a parameter.

barrier; (3) that no image-force lowering of the edge of the potential barrier on the cathode side is considered; and (4) that the electron effective mass does not depend on the electron energy. The measured curves are parallel to the calculated curves over a wide range of current. The barrier height obtained by comparing the measured and calculated curves is around 680 meV at room temperature and 750 meV at 77 K, which is slightly below the Γ-conduction band discontinuity at the $Al_{0.7}Ga_{0.3}As$–GaAs heterojunction. The actual barrier height, however, is determined by the combination of the three factors: the bandgap discontinuity or the difference in electron affinity, the built-in potential or the difference in Fermi energy, and the width of the barrier through which the electrons pass. The observed barrier height, therefore, includes some information about the amount of electro-active impurities in the barrier layer although we will not discuss this in more detail. It is significant that the observed current can be expressed as the sum of the thermionic emission and the tunneling emission without introducing any unknown current component over a wide range of the total current and the agreement between the barrier heights obtained from two series of data (temperature dependence and voltage dependence of the current) was good.

4.6. Summary

MOCVD growth of ultra-thin AlGaAs/GaAs stacked layers has been discussed. In order to grow atomic layer superlattices it is necessary to eliminate the dead space in the reactor and to control precisely the pressure difference between the reactor-line and the vent-line. Decrease of the growth rate is not always a necessary condition. A few systems have already succeeded in achieving these conditions.

The AlGaAs/GaAs heterojunction width has been measured by ion sputtering Auger electron spectroscopy and transmission electron microscopy. The depth profile of the superlattice obtained by AES showed that the true width of the AlAs/GaAs interface was hidden behind the limit of the system resolution. The TEM lattice image of the AlAs/GaAs superlattice showed that the hetero-interface changed from GaAs to AlAs within one monolayer. Furthermore, the $(AlAs)_5$–$(GaAs)_2$ superlattice exhibited TED satellite spots, each of which was spaced exactly $\frac{1}{7}$ of the interatomic distance in the reciprocal lattice space. Photoluminescence of single and coupled quantum wells was observed. The two-dimensional free excitons in the wells were formed at 4.2 K, while at 75 K the luminescence originated from the recombination of free electrons with heavy holes confined in the well. The photoluminescence line width of narrow wells is much smaller than expected from a model of the formation of one atomic-step island. Coupling of the wavefunctions in closely spaced (20 Å) quantum wells could be seen by means of photoluminescence spectroscopy. Good agreement of the energy location of both the peak and shoulders with the calculation confirms the validity of this coupled-well energy level scheme.

Vertical transport diodes consisting of a 200 Å thick AlGaAs barrier layer sandwiched between two n^+GaAs were grown by MOCVD. The diode current was defined as the sum of the thermionic emission and tunneling emission currents through the AlGaAs barrier with a reasonable barrier height.

MOCVD has been shown to have the capability of growing ultra-thin AlGaAs/GaAs layers with uniform and atomically abrupt heterojunctions. The structural, optical and electrical qualities of these stacked crystals are equal to or superior to that grown by other growth methods.

Acknowledgements

This work was performed under the management of the R&D Association for Future Electron Devices as a part of the R&D Project of Basic Technology for Future Industries sponsored by the Agency of Industrial Science and Technology, MITI.

References

1) H. Holonyak, Jr., R. M. Kolbas, R. D. Dupuis, and P. D. Dupkus: Appl. Phys. Lett., 33 (1978) 73.
2) P. M. Frijlink and J. Maluenda: Japan. J. Appl. Phys. Lett., 21 (1982) L574.
3) J. J. Coleman, P. D. Dapkus, W. D. Laidig, B. A. Vojak, and N. Holonyak, Jr.: Appl. Phys. Lett. 38 (1981) 63..
4) H. Kawai, K. Kaneko, and N. Watanabe: J. Appl. Phys., 55 (1984) 463.
5) R. D. Dupuis and P. D. Dupkus: Appl. Phys. Lett., 34 (1979) 335.
6) R. J. M. Griffiths, N. G. Chew, A. G. Cullis, and G. C. Joyce: Electron. Lett., 19 (1983) 988.
7) J. M. Brown, N. Holonyak, M. J. Ludowise, W. T. Dietze, and C. R. Lewis: Electron. Lett., 20 (1984) 204.
8) M. R. Leys, C. V. Opdrop, M. P. A. Viegers, H. J. Talen, and V. D. Mheen: J. Cryst. Growth, 68 (1984) 431.
9) M. Erman and P. M. Frijlink: Appl. Phys. Lett., 43 (1983) 285.
10) L. Goldstein, Y. Suzuki, S. Tarucha, K. Horikoshi, and H. Okamoto: 30th Spring meeting of Japan. Soc. Appl. Phys. (1983) 514. [in Japanese]
11) C. Weisbush, R. Dingle, A. C. Gossard, and W. Wiegman; Solid State Commun., 38 (1981) 709.
12) H. Kawai, K. Kanako, and N. Watanabe: J. Cryst. Growth, 68 (1984) 406.
13) S. D. Hersee, J. P. Hirtz, M. Baldy, and J. P. Duchemin: Electron. Lett. 18 (1982) 1076.
14) J. Maluenda and P. M. Frijlink: J. Vac. Sci. Technol., B1 (1983) 334.
15) T. Usagawa, Y. Ono, S. Kawase, Y. Katayama, and S. Takahashi: Extended Abst. of 15th Conf. on Solid State Devices and Materials (1983) p. 289.
16) J. P. Andre, A. Briere, M. Rocchi, M. Riet: J. Cryst. Growth, 68 (1984) 445.
17) H. M. Manasevit: Appl. Phys. Lett., 12 (1968) 156.
18) S. J. Bass: J. Cryst. Growth, 31 (1975) 172.
19) C. M. Garner, C. Y. Su, Y. D. Shen, C. S. Lee, G. L. Pearson, W. E. Spicer, D. D. Edwall, D. Miller, and J. S. Harris, Jr.: J. Appl. Phys., 50 (1979) 3383.
20) J. S. Johannessen, W. E. Spicer, and Y. E. Strausser: J. Vac. Sci. Technol., 13 (1976) 849.
21) P. M. Petroff: J. Vac. Sci. Technol., 14 (1977) 973.
22) P. M. Petroff, A. C. Gossard, W. Weigmann, and A. Savage: J. Cryst. Growth, 44 (1978) 5.
23) K. Kajiwara, K. Kawai, K. Kaneko, and N. Watanabe: Jpn. J. Appl. Phys. Lett., 24 (1985) 485.
24) R. C. Miller, D. A. Kleinman, W. A. Nordland, Jr., and A. C. Gossard: Phys. Rev., B22 (1980) 863.
25) R. Dingle, in Festkolperprobleme, Advances in Solid State Physics, Ed. H. J. Queisser (Pergamon/Vieweg, Braunschweig, 1975) vol. XV, p. 21.
26) P. Dawson, G. Duggan, H. I. Ralph, and K. Woodbridge: Phys. Rev., B28 (1983) 7381.
27) B. A. Vojak, N. Holonyak, Jr., W. D. Laidig, K. Hess, J. J. Colemann, and P. D. Dupkus: Solid State Commun., 35 (1980) 477.
28) R. C. Miller and W. T. Tsang: Appl. Phys. Lett., 39 (1981) 334.
29) R. Dingle, A. C. Gossard, and W. Wiegmann: Phys. Rev. Lett., 34 (1975) 1327.
30) R. C. Miller, A. C. Gossard, W. T. Tsang, and O. Munteanu: Phys. Rev., B25 (1982) 3871.
31) R. C. Miller, D. A. Kleinman, W. T. Tsang, and A. C. Gossard: Phys. Rev., B24 (1981) 1134.
32) J. Singh, K. K. Bajaj, and S. Chaudhuri: Appl. Phys. Lett., 44 (1984) 805.
33) I. Hase, H. Kawai, and K. Kaneko: Electron. Lett., 20 (1984) 496.
34) L. Esaki and R. Tsu: IBM Research Note RC-2418 (1969).
35) D. Delagebeaudeuf, P. Delescluse, P. Etienne, J. Massies, M. Laviron, J. Chaplart, and T. Linh: Electron. Lett., 18 (1983) 85.
36) P. M. Solomon, T. W. Hickmott, H. Morkoc, and R. Fischer: Appl. Phys. Lett., 42 (1982) 821.

5 MOLECULAR-BEAM EPITAXY OF II–VI COMPOUNDS

Takafumi YAO†

Abstract

A systematic study on molecular-beam epitaxy (MBE) of undoped ZnSe as a typical case of II–VI compounds has been made. The crystallinity, surface morphology, electrical properties, and photoluminescence (PL) properties of undoped ZnSe films grown under various substrate temperatures and molecular-beam fluxes have been extensively investigated. Electrical and PL properties depend strongly on the stoichiometry of the epilayer. Low-resistivity ZnSe films are obtained only when the substrate temperature is between 280°C and 370°C, and the molecular-beam flux ratio of the constituent elements is about 1. High quality undoped ZnSe films are grown at 280°C and at unity molecular-beam flux ratio of the constituents. The dominant donor species in undoped ZnSe films are considered to be native donors such as Se vacancy or its associated complex defects.

Keywords: Molecular-beam Epitaxy, ZnSe, II–VI Compounds

5.1. Introduction

Recently, major efforts have been devoted to realizing visible light-emitting devices from the green to near-UV region with the wide bandgap II–VI compounds.[1)] The Zn chalcogenides have a direct wide energy bandgap between 2.26 eV and 3.76 eV. Among them, ZnSe and ZnS are especially important for blue light-emitting devices, since the near-bandgap emission of ZnSe occurs at 4600 Å and that of ZnS at 3400 Å.[1)]

Most of the works on the growth of II–VI compounds so far have been concentrated on bulk crystals which are grown under thermal equilibrium conditions and generally require very high growth temperatures.[2)] With high growth temperatures, both the contamination from background and the deviation from stoichiometry easily occur during the crystal growth. Both of these factors produce a degraded crystallinity: as-grown bulk Zn chalcogenides show very high resistivity (10^{12} Ω cm) and very weak luminescence.

ZnTe is generally p-type, while ZnS and ZnSe have n-type conduction. It is difficult to convert the conduction type, because of the self-compensation effect caused by induced native defects on incorporation of impurities[3)] and the presence of residual impurities, which makes it difficult to control the conductivity.

The low growth temperature is considered to be a necessary condition in order to achieve high-quality II–VI compounds. The low growth temperature in MBE will help realize low concentration of non-stoichiometric defects in the epilayers and suppress cross-diffusion from the substrate. Furthermore, precise control of alloy composition in mixed crystals with high uniformity can be performed by controlling the constituent

† Electrotechnical Laboratory, Umezono 1-1-4, Sakura-mura, Niihari-gun, Ibaraki 305.

JARECT Vol. 19, Semiconductor Technologies (1986), *J. Nishizawa (ed.)*
OHMSHA, LTD. and North-Holland

molecular-beam fluxes. All of these aspects of MBE make it an important technique for the growth of II–VI compounds.

The purpose of this chapter is to elucidate the general characteristics of MBE growth of II–VI compounds, and then to describe the evaluated properties of MBE-grown epilayers. As a typical example, we will concentrate on ZnSe, since systematic studies on optical and electrical properties of MBE-grown ZnSe have been done.

5.2. Molecular-beam epitaxy

5.2.1. *Growth conditions*

MBE growth has been conducted in a conventional MBE machine which is composed of a growth chamber and a substrate-introduction chamber as shown schematically in Fig. 5.1. The substrate used for ZnSe epitaxy was (100)GaAs. The substrate was degreased, etched in a solution of $H_2SO_4:H_2O_2:H_2O$ (20:1:1), and mounted on a molybdenum heating block. Before commencing growth, the substrate was thermally cleaned at 630°C for 1 min. The substrate temperature was varied between 250°C and 410°C. The molecular-beam flux ratio $k_{Zn}J_{Zn}/k_{Se}J_{Se}$ was varied between 0.2 and 5, where k_{Zn} and k_{Se} are the sticking probabilities of the Zn and Se molecular beams, and J_{Zn} and J_{Se} are the molecular-beam fluxes of Zn and Se, respectively.[14] The growth rate was about 0.5 μm/h. The thickness of the epilayer was larger than 3 μm, which is sufficiently thick for characterization of the epilayer.[4]

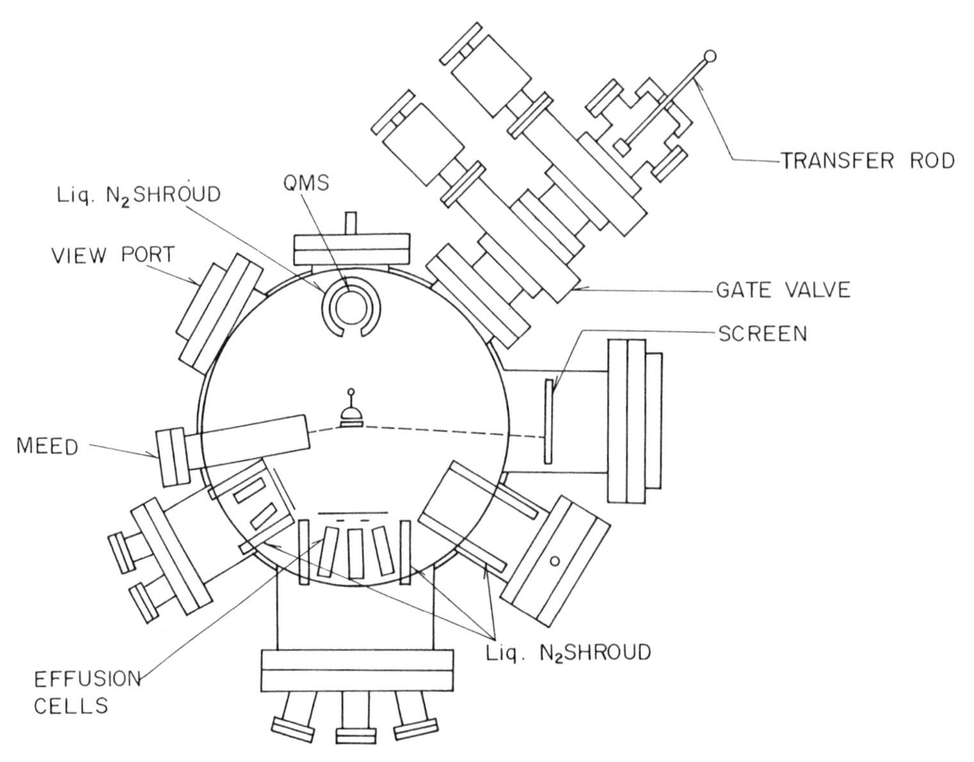

Fig. 5.1 Top view of the MBE system used for the growth of II–VI compounds.

5.2.2. *Growth rate*

The growth rate of the deposited film depends greatly both on the impinging molecular-beam fluxes and on the substrate temperature. Fig. 5.2 shows the measured growth rate of ZnSe grown at 310°C as a function of J_{Se}, when J_{Zn} is kept almost at a constant value.[5)] The molecular-beam fluxes are expressed in vapor pressure (Torr) measured with an ion gauge located near the substrate holder. The dotted line indicates the point of $k_{Se}J_{Se} \sim k_{Zn}J_{Zn}$. The growth rate increases almost in proportion to J_{Se} when $J_{Se} \ll J_{Zn}$, and tends to saturate when $J_{Se} \gg J_{Zn}$. On the other hand, when J_{Se} is kept at a constant value and J_{Zn} is varied, the growth rate increases almost proportionally to J_{Zn} at low J_{Zn} value and saturates at a high J_{Zn} value.[5)] When $k_{Zn}J_{Zn}/k_{Se}J_{Se} < 1$, the stoichiometry of ZnSe would be slightly on the Se rich side, while for $k_{Zn}J_{Zn}/k_{Se}J_{Se} > 1$, the stoichiometry would be slightly on the Zn-rich side.

The measured growth rate of ZnSe films grown on GaAs substrate is shown in Fig. 5.3 as a function of the substrate temperature,[6)] where the incoming flux ratio $k_{Zn}J_{Zn}/k_{Se}J_{Se}$ is set about unity for the substrate temperature of 280°C. The growth rate is almost constant between 280°C and 340°C. Above this temperature range it decreases, while it increases below this temperature range. Such a tendency is due to the variation of the sticking probabilities of Zn and Se; the sticking probability of Se would increase below 250°C which is responsible for the increase of the growth rate below 250°C, while the sticking probability of Se and Zn would decrease above 350°C, which is responsible for the decrease

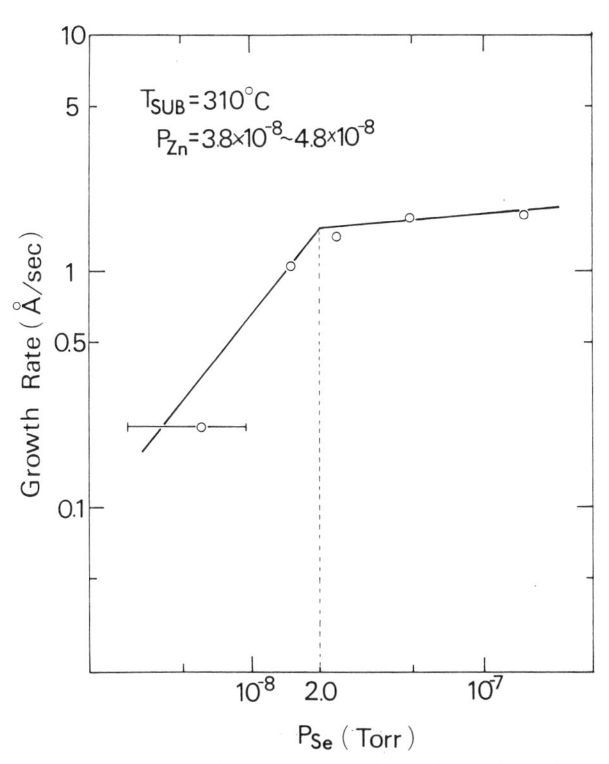

Fig. 5.2 Growth rate of MBE ZnSe against the Se molecular-beam flux, where the Zn molecular-beam flux is kept constant and the substrate temperature is 310°C.

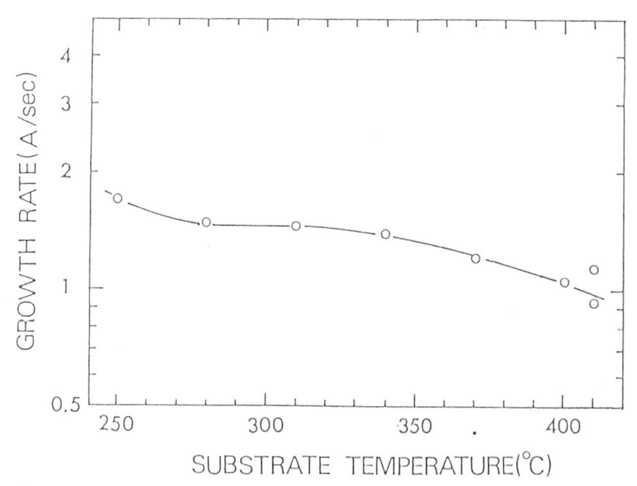

Fig. 5.3 Growth rate of MBE ZnSe against the substrate temperature for constant impinging molecular-beam fluxes. The flux ratio is $k_{Zn}J_{Zn}/k_{Se}J_{Se} = 1$. [after T. Yao et al.[6]]

of the growth rate above 350°C. The variation of k_{Se} would be much steeper than that of k_{Zn} in this temperature range because of the higher Se vapor pressure. Therefore, the film stoichiometry would be slightly on the Se-rich side below 250°C, while slightly on the Zn-rich side above 350°C.

5.2.3. *Surface morphology and RHEED observation*

Crystallinity of ZnSe epilayers has been assessed with reflection high-energy electron diffraction (RHEED) at 50 keV and surface morphology with a Nomarski contrast microscope. Fig. 5.4 shows the RHEED patterns and the surface morphology of MBE ZnSe layers grown at 280°C for various $k_{Zn}J_{Zn}/k_{Se}J_{Se}$ values.[7] When $k_{Zn}J_{Zn}/k_{Se}J_{Se} \sim 1$, the RHEED pattern shows a streaked pattern together with Kikuchi lines and the epilayer shows a smooth surface morphology. As the value of $k_{Zn}J_{Zn}/k_{Se}J_{Se}$ decreases from unity, the surface of the epilayer has a mat finish and many rectangular-shaped hillocks appear on the surface. Consequently, the RHEED pattern shows a diffused spotty pattern. As the Zn beam flux increases, the epilayer exhibits a smooth surface morphology, while the RHEED pattern shows a spotty pattern, which is indicative of the presence of microscale roughness over the surface.

Fig. 5.5 shows the RHEED patterns and surface morphology of MBE ZnSe grown at various substrate temperatures.[7] The flux ratio of J_{Zn}/J_{Se} was kept constant throughout the whole substrate temperature range and the value of $k_{Zn}J_{Zn}/k_{Se}J_{Se}$ was set such that it is about unity at the substrate temperature of 280°C. Below 250°C, there appear often rectangular-shaped hillocks on the surface and the corresponding RHEED pattern shows an elongated spotty pattern. As the substrate temperature is raised above 280°C, the surface becomes very smooth and the corresponding RHEED pattern shows a streaky pattern with Kikuchi lines. As the substrate temperatures is further increased above 340°C, the surface morphology is still smooth though the corresponding RHEED pattern shows a spotty pattern. Such a variation of the surface morphology can be understood in terms of the effect of the flux ratio on the crystallinity and the surface morphology: At

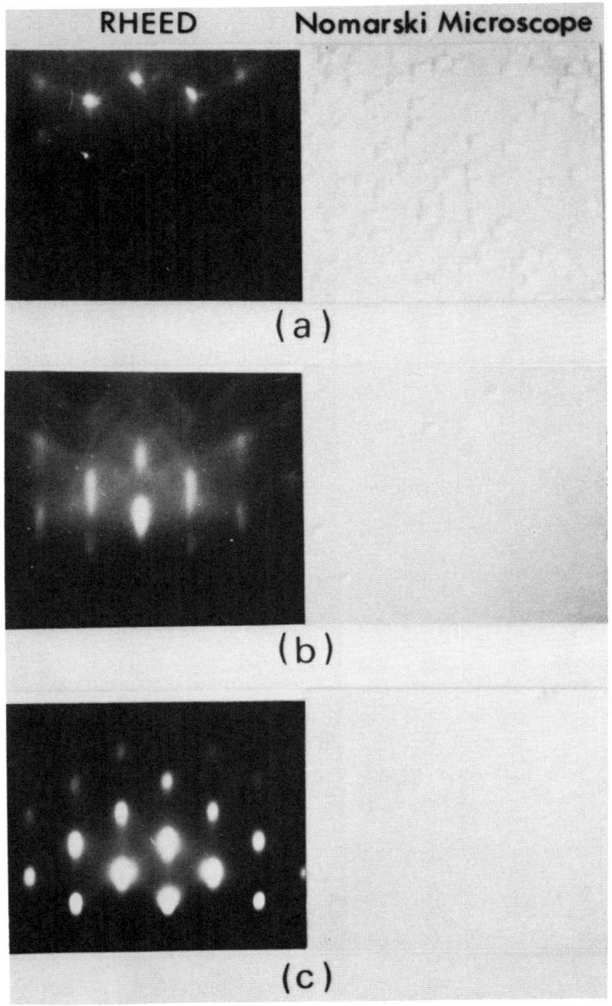

Fig. 5.4 RHEED patterns and surface morphology of undoped ZnSe at 280°C for various molecular-beam flux ratios: $k_{Zn}J_{Zn}/k_{Se}J_{Se}$ = (a) 2, (b) 1, (c) 0.5.

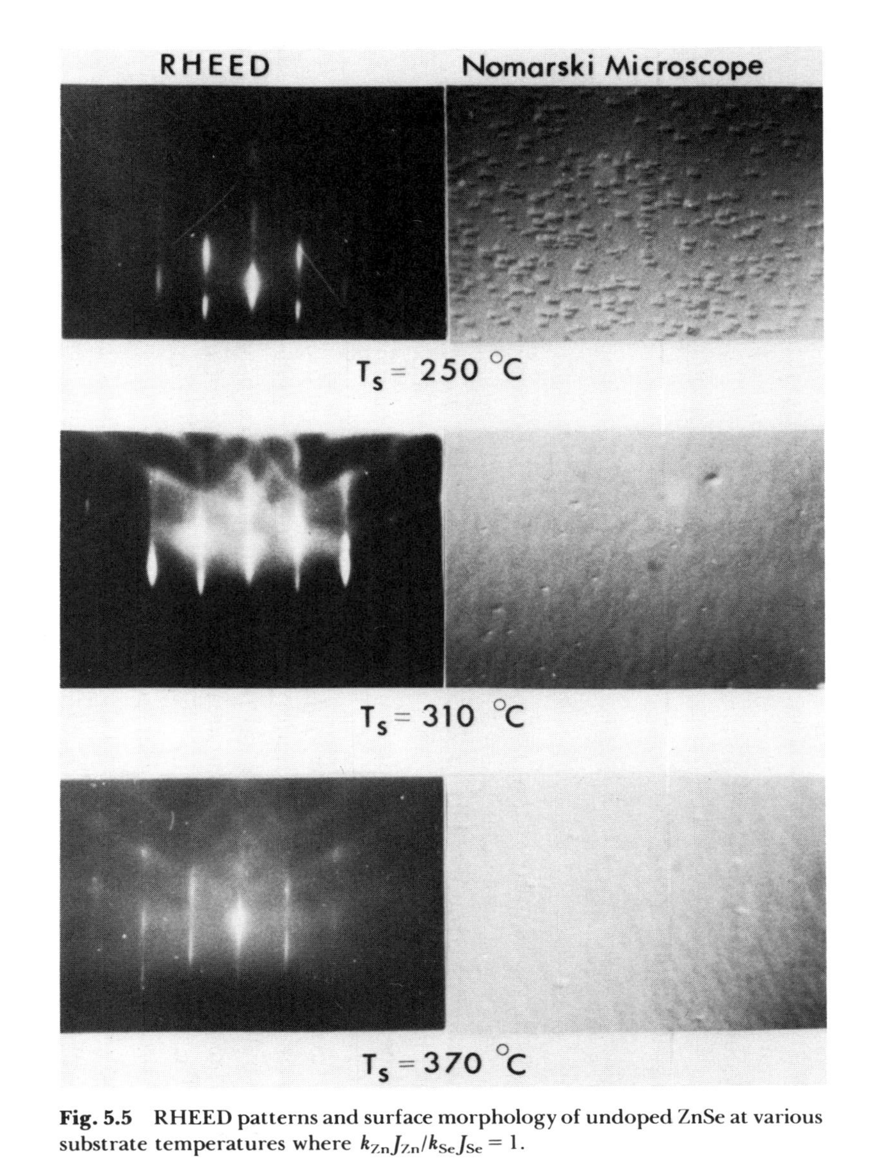

Fig. 5.5 RHEED patterns and surface morphology of undoped ZnSe at various substrate temperatures where $k_{Zn}J_{Zn}/k_{Se}J_{Se} = 1$.

lower substrate temperatures than 280°C, the value of $k_{Zn}J_{Zn}/k_{Se}J_{Se}$ would decrease from unity, and the surface of the epilayers exhibit rectangular-shaped hillocks. In contrast, at substrate temperatures higher than 280°C, the value of $k_{Zn}J_{Zn}/k_{Se}J_{Se}$ becomes larger than unity, and a smooth surface is obtained.

5.3. Evaluation of epilayers

5.3.1. *Electrical properties*

The electrical properties of the epilayers are influenced considerably by the substrate temperature and molecular-beam flux. Fig. 5.6 shows the resistivity, electron concentration, and electron mobility of ZnSe films grown at 280°C as a function of molecular-beam flux ratio $k_{Zn}J_{Zn}/k_{Se}J_{Se}$.[7] The MBE ZnSe films become low-resistive (~1 ohm cm) only when the molecular-beam flux ratio is about unity. The resistivity of the film increases abruptly up to 10^5 ohm cm when the value of $k_{Zn}J_{Zn}/k_{Se}J_{Se}$ decreases below 0.5, while the resistivity increases gradually for $k_{Zn}J_{Zn}/k_{Se}J_{Se} > 1$. The electron concentration and mobility seem to reach at their maximum values at around unity molecular-beam flux ratio. Such a dependence of the electrical properties on the molecular-beam flux ratio implies that the film stoichiometry dominates the electrical properties of undoped ZnSe.

Fig. 5.7 shows the substrate temperature dependence of the resistivity, electron concentration, and electron mobility at room temperature.[8] The value of $k_{Zn}J_{Zn}/k_{Se}J_{Se}$ is set such that it is unity at the substrate temperature of 280°C, and the values of J_{Zn} and J_{Se} are kept constant throughout the whole range of substrate temperatures. The epilayer is low-resistive (resistivity ~1 ohm cm) for the substrate temperatures between 280°C and

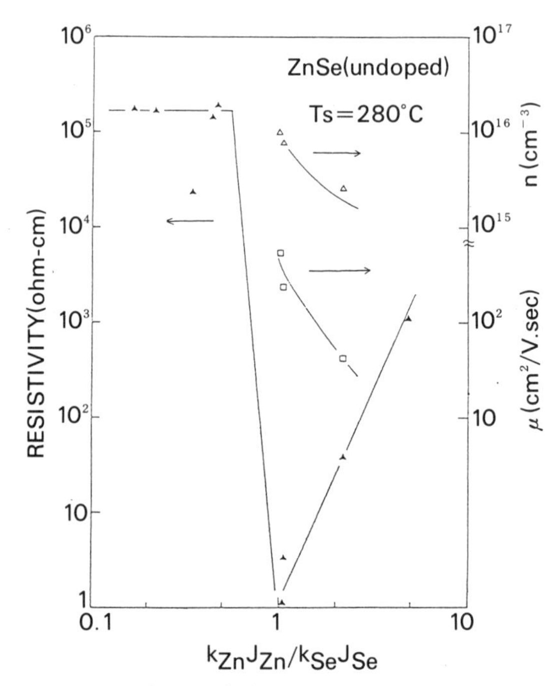

Fig. 5.6 Resistivity, electron concentration, and electron mobility of undoped ZnSe at room temperature against the molecular-beam flux ratio of the constituent elements when the substrate temperature is 280°C.

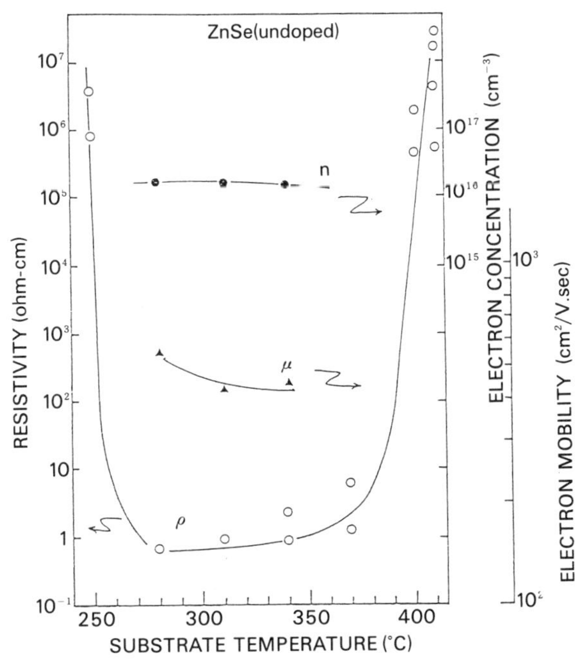

Fig. 5.7 Resistivity, electron concentration, and electron mobility of undoped ZnSe at room temperature against the substrate temperature when $k_{Zn}J_{Zn}/k_{Se}J_{Se} = 1$. [after T. Yao et al.[6]]

370°C, while high-resistivity films (resistivity $\sim 10^6$ ohm cm) grow below 250°C or above 400°C. The electron concentration of the epilayer grown at 280°C is $1.7 \times 10^{16}\ \mathrm{cm}^{-3}$. The mobility reaches its maximum value at this substrate temperature and is as high as 550 cm^2/V s which is even larger than the calculated mobility value of polar optical phonon scattering ($\sim$470 cm^2/V s). As the substrate temperature increases above 280°C, both the electron concentration and the mobility decrease gradually.

5.3.2. *Photoluminescence properties*

Fig. 5.8 shows PL spectrum at 4.2 K from the epilayer grown at 280°C and $k_{Zn}J_{Zn}/k_{Se}J_{Se} \sim 1$.[6] The spectrum consists of dominant bound-exciton lines trapped at neutral donors (I_x) at around 2.8 eV and other very weak emission bands: a donor–acceptor (DA) pair emission band at 2.7 eV, an unidentified emission band at 2.5 eV, and the self-activated (SA) emission band at 2.0 eV which is a radiative transition at the SA center composed of a Zn vacancy and a donor impurity. The associated donor and acceptor responsible for the DA pair emission are related to Ga and As impurities slightly diffusing from the GaAs substrate. The extremely weak intensity of the SA and the DA pair emission bands and strong excitonic emission lines indicate that MBE ZnSe contains a very small concentration of Zn vacancy and residual impurities. Details of the PL spectrum in the exciton emission region are shown in the inset. Free-exciton emission lines (E_x) appear at 2.805 eV and 2.801 eV which are ascribed to the splitting of the free-exciton emission due to residual strain in the epilayer induced by the small lattice misfit and the difference in expansion coefficient between the GaAs substrate and the ZnSe overgrowth.[8,9] A weak I_2 line appears at 2.798 eV, which is bound-exciton emission at neutral

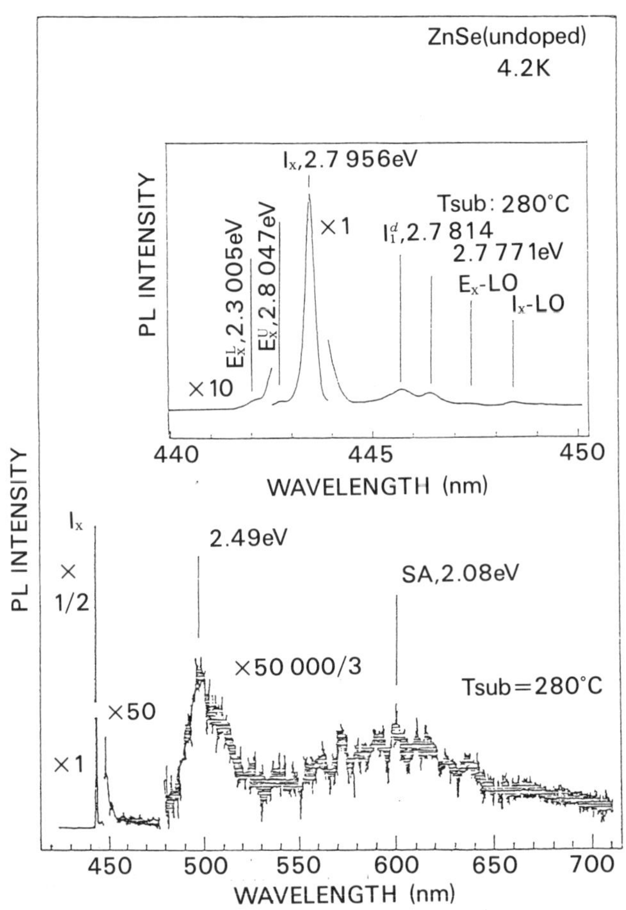

Fig. 5.8 Low-temperature PL spectrum of undoped ZnSe grown at 280°C and $k_{Zn}J_{Zn}/k_{Se}J_{Se} = 1$. The inset shows details of the exciton emission region.

donors presumably associated with the diffused Ga. The principal bound-exciton emission line at a neutral donor (I_x) appears at 2.796 eV. The emission lines due to the bound excitons at neutral acceptors such as the so-called I_1^s and I_1^d lines[10] are rarely observed. The weakness of the bound-exciton emission associated with acceptor impurities or Zn vacancy again indicates that MBE ZnSe is almost free from acceptor impurities and contains a small concentration of V_{Zn}.

The PL properties of MBE ZnSe vary considerably with the substrate temperature and the molecular-beam intensity ratio. Fig. 5.9(a) shows the dependence of emission intensities of I_x, E_x, and SA emission lines on the molecular-beam flux ratio.[7] The intensities of the I_x and the E_x emission lines are the strongest at around $k_{Zn}J_{Zn}/k_{Se}J_{Se} \sim 1$, while that of the SA emission band is the weakest at this molecular-beam flux ratio. This tendency is closely correlated with the variation of conductivity (cf. Fig. 5.6). Fig. 5.9(b) shows the intensities of the I_x, E_x, and SA emission lines against the substrate temperature.[8] The I_x and E_x emission lines vary such that the emission intensity is weaker in high-resistivity films and stronger in low-resistivity films while the SA emission intensity is enhanced in high-resistivity films and is suppressed in low-resistivity films. This tendency is also closely

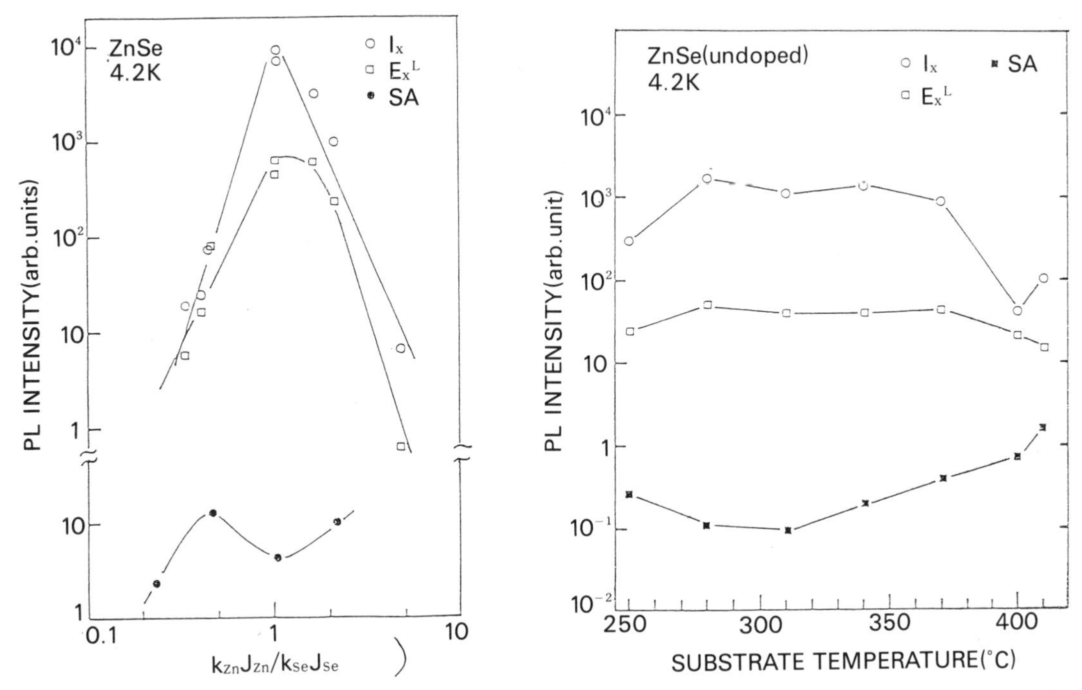

Fig. 5.9 Peak intensity variation of the I_x, E_x, and SA emission lines from undoped ZnSe against (a) the molecular-beam flux ratio when the substrate temperature is 280°C, and (b) the substrate temperature when $k_{Zn}J_{Zn}/k_{Se}J_{Se} = 1$. [after T. Yao et al.[6]]

correlated with the conductivity (cf. Fig. 5.7). Such a close correlation suggests that the principal I_x line is associated with donors whose concentration is sensitive to the film stoichiometry.

5.4. Discussion

5.4.1. *Optimum growth conditions*

The optimum growth condition was obtained from the assessment with PL and electrical measurements. As the crystallinity of the undoped epilayer is improved, intrinsic emission lines due to radiative recombination of excitons will be enhanced while deep emission intensity such as the SA emission will be suppressed. Strong intrinsic emission and weak deep emission implies a very small concentration of impurities and complex defects. As can be seen from Fig. 5.9(b), the intensity of the exciton emission line E_x is very strong when the substrate temperature is between 280°C and 370°C. The SA emission intensity becomes weakest for the epilayers grown at the substrate temperatures between 280°C and 310°C. As shown in Fig. 5.9(a), the exciton emission intensity becomes strongest for the epilayers grown at the molecular-beam flux ratio of $k_{Zn}J_{Zn}/k_{Se}J_{Se} \sim 1$, at which condition the SA emission intensity becomes the weakest.

As the crystallinity of the epilayer is improved, the electron mobility of the epilayer will be enhanced. As can be seen from Fig. 5.7, the electron mobility reaches its maximum value for the films grown at 280°C. From the dependence of the electron mobility on the molecular-beam flux ratio, the electron mobility becomes maximum when $k_{Zn}J_{Zn}/k_{Se}J_{Se} \sim 1$. Furthermore, the MBE ZnSe exhibits a very smooth surface morphology and good crystallinity as shown in Figs. 5.4 and 5.5, when the substrate temperature is ≳280°C and $k_{Zn}J_{Zn}/k_{Se}J_{Se} \gtrsim 1$. Therefore, the optimum growth conditions for high-quality ZnSe having a smooth surface morphology are: the substrate temperature is 280°C and $k_{Zn}J_{Zn}/k_{Se}J_{Se} \sim 1$. In other words, the stoichiometric epilayer is of high quality, and even a slight deviation from stoichiometry in the epilayer causes degraded crystallinity.

It is interesting to evaluate the ZnSe epilayer grown at the optimum growth conditions in terms of the PL and electrical properties. Fig. 5.10 shows a typical PL spectrum of MBE ZnSe grown under the optimized growth condition. The spectrum was measured at 300 K by using the 3250 Å line from a He–Cd laser as the exciting light with an excitation power of 200 mW/cm^2, a weak excitation condition.[11)] Here we have placed emphasis on the PL spectrum at room temperature, since, as the exciting light intensity increases, the intensity of the near-bandedge (NBE) emission becomes stronger, while deep emission tends to saturate.[12)] Moreover, the NBE emission is strongly enhanced compared with the deep emission band on lowering the temperature.[11)] The PL spectrum of MBE ZnSe consists of a dominant NBE emission band at around 460 nm and very weak deep emission bands. The ZnSe films grown at the optimized growth condition exhibit the peak intensity ratio (R) between the narrow NBE and the broad deep center luminescence as high as 40 even at room temperature. Since the deep emission band is generally related with complex defects or impurities, the high R value in MBE ZnSe is attributed to a low concentration of these crystalline defects. It has been demonstrated that this R value is the highest among the epitaxial ZnSe films.

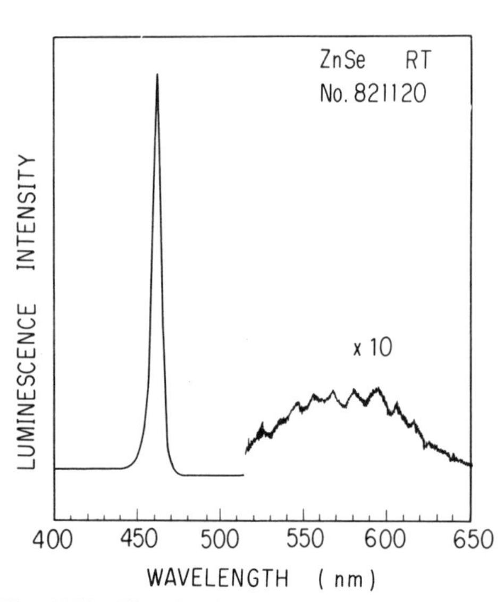

Fig. 5.10 Photoluminescence spectrum of MBE ZnSe grown under the optimized growth condition. The spectrum was measured at 300 K by using the 3250 Å line from a He–Cd laser as the exciting light with an excitation power of 200 mW/cm^2. [after T. Yao et al.[11)]]

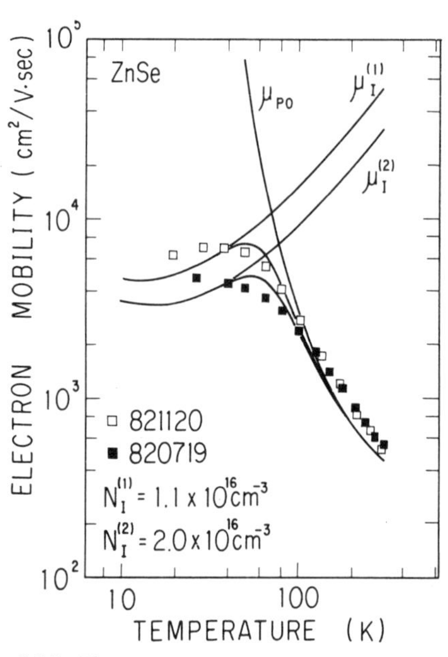

Fig. 5.11 Temperature dependence of the electron mobility of MBE ZnSe thin films and the calculated electron mobility. μ_{PO} is the polar optical phonon scattering mobility and μ_I is the charged defects scattering mobility. [after T. Yao et al.[11)]]

Fig. 5.11 shows the temperature dependence of the electron mobility of ZnSe epilayers.[11] The room-temperature mobility values are 530–540 $cm^2/V\,s$. As the temperature decreases below room temperature, the mobility increases because of the decrease in the longitudinal optical phonon scattering. The mobility shows a saturation behaviour or a maximum value at around 50 K which indicates that the ionized defect scattering participates at low temperatures. The maximum electron mobility of ZnSe epilayers grown at 280°C is $(4.7–6.9)\times10^3$ $cm^2/V\,s$ which is the highest electron mobility ever reported for ZnSe epilayers. It should be mentioned here that the electron mobility values of ZnSe epilayer grown at 310 and 340°C show maximum values of 3.4×10^3 $cm^2/V\,s$ and 3.0×10^3 $cm^2/V\,s$, respectively. The maximum electron mobility decreases with the increase of the substrate temperature. It is suggested that the ionized defect scattering increases with the substrate temperatures.

The solid lines in Fig. 5.11 show the calculated mobility via polar optical phonon scattering (μ_{PO}), scattering by charged defects (μ_I), and the resultant mobility ($1/\mu = 1/\mu_{PO} + 1/\mu_I$). The μ_I values were calculated by using the Conwell and Weisskopf formula[15] assuming singly charged defects. The concentration of charged defects (N_I) was chosen so as to make the calculated mobility values fit the experimental values. The estimated N_I value in MBE ZnSe is as low as $(1–2)\times10^{16}$ cm^{-3}.

Why can MBE grow such high-quality ZnSe films? This is mainly due to the low growth temperature in the MBE growth. In fact, the substrate temperature of MBE ZnSe is the lowest among the epitaxial growth techniques. We have shown that the maximum mobility values of MBE ZnSe decrease with the increase of the substrate temperature above 280°C. Furthermore, the flux ratio $k_{Zn}J_{Zn}/k_{Se}J_{Se}$ also influences the properties of MBE ZnSe. MBE ZnSe grown at 280°C has low resistivity and high electron mobility only when the flux ratio is about unity. This stresses the importance of the stoichiometric growth.

5.4.2. *Origin of donor in undoped ZnSe*

Since the conductivity is sensitive to stoichiometry, one can consider that the donor responsible for the electric conduction in undoped ZnSe would arise from intrinsic point defects, that is Se vacancy (V_{Se}), interstitial Zn(Zn_i), or their associated complex defects. According to Watkins,[13] however, the interstitial Zn would not be stable at the growth temperature. We will consider the Se vacancy or its associated complex.

If the native donor is related to V_{Se}, the dependence of the conductivity on the substrate temperature and the molecular-beam flux ratio will be explained as follows: As mentioned in the previous section, the ZnSe films grown at the substrate temperatures below 250°C are slightly on the Se-rich side, so that V_{Se} decreases. On the other hand, the ZnSe films grown at the substrate temperatures higher than 400°C contains a high concentration of V_{Zn}, V_{Se}, or their associated complex defects, so that deep centers would be induced in the bandgap and compensation of native donors would be enhanced. In fact, deep center emission is enhanced and total emission intensity is decreased considerably in ZnSe films grown above 400°C as shown in Fig. 5.9(b). Similarly, the increase in resistivity for ZnSe films grown at the molecular-beam flux ratio of $k_{Zn}J_{Zn}/k_{Se}J_{Se} > 1$ would be ascribed to the decrease in concentration of V_{Se}. On the other hand, the increase in resistivity of ZnSe films in the case of $k_{Zn}J_{Zn}/k_{Se}J_{Se} < 1$ would be due to the compensation of native donors due to deep centers generated by the presence of excessive V_{Zn}. In fact, as shown in Fig. 5.9(a), the deep center emission increases and the total emission intensity decreases.

The intensity of the I_x emission line is closely correlated with the conductivity of ZnSe films. Since the donor species are associated with native defects, the I_x line would be related to the native donors. As for the dependence of the I_x line on the substrate temperature and the molecular-beam flux ratio, similar arguments as the resistivity would explain the observed tendency.

5.5. Conclusions

We have systematically studied MBE growth of II–VI compounds with special emphasis on undoped ZnSe. The crystallinity, surface morphology, electrical properties, and PL properties have been extensively investigated. We have elucidated the growth conditions which enable one to grow low-resistivity ZnSe by MBE: The substrate temperature is between 280°C and 370°C, and the molecular-beam flux ratio $k_{Zn}J_{Zn}/k_{Se}J_{Se}$ is about 1. We have established the optimum growth conditions: The substrate temperature is 280°C and the molecular-beam flux ratio $k_{Zn}J_{Zn}/k_{Se}J_{Se} \sim 1$, i.e. stoichiometric growth condition. The ZnSe film grown at the optimum growth condition exhibits very smooth surface morphology and the quality of the film is highest among epitaxial ZnSe films grown by other various epitaxial techniques.[11)]

There is a close correlation between the electrical and PL properties, which are very sensitive to the film stoichiometry. The donor species in undoped ZnSe films are considered to be native donors, such as V_{Zn} or its associated complex defects. The concentrations of such native donors are sensitive to the film stoichiometry. The dominant bound-exciton emission is related to the native donors. The emission line is the most intense and the total emission intensity is the strongest in the stoichiometric ZnSe films.

Acknowledgements

The author is grateful to Dr. Tsunemasa Taguchi of Osaka University for a useful discussion.

References

1) Y. S. Park and B. K. Shin: in Electroluminescence, Ed. J. I. Pankove (Springer-Verlag, New York, 1977).
2) B. Ray: II–VI Compounds (Pergamon, Oxford, 1969).
3) J. C. Philips: in Bonds and Bands in Semiconductors (Academic Press, New York, 1973).
4) T. Yao, Y. Makita, and S. Maekawa: Japan. J. Appl. Phys., 20 (1981) L741.
5) T. Yao, Y. Miyoshi, Y. Makita, and S. Maekwa: Japan. J. Appl. Phys., 16 (1977) 369.
6) T. Yao, M. Ogura, S. Matsuoka, and T. Morishita: Japan. J. Appl. Phys., 22 (1983) L144.
7) T. Yao: unpublished.
8) P. J. Dean, A. D. Pitt, P. J. Wright, M. L. Young, and B. Cockayne: Physica 116B (1983) 508.
9) T. Yao: unpublished.
10) J. L. Merz, K. Nassau, and J. Shiever: Phys. Rev., B8 (1973) 1444.
11) T. Yao, M. Ogura, S. Matsuoka, and T. Morishita: Appl. Phys. Lett., 43 (1983) 499.
12) S. Fujita, H. Mimoto, and T. Noguchi: J. Appl. Phys., 50 (1979) 1079.
13) G. D. Watkins: Lattice Defects in Semiconductors 1974 (Inst. Phys. Conf. Ser. No. 23, 1975) p. 338.
14) T. Yao and S. Maekawa: J. Cryst. Growth, 53 (1981) 423.
15) R. A. Smith: Semiconductors (Cambridge University Press, Cambridge, 1968).

6 MID-GAP ELECTRON TRAPS (EL2 FAMILY) IN GaAs

Toshiaki IKOMA†

Abstract

The main mid-gap electron trap called EL2 plays a dominant role in compensating shallow acceptors in undoped semi-insulating GaAs, which is now widely used as a substrate for high-speed integrated circuits. Although the properties and origin of this trap have been intensively investigated by many researchers, they still remain unclear. The author's group have experimentally shown that EL2 is not a single level but consists of a "family". This chapter proposes a new model for EL2, namely arsenic atom aggregates, to explain the variation of the thermal and optical properties of the EL2 family and their change with heat treatment and reaction with metals.

Keywords: Deep Level, EL2, GaAs

6.1. Introduction

The mid-gap electron trap, whose energy level and capture cross-section were first measured by the present author's group in 1974[1] plays a dominant role in compensating shallow acceptors in undoped semi-insulating GaAs, which is now widely used as a substrate for high-speed integrated circuits. This level is often called EL2 after Martin et al.,[2] who collected all data on deep levels in GaAs and classified them according to their emission rate versus temperature characteristics. They named mid-gap levels found at University of Tokyo,[1] Bell Laboratory[3] and Laboratoire d'Electronique et de Physique Appliquée,[4] ET1, EB2 and EL2, respectively and judged that they should be the same level. The name of EL2 has become popular because of a large contribution of French groups to the study of this level. Recently, Taniguchi and Ikoma have shown that EL2 is not a single level but consists of a family of similar levels.[5]

It is important to investigate the properties of these traps and to elucidate their origins. In the early years, EL2 was believed to be oxygen-related, but this model was refuted in 1979.[6] However, it has been shown that high-concentration doping of oxygen during Bridgman growth induces a mid-gap level, which is similar to EL2.[7]

After the oxygen-related model was refuted, the As_{Ga} antisite model has been investigated. The existence of such antisite defects in GaAs was predicted by Van Vechten[8] and experimentally discovered with the electron spin resonance (ESR) measurements by Wagner et al.[9] Later on, Kaufmann et al.[10] suggested that EL2 was the As_{Ga} antisite defect. Since then, many researchers have investigated the antisite defect by using ESR and similar techniques. Very recently, the detailed experiments carried out by

† Research Center for Function-Oriented Electronics, Institute of Industrial Science, University of Tokyo, 7-22-1, Roppongi, Minatoku, Tokyo 106.

JARECT Vol. 19, *Semiconductor Technologies* (1986), *J. Nishizawa (ed.)*
OHMSHA, LTD. and North-Holland

Meyer et al.[11] using the optical detection of electron spin resonance (ODESR) and the optical detection of electron nuclear double resonance (ODENDOR) showed that the isolated antisite defect was not EL2. They also showed that a variety of As_{Ga} antisite defects existed. Goltzene et al.[12] also suggested that the antisite defect was associated with the gallium vacancy.

The difficulty of identifying the origin of EL2 lies in the fact that the data reported by various authors are in many cases different and sometimes contradictory. Meantime, the present authors showed, by measuring DLTS spectra[5] photocapacitance[13] and photoquenching spectra[14] that the mid-gap traps detected in various GaAs crystals are not the same level. They have pointed out that there exist more than two mid-gap levels which have very similar properties. These levels were named the EL2 family.[14]

The existence of the EL2 family has been supported by many authors since then. Therefore, a question arises as to what kinds of point defect are involved in the EL2 family. Fillard et al.[15] extended Levinson's model[16] and proposed that EL2 should be an association of a deep state with different shallow states in order to account for the variation that they observed. Recently Martin et al.[17] proposed a hypothetical model to explain how the U-shaped defect-clustering dissociates into an "isolated" EL2 (As_{Ga}) in an as-neutron-irradiated sample. In their model, they assumed that the "isolated" EL2 could be associated with other neighboring defects at various distances.

Since a variety of mid-gap levels exist in different crystals, it is necessary to specify very carefully in what kind of GaAs crystal the level is detected and how the thermal history of that sample is, when examining the origins of mid-gap levels.

In this chapter the variation of properties of mid-gap levels in various crystals is shown, and their changes with heat treatment and by depositing metals are discussed. Then specific features of the EL2 family are deduced. From these features, the origin of the EL2 family is speculated.

6.2. Process dependence of EL2 properties

6.2.1. *Creation of EL2*

EL2 exists in bulk [both liquid encapsulated Czochralski (LEC) and horizontal Bridgman (HB)] and vapor-phase epitaxy [halide, hydride and organo-metallic (OMVPE)] GaAs but neither in liquid-phase epitaxy (LPE) nor molecular-beam epitaxy (MBE) GaAs. The non-existence of EL2 in LPE GaAs is explained by the fact that a crystal is grown under Ga-rich conditions, since EL2 is related to excess arsenic. However, there is no reasonable explanation why an MBE crystal does not contain EL2 even if it is grown under arsenic overpressure.

It has been reported that the density of EL2 increases with increase of the As to Ga ratio in a nearly stoichiometric melt in LEC growth[18] and with increase of the As to Ga mole fraction in OMVPE growth.[19,20] We have observed that increase of the 0.35 eV trap (EL6) was greater than that of EL2 in a Czochralski GaAs when applying arsenic pressure.[21] EL6 is another common electron trap in both HB and LEC materials, and sometimes the density is higher than that of EL2. Therefore, when the ESR signal is analyzed in terms of existing traps in LEC GaAs, EL6 cannot be ignored.

EL2 can be created by ion implantation and successive low-temperature annealing.[22]

In an as-implanted sample, the DLTS exhibited a U-shaped spectrum, the peak of which was around 260 K. After annealing at 400°C for 15 min, the U-shaped peak disappeared quickly and a small peak which is similar to EL2 appeared. This peak became very large after 700°C annealing. The position of the peak was almost the same as the peak of EL2 in HB GaAs. However, on close examination, it was found that the peak temperature was slightly lower than the peak of EL2 in the before-implantation sample. Such an annealing behavior was also observed for O, Ga, As, Si, and N implantation.[22] All these levels show the photoquenching effect which is regarded as a fingerprint of the EL2 family.[14] They are thermally unstable and annihilated by a longer time or higher-temperature annealing. Appearance of the similar U-shaped peak and dissociation into an EL2-like level were also observed in a neutron-irradiated sample.[23]

Plastic deformation at around 400°C also created EL2.[24] However, this is not always true. It should be noted that electron-beam (EB) irradiation did not create EL2 even after annealing. The 0.75 eV level whose DLTS peak appeared at the same position as that of one of the EL2 family (ETX-5[12]) was created by EB irradiation and annealing. However this level did not show photoquenching and hence is not a member of the EL2 family. Neutron irradiation, which is known to increase the antisite ESR signal[25] and induce the U-shaped DLTS peak, does not create a clear EL2 peak in the DLTS spectrum at least in as-irradiated samples.[26] After annealing, the EL2-like peak was observed[17] but an experimental verification whether this level belongs to the EL2 family is not clear.

6.2.2. *Annihilation of EL2*

It is well known that the density of EL2 decreases after annealing in hydrogen atmosphere. Its dependence on annealing temperature is different in HB and LEC crystals. Hydrogen plasma enhances the decrease of EL2.[27] In LEC crystals annealing induces the change of the electron emission rate.[5] Ion implantation also annihilates EL2 as shown in Fig. 6.1, which shows how EL2 was annihilated by oxygen implantation in an as-irradiated HB wafer, which originally contained EL2. EL2 reappeared after annealing.

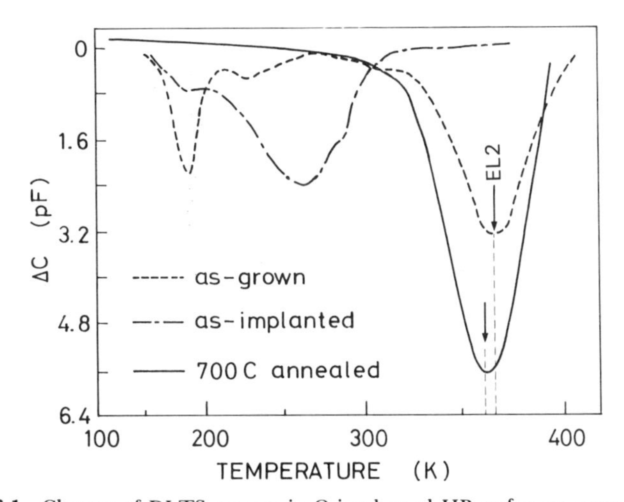

Fig. 6.1 Change of DLTS spectra in O-implanted HB wafer on annealing.

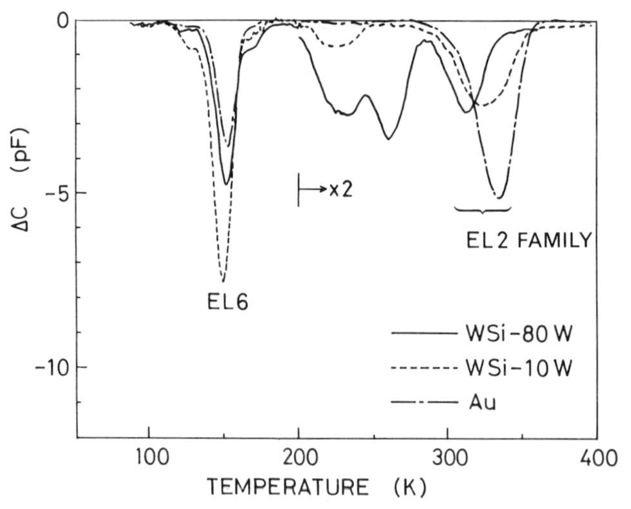

Fig. 6.2 DLTS spectra for WSi and Au Schottky contracts.

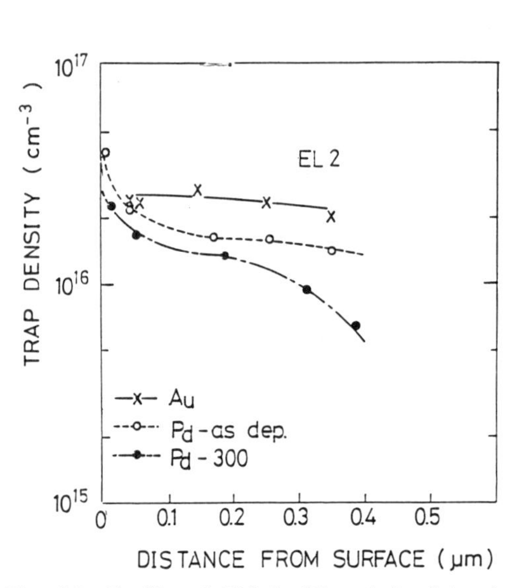

Fig. 6.3 Profiles of EL2 in Pd and Au Schottky contacts.

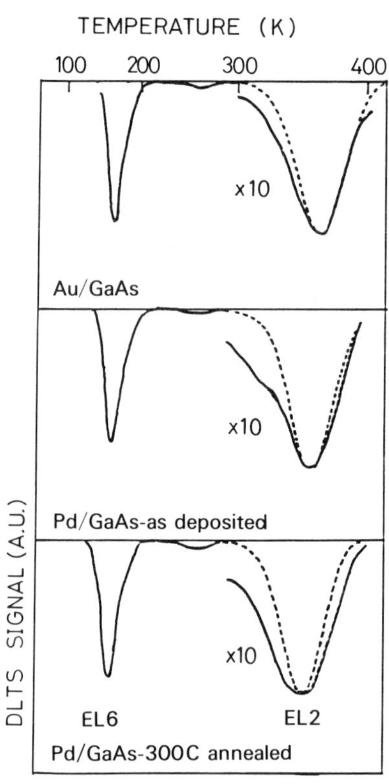

Fig. 6.4 DLTS spectra near the interfaces of Au/GaAs, as-deposited Pd/GaAs, and 300°C annealed Pd/GaAs systems. Dotted lines are the calculated curves for single level. Bias voltage: 0 V. Injection pulse: 0.5 V.

It should be noted, however, that the peak temperature of the DLTS spectrum is different before and after implantation. This is discussed in more detail later.

Recently we have found that EL2 is annihilated or changes its properties with deposition of WSi as a Schottky metal.[28] Fig. 6.2 shows the DLTS spectra measured in WSi/GaAs Schottky contacts which were fabricated at different sputtering input powers. For reference, the DLTS spectrum for a Au/GaAs Schottky contact made on the same wafer is included. For the samples fabricated by sputtering, the density of EL2 decreased and the peak temperature shifted to lower temperatures. The shift of peak temperature is larger for higher sputtering powers.

The properties of EL2 were also changed by Pd evaporation. Pd is known to react with GaAs at low temperature and form metallic compounds. The distributions of EL2 density in Pd/GaAs Schottky contacts are shown in Fig. 6.3 as well as the distribution in a Au/GaAs Schottky contact for reference. It is clear that Pd evaporation and successive low-temperature annealing changes the density distribution. The emission rate also increased, resulting in a lower temperature shift of the DLTS peak. A similar effect was also observed for Al/GaAs Schottky contacts.[29]

Furthermore, we have found that the DLTS spectra measured at the interface were broadened as shown in Fig. 6.4. This broadening was observed even in a Au/GaAs Schottky contact but most remarkably in the Pd/GaAs contact annealed at 300°C. This broadening suggests that EL2 at the interface cannot be regarded as a single energy level but has level broadening. This broadening is extended as deep as 50 nm from the metallurgical interface. This broadened energy states showed the photoquenching effect. All these experimental facts show that EL2 reacts with metals selectively at very low temperature.

6.3. Features of the EL2 family

6.3.1. *DLTS*

A DLTS spectrum reflects the emission rate versus temperature characteristic of the level of interest. Therefore, from a peak temperature shift with change of the rate window are deduced the activation energy and capture cross-section, with which the level is classified. However, in DLTS measurements, accurate determination of temperature is rather difficult, and data obtained on the emission rate show a wide scatter between researchers. Therefore, in the past, levels lying near the mid-gap have been regarded as being the same level and called EL2. However, when we examine the data in the past more closely, we can find some levels are actually different. In 1977, Ikoma and Takikawa[30] found a step-like increase of the density of the mid-gap level in a VPE layer as shown in Fig. 6.5. This step-like increase was found to be due to the appearance of another level (#11) which could not be separated by DLTS but could be distinguished by an optical measurement; the threshold energy for photocapacitance was 0.82 eV for #10 and 0.75 eV for #11 in the figure. In 1977, Mircea[31] also reported the existence of two levels, the DLTS peaks of which were separated by 5–6°C. Ozeki et al.[32] found two mid-gap levels in their VPE layers which had different activation energies depending on the growth temperatures. A similar effect was observed in OMVPE layers.[20]

We have shown that oxygen implantation and annealing induces a mid-gap level which has a photoionization threshold energy different from that of EL2 in HB GaAs.[13]

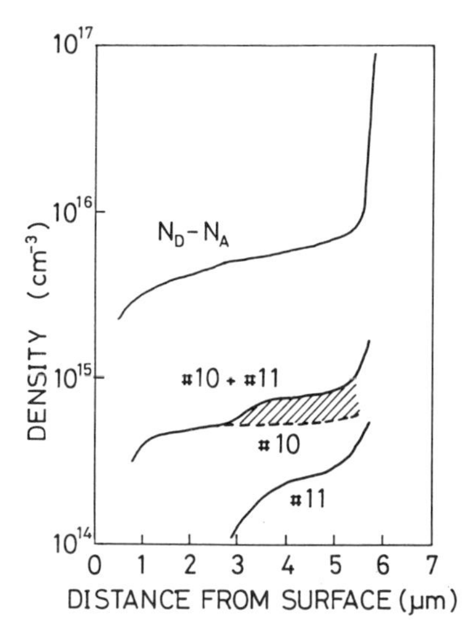

Fig. 6.5 Profiles of mid-gap traps showing the existence of two levels (#10 and #11).

After low-temperature annealing of an oxygen-implanted sample, two mid-gap levels exist simultaneously as shown in Fig. 6.6, which shows the DLTS spectrum measured in an HB wafer oxygen-implanted and annealed at 400°C for 15 min. As shown in the figure, the broadened peak could be fitted by assuming two levels; one existed in the original HB wafer and the other a newly generated level. This indicates that during the annealing stage, one mid-gap level diffused out from the bulk into the surface region and another mid-gap level was created by dissociation of defect clusterings formed by implantation. Such simultaneous existence of two mid-gap levels were also found in as-grown LEC[33] and oxygen-doped HB GaAs.[7]

In LEC GaAs we have observed two groups of mid-gap levels:[12] one is stable and has the same electron emission rate as that of EL2 in HB GaAs and the other changes its emission rate and density with low-temperature annealing. Furthermore, we have found

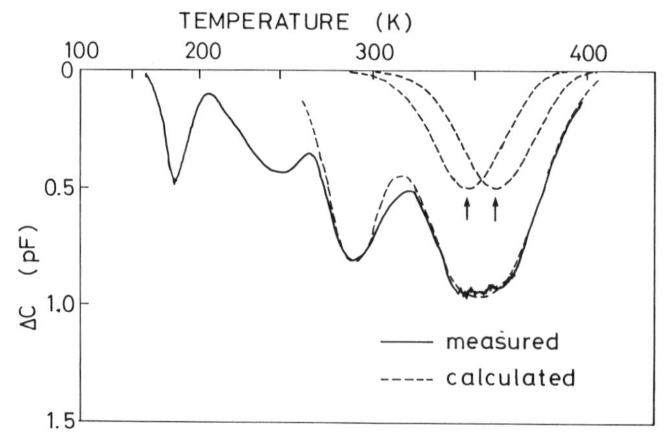

Fig. 6.6 DLTS spectrum for O-implanted and annealed HB GaAs wafer. The measured curve is fitted assuming the existence of two mid-gap levels.

that the mid-gap levels in the front and tail sections of an LEC ingot are very different; the difference of the DLTS peak temperatures was about 25 K. This indicates that the mid-gap level in the front section is annealed during the cooling process after growth and, therefore, the emission rate is increased and the density is reduced. This result is consistent with that of the annealing experiment. Different properties of a mid-gap level in LEC GaAs from those in VPE and HB GaAs were also reported by G. P. Li et al.[34)]

From the aforementioned facts we can conclude that a variety of mid-gap levels, which have different electron emission rates and are distributed non-uniformly, exist in LEC GaAs. This causes non-uniformity in an undoped semi-insulating substrate. Accordingly, reduction of the mid-gap level density with annealing is strongly non-uniform in an LEC wafer.

6.3.2. *Photoquenching effect*

The photoquenching effect is a very peculiar characteristic and hence regarded as a fingerprint for the EL2 family. We have found that all the mid-gap levels detected in our samples and referred to in the above sections showed the photoquenching effect in junction capacitance except the one created by EB irradiation as mentioned before. This indicates that all these mid-gap levels which have different activation energies and capture cross-sections should have similar atomic structures and be classified in the same group; the EL2 family.

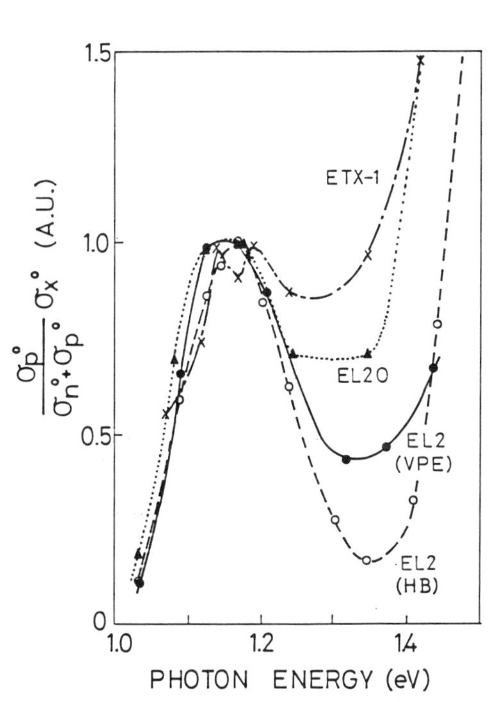

Fig. 6.7 Photoquenching spectra for four mid-gap levels belonging to the EL2 family.

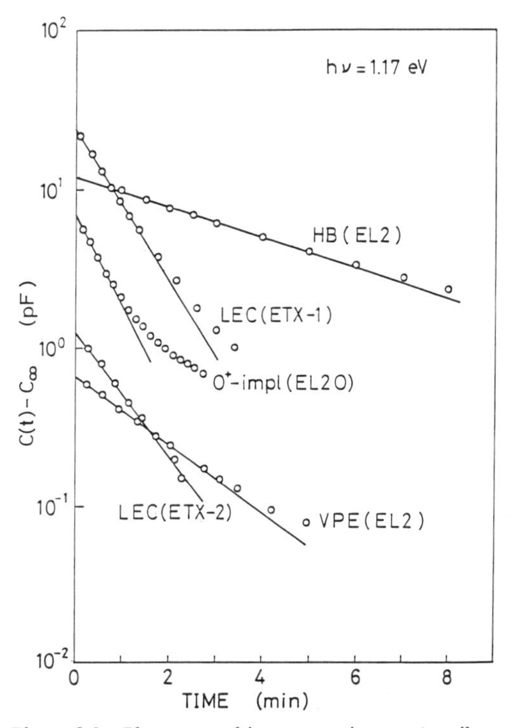

Fig. 6.8 Photoquenching transients (semilogarithmic plots of $C(t)$–C_∞ versus time) for five mid-gap electron traps.

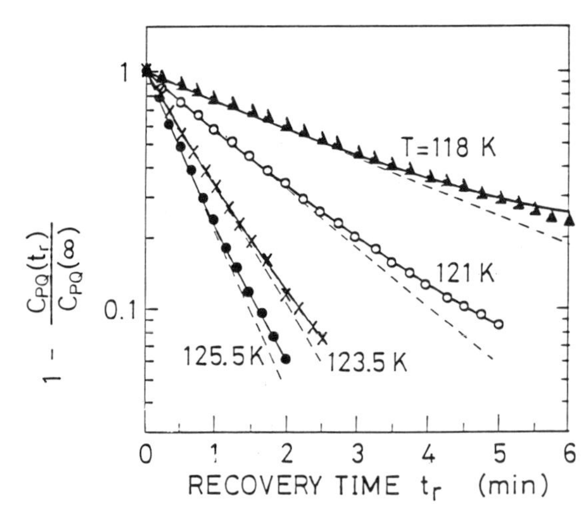

Fig. 6.9 Transients of the thermal recovery process from the metastable state to the normal state at ETX-1. $C_{PQ}(t_r)$ denotes the photoquenched capacitance corresponding to the recovery during the period of t_r.

It is important, however, to note that the time constants of transition from the normal to metastable states are different for these mid-gap levels and also the photoquenching spectra have a variation as shown in Fig. 6.7, which shows normalized photoquenching spectra for four levels in various crystals. We have found that the photoquenching efficiency increases for photon energies higher than 1.4 eV, which was not reported in the original paper by Vincent et al.[35] Such photoquenching spectra were also reported by Hasegawa.[36]

It is interesting to note that the capacitance transient due to quenching is non-exponential as shown in Fig. 6.8[37]. The non-exponentiality is most remarkable for the oxygen-implantation induced level. The recovery from the metastable state to the normal state is also non-exponential as shown in Fig. 6.9. We have measured and analyzed these transients at different temperatures. Using the model of Vincent et al., we can estimate that the barrier height for the photoquenching transition has a variation. These observations suggest that there exist multiple metastable states and the EL2 family has various electron–lattice couplings.[37]

6.3.3. *Photoluminescence*

Typical photoluminescence bands observed in semi-insulating GaAs are around 0.64, 0.68 and 0.78 eV. The correlation of these emission bands with EL2 has been studied but not certainly confirmed. The photoquenching effect is usually examined in order to correlate these bands with EL2. However, the experimental results differ among researchers. The photoquenching effect was observed for the 0.64 eV band by Leyral et al.[38] and Shanabrook et al.[39] but recently the enhancement of this emission band was reported after illumination by a YAG laser.[40] We have observed no correlation of this emission band with the EL2 density. According to Tajima,[41] the photoquenching of this band depends on samples, and he sometimes observed partial photoquenching for some of the EL2 levels. Recently photoquenching of the 0.78 eV band was reported.[42] However, this emission band is not always observed in samples which contain EL2.

6.4. Model and discussion

Some features of the EL2 family are summarized as follows:

(1) EL2 is associated with excess arsenic.
(2) EL2 appears at an intermediate stage of annealing, during which clusters of native defects created by heavy-particle bombardment dissociate and migrate to form more stable forms.
(3) EL2 has a large lattice distortion as revealed by the configuration-coordinate picture and also has multiple metastable states which are caused by an atomic structural change due to change of charged states. This implies that the atomic bonds surrounding the defect should be "soft" and that the defect would not be a simple point defect like an isolated antisite but a cluster or an association of point defects.
(4) EL2 has a variation in thermal and optical properties, which suggests that the defect is a sizeable defect consisting of a cluster or an association of point defects.
(5) EL2 changes its form on low-temperature annealing and on reaction with Schottky metals at very low temperature. This suggests that the constituents of the defect are fast diffusers.
(6) EL2 can appear as broadened energy states at the interface between metal and GaAs. This also indicates that the defect would have a complex form.

The models proposed so far are an isolated As_{Ga} (Kaminska[43]), an As_{Ga}–V_{As} complex (Lagowski et al.[44]), a V_{Ga}–As_{Ga}–V_{Ga} complex (Van Vechten[8]), a V_{Ga}–As_{Ga}–V_{As} complex (Yuanxi[45]) and an As_{Ga}–V_{Ga} complex (Goltzene et al.[11]). Recently Levinson[16] proposed a new model in which EL2 is a lattice defect which has a dipole interaction with an ionized impurity. Martin et al.[17] has extended this idea and considered that EL2 is an antisite defect associated with other lattice defects located at various distances from the core defect.

The present author's group has proposed that the EL2 family consists of arsenic atom aggregates[14,46], which perturb a periodical potential in the GaAs network. The size of aggregates can vary. The smallest one might be a five-atom cluster which could be an As_{Ga} antisite. Large ones might have a size of the order of 10 nm or larger. These defects could be regarded as a complex of an As_{Ga} antisite with interstitial arsenic atoms, amorphous arsenic atoms or microcrystalline arsenic precipitation. The involvement of interstitial arsenic atoms is conceived from point (5) and is very important. Interstitial atoms migrate faster than vacancies. In bulk GaAs, the existence of excess arsenic atoms was suggested.[47,48]

In an amorphous-phase or microcrystalline arsenic aggregate, lattice distortion is considered to be large, and the electron–lattice coupling can vary according to a possible variety of atomic structures. Furthermore, a configurational change due to charged states (transition from normal to metastable states) can take place in such an aggregate. Actually, in amorphous arsenic, the quenching of photoluminescence was observed on illumination with 1.1 eV light.[49]

The change of EL2 by deposition of certain metals is explained as follows. When a Schottky metal reacts with arsenic to form an intermetallic compound, interstitial arsenic atoms diffuse quickly to the surface to induce dissociation of arsenic aggregates near the surface. This results in reduction of the EL2 density near the surface. The reappearance of

EL2 after annealing might be ascribed to diffusion of excess arsenic atoms from the bulk region into the vicinity of the interface to again form arsenic aggregates. Antisite arsenic atoms may play a similar role as interstitial arsenic in this process.

The idea that atomic aggregates perturb the periodical crystal potential to generate localized states is somewhat similar to the generation of surface states. The similarity of the EL2 level with surface states in GaAs can be pointed out.

6.5. Conclusion

A new model for the main mid-gap electron traps (the EL2 family), namely arsenic atoms aggregates, was proposed to explain the variation of the thermal and optical properties of the EL2 family and their change with heat treatment and reaction with metals. To understand how arsenic aggregates create localized states near the mid-gap in the forbidden band and why they show very peculiar characteristics such as photoquenching effect, further extensive studies, both experimental and theoretical, are needed.

Acknowledgements

This chapter contains many data which were taken in the author's laboratory at University of Tokyo in the past. The author would like to thank Drs. Kazuo Sakai, Tsugunori Okumura, Hiroshi Gotoh, Masahiko Takikawa, Mitsuhiro Taniguchi and Mr. Yasunori Mochizuki for their contributions to this study.

References

1) K. Sakai and T. Ikoma: Appl. Phys., 5 (1974) 165.
2) G. M. Martin, R. C. Mitonneau, and A. Mircea: Electron. Lett., 13 (1977) 191.
3) D. V. Lang and R. A. Logan: J. Electron. Mater., 4 (1975) 1053.
4) A. Mircea and A. Mitonneau: Appl. Phys., 8 (1975) 15.
5) M. Taniguchi and T. Ikoma: J. Appl. Phys., 54 (1983) 6448.
6) A. M. Huber, N. T. Linh, M. Valladon, J. L. Debrun, G. M. Martin, A. Mitonneau, and M. Mircea: J. Appl. Phys., 50 (1979) 4022.
7) J. Lagowski, D. G. Lin, T. Aoyama, and H. C. Gatos: Appl. Phys. Lett., 44 (1984) 336.
8) J. A. Van Vechten: J. Electrochem. Soc., 122 (1975) 423.
9) R. J. Wagner, J. J. Krebs, G. H. Stauss, and A. M. White: Solid State Commun., 36 (1980) 15.
10) U. Kaufmann and J. Schneider: Advances in Electronics and Electron Physics 58 (Academic Press, New York, 1982) p. 81.
11) B. K. Meyer, D. M. Hofmann, F. Lohse, and J. M. Spaeth: 13th Int. Conf. on Defects in Semiconductors (Coronado, USA, 1984) p. 921.
12) A. Goltzene, B. K. Meyer, and C. Schwab: 13th Int. Conf. on Defects in Semiconductors (Coronado, USA, 1984).
13) M. Taniguchi and T. Ikoma: Semi-Insulating III–V Materials (Evian), Ed. S. Makram-Ebeid, (Shiva, Nantwich, UK, 1982) p. 283.
14) M. Taniguchi and T. Ikoma: Appl. Phys. Lett., 45 (1984) 69.
15) J. P. Fillard, J. Bonnafe, and M. Castagne: Solid State Commun., 52 (1984) 855.
16) Levinson: Phys. Rev., B28 (1983) 635.
17) G. M. Martin, E. Esteve, P. Langlade and S. Makram-Ebeid: private communication.
18) D. E. Holmes, R. T. Chen, K. R. Elliott, and C. G. Kirkpatrick: Appl. Phys. Lett., 40 (1982) 46.
19) L. Samuelson, P. Omling, H. Titze, and H. Grimmeiss, J. Cryst. Growth, 55 (1981) 164.
20) M. O. Watanabe, M. Tanaka, T. Nakanishi, and Y. Zohta: Japan. J. Appl. Phys., 20 (1981) L249.
21) M. Takikawa, T. Ikoma and K. Tomizaw: presented at the 41st meeting of Japan Society of Applied Physics (1980) 18p-A-7.

22) M. Taniguchi and T. Ikoma: Inst. Phys. Conf. Ser. 65 (1983) p. 65.
23) G. M. Martin and S. Makram Ebeid: Physica, 116B (1983) 371.
24) T. Ishida, K. Maeda, and S. Takeuchi: Appl. Phys., 21 (1980) 257.
25) E. R. Weber, H. Ennen, U. Kaumann, J. Windscheif, J. Schneider, and T. Wosinski: J. Appl. Phys., 53 (1982) 6140.
26) R. Magno, J. G. Giessner and M. Spencer: 13th Int. Conf. on Defects in Semiconductors (Coronado, USA, 1984) p. 981.
27) J. Lagowski, M. Kaminska, J. M. Parsey, H. C. Gatos, and M. Lichtensteiger: Appl. Phys. Lett., 41 (1982) 1078.
28) T. Makimoto, M. Taniguchi, T. Ogiwara, T. Ikoma, and T. Okumura: Int. Conf. on Solid State Devices and Materials (Kobe, Japan, 1984) D-3-2.
29) A. Yahata and M. Nakajima: Japan. J. Appl. Phys., 23 (1984) 313.
30) T. Ikoma and M. Takikawa: unpublished work; see also T. Ikoma and Y. Kurihara: Seisan Kenkyu, 31 (1980) 555. [in Japanese]
31) A. Mircea, R. C. Mitonneau, A. Hallis, and M. Jaros: Phys. Rev., B16 (1977) 3665.
32) M. Ozeki, J. Komeno, A. Shibatomi, and S. Ohkawa: J. Appl. Phys., 50 (1979) 4808.
33) M. Taniguchi and T. Ikoma: unpublished work.
34) G. P. Li, Y. Wu and K. L. Wang: 13th Int. Conf. on Defects in Semiconductors (Coronado, USA, 1984) p. 951.
35) G. Vincent and P. Bois: Solid State Commun., 27 (1978) 731.
36) F. Hasegawa: private communication.
37) M. Taniguchi, Y. Mochizuki, and T. Ikoma: Semi-Insulating III–V Materials, (Kah-nee-ta) Ed. D. C. Look and J. S. Blakemore (Shiva, Nantwich, UK, 1984) p. 231.
38) P. Leyral, G. Vincent, A. Nouailhat, and G. Guillot: Solid State Commun., 42 (1982) 67.
39) B. V. Shanabrook, P. B. Klein, E. M. Swiggard, and S. G. Bishop: J. Appl. Phys., 54 (1983) 336.
40) D. Paget and P. B. Klein: 13th Int. Conf. on Defects in Semiconductors (Coronado, USA, 1984) p. 959.
41) M. Tajima: private communication.
42) P. W. Yu: 17th Int. Conf. on Phys. of Semiconductors (San Francisco, USA, 1984) p. 747.
43) M. Kaminska: 17th Int. Conf. on Phys. of Semiconductors (San Francisco, USA, 1984) p. 741.
44) J. Lagowski, M. Kaminska, J. M. Parsey, H. C. Gatos, and W. Walukiewicz: Inst. Phys. Conf. Ser. 65 (1982) p. 41.
45) Z. Yuanxi, Z. Jicheng, L. Yiang, W. Kuangyu, H. Binghua, L. Bingfang, L. Cuncai, L. Liansheng, and S. Jiuan: 13th Int. Conf. on Defects in Semiconductors (Coronado, USA, 1984) p. 1021.
46) T. Ikoma, M. Taniguchi, and Y. Mochizuki: Gallium Arsenide and Related Compounds 1984 (Inst. Phys. Conf. Ser. No. 74, 1985) p. 65.
47) H. Gant, L. Koenders, F. Bartels, and W. Monch: Appl. Phys. Lett., 43 (1983) 1032.
48) I. Fujimoto: Japan. J. Appl. Phys., 23 (1984) L-287.
49) S. G. Bishop, V. Storm and P. C. Taylor: Solid State Commun., 18 (1976) 573.

7 QUANTUM MECHANICAL SIZE-EFFECT MODULATION LIGHT SOURCES WITH VERY HIGH SPEED CAPABILITY

Masamichi YAMANISHI†

Abstract

The state of the art is presented for recent developments in a new light-emitting device with very high speed capability, in which the application of an electrostatic field perpendicular to the active layer will produce a spatial separation of the carriers in the quantum well, providing a unique method of modulating the light output. The theoretical estimation shows a possibility of promising lasing and LED mode operations of the proposed devices at 80 K and 300 K, respectively. The experimental results on the field-induced photoluminescence quenching in a GaAlAs single quantum well at high temperatures suggest that the proposed devices will be constructed in the near future.

Keywords: Quantum Well, Semiconductor Light Source, Electric Field Effect

7.1. Introduction

Direct modulation of injection current can be used most conveniently to modulate the light output intensity from a semiconductor laser or light-emitting diode (LED).[1] The switching speed in the laser diode or LED modulated by the injection current is limited essentially by the recombination lifetime of the injected carriers in the active layer. In other words, the switching speeds in the light sources are dominated by the variation speeds in the carrier densities in the active layers. A new device concept[2,3] has been proposed to solve such a problem. In the proposed devices, the photon emission rate is controlled by the gate voltage mainly through changes in the spatial distributions of the carriers while the carrier concentration per unit area (surface carrier density) is kept more or less constant. Consequently, the switching speeds in the devices will be determined from the time (~10 ps) that the carriers take to transit in the very thin active layers. That is, the speeds will be free from the recombination lifetime limitation. Furthermore, the functions of field-effect control and of light emission are fully integrated only around the active layer in a single device, unlike the situation in a conventional integrated electro-optic circuit in which a light source and field-effect transistor or bipolar transistor are fabricated separately on a substrate.[4]

In respect of testing the operating principle of the proposed devices, Mendez et al.[5] and Miller and Gossard[6] reported the field-induced quenching of the photoluminescence (PL) from GaAlAs multi-quantum wells (MQWs) at cryogenic temperatures as well as shifts in the PL peak energies. Also, an optical loss modulation caused by field-induced changes

† Department of Physical Electronics, Faculty of Engineering, Hiroshima University, Shitami, Saijocho, Higashihiroshima 724

JARECT Vol. 19, Semiconductor Technologies (1986), *J. Nishizawa (ed.)*
OHMSHA, LTD. and North-Holland

in carrier distributions in a GaAlAs MQW was demonstrated by Wood et al.,[7] showing a high speed capability of the loss modulation. Very recently, we have observed the quenching of PL from a GaAlAs single quantum well (SQW) by an electric field at room temperature.[8]

This chapter will review work on the new light-emitting devices, concentrating mainly on the author's own work.[2,3,8,9]

7.2. Operating principle and design theory

Fig. 7.1 shows the proposed device structure[2] where electrons and holes are injected into the very thin active layer from the n- and p-type clading $Ga_{1-x}Al_xAs$ regions, respectively, and electric fields perpendicular to the injection current flows are applied through insulating $Ga_{1-y}Al_yAs$ ($y > x$) layers. When a sufficient positive voltage is applied to the upper gate electrode (Gate 1) and a negative voltage of similar magnitude is applied to the lower gate electrode (Gate 2), the injected electrons and holes accumulate at the upper and lower sides, respectively, in the active layer. The spatial separation of the carriers may bring about a reduction in the momentum matrix elements related to electron–hole recombination, i.e. a reduction in photon emission rates. In order to obtain a significant change in the emission rates, the thickness of the active layer might be chosen to be around 100 Å for the following reason. For an active layer much thicker than 100 Å and a certain value of carrier concentration (10^{17}–10^{18} cm^{-3}) in the active layer, an electric field lower than the breakdown field in the insulating $Ga_{1-y}Al_yAs$ layers ($\sim 5 \times 10^5$ V/cm) may not be sufficient to separate the electrons from the holes. Conversely, when the active layer is thinner than the screening length, several tens of ängstrom, the wavefunctions of the accumulated electrons and holes may overlap spatially, resulting in no significant change in the matrix elements. The behaviors of the electrons and holes in the active layer with a thickness of around 100 Å should be treated quantum mechanically so that the proposed devices have been named size-effect modulation laser device (SEM/LD) or light-emitting device (SEM/LED). Also, in the proposed devices, changes in the quantum states of the carriers play an important role in modulating the emission rate and wavelength. In this sense, the device is one of the wavefunction engineering devices.[10]

The wavefunctions associated with electrons and holes in the active layers were estimated by solving self-consistently the Schrödinger and Poisson equations, predicting

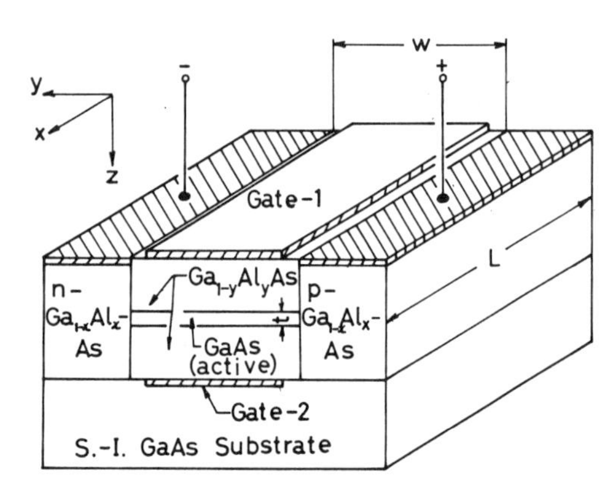

Fig. 7.1 An example of the proposed light sources. [after M. Yamanishi et al.[2]]

the changes in the gain and spontaneous emission spectra for the gate fields. Firstly, we shall concentrate on making a theoretical estimate of the characteristics of the devices at 80 K, showing a problem related to the field-induced changes in the surface carrier density in the active layer. Secondly, a new scheme will be proposed to eliminate the problem and, then, a theoretical estimate of the characteristics of the devices operating in LED modes at room temperature will be made.

Consider a double well structure,[3] as a specific example of the active region, which consists of active GaAs (40 Å), coupling barrier $Ga_{0.85}Al_{0.15}As$ (30 Å) and confining $Ga_{0.1}Al_{0.9}As$ layers, as shown in Fig. 7.2. In the analysis, it was assumed that the p–n junction consisting of the p-$Ga_{1-x}Al_xAs$, the active region and the n-$Ga_{1-x}Al_xAs$ is connected to a constant voltage source to inject carriers into the active region, and the contributions of light holes to the emission characteristics were neglected.

Fig. 7.3(a) and (b) show the estimated wavefunctions for zero gate field and an applied gate field F_s of 3×10^5 V/cm, respectively, at 80 K under a constant applied junction voltage of 1.68 V which gives a surface carrier density of 1.42×10^{12} cm^{-2} for zero gate field. The gain spectra in the device at 80 K are shown in Fig. 7.4. A significant reduction in the peak gains due to the lowest state transitions (1e–1h) can be expected with increasing field up to 3.5×10^5 V/cm, while an increase in the peak gains due to the transitions including higher states (1e–2h, 2e–1h) appears over the field. The latter fact is brought about by an asymmetry of the potential, i.e. a field-induced break of the selection rule. The peak gain together with the total spontaneous emission rate are quenched significantly by the applied field while the carrier density is kept almost constant for a field less than 2×10^5 V/cm as shown in Fig. 7.5. However, the carrier density increases with increasing fields larger than 2×10^5 V/cm.

The switching characteristics expected from the results in Fig. 7.5 are drawn schematically in Fig. 7.6. After the gate voltage is removed, a rapid increase in the emission rates will be caused by changes in the spatial distribution of the carriers. The rise time will be determined from the transit time of carriers across the active layer, the transverse relaxation time related to electron–hole recombination (T_2 time) and the *CR* time-constant, which is the product of the gate capacitance and the circuit resistance. Usually, the transit time and T_2 time are of the order of 0.1 ps while the *CR* time-constant

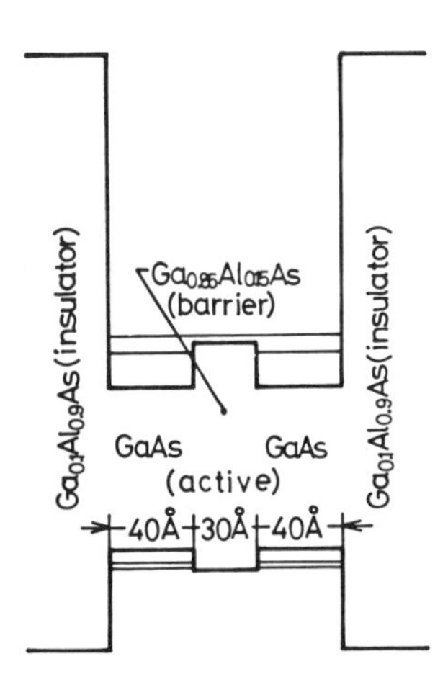

Fig. 7.2 A double well structure as a specific example of the active layers for a lasing mode operation at 80 K. [after M. Yamanishi et al.[3]]

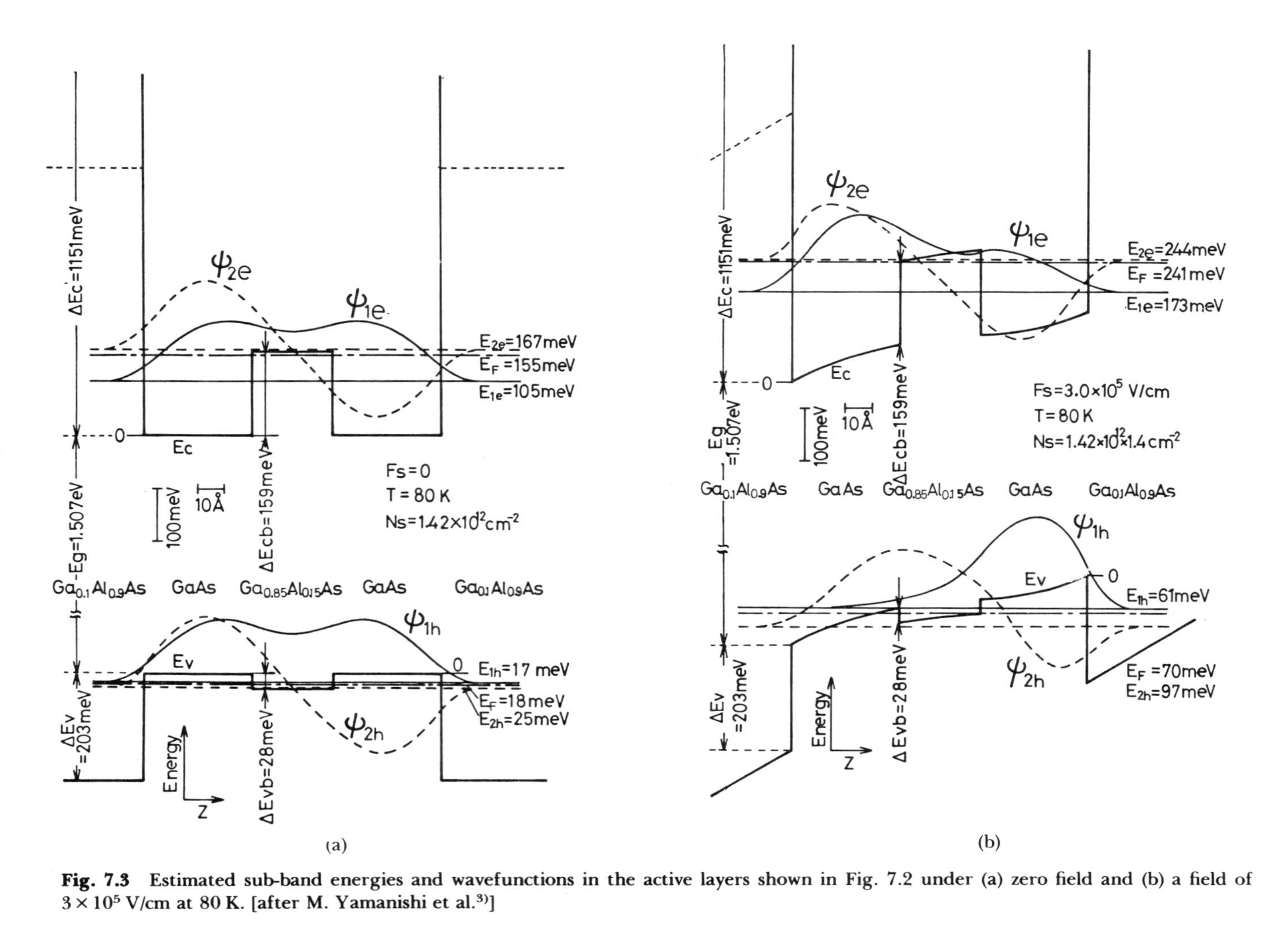

Fig. 7.3 Estimated sub-band energies and wavefunctions in the active layers shown in Fig. 7.2 under (a) zero field and (b) a field of 3×10^5 V/cm at 80 K. [after M. Yamanishi et al.[3]]

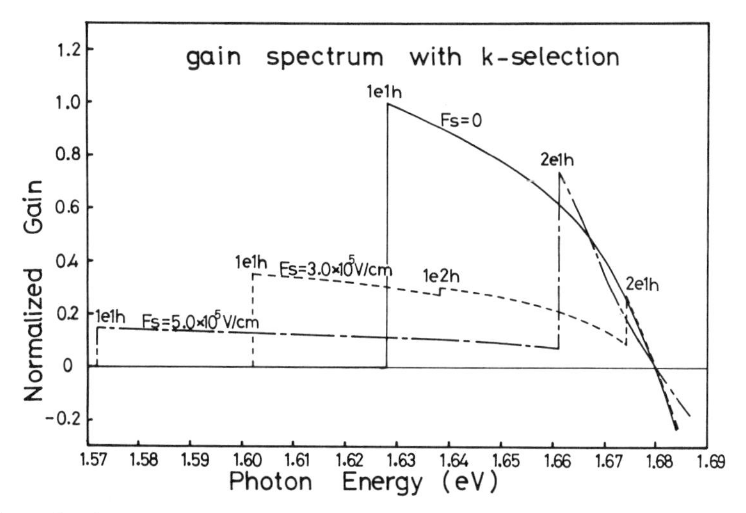

Fig. 7.4 Estimated gain spectra at 80 K in the device with the active layers shown in Fig. 7.2. [after M. Yamanishi et al.[3)]]

could be designed to be around 10 ps. Therefore, the switching speed will be limited by the *CR* time-constant. Following the very fast switching in the emission rates, the relatively slow decrease in the emission rates caused by the changes in the carrier density will be observed. The decay rate will be determined from the reconbination lifetime of the carriers. In particular, if the p–n junction is connected to a constant current source like a conventional laser diode or LED and if the radiative processes dominate over recombination, then the spontaneous emission rate, when compared under steady-state condition, may be constant, as shown by the dash-dotted line in Fig. 7.6. This problem associated with the changes in the carrier density can be resolved with a new scheme which will be discussed in what follows.

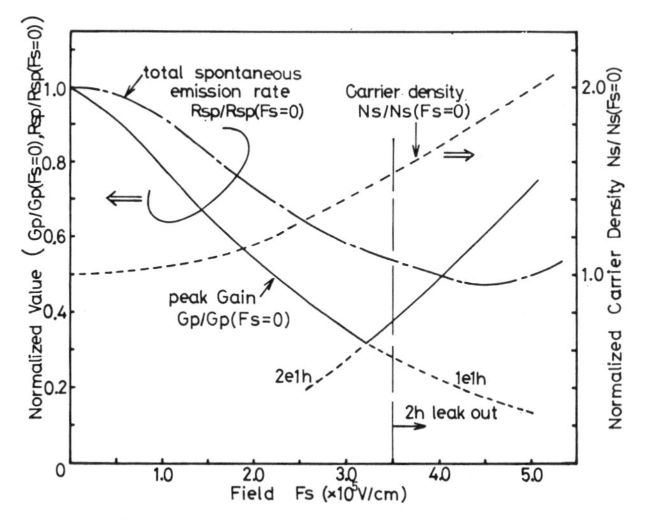

Fig. 7.5 Estimated peak gain, total spontaneous emission intensity and carrier concentration N_s as a function of applied field at 80 K in the device with the active layers shown in Fig. 7.2. [after M. Yamanishi et al.[3)]]

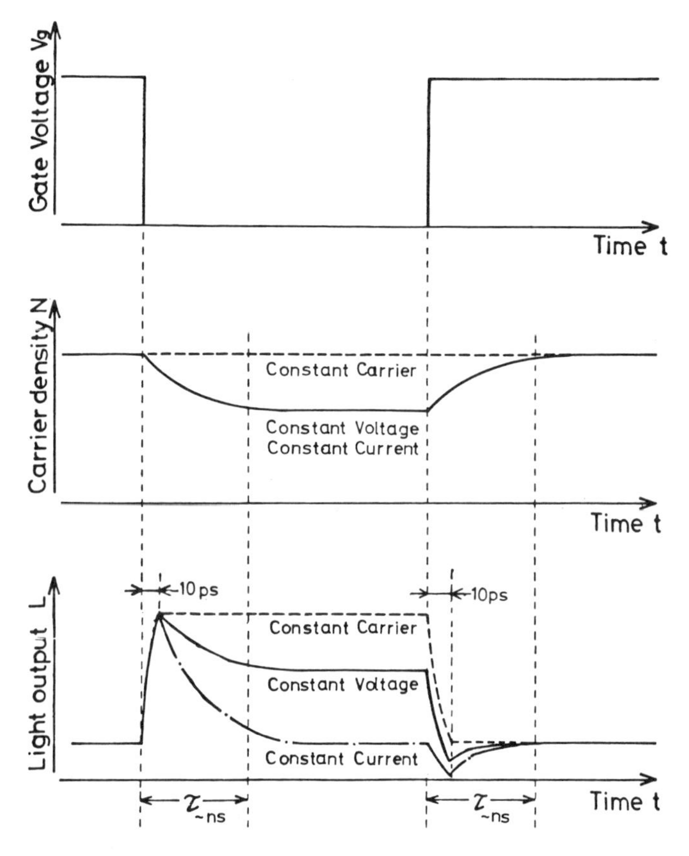

Fig. 7.6 Schematic diagram of the switching characteristics of the proposed devices in LED mode operations. [after M. Yamanishi et al.[9]]

In addition to the problem, there exists another problem related to free-carrier screening of the electric field. At higher temperatures, a higher carrier density is required to realize a population inversion, i.e. an optical gain. As a result, the electric field in the active layer is screened significantly by the high density of carriers if we expect to have a lasing mode operation at room temperature. Therefore we will focus on the estimation of the characteristics of the devices operating an LED mode at room temperature. The above-mentioned problem associated with the changes in the carrier density will be resolved by controlling the injection current with the application of the gate voltage to an external field-effect transistor (FET), as shown in Fig. 7.7.[9] It should be noted that the switching time of the injection current is required to be of the order of the recombination lifetime. Thus, one may be able to realize easily a constant charge operation of the SEM/LED with the proposed scheme.

Fig. 7.8 shows the estimated spontaneous emission rate as a function of the gate field F_s (field in the insulating layer of the SEM/LED) at 300 K. On this estimation, the structures of the active layers shown in the inset of Fig. 7.8 as well as the constant charge operation were postulated. In the cases of the structures B and C, the electrons can be confined effectively in the 50 Å GaAs wells while the holes can be accumulated easily near the interface between the $Ga_{0.85}Al_{0.15}As$ layer and $Ga_{0.1}Al_{0.9}As$ barrier when positive and

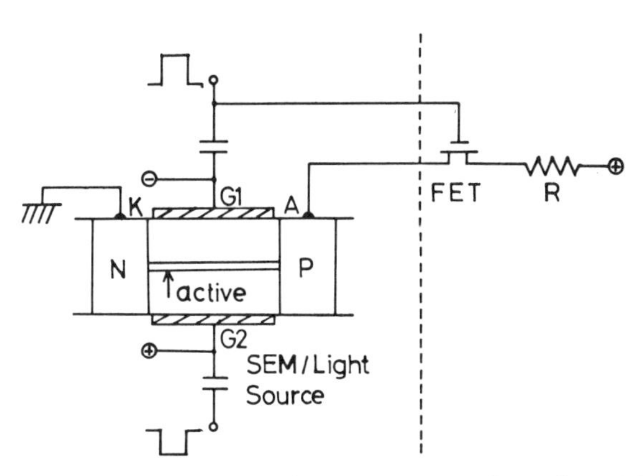

Fig. 7.7 Proposed circuit for a constant carrier number operation. [after M. Yamanishi et al.[9)]]

negative voltages are applied to the gate electrodes on the sides of the $Ga_{0.7}Al_{0.3}As$ and $Ga_{0.1}Al_{0.9}As$ barriers, respectively. This is because the barrier height (~150 meV) between GaAs and $Ga_{0.85}Al_{0.15}As$ in the conduction band is higher than that (~26 meV) in the valence band. Thus, for the structure C in which the thickness of the $Ga_{0.85}Al_{0.15}As$ layer, L_z, is 200 Å, we can expect a significant quenching of the emission rate by applying a field of 5×10^4 V/cm in the insulating $Ga_{0.7}Al_{0.3}As$ layer. Fig. 7.9 shows the estimated

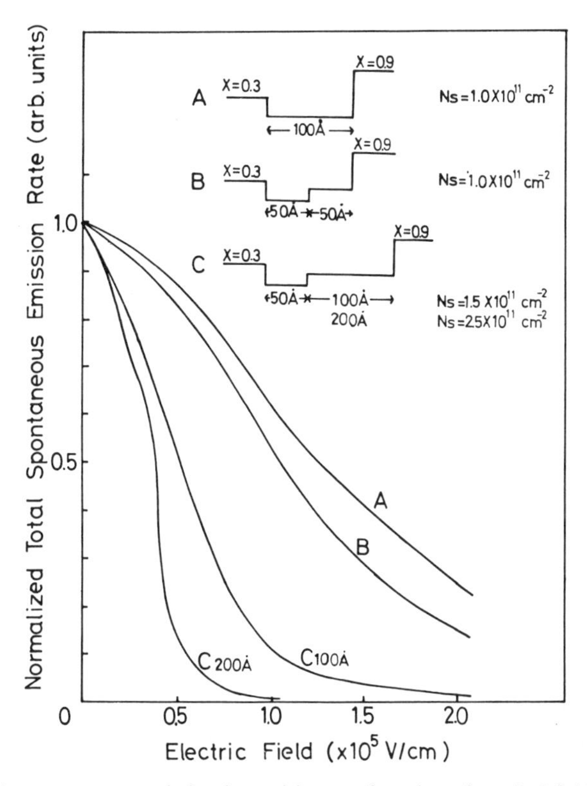

Fig. 7.8 Estimated total spontaneous emission intensities as a function of applied field at 300 K for the devices of which active layer structures are shown in the inset. Constant values of the surface carrier densities N_s are shown in the figure. [after M. Yamanishi et al.[9)]]

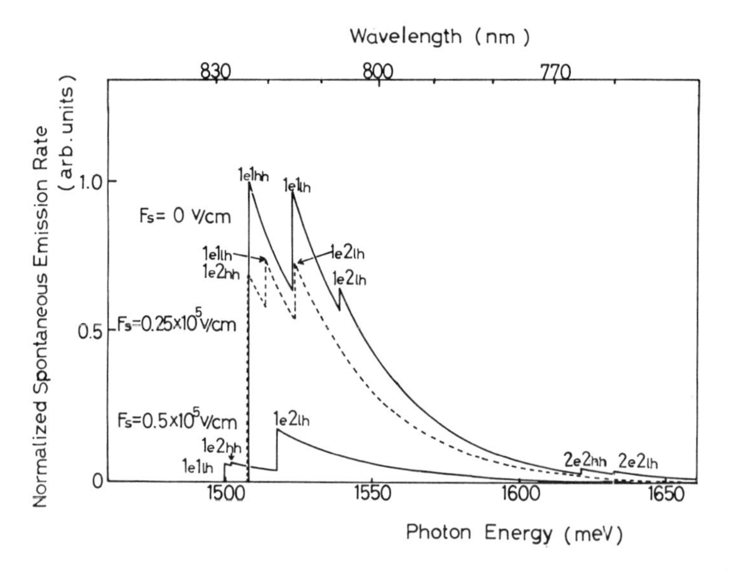

Fig. 7.9 Estimated spontaneous emission spectra for various fields at 300 K for the device (structure C with $L_z' = 200$ Å). [after M. Yamanishi et al.[9]]

spontaneous emission spectra for various electric fields in the device with the structure C ($L_z' = 200$ Å).

In addition to the amplitude modulation of the emitted light by the gate voltage, one may expect the wavelength tunable operation due to the shifts of the sub-band energies by the gate voltage in the present device. This tunable characteristic is apparently similar to that in a nipi superlattice device.[11] However, the tunable capability in the present device is caused by the change in the spatial distribution of the carriers induced by the static field while that in a nipi superlattice device is caused by the free-carrier screening effect for the built-in potential, induced by the change in the carrier number. There is an essential difference between them with respect to the switching speed.

7.3. Quenching of photoluminescence from GaAlAs single quantum well by electric field

As already mentioned, photoluminescence (PL) measurements at cryogenic temperatures on GaAlAs multi-quantum well (MQW) structures subject to an electric field were reported by Mendez et al.[5] and by Miller and Gossard.[6] They observed drastic quenching of the PL intensities by a factor of ~100 by the electric field (~10^4 V/cm) as well as shifts in the PL peak energies. In ref. 5, the shifts of the peak energy to the lower-energy side were ascribed to the field-induced separation of the electrons and holes, and the perturbation of the energy levels of the confined particles in the MQW samples, while in ref. 6, a shift in the PL from one transition to another at higher energy with increasing field (opposite in direction to the continuous shift with field reported in ref. 5) was observed. This suggests that different mechanisms are operative in the two cases. In addition, the observed quenching rates of the PL intensities were significantly larger than those expected from the carrier separation model. It seems that some complications were brought about by the coupling between the wells in the MQWs and by the existence of non-radiative processes. Furthermore, from the viewpoint of the application of the field-induced quenching of the PL to the very high-speed light-emitting device,[2,3,9] it is urgently required to know the influence of the electric field on the PL at higher temperature. The author's experimental results on the PL from a GaAlAs single quantum well (SQW) subject to an electric field at room temperature and at liquid nitrogen temperature, will be described briefly.

The experimental arrangements and sample configuration are shown in Fig. 7.10. The GaAs/GaAlAs layers were grown on the semi-insulating GaAs substrate by the MBE technique. The GaAs well was sandwiched between the GaAlAs barrier of which the Al mole fraction $x \sim 0.7$ was high enough to prevent carrier leakage by an electric field ($\sim 10^5$ V/cm). We adopted the SQW to avoid the complications involved in the MQW structures. A 400 μm diameter hole was etched through the opaque substrate by selective chemical etching. Finally, metal contact layers were deposited on the front and back surfaces. The samples were pumped with the 6328 Å laser line corresponding to a photon energy of 1.96 eV, which is sufficiently lower than the bandgap energy of the GaAlAs barriers (2.05 eV at 300 K and 2.15 eV at 80 K) that the GaAs well may be excited selectively. The excitation power density was around 20 W/cm^2 giving an estimated carrier density of the order of 10^{17} cm^{-3}.

Figs. 7.11 and 7.12 show the effect of the electric field on the PL from the GaAlAs SQW at 300 K and 80 K, respectively. These spectra were mainly due to electron–hole recombination in the SQW. The observed structures in the PL spectra were not intrinsic but extrinsic ones which were caused by Farby–Perot resonance due to the front and back surfaces of the sample. We observed significant quenchings of the PL intensities (by a factor of ~ 10 for an electric field of $\sim 10^5$ V/cm) as well as shifts in the PL energies (~ 20 meV) to the lower-energy side with increasing reverse bias voltage. The quenching rates were independent of the excitation power (4–20 W/cm^2).

We measured the photocurrent to reveal the reasons for the PL quenching. At room temperature, the photocurrent saturated gradually with increasing reverse bias voltage while the PL intensity decreased rapidly with increasing voltage, as shown in Fig. 7.13. At liquid nitrogen temperature, the observed photocurrents less than 1 μA were much smaller than the effective excitation current of 100 μA corresponding to 10% of the

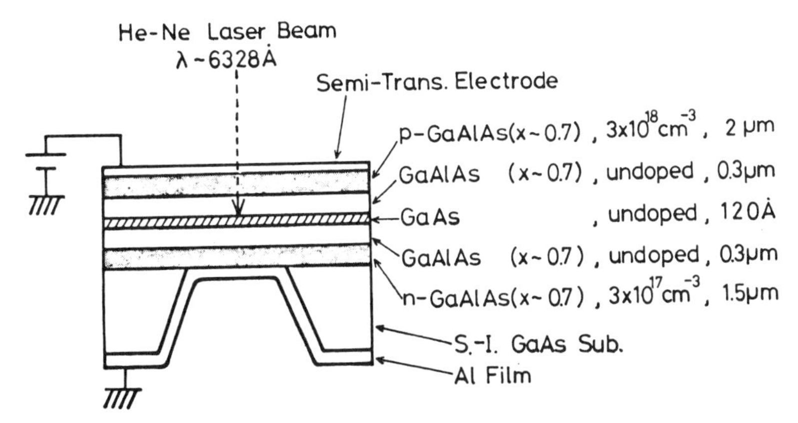

Fig. 7.10 Sample configuration and experimental arrangements. [after M. Yamanishi et al.[8]]

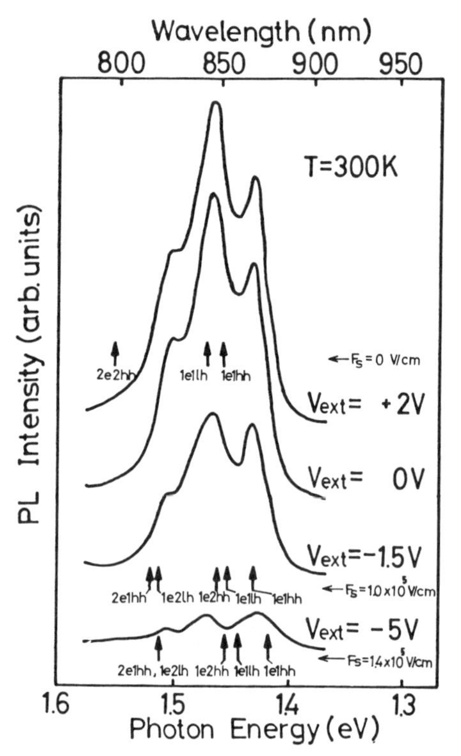

Fig. 7.11 Photoluminescence spectra for the sample shown in Fig. 7.10 at 300 K. The arrows show the optical transition energies estimated on the basis of the field-induced electron–hole separation model. [after M. Yamanishi et al.[8]]

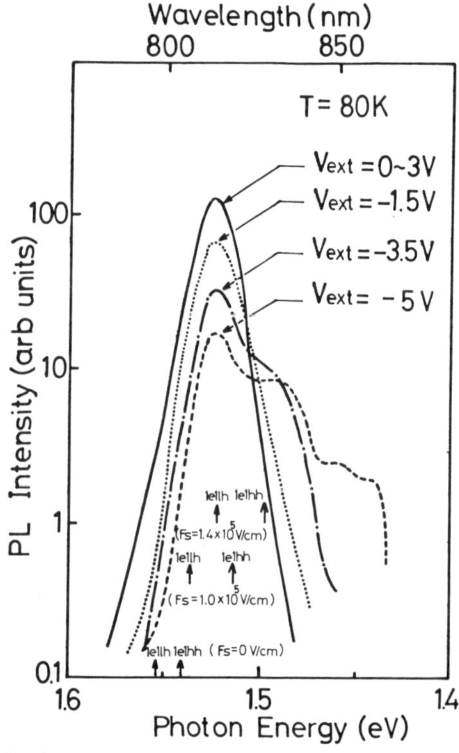

Fig. 7.12 Photoluminescence spectra at 80 K for the same sample as that in Fig. 7.11. The arrows show the theoretical transition energies. [after M. Yamanishi et al.[8]]

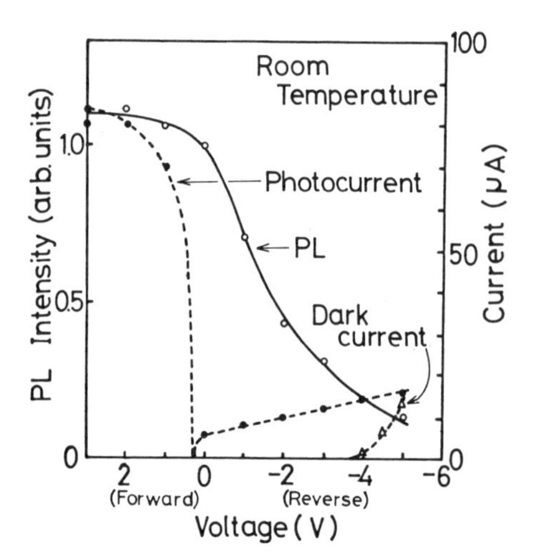

Fig. 7.13 Photoluminescence intensity and photo- and dark-currents at 300 K for the same sample as that in Fig. 7.11. [after M. Yamanishi et al.[8]]

He–Ne laser power (~2 mW). These seem to exclude field-induced carrier leakages from PL quenching mechanisms. We estimated the quenching rates of the PL intensities based on the carrier polarization model, postulating that the surface carrier concentrations (the carrier numbers) in the well are unchanged on application of the electric field. The experimental quenching rates were explained apparently in terms of the model. However, as discussed in Section 7.2, if the radiative processes dominate over recombination, then the total PL intensity (integrated over the photon energy) should be unchangeable in the steady state since the photon energy 1.96 eV of the excitation light is significantly higher than the bandgap energy of GaAs and, therefore, the absorption coefficient for the excitation light in the GaAs well may not be affected significantly by the applied field. One of the most plausible mechanisms for the PL quenching will be a change in the ratio of non-radiative to radiative recombination rates due to electron–hole separation in the SQW induced by the static field. On this matter, more work is still required to clarify the mechanisms for the PL quenching.

On the other hand, the shifts in the energies can be explained successfully in terms of the field-induced electron–hole separation and the perturbation of the energy levels of the confined particles in the well, which have been discussed in Section 7.2. We estimated the optical transition energies, based on the carrier separation model. The estimated transition energies are shown as arrows with symbols for sub-bands in Figs. 7.11 and 7.12. There is a fair correspondence between the observed spectra and estimated transition energies. The detailed descriptions of our works on the influence of an electric field on the PL intensities and spectra will appear elsewhere.

7.4. Conclusion

Theoretical and experimental work on a new field-effect light source concerned mainly with the work of the author's group have been reviewed. The theoretical estimate has shown a promising lasing and LED mode of operation of the devices at 80 K and 300 K, respectively. The possible advantages of the light sources, especially their high-speed capabilities, have been discussed. It has been shown that the number of carriers in the active layers in the devices is kept completely constant, utilizing a supplemental external switch such as a field-effect transistor or bipolar transistor. The experiments on the effect of an electric field for the photoluminescence from GaAlAs single quantum well have encouraged us to expect a realization of the proposed devices. The shifts in the luminescence energies associated with applications of electric fields have been successfully explained in terms of the field-induced electron–hole separation model which is the basic mechanism of the proposed devices. However, for the quenching of the luminescence intensities, more work, particularly on non-radiative processes, is required to clarify the mechanism.

Acknowledgements

The author wishes to thank Assoc. Prof. Ikuo Suemune and Assoc. Prof. Tadatsugu Minami (Kanazawa Institute of Technology) for their helpful discussions and collaboration, and graduate students Yasuo Kan, Hiroaki Yamamoto, Kazuo Ishiyama and Yuichi Usami for their collaboration in the course of the present work. He also wishes to thank Dr. Takahiko Misugi and Dr. Shigenobu Yamakoshi (Fujitsu Labs.) for supplying the MBE-grown sample used in the present experiments.

References

1) For instance, H. Kressel, and J. K. Butler: Semiconductor Lasers and Heterojunction LEDs (Academic Press, New York, 1977) Ch. 17.
2) M. Yamanishi and I. Suemune, Japan. J. Appl. Phys., 22 (1983) L22.
3) M. Yamanishi, T. Minami, and I. Suemune: extended abstr. of 15th Conf. on Solid State Devices and Materials (Tokyo, 1983) p. 325.
4) For a review of electro-optic ICs, see N. Bar-Chaim, I. Ury, and A. Yariv: IEEE Spectrum, 19 (1982) No. 5, p. 38.
5) E. E. Mendez, G. Bastard, L. L. Chang, L. Esaki, H. Morkoc, and R. Fisher, Phys. Rev., B26 (1982) 7101.
6) R. C. Miller and A. C. Gossard, Appl. Phys. Lett., 43 (1983) 954.
7) T. H. Wood, C. A. Burrus, D. A. B. Miller, D. S. Chemla, T. C. Damen, A. C. Gossard, and W. Wiegmann, Appl. Phys. Lett., 44 (1984) 16.
8) M. Yamanishi, Y. Kan, T. Minami, I. Suemune, H. Yamamoto, and Y. Usami, Superlattices and Microstructures, 1 (1985) 111.
9) M. Yamanishi, H. Yamamoto, and I. Suemune, Superlattice and Microstructures, 1 (1985) 335.
10) H. Sakaki, extended abstr. of 15th Conf. on Solid State Devices and Materials (Tokyo, 1983) p. 3.
11) H. Jung, H. Künzel, G. H. Döhler, and K. Ploog, J. Appl. Phys., 54 (1983) 6965; many references of the Max Planck Institüt group on nipi superlattices are cited therein.

8 InP ENHANCEMENT-MODE MISFETs FOR NEW MICROWAVE DEVICES

Tomohiro ITOH and Keiichi OHATA†

Abstract

CVD SiO_2/InP MIS diodes have good characteristics suitable for the accumulation-mode operation. They show no frequency dispersion in accumulation capacitance and minimum surface state density near the conduction band edge. Self-aligned gate enhancement-mode InP/SiO_2 MISFETs with short channel length have been successfully fabricated on an Fe-doped semi-insulating substrate. They exhibit very high transconductance, as high as 220 mS/mm, and good X-band operation, especially marked high power-output characteristics. The minimum noise figure at 4 GHz is 1.87 dB with 10.0 dB associated gain. Power-outputs 1.17 W/mm and 1.0 W/mm are obtained at 6.5 and 11.5 GHz, respectively. A maximum power-added efficiency of 43.5% is attained at 6.5 GHz.

Keywords: InP, MISFET, Microwave Devices

8.1. Introduction

High-frequency and high-speed devices are the principal devices mostly used in communication equipment and computers. GaAs MESFETs have been developed as practical low-noise and high-power devices. Their integration in high-speed circuits is being conducted in many laboratories.

However, there are various disadvantages originating from the Schottky gate structure, such as small tolerances in applicable gate voltages, threshold voltages, device structures, and fabrication process. MIS (Metal–Insulator–Semiconductor) or insulated gate structures can eliminate these disadvantages. Especially, enhancement-mode MISFETs are useful for digital integrated circuits.

InP has higher potentialities than GaAs for high-speed devices, because of the higher peak and saturation velocities of electrons. Furthermore, it has favorable MIS interfacial properties, which allows the formation of n-channel inversion or accumulation layers on p-type or semi-insulating InP substrates.[1–3]

There are many reports on investigation and improvement of InP MIS characteristics and MISFET performance,[4–9] and an inverter circuit has been reported.[8] However, successful high-speed and high-frequency operation for InP MISFETs has not yet been reported.

In addition to the above advantages over GaAs, InP has such properties as higher thermal conductivity and lower ionization coefficients. These properties, as well as high breakdown voltage due to insulated gate structure, are suitable for high-power devices.

† Microelectronics Research Laboratories, NEC Corporation, 4-1-1, Miyazaki, Miyamae-ku, Kawasaki, 213.

JARECT Vol. 19, *Semiconductor Technologies* (1986), *J. Nishizawa (ed.)*
OHMSHA, LTD. and North-Holland

The purpose of this paper is to clarify the ability of enhancement-mode InP MISFETs for high-speed application. In this work, a CVD (chemical vapor deposited) SiO_2 film was used as the gate insulator because it has good insulating properties and is easily deposited at low temperature. The condition to obtain a good MIS system was examined, and self-aligned gate enhancement-mode InP MISFETs were successfully fabricated on an Fe-doped semi-insulating substrate. The fabricated MISFETs exhibited successful X-band operation for the first time and higher power-output capability than that for GaAs MESFETs.

8.2. Investigation of MIS system

The deposition temperature of SiO_2 may strongly affect both film quality and interface properties between SiO_2 and InP. A few papers have reported on the effects of the deposition temperature.[10,11] However, generalized investigations on both electrical characteristics and interfacial behavior over a wide range of deposition temperature have not been conducted. In this section, the dependence of MIS diode properties on the SiO_2 deposition temperature was investigated in order to obtain the best condition for good MIS characteristics.

MIS diodes were fabricated on n-type InP (100) substrates with carrier concentration of $4.8 \times 10^{16}\,cm^{-3}$. First ohmic contacts were formed on the back with Ni/Au–Ge in order to avoid a high-temperature process after deposition of the insulating film. Then the substrates were rinsed in organic solvents and chemically etched in $H_2SO_4 : H_2O_2 : H_2O = 3 : 1 : 1$ solution prior to SiO_2 deposition. A 600–900 Å thick CVD–SiO_2 film was deposited as the insulator of the MIS diodes. The deposition temperature varied from 300°C to 420°C. Finally, Al circular electrodes were evaporated on the SiO_2 films.

MIS diodes were evaluated by measuring the resistivity of the insulator, its breakdown strength and C–V curves at room temperature and at 77 K, and estimating the surface state density. The interfacial behavior between the SiO_2 film and InP was also examined by AES (Auger electron spectroscopy) profile measurement with the Ar ion sputter technique.

Resistivities of SiO_2 films deposited at various temperatures were evaluated from the I–V characteristics of MIS diodes. The resistivity of the film had a maximum value for the deposition temperature of 340–360°C. Below this temperature, the resistivity decreased considerably, indicating poorer quality SiO_2 film. An insulating film with resistivity of $7 \times 10^{15}\,\Omega\,cm$ and the breakdown strength of 7.7×10^6 V/cm was obtained at the deposition temperature of 340–360°C.

Fig. 8.1 shows the C–V characteristics of a MIS diode deposited at 360°C. The measurement was carried out at 1 kHz, 10 kHz, 100 kHz and 1 MHz at room temperature. The bias voltage was applied in the range of ±10 V at a repetition of 0.0015 Hz. At frequencies about 10 kHz, there were clockwise, 1.5 to 2.0 V wide, carrier injection type hystereses. It is noticed that a frequency dispersion was scarcely observed in the C–V curves at the accumulation side.

The surface density of states was evaluated from the C–V characteristics at 1 MHz by the Terman method.[12] Fig. 8.2 shows the dependence of the surface density of states on the surface potential for MIS diodes deposited at 300°C, 340°C and 420°C. The surface density of states had a minimum value at around the flat band (0.06 eV below the

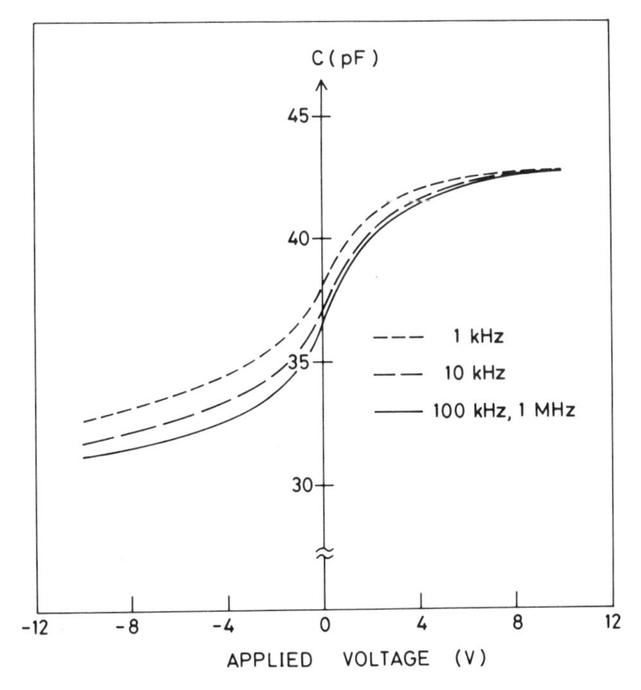

Fig. 8.1 *C–V* curves of a MIS diode.

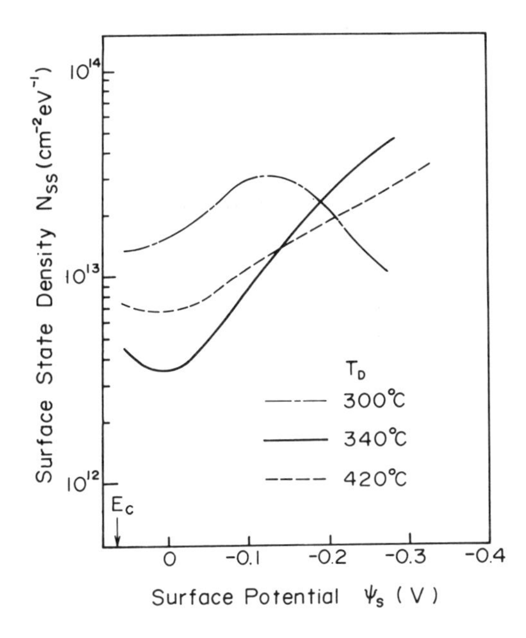

Fig. 8.2 Dependence of surface density of states on surface potential. [after K. Ohata et al.[7)]]

conduction band edge), and increased considerably towards the mid-gap. *C–V* measurements at 77 K showed much better characteristics and the minimum surface density of states was one order of magnitude less than the value at room temperature. MIS characteristics obtained in this experiment, that is no frequency dispersion in the accumulation capacitance and the minimum surface density of states near the conduction band edge, are suitable for devices in accumulation-mode operation.

AES measurements[7] also showed good MIS interfacial properties deposited at the temperature of 340–360°C. The device deposited at 340°C showed a very sharp interface transition, in which the transition points for all atoms coincide. Phosphorus or indium pile-up were not observed and native oxide was scarcely seen.

From these results, it is shown that the best SiO_2 film for the MIS device with good electrical characteristics as well as good interfacial properties can be obtained at the deposition temperature of around 350°C.

8.3. MISFET structure and fabrication

On the basis of the results of the MIS diode investigation,[7] self-aligned gate n-channel enhancement-mode InP/SiO_2 MISFETs were successfully fabricated on an Fe-doped semi-insulating (SI) substrate.[13,14] The SI substrates were grown in NEC Laboratories. A cross-sectional view of an InP MISFET is shown in Fig. 8.3. An n^+-InP layer, grown by vapor-phase epitaxy, was used for the source and drain regions.

The fabrication process for the enhancement-mode InP MISFETs is outlined in Fig. 8.4. Using a CVD SiO_2 film as a mask, the n^+ layer was etched off to form an n-channel region on the SI substrate (Fig. 8.4a). Source and drain electrodes were formed with Ni/Au–Ge alloyed at 390°C on the n^+ layer (Fig. 8.4b). After mild SI substrate surface etching in a mixture of H_2SO_4, H_2O_2, and H_2O with a 3 : 1 : 1 volume ratio, a CVD SiO_2 gate insulator was deposited at 350°C, as described earlier (Fig. 8.4c). A self-aligned gate electrode was made with 0.3 μm thick Al by using the SiO_2 film again as an evaporation mask (Fig. 8.4d,e). Finally, Au/Pt/Ti bonding pads for ohmic contacts were formed (Fig. 8.4f), completing the device processing.

In the fabrication process as shown in Fig. 8.4(a), a preferential etching technique to form trapezoidal shape was employed for the n^+ layer etching. This technique allowed the channel length to be shorter than the window of the SiO_2 mask. Therefore, even if the mask window was reduced by the deposition of a gate insulator, the gate electrode could cover the channel completely, resulting in good enhancement-mode operation.

A cross-sectional SEM view of the fabricated MISFET is shown in Fig. 8.5. By using the self-alignment technique for forming the gate electrode as described above, the area

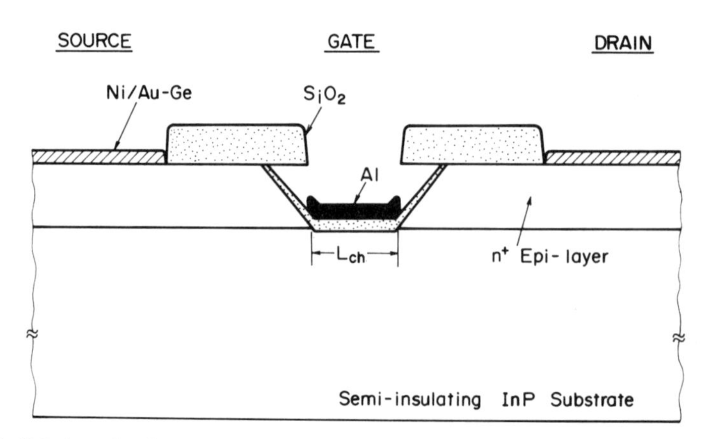

Fig. 8.3 Fabricated enhancement-mode InP MISFET structure. [after T. Itoh et al.[14]]

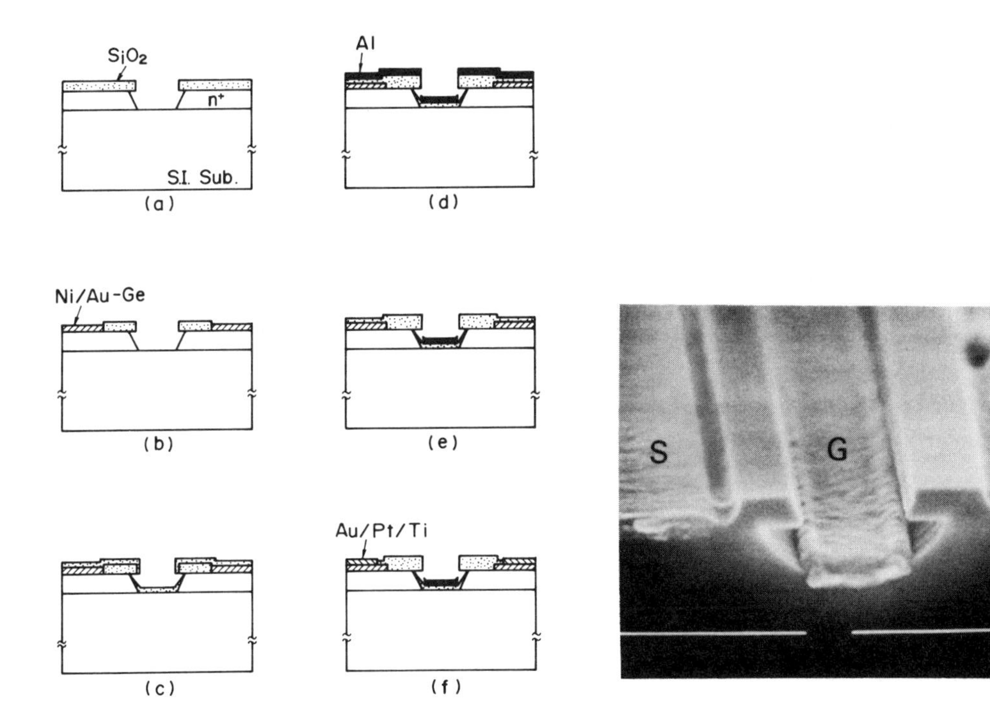

Fig. 8.4 Enhancement-mode InP MISFET fabrication process. [after T. Itoh et al.[14)]]

Fig. 8.5 Cross-sectional view of the fabricated InP MISFET. The marker indicates the 0.5 μm length. [after T. Itoh et al.[14)]]

of overlap for the gate metal and the n^+ contact layer through the SiO_2 gate insulator could be reduced, resulting in small parasitic capacitances. The channel width is 280 μm, channel length is about 0.8 μm and gate electrode length is about 1.2 μm. The gate insulator is about 650 Å thick. The device has two gate pads.

The InP MISFETs were assembled in a ceramic package and a 50 Ω microstrip carrier, and were evaluated by measuring DC characteristics and microwave performance.

8.4. DC characteristics

8.4.1 *Current–voltage characteristics*

Fig. 8.6 shows drain-current–voltage characteristics for a fabricated MISFET. The MISFETs showed good enhancement-mode characteristics with a very high transconductance. The transconductance was typically 160–170 mS/mm, and some MISFET's exhibited very high transconductance, as high as 220 mS/mm, which is about three times higher than that observed in enhancement-mode InP MISFETs reported thus far.[2,7)] This high transconductance could be achieved by reducing the channel length and the gate insulator thickness, and using a high-quality semi-insulating substrate, and is due to good MIS interfacial properties, as described in Section 8.2.

It is also due to the high electron velocity in InP. The average electron velocity in the fabricated InP MISFETs was evaluated, and the result was compared with values for some other FETs. The average electron velocity v_a in short-channel MISFET and MESFET

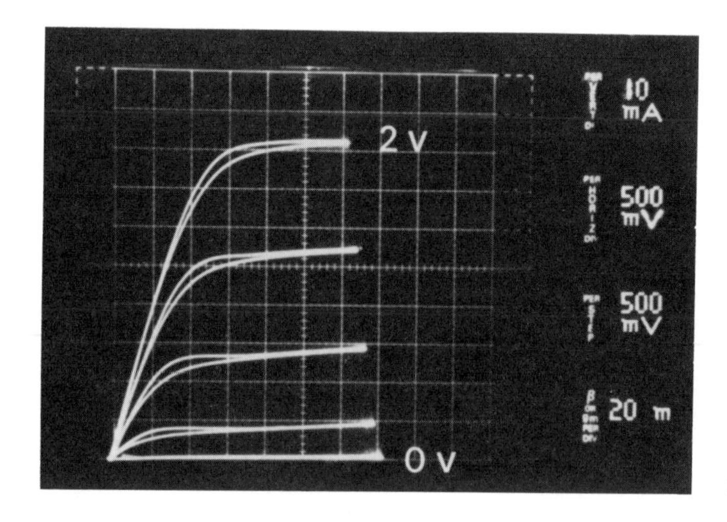

Fig. 8.6 Drain-current–voltage characteristics for an enhancement-mode InP MISFET. [after T. Itoh et al.[14)]]

structures can be evaluated by the following equations:

$$g_m = C_i v_a Z \qquad \text{(MISFET)}$$

$$g_m = Z v_a \left(\frac{\epsilon q N_D}{2(\phi_B - V_G)} \right)^{1/2} \qquad \text{(MESFET)}$$

where g_m is the transconductance in the saturation region, C_i is the gate insulator capacitance per unit area, Z is the channel width, ϵ is the permittivity of the semiconductor, q is the electronic charge, N_D is the ionized donor concentration, ϕ_B is the Schottky built-in potential, and V_G is the gate voltage. By using the above equations, average electron velocities were calculated for the present InP MISFETs and some other FETs. The results are shown in Table 8.1. For the InP MISFETs, a typical transconductance of

Table 8.1 Comparison between average electron velocities in various FETs. [after T. Itoh et al.[14)]]

Material	Device	Gate (Channel) Length $L_g(L_{ch})$ (μm)	Average Velocity V_a (10^7 cm/s)
Si	N-MOSFET	1.5 (1.3)	0.6
GaAs	D-MESFET	1.0	1.0
GaAs	D-MESFET	0.5	1.4
InP	E-MISFET	1.2 (0.8)	2.5
AlGaAs/GaAs	E-TEGFET(RT)	0.8	1.2
AlGaAs/GaAs	E-HEMT(77K)	2.0	2.5

160 mS/mm gave the average electron velocity of 2.5×10^7 cm/s, which is about twice as high as that in GaAs MESFETs[15] [$v_a \cong (1.0–1.4) \times 10^7$ cm/s], and much higher than that in Si NMOSFETs[16] ($v_a \cong 0.6 \times 10^7$ cm/s). Furthermore, this average velocity in InP enhancement-mode MISFETs is almost the same as that in modulation-doped AlGaAs/GaAs heterojunction FETs at 77 K.[17,18]

These results indicate that InP enhancement-mode MISFETs can provide very high transconductance, even at room temperature, and are very favorable for use in high-frequency and high-speed devices, including integrated circuits.

8.4.2. *Drift characteristics*

The drain current drift after the gate voltage is applied is a serious problem to be solved for InP MISFETs.[19,20] Especially, the long-term drift is unfavorable for practical devices.

Fig. 8.7 shows the long-term drain-current drift for the fabricated InP MISFETs with drain voltage of 4 V. The drain current was measured after the application of a 3 V step gate voltage. The short-term drift in a period shorter than 1 s from turning on for the present MISFET is relatively small (less than a 20% decrease in drain current compared with the real initial value). In Fig. 8.7, from a practical point of view, the drain current is normalized by the value at 1 s after turning on. Some other results reported so far are also shown.[19–21] The drift of the present InP MISFETs is relatively small and mostly the best level among reported results. However, a few tenths of a percent decrease in drain current was observed after 10^3 s from turning on, although the drain current tended to be stable later.

Although the fabricated InP MISFETs showed DC drift, as described above, the drifts in RF performance were fairly small. This is due to small changes in equivalent circuit parameters. For example, the gate capacitance is hardly affected by the DC drain current

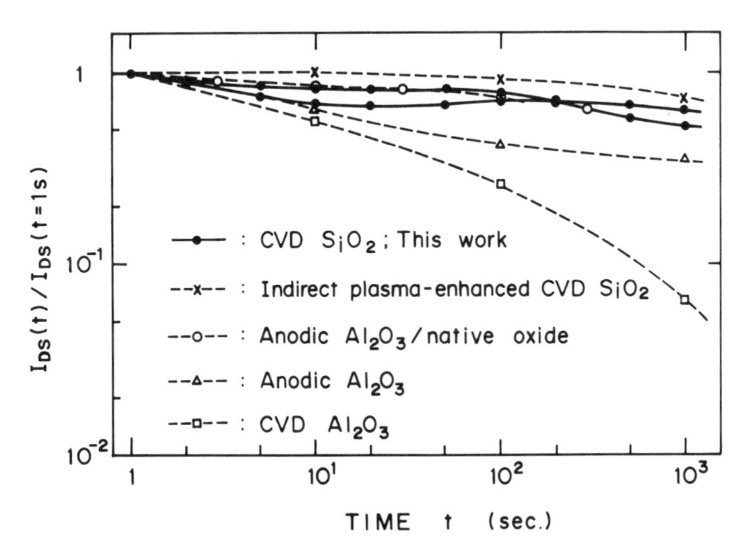

Fig. 8.7 Long-term drift characteristics for drain current after switching on the gate voltage. [after T. Itoh et al.[14]]

drift. RF parameter stability is one of the advantages of enhancement-mode MISFETs, in contrast with depletion-mode MISFETs and MESFETs, where the gate capacitance is greatly affected by the change in the gate depletion layer width.

8.5. Microwave performance

8.5.1. *Small-signal characteristics*

Power gain and noise figures for the fabricated MISFETs were measured in the 4–12 GHz frequency range.

Fig. 8.8 shows the drain-voltage dependence of the maximum power gain and the drain current at 4 GHz. The maximum power gain G_{max} increases steeply with increase in the drain voltage, and starts to saturate at relatively small drain voltage of about 4 V.

The drain-current dependence of the minimum noise figure NF_{min} and the associated gain G_a at 12 GHz is shown in Fig. 8.9. As in GaAs MESFETs, the minimum noise figure NF_{min} takes the minimum value at small drain current (in this case, at about 10 mA), and increases with increasing drain current, as shown in Fig. 8.9.

Fig. 8.10 shows the frequency dependence of the power gain obtained for the fabricated MISFETs. The results so far observed for depletion-mode InP MISFETs (broken lines) and enhancement-mode InP MISFETs (solid lines) are also shown in this figure.[4–7] The results for D-MISFETs and E-MISFET(a) were reported by the Naval Ocean Systems Center (NOSC) group,[1,2] and that for E-MISFET(b) were reported by the authors.[7]

As shown in Fig. 8.10, the microwave performance for the fabricated MISFETs were drastically improved over InP MISFETs so far reported. To the authors' knowledge, good X-band operation has been achieved for the first time in MISFET structures.

The maximum gains obtained at 6.5, 8, and 11.5 GHz were 11.4, 9.8, and 7.2 dB,

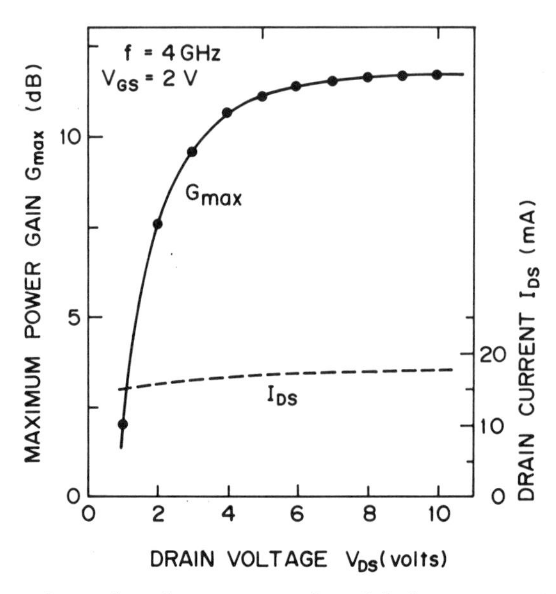

Fig. 8.8 Drain-voltage dependence of maximum power gain and drain current at 4 GHz. [after T. Itoh et al.[14]]

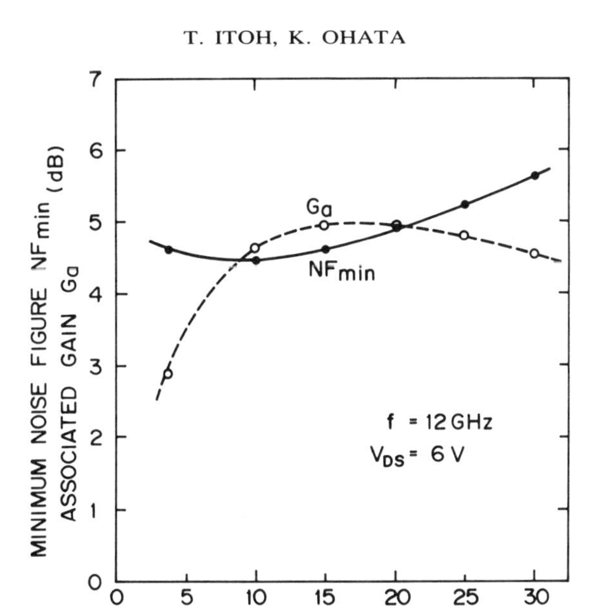

Fig. 8.9 Drain-current dependence of minimum noise figure and associated gain at 12 GHz. [after T. Itoh et al.[14]]

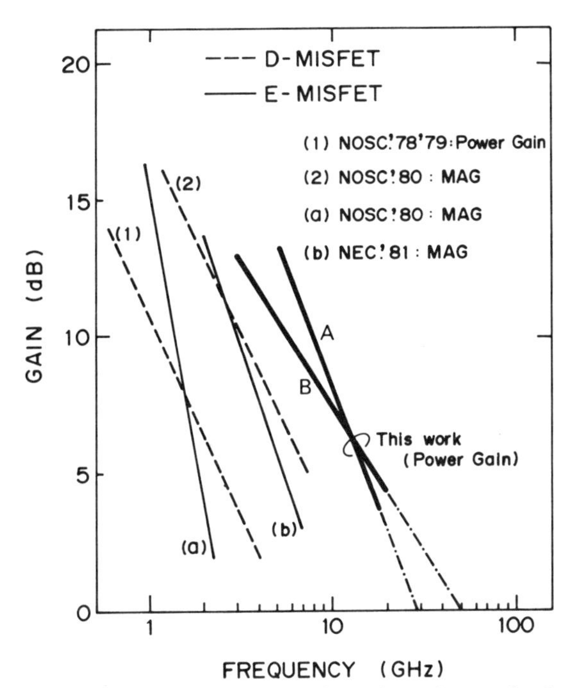

Fig. 8.10 Frequency dependence of power gain. Solid lines show the results for enhancement-mode InP-MISFETs and broken lines show the results for depletion-mode InP MISFETs. [after T. Itoh et al.[14]]

respectively. The present MISFETs typically exhibited 6 dB/oct reduction in power gain with frequency, as denoted by A in Fig. 8.10. However, a more moderate power reduction with frequency, denoted by B, was observed in some MISFETs which gave an extrapolated cut-off frequency of as high as 50 GHz.

The minimum noise figure NF_{min} and associated gain G_a obtained at 4, 8, and 12 GHz

Table 8.2 Summary of minimum noise figures and associated gains at 4, 8, and 12 GHz. [after T. Itoh et al.[14]]

	4 GHz	8 GHz	12 GHz
NF_{min}	1.87 dB	3.2 dB	4.4 dB
G_a	10.0 dB	7.5 dB	4.7 dB

are summarized in Table 8.2. In the previous work,[7] 4.22 dB NF_{min} with 4.0 dB G_a was obtained at 4 GHz. However, for the present MISFETs, almost the same performance was attained at 12 GHz. At 4 GHz, the MISFET gave 1.87 dB NF_{min} with 10.0 dB G_a, which compares favorably with that for flat-type 1 μm gate GaAs MESFETs.

8.5.2. *Power-output capability*

InP MISFETs would be suitable for high-power devices in respect to applicable drain and gate voltage and material properties, as discussed before. Hence, the large-signal

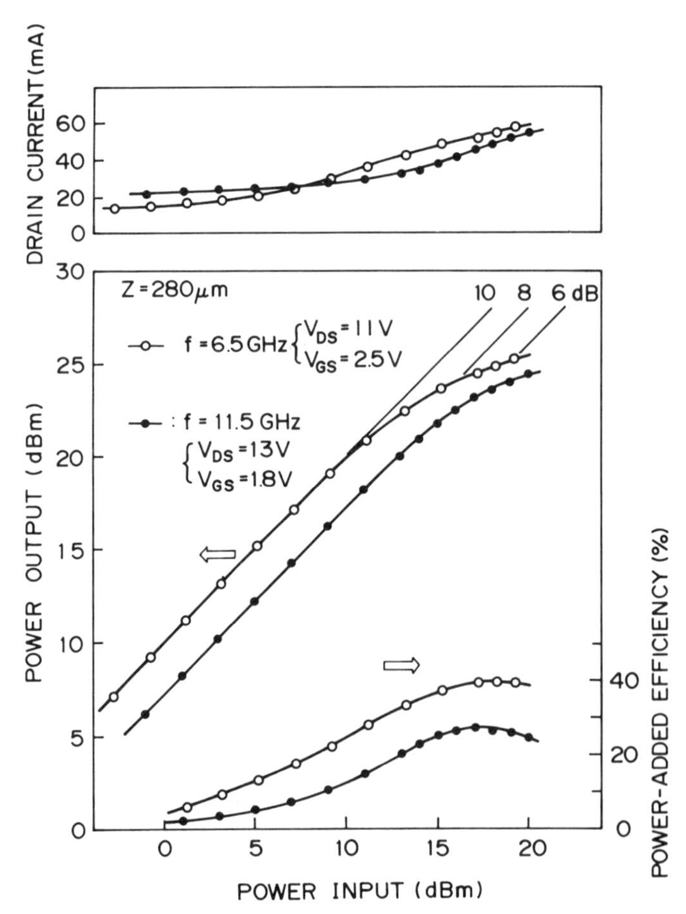

Fig. 8.11 Power input-to-output responses at 6.5 GHz and 11.5 GHz, respectively. [after T. Itoh et al.[14]]

microwave performance for the fabricated InP MISFETs was evaluated. Fig. 8.11 shows the power input-to-output responses at 6.5 and 11.5 GHz. The MISFETs exhibited very good linearity in the input-to-output responses at 6.5 and 11.5 GHz. The MISFETs exhibited very good linearity in the input-to-output power relations with high power-added efficiencies. It should be noted that very large output power per unit gate width was achieved. The maximum power outputs were 327 mW (1.17 W/mm) at 6.5 GHz and 280 mV (1.0 W/mm) at 11.5 GHz. Fig. 8.12 shows the drain-voltage dependence of large signal performance at 6.5 GHz, where the power output with 6 dB gain: P_o (6 dB gain), the power output at 1 dB gain compression point: P_o (1 dB comp.), the small-signal gain G_{ss}, linear gain G_L, and power-added efficiency at 1 dB gain compression point: η_{add} (1 dB comp.) are shown. Output powers increased markedly with an increase in the drain voltage. The power gains increased slightly with larger drain bias. On the other hand, the optimum drain voltage for maximum power-added efficiencies was about 9 V.

The high-power performances of the fabricated InP MISFETs at 6.5, 8, and 11.5 GHz are summarized in Table 8.3. Power outputs of more than 1 W per mm gate width were attained through X-band, which are superior to the values of 0.5–0.8 W/mm for GaAs MESFETs[22] and 0.3 W/mm for GaAs MOSFETs.[23] The power-added efficiencies were also considerably high, and η_{add} of 43.5% was observed at 6.5 GHz. The output power per unit gate width below 8 GHz is mostly the same as that reported for the InP D-MESFET.[24] However, the efficiency for the fabricated MISFETs is much higher. It is also noted that the gate current was not observed, even at high input power level. This is possibly due to the insulated gate structure.

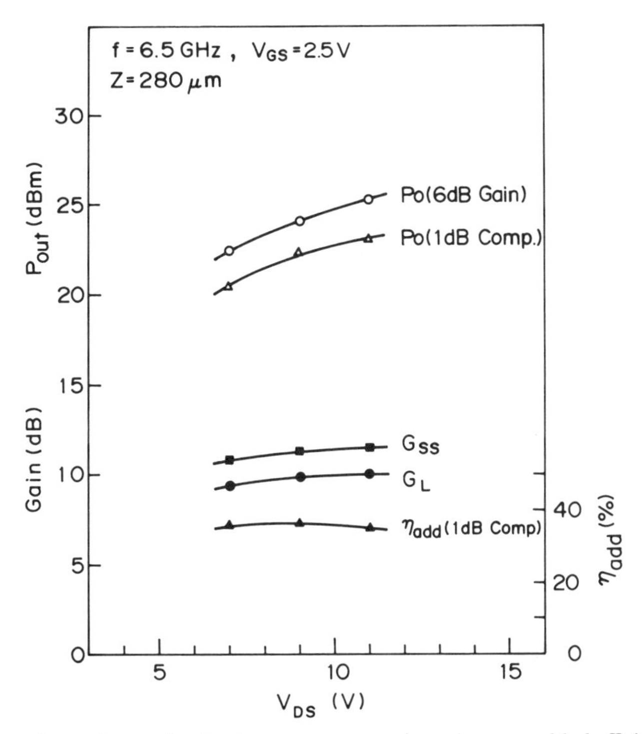

Fig. 8.12 Drain-voltage dependence of output power, power gain and power-added efficiency at 6.5 GHz. [after T. Itoh et al.[14]]

Table 8.3. InP MISFET output-power performance at 6.5, 8, and 11.5 GHz. [after T. Itoh et al.[14)]]

	6.5 GHz	8 GHz	11.5 GHz
G_L (dB)	10.0	8.0	6.8
P_{out} (mW)	327 (1.17 W/mm)	295 (1.05 W/mm)	280 (1.0 W/mm)
η_{add} (%)	38.9	40.7	26
G_a (dB)	6.0	5.2	4.5
$\eta_{add\ max}$ (%)	43.5	42.7	30.2
$G_{ss\ max}$ (dB)	11.4	9.8	7.2

These power output characteristics demonstrate the excellent power output capability of the InP MISFETs. Furthermore, the MISFETs fabricated here were not designed for power devices and the substrate was thick (~200 μm), so further improved performance can be expected. The InP MISFETs are expected to have application in high-efficiency, high-power devices.

8.6. Conclusion

CVD–SiO_2/InP MIS diodes and enhancement-mode InP MISFETs were fabricated and evaluated to clarify the feasibility of their high-speed applications. MIS diodes on n-InP substrates were investigated to find the optimum condition for a good MIS system by varying the deposition temperature of SiO_2 films. The MIS devices deposited at 340–360°C exhibited the best characteristics as well as very sharp interface transition from SiO_2 to InP. They showed no frequency dispersion in the *C–V* curves at the accumulation side and had minimum surface density of states near the conduction band edge. These characteristics are suitable for enhancement-mode operation.

N-channel enhancement-mode InP MISFET's with short channel length were successfully fabricated on an Fe-doped semi-insulating substrate by using a self-alignment technique. Very high transconductance, as high as 220 mS/mm, was obtained for MISFETs with a CVD SiO_2 gate insulator about 650 Å thick. They exhibited successful X-band operation and higher power output capability than that for GaAs MESFETs.

Further advanced performance characteristics will be achieved by device structure optimization, and by the improvement in the MIS interfacial properties.

Acknowledgements

The authors wish to thank Hisao Watanabe for supplying semi-insulating InP substrates, and Akira Usui and Dr. Takashi Mizutani for epitaxial growth of n^+-InP layers. They also wish to thank Yoichi Aono for helpful suggestions regarding RF large signal measurements. Thanks are also due to Dr. Yoichiro Takayama, Dr. Hisatsune Watanabe and Masaki Ogawa for valuable suggestions and discussions, and to Dr. Hidehiko Katoh and Dr. Hiroki Muta for their continuous support during this work and their warm encouragement.

References

1) D. L. Lile, D. A. Collins, L. G. Meiners, and L. Messick: Electron. Lett., 14 (1978) 657.
2) T. Kawakami and M. Okamura: Electron. Lett., 15 (1979) 502.
3) T. Kawakami and M. Okamura: Electron. Lett., 15 (1979) 743.
4) L. Messick, D. L. Lile, and A. R. Clawson: Appl. Phys. Lett., 32 (1978) 494.
5) L. Messick: Solid State Electron., 23 (1980) 551.
6) D. L. Lile, D. A. Collins, L. G. Meiners, and M. J. Taylor: Gallium Arsenide and Related Compounds 1980 (Inst. Phys. Conf. Ser. No. 56, 1981) p. 493.
7) K. Ohata, T. Itoh, H. Watanabe, T. Mizutani, and Y. Takayama: Gallium Arsenide and Related Compounds 1981 (Inst. Phys. Conf. Ser. No. 63, 1982) p. 353.
8) L. Messick: IEEE Trans. Electron Devices, ED-28 (1981) 218.
9) D. L. Lile and D. A. Collins: IEEE Trans. Electron Devices, ED-29 (1982) 842.
10) J. Stannard: J. Vac. Sci. Technol., 16 (1979) 1462.
11) C. W. Wilmsen, J. F. Wagner, and J. Stannard: Insulating Films on Semiconductors 1979 (Inst. Phys. Conf. Ser. No. 50, 1980) p. 251.
12) L. M. Terman: Solid State Electron., 5 (1962) 285.
13) T. Itoh and K. Ohata: 1982 Int. Electron Devices Meeting, Dig. Tech. Papers (1982) p. 774.
14) T. Itoh and K. Ohata: IEEE Trans. Electron Devices, ED-30 (1983) 811.
15) K. Ohata, H. Itoh, F. Hasagawa, and Y. Fujiki: IEEE Trans. Electron Devices, ED-27 (1980) 1079.
16) M. Morimoto, M. Sugimoto, K. Terada, K. Takahashi, T. Ishijima, H. Muta, and S. Suzuki: Japan. J. Appl. Phys., Suppl. 20-1 (1981) 123.
17) D. Delagabeaudeuf, M. Laviron, P. Delescluse, P. N. Tung, J. Chaplart, and N. T. Linh: Electron Lett., 18 (1982) 103.
18) T. Mimura, S. Hiyamizu, K. Joshin, and K. Hikosaka: Japan. J. Appl. Phys., 20 (1981) L317.
19) M. Okamura and T. Kobayashi: Japan. J. Appl. Phys. 19 (1980) 2143.
20) D. Lile, M. Taylor, and L. Meiners: presented at the 1982 Int. Conf. Solid State Devices (Tokyo).
21) T. Sawada and H. Hasegawa: Electron. Lett., 18 (1982) 742.
22) S. H. Wemple, P. G. Flahive, C. L. Allyn, W. O. Schlosser and D. E. Iglesias: Gallium Arsenide and Related Compounds 1981 (Inst. Phys. Conf. Ser. No. 63, 1982) p. 437.
23) T. Mimura, K. Odani, N. Yokoyama, Y. Nakamura, and M. Fukuta: IEEE Trans. Electron Devices, ED-25 (1978) 573.
24) M. Armand, J. Chevieer, and N. T. Linh: Electron. Lett., 16 (1980) 906.

9 AlGaAs TRANSVERSE JUNCTION STRIPE LASERS

Hirofumi NAMIZAKI† and Saburo TAKAMIYA*

Abstract

Transverse junction stripe (TJS) lasers have a unique current-confining and mode-controlling scheme. Since the first realization of single-mode oscillation, many improvements have been made, especially TJS on semi-insulating substrate and CRANK structure. Superior characteristics such as low threshold current, fundamental transverse and single longitudinal mode oscillation, high-speed modulation capability and high reliability have been realized.

Keywords: Semiconductor Laser, AlGaAs Laser, TJS Structure

9.1. Introduction

Since the realization of CW operation of semiconductor lasers at room temperature in 1970,[1] mode control and lifetime of the devices have been major problems to be overcome. The transverse junction stripe (TJS) laser[2] was the first laser structure in which the transverse mode is fully controlled into the fundamental mode by a refractive index guide, resulting in a single longitudinal mode oscillation.[3,4] Today, many single-mode laser structures have been developed such as the channeled substrate planar (CSP) structure,[5] the constricted double-heterojunction (CDH) structure,[6] the terraced substrate (TS) structure,[7] the buried heterostructure with buried optical guide (BH-BOG),[8] the V-channeled substrate inner stripe (VSIS) structure,[9] etc. TJS lasers have the advantages of reproducibly obtained fine mode stability and high reliability and are characterized by low threshold current, stable single mode oscillation, high-speed modulation capability and long operation life, so they are used in a variety of application fields.

This chapter describes the basic scheme of the TJS structure, improvements such as TJS on semi-insulating GaAs substrate and CRANK structure, and state of-the-art characteristics and operation life of the devices.

9.2. Basic structure

The basic structure of the TJS laser is shown in Fig. 9.1. A p–n junction traverses closely separated double-heterojunctions. The structure is made by multi-layer epitaxial growth and selective impurity diffusion. The bandgap energy of the center layer, the active layer, E_{g1}, is smaller than that of the outer cladding layers, E_{g2}. The p–n junction consists of two kinds of homojunction, one of narrow-bandgap material in the active layer and the other of wider-bandgap material outside. The diffusion potential of a p–n

† LSI Research and Development Laboratory, Mitsubishi Electric Corporation, 4-1, Mizuhara Itami 664.
* Kitaitami Works, Mitsubishi Electric Corporation, 4-1, Mizuhara Itami 664.

JARECT Vol. 19, *Semiconductor Technologies* (1986), *J. Nishizawa (ed.)*
OHMSHA, LTD. and North-Holland

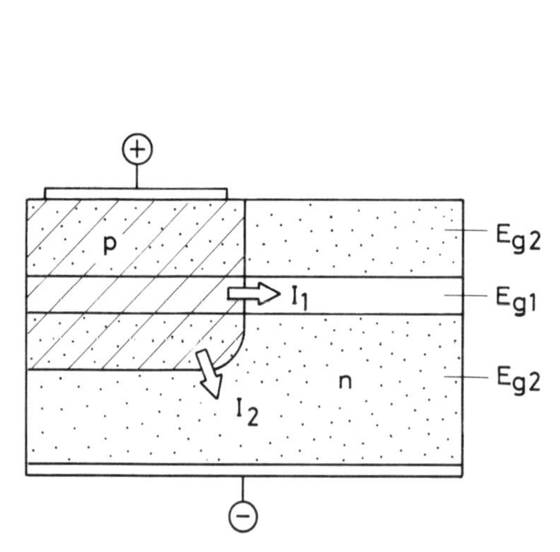

Fig. 9.1 Basic structure of the TJS laser.

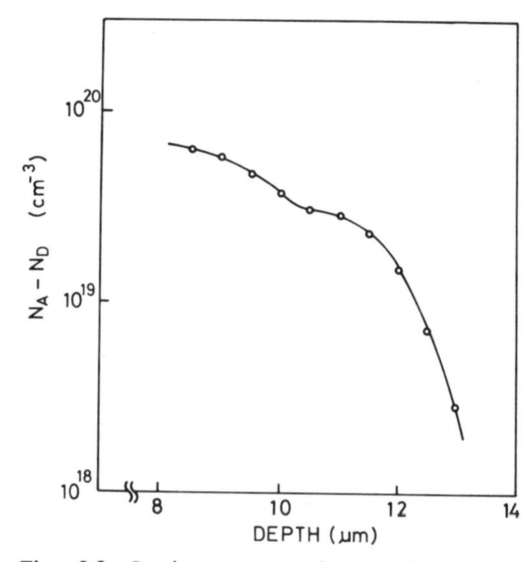

Fig. 9.2 Carrier concentration profile in Zn-diffused GaAs.

junction is roughly proportional to the bandgap energy of the material, and current tends to flow through a p–n junction with lower diffusion potential. Thus, current is concentrated into the p–n junction in the active layer. Current densities can be estimated from the following relation with subscripts $i = 1$ and 2 corresponding to the p–n junction in the active layer and that outside, respectively:[10,3)]

$$\Delta E_g - (kT/q)\ln(J_1/J_2) = J_2R_2 - J_1R_1, \tag{9.1}$$

where ΔE_g is the bandgap energy difference, k is the Boltzmann constant, q is the electronic charge and T is the absolute temperature. R_1 and R_2 are the reduced resistances between electrodes for unit area of the p–n junctions. The total diode current is expressed as

$$I = J_1S_1 + J_2S_2, \tag{9.2}$$

where S_1 and S_2 are the areas of the two p–n junctions. To minimize the ineffective current for lasing, J_2 should be far smaller than J_1, because S_1 is usually of the order of 1/100 of S_2. J_1/J_2 is temperature sensitive as is seen from Eq. (9.1), and above a critical temperature this ratio decreases drastically. To assure stable operation at high temperature, ΔE_g should be sufficiently large and R_1 should be small enough compared with R_2. For active layer material, GaAs or AlGaAs with small Al content is used and AlGaAs with relatively high content of Al is used for the cladding layers. The bandgap energy of AlGaAs is related to the Al content x by[11)]

$$\begin{aligned} E_g &= 1.424 + 1.247x && (0 \le x < 0.45) \\ &= 1.900 + 0.125x + 0.143x^2 && (0.45 \le x \le 1). \end{aligned} \tag{9.3}$$

When the difference in Al content Δx is 0.3, the energy difference ΔE_g can be 0.37 eV. However, when the X indirect gap becomes smaller than the direct gap, that is $x_2 > 0.45$, ΔE_g becomes smaller for the same value of Δx.

Electrons injected through a p–n junction into the active layer recombine with holes radiatively and laser action occurs along the p–n junction plane. To get a low threshold current, it is important to achieve effective confinement of carriers and optical field and high interaction between them. The TJS laser has successfully realized these using a double heterostructure in the direction perpendicular to the grown layers and homojunction in the direction parallel to them. More importantly, the transverse mode is controlled by the p–n junction utilizing the refractive index guide formed by the impurity distribution in the direction parallel to the layers. Although the precise index distribution in the actual structure is not clear, a rough estimate of the distribution can be deduced by examining the variation of the carrier concentration in flat Zn-diffused GaAs samples. An example of such distribution is shown in Fig. 9.2. The refractive index could be obtained from the following empirical relation when the net carrier concentration is around $1 \times 10^{19}\,\mathrm{cm}^{-3}$:[12]

$$n \simeq 3.6 - 1.7 \times 10^{-21}(N_\mathrm{A} - N_\mathrm{D}). \tag{9.4}$$

Taking into account the change of index with injected carriers, a guide of p^+–p–n structure is supposedly formed in which the width is about 2 μm and index difference is $(2–5) \times 10^{-3}$.

9.3. Device fabrication and improvements

In the early stage of development, the TJS structure was made on a n-type GaAs substrate.[2–4] Low threshold current less than 50 mA and single mode operation were achieved with this structure. However, the temperature dependence above room temperature was rather large. Many attempts were made to reduce the ineffective current which flows through the p–n junctions in the cladding layers. Additional p–n junctions were introduced to block the leakage current using p-type substrate,[13] or buried p-type region to make the leakage current path narrower.[14] Finally, semi-insulating GaAs substrate is used so that the area of the homojunctions in the cladding layers is minimized.[15–17]

The actual structure with semi-insulating GaAs substrate is shown in Fig. 9.3; it is fabricated as follows. LPE is used for multilayer epitaxial growth of Te-doped GaAs/Al-

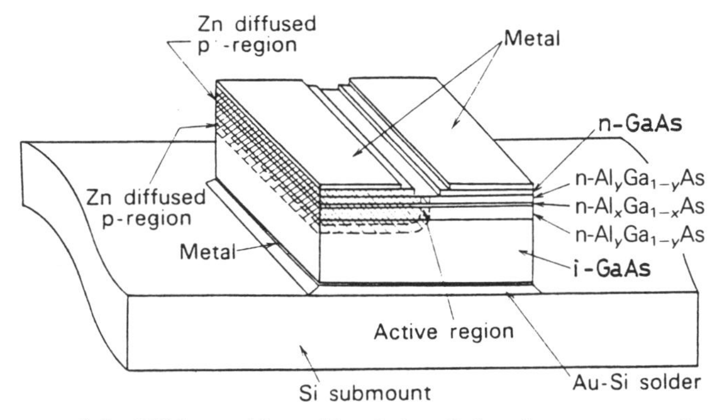

Fig. 9.3 Actual structure of the TJS laser with semi-insulating GaAs substrate mounted on Si submount. [after S. Takamiya[43]]

GaAs. The thickness of the active layer is about 0.1 μm, and the Al content in the active layer is 0 to 0.2 depending on the desired lasing wavelength. The doping concentration in the active layer is $(2–5) \times 10^{18}$ cm^{-3}. Two-step Zn diffusion is done selectively, to form the impurity distribution, with Si_3N_4 films as a diffusion mask. The width of the active region is automatically controlled by diffusion of Zn and reproducible characteristics can be obtained provided a good epitaxial wafer is prepared. After evaporating metals for ohmic contacts, isolation between n- and p-type regions is achieved by etching the metal and the top GaAs contact layer. The cavity length is 200–300 μm and the thickness of the chip is about 100 μm. The mirror facets are coated with Si_3N_4[18)] or Al_2O_3[19)] films so as to prevent facet damage. The chips are bonded on a Si heatsink using Au–Si solder in a junction-up configuration. Elimination of soft solders such as In gives high reliability.[20)] The thickness of the Si heatsink is optimized to be 150 μm so as to minimize the strain inside the laser chip.[21,22)]

Further improvement, with respect to high-power operation and high reliability, has been achieved by introducing the CRANK structure[23,24)] in which mirror damage is eliminated. The CRANK structure, which is shown in Fig. 9.4, is similarly fabricated. The difference from the conventional structure is that Zn is diffused in the shape of a crank and the two sides of the straight lasing region remain undiffused. The cavity length is 300 μm and the length of the crank web is 10–15 μm. It is well known that AlGaAs lasers have a problem of facet degradation. The intense optical flux causes catastrophic mirror damage when driven at high output power. Even with low power density, optically accelerated facet oxidation occurs in long-term CW operation. These degradations are due to absorption of light and probably to carrier recombination near the mirror facets. In the CRANK structure the laser beam guided by the p^+–p–n waveguide at the center portion comes out into the undiffused n-type region where no carrier recombination occurs because the length of the n-type region is much longer than the carrier diffusion length. The n-type region has a wider bandgap than that of the heavily Zn-diffused p-type active region. So, the laser beam suffers little loss due to absorption in the n-type region. For these reasons, stability of the mirror facets has been much improved in the CRANK structure compared with the conventional straight lasers.

Lee et al.[25)] showed the possibility of integration of the TJS laser with a Gunn device

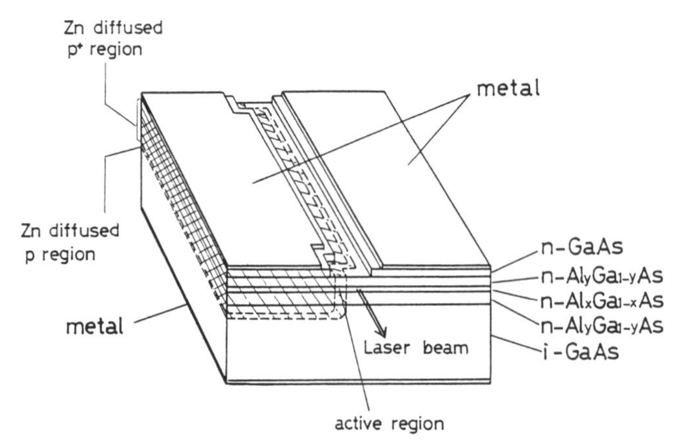

Fig. 9.4 CRANK TJS structure. [after H. Kumabe et al.[24)]]

and FET monolithically on a semi-insulating GaAs substrate. Tsang[26] utilized the structure for multiple wavelength laser sources with multiple active layers with different Al contents. A DFB-TJS laser has been realized by Kawanishi et al.[27] These experiments have shown the feasibility of the structure used in the integrated optics. Although at an experimental stage, MBE[28,29] and, very recently, MOCVD[30] have been successfully used to fabricate TJS wafers. In these cases, Sn, Si or Se was used instead of Te as the n-type dopant. The lasers made by MBE and MOCVD have been shown to have similar characteristics as those made by LPE.

9.4. Characteristics

The typical threshold current at room temperature is about 20 mA and the external quantum efficiency is 20–30% per facet. The standard light output for straight TJS lasers is 3 mW/facet, and 5 to 15 mW/facet for CRANK TJS lasers depending on the field of application. The threshold current density is 40–100 kA/cm^2. The relatively high threshold current density is due to the fact that in TJS lasers the current is injected through the narrower side of the active volume. However, the carrier density at threshold is not so different from that in the usual double-heterostructure lasers. The threshold current density for lasers with $L = 300\ \mu$m is expressed by the approximate empirical relation:

$$J_{th} = 10(1/\Gamma) + 28\,(\text{kA/cm}^2), \tag{9.5}$$

where Γ is the optical confinement factor. With the experimentally obtained values of internal loss $\alpha = 25\ \text{cm}^{-1}$ and carrier lifetime at threshold $\tau = 0.7$ ns, one can find an approximate relation between the gain g and average carrier density N as

$$g = 2.8 \times 10^{-16} N - 170\,(\text{cm}^{-1}). \tag{9.6}$$

The calculated dependences of I_{th}, J_{th} and N_{th} on the thickness of the active region are shown in Fig. 9.5. The CRANK structure has a slightly larger value of J_{th} due to the

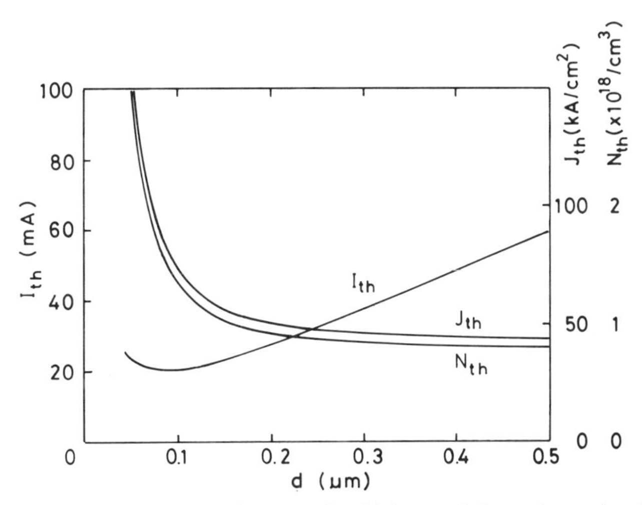

Fig. 9.5 Calculated dependence of I_{th}, J_{th} and N_{th} on the thickness of the active region for straight TJS lasers with $L = 300\ \mu$m and $\Delta x = 0.3$.

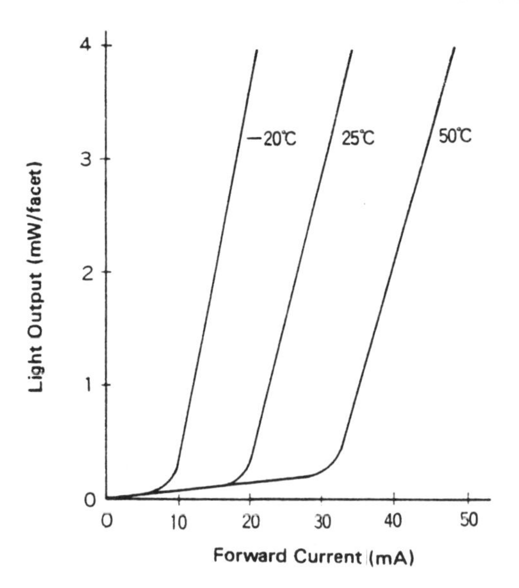

Fig. 9.6 Typical light output versus current characteristics.

reduction of the effective mirror reflectivity, but the increase is less than 20% when the length of the crank web is less than 15 μm.[24] The threshold current varies with temperature in proportion to T^3 from room temperature down to liquid nitrogen temperature. Around room temperature, this dependence can be transformed to the familiar relation $\exp(T/T_0)$ with $T_0 \simeq 100$ K. Fig. 9.6 shows a typical light-output–current relation at various temperatures.

The fundamental transverse mode is usually obtained and is maintained within CW operation limits. Near-field and far-field patterns are shown in Fig. 9.7. The full width at half maximum of the near-field in the direction parallel to the layers is about or slightly less than 2 μm, and those of the far-field are typically 12 and 35 degrees in the directions parallel and perpendicular to the layers, respectively. Although undulations are sometimes seen on the n-type side of the far-field patterns, higher-order transverse modes are scarcely observed in thousands of lasers, which can be attributed to the stability and reproducibility of index guide produced by the Zn diffusion. Straight TJS lasers have little astigmatism, while the CRANK TJS lasers have astigmatism of about 3–4 μm, which is in good agreement with the theoretically expected value for lasers with 10–15 μm crank webs.

Today, it is widely accepted that well controlled index guide lasers tend to lase in a single longitudinal mode in contrast to multimode operation in narrow stripe gain guide lasers. The TJS laser with the index guide, lases in a single longitudinal mode as is shown in Fig. 9.8. Although gain suppression is predicted by theory in some conditions[31,32] and is observed in many laser structures, such as in the CSP laser,[33] it is less apparent in the TJS lasers. Basically, the spectral behavior of the TJS lasers can be understood in terms of homogeneous broadening which may be due to the heavily doped active region.[34] Around threshold, several longitudinal modes are seen in the spectrum. At the output power of above 1 mW, the spectrum becomes single as the main mode power increases, while the powers in adjacent modes stay almost unchanged as shown in Fig. 9.9.[35] Stable single-mode operation has been confirmed up to the maximum CW output power, up to

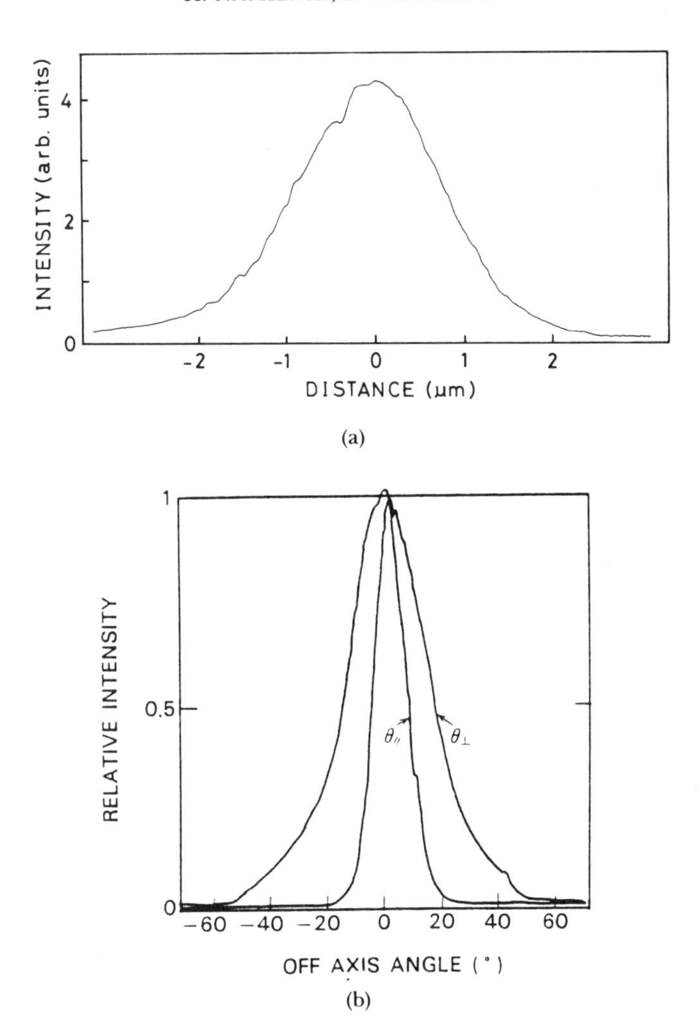

Fig. 9.7 (a) Near-field distribution in the direction parallel to the layers. (b) Far-field distributions.

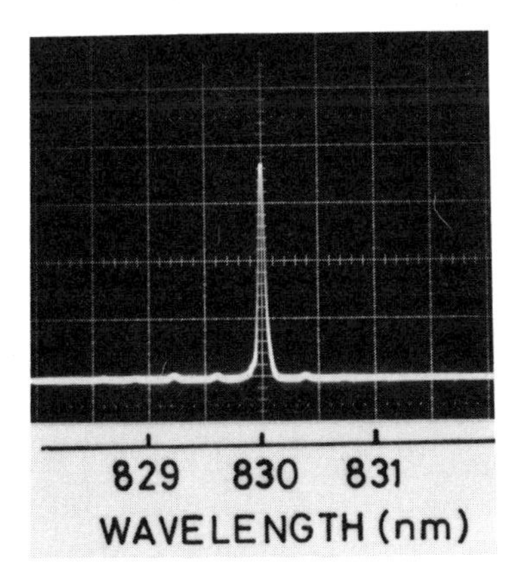

Fig. 9.8 Typical lasing spectrum of the TJS laser.

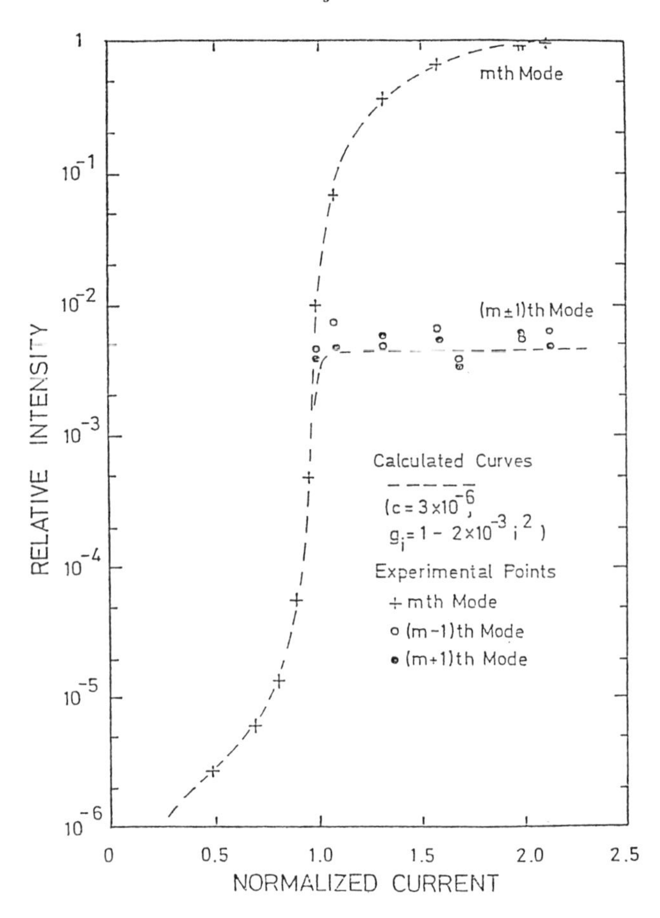

Fig. 9.9 Longitudinal mode intensity versus normalized input current. [after W. Susaki et al.[35]]

100°C[18] and after more than 10,000 h of CW operation.[36] The line width was carefully measured by Werford and Mooradian[37] to be inversely proportional to the output power and is a few tens of MHz at 3 mW/facet.

The lasing wavelength can be varied with the Al content in the active region. Because of the heavy impurity doping, the lasing energy is smaller than the bandgap energy of the material. The difference between these is experimentally estimated to be about 58 meV.[38] This enables a longer wavelength than that corresponding to the bandgap energy of GaAs, and the TJS laser covers the entire range of the 0.8 μm band. Visible lasers are also fabricated with this structure. Reliable 780 nm lasers have been realized, and 760 nm lasers have been obtained experimentally without much increase of threshold current.[39]

The TJS laser has a wide modulation bandwidth,[40] partly because its resonance frequency is high due to the short spontaneous carrier lifetime and partly because the parasitic capacitance is small due to usage of semi-insulating substrate. An example of frequency response is shown in Fig. 9.10. When the bias current is 1.3 times the threshold current, a flat response over 2 GHz is obtained. The resonance frequencies can be explained fairly well by the following equation with photon lifetime τ_p of about 1.5 ps:

$$f_r = (1/2\pi)(1/\tau_p\tau_s)^{1/2}(I/I_{th} - 1)^{1/2}. \tag{9.7}$$

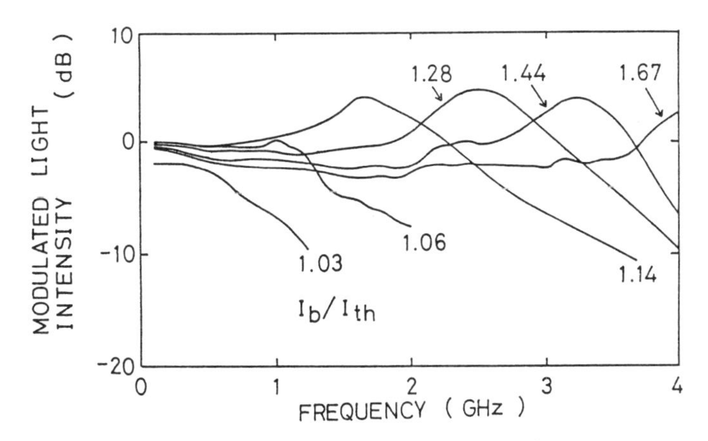

Fig. 9.10 Frequency response of a TJS laser. [after W. Susaki et al.[40]]

The laser has a response time of less than 150 ps, and pulse modulation over 1 Gbit/s is possible. Tell and Eng[41] reported 8 Gbit/s PCM direct modulation with this laser.

9.5. Reliability

The operating life has been drastically improved during the last decade, from less than 1000 h at the initial stage to an estimated 10^6 h at present. Technical contributions are high-quality, especially oxygen-free, crystal growth,[42] elimination of soft solders, stress compensation inside the laser chips and facet passivation with dielectric films such as Si_3N_4 or Al_2O_3.

To estimate such a long life as 10^6 h, life tests at elevated temperatures are commonly used in order to evaluate the activation energy of degradation. Fig. 9.11 shows an example of such a life test, in which diodes were operated with constant light output power of 3 mW/facet at case temperatures of 50, 60, 70, 80 and 90°C. The mean time to failure on

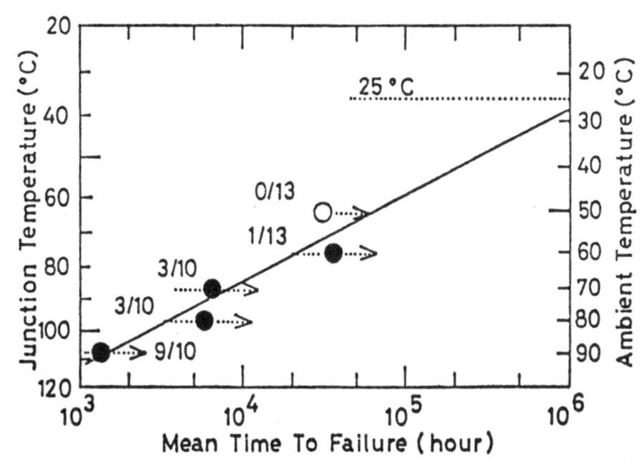

Fig. 9.11 Temperature dependence of mean time to failure. [after S. Nita et al.[25]]

the abcissa is defined as the sum of the operation time divided by the number of failed devices. The junction temperatures were estimated from the case temperature, power consumption and the thermal resistance. The activation energy of the TJS laser can be obtained from the figure to be about 1.0 eV, and the estimated lifetime at room temperature is more than 10^6 h.[20)]

The failure rate of the TJS lasers was estimated with a total of 180 devices, which were operated at temperatures of 50 to 70°C for 2000 to 4000 hours with light output of 3 mW/facet. Using the activation energy estimated above, component-hours at 50°C is calculated to be 2053×10^3. The number of failed samples, in which the operation current exceeds 1.5 times the initial value, was only one out of 180 devices. Supposing a random failure, the failure rate can be estimated to be around 500 Fit. This means that, when an operating condition of 3 mW/facet, 50°C and 1000 h is considered, the failure rate is about 0.05%.[43)]

In spite of facet coating, a gradual degradation mode due to facet oxidation was not completely eliminated in conventional short-wavelength lasers. Usually, operating currents gradually increased at a rate of 1 to a few %/kh. The CRANK TJS lasers eliminated the gradual increase completely.[44)] Fig. 9.12 shows the result of an operating life test of CRANK TJS lasers. Diodes were operated at 50°C with output powers of 5 mW/facet and 15 mW/facet, respectively. The gradual increase can hardly be seen and very stable operation is confirmed for more than 9000 and 5000 h, respectively. This can be attributed to the facet stability of the CRANK structure described above.

Surge endurance is also improve by the CRANK structure. The short-wavelength lasers were very fragile for current surge due to catastrophic mirror damage. Diodes were often given mortal damage by surge currents through careless treatment or troubles in test equipment. By eliminating the absorption at the mirror facets, the CRANK structure has greatly improved the surge endurance.[44)] Fig. 9.13 shows typical light output versus current relations of the CRANK TJS laser in various pulsed operation conditions and also that of a straight TJS laser in the case of a 1 μs pulse width for comparison. In straight

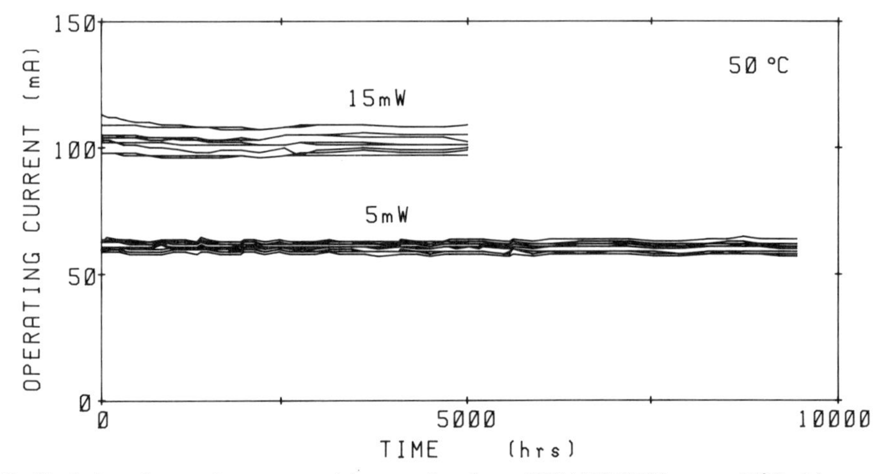

Fig. 9.12 Variation of operation current with operation time of CRANK TJS lasers at 50°C with output power of 5 mW/facet and 15 mW/facet.

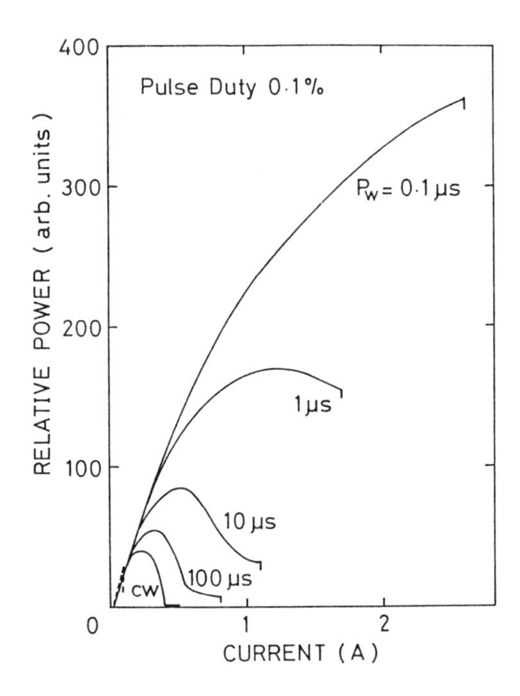

Fig. 9.13 Typical light output versus current relation of the CRANK TJS laser under CW and pulsed conditions. For comparison, that of a straight TJS laser driven under 1 μs wide pulses is shown with a dotted line. [after Y. Kadota et al.[44]]

TJS lasers, degradation occurs at the maximum output which is determined by the catastrophic mirror damage. When the pulse width is 1 μs, this power level is about 30 mW, corresponding to a power density of about 3×10^{6} W/cm^{2}. The CRANK TJS lasers, which eliminate mirror absorption, show no mirror damage and endure much higher power levels. More importantly, the current levels at which the CRANK lasers degrade are nearly 10 times as high as those of straight TJS lasers. This ensures reliable operation further in actual usage.

9.6. Conclusion

The transverse junction stripe laser occupies a peculiar position among laser structures. Unlike the case in most stripe lasers, carriers are injected laterally to the double-heterojunctions into the active region whose width is usually much larger than its thickness. The structure can be also considered as a very thin p–n homojunction laser sandwiched between cladding layers. This is the reason why some of the characteristics resemble those of homojunction lasers, while most of the characteristics are apparently shared with the usual double heterostructure lasers. Reproducible production is due to the fine controllability of the active region made by the Zn diffusion. Superior characteristics such as low threshold current, stable single-mode oscillation, high-speed modulation capability and long life enable the lasers to be widely used in many application fields.

References

1) I. Hayashi, M. B. Panishi, P. W. Foy, and S. Sumski: Appl. Phys. Lett., 17 (1970) 109.
2) H. Namizaki, H. Kan, M. Ishii, and A. Ito: J. Appl. Phys., 45 (1974) 2785.
3) H. Namizaki: IEEE J. Quantum Electron., QE-11 (1975) 427.

4) H. Namizaki: Trans. IECE Japan, E59 (1976) 8.
5) K. Aiki, M. Nakamura, T. Kuroda, J. Umeda, R. Ito, N. Chinone, and M. Maeda: IEEE J. Quantum Electron., QE-14 (1978) 89.
6) D. Botez: Appl. Phys. Lett., 33 (1978) 872.
7) T. Sugino, M. Wada, H. Shimizu, K. Itoh, and I. Teramoto: Appl. Phys. Lett., 34 (1979) 270.
8) N. Chinone, K. Saito, R. Ito, K. Aiki, and N. Shige: Appl. Phys. Lett., 35 (1979) 513.
9) S. Yamamoto, H. Hayashi, S. Yano, T. Sakurai, and T. Hijikata: Appl. Phys. Lett., 40 (1982) 372.
10) H. Namizaki, H. Kan, M. Ishii, A. Ito, and W. Susaki: Japan. J. Appl. Phys., 13 (1974) 1618.
11) H. C. Casey, Jr. and M. B. Panish: Heterostructure Lasers, Part A (Academic Press, New York, 1978) p. 193.
12) D. D. Sell, H. C. Casey, Jr., and K. W. Wecht: J. Appl. Phys., 45 (1974) 2650.
13) W. Susaki, T. Tanaka, H. Kan, and M. Ishii: IEEE J. Quantum Electron., QE-13 (1977) 587.
14) E. Oomura, R. Hirano, T. Tanaka, M. Ishii, and W. Susaki: IEEE J. Quantum Electron., QE-14 (1978) 460.
15) C. P. Lee, S. Margalit, I. Ury, and A. Yariv: Appl. Phys. Lett., 32 (1978) 410.
16) H. Kumabe, T. Tanaka, H. Namizaki, M. Ishii, and W. Susaki: Appl. Phys. Lett., 33 (1978) 38.
17) H. Kumabe, T. Tanaka, H. Namizaki, S. Takamiya, M. Ishii, and W. Susaki: Japan. J. Appl. Phys., 18 (1979) Suppl. 18-1, 371.
18) H. Namizaki, S. Takamiya, M. Ishii, and W. Susaki: J. Appl. Phys., 50 (1979) 3743.
19) I. Ladany, M. Ettenberg, H. F. Lockwood, and H. Kressel: Appl. Phys. Lett., 30 (1977) 87.
20) S. Nita, H. Namizaki, S. Takamiya, and W. Susaki: IEEE J. Quantum Electron., QE-15 (1979) 1208.
21) H. Kumabe, K. Isshiki, H. Namizaki, T. Nishioka, H. Koyama, and S. Takamiya: 15th Conf. Solid State Devices & Materials, (Tokyo, 1983).
22) H. Koyama, T. Nishioka, K. Isshiki, H. Namizaki, and S. Kawazu: Appl. Phys. Lett., 43 (1983) 733.
23) S. Takamiya, Y. Seiwa, T. Tanaka, T. Sogo, H. Namizaki, and K. Shirahata: 7th IEEE Semiconductor Laser Conf. (Brighton, 1980).
24) H. Kumabe, T. Tanaka, S. Nita, Y. Seiwa, T. Sogo, and S. Takamiya: Japan. J. Appl. Phys., 21 (1982) Suppl. 21-1, 347.
25) C. P. Lee, S. Margalit, and A. Yariv: IEEE Trans. Electron Devices, ED-25 (1978) 1250.
26) W. T. Tsang: Appl. Phys. Lett., 36 (1980) 441.
27) H. Kawanishi, M. J. Hafich, R. A. Skogman, B. S. Lenz, and P. E. Petersen: J. Appl. Phys., 52 (1981) 4447.
28) T. P. Lee, C. A. Burrus, and A. Y. Cho: Electron. Lett., 16 (1980) 510.
29) K. Mitsunga, K. Fujiwara, M. Nunoshita, and T. Nakayama: 5th Molecular Beam Epitaxy Workshop, (Atlanta, GA, 1983).
30) H. Kumabe, J. Ohsawa, K. Issiki, N. Kaneno, H. Namizaki, K. Ikeda, S. Takamiya, and W. Susaki: 9th IEEE Semiconductor Laser Conf. (Rio de Janeiro, 1984).
31) M. Yamada and Y. Suematsu: IEEE J. Quantum Electron., QE-15 (1979) 743.
32) R. F. Kazarinov, C. H. Henry, and R. A. Logan: J. Appl. Phys., 53 (1982) 4631.
33) M. Nakamura, K. Aiki, N. Chinone, R. Ito, and J. Umeda: J. Appl. Phys., 49 (1978) 4644.
34) M. Yamada and Y. Suematsu: Japan. J. Appl. Phys., 18 (1979) Suppl. 18-1, 347.
35) W. Susaki, E. Oomura, K. Ikeda, M. Ishii, and K. Shirahata: Proc. 3rd European Conf. on Optical Communication, (Berlin, 1977) p. 123.
36) K. Ikeda, H. Kan, E. Oomura, K. Matsui, M. Ishii, and W. Susaki: Trans IECE Japan, E61 (1978) 136.
37) D. Welford and A. Mooradian: Appl. Phys. Lett., 40 (1982) 560.
38) H. Namizaki, H. Kumabe, and W. Susaki: IEEE J. Quantum Electron., QE-17 (1981) 799.
39) H. Kumabe, H. Namizaki, and W. Susaki: IECE Japan, Tech. Group Electron Devices, ED79-79 (1979) 61. [in Japanese]
40) W. Susaki, K. Ikeda, and K. Shirahata: Techn. Digest of 100C '81 (1981) p. 46.
41) R. Tell and S. T. Eng: Electron. Lett., 16 (1980) 497.
42) M. Ishii, H. Kan, W. Susaki and Y. Ogata: Appl. Phys. Lett., 29 (1976) 375.
43) S. Takamiya: JARECT, Optical Devices and Fibers 1983, ed. Y. Suematsu (OHM, Tokyo, and North-Holland, Amsterdam, 1983) p. 81.
44) Y. Kadota, K. Chino, Y. Onodera, H. Namizaki, and S. Takamiya: IEEE J. Quantum Electron, QE-20 (1984) 1247.

10 GROWTH OF Si AND GaAs CRYSTALS IN THE PRESENCE OF A MAGNETIC FIELD

Keigo HOSHIKAWA and Jiro OSAKA†

Abstract

This chapter briefly reviews the techniques developed in Japan during the past few years for growing Si and GaAs crystals in the presence of a magnetic field. Since Hoshi et al. reported CZ crystal growth of Si in the presence of a transverse magnetic field in 1980, extensive studies have been carried out by several groups. A new technique for the application of a magnetic field has been developed using a small coil to apply a vertical magnetic field to the melt. A new apparatus to apply a magnetic field using a super-conducting magnet is also being developed. Crystals with a homogeneous dopant distribution are realized by suppressing thermal convection and improving thermal symmetry by optimizing the magnetic field strength and crystal and crucible rotation conditions. The oxygen concentration in CZ-grown Si crystal is controlled to a large extent by the applied magnetic field. The potential for controlling the native defect density and dislocation density in LEC-grown GaAs is also discussed.

Keywords: CZ-Si, LEC-GaAs, MCZ, VM-CZ, MLEC, Magnetic Field

10.1. Introduction

In Japan there has been a considerable research effort into techniques for applying a magnetic field to improve the quality of Czochralski Si (CZ-Si) and Liquid Encapsulated Czochralski GaAs (LEC-GaAs) crystals. The application of a magnetic field in crystal growth was first attempted a number of years ago. If the discussion is restricted to the CZ process, there was a report by Witt et al.[1)] as early as 1970 on the growth of InSb crystals in the presence of a magnetic field. This early study showed that, although thermal convection in the melt was suppressed and irregular growth striations were reduced by application of the magnetic field, rotational striations caused by crystal rotation appeared. Therefore, the findings of Witt et al. did not yield a method for growing homogeneous crystals nor present an appealing growth technique. Furthermore, the apparatus used required a large electromagnet which was difficult to adapt to the growth of large crystals. Thus, there were no reports of further studies for quite a while.

Recently, CZ-Si crystals grown in a transverse magnetic field were reported by Hoshi et al.[2,3)] Their study attracted a lot of attention for its claim that they were able to control over a wide range the oxygen concentration which has a definite relationship to the crystal quality. Since their study was reported, there have been many reports[4–8)] of developments on growth techniques in the presence of a magnetic field.

† Atsugi Laboratories, NTT Electrical Communications Laboratories, 3-1 Morinosato Wakamiya, Atsugi-shi, Kanagawa 243-01.

JARECT Vol. 19, *Semiconductor Technologies* (1986), *J. Nishizawa* (*ed.*)
OHMSHA, LTD. and North-Holland

This chapter briefly reviews the techniques developed in Japan during the past few years for growing CZ–Si and LEC–GaAs crystals in the presence of a magnetic field.

10.2. Transverse and vertical magnetic fields: apparatus and effects

10.2.1. *Transverse magnetic field*

The CZ-Si (MCZ-Si) crystal growth apparatus with transverse applied magnetic field used by Suzuki et al.[3] is shown schematically in Fig. 10.1. This apparatus uses a large electromagnet with an iron yoke and is thus essentially the same as that used by Witt et al.[1] in their study of InSb crystal growth. In the study of Suzuki et al. a transverse magnetic field of 3700 Oe was generated in a 400 mm pole gap and a conventional CZ puller was positioned in the middle of the pole gap. Temperature fluctuations in 4.5 kg of molten silicon in a 200 mm diameter silica crucible were measured to investigate the effects of the magnetic field of the suppression of thermal convection. The results, shown in Fig. 10.2, indicate that when a magnetic field of 1500 Oe is applied temperature fluctuations are reduced to about 1/10. The suppression of thermal convection in an electrically conductive melt resulting from the application of a magnetic field can be explained qualitatively by Lenz's Law. More recently, a scaled-up apparatus has been developed and the production of 5-inch diameter CZ-Si crystals has become possible using this method.

10.2.2. *Vertical magnetic field*

Although a detailed report is not yet available concerning the crystal growth apparatus used in transverse magnetic field applications, it can be assumed that the

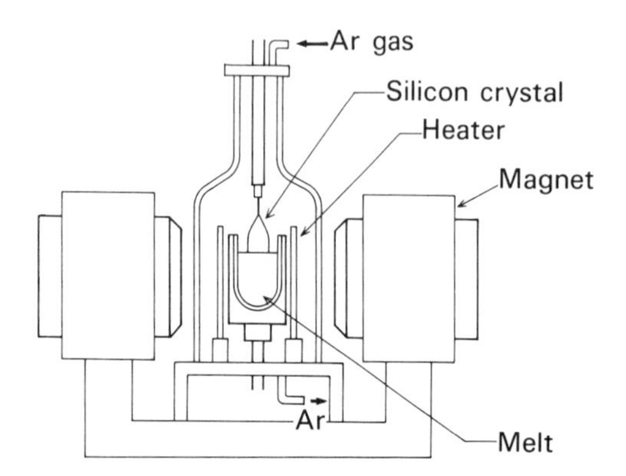

Fig. 10.1 Schematic diagram of the CZ-Si growth apparatus with transverse applied magnetic field.

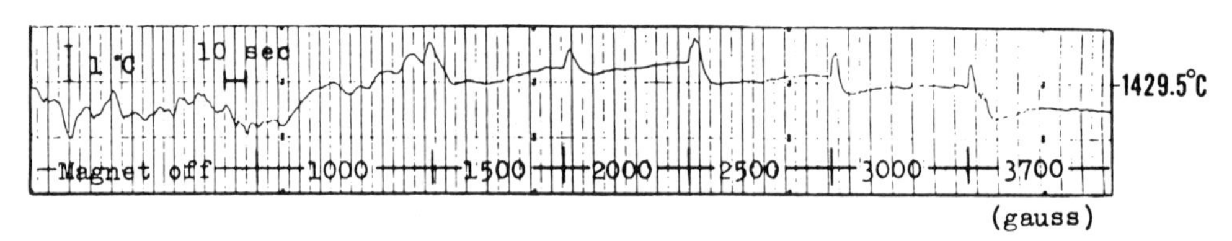

Fig. 10.2 Effect of a magnetic field on temperature fluctuations in the silicon melt at a position 10 mm below the center region of the melt surface.[3]

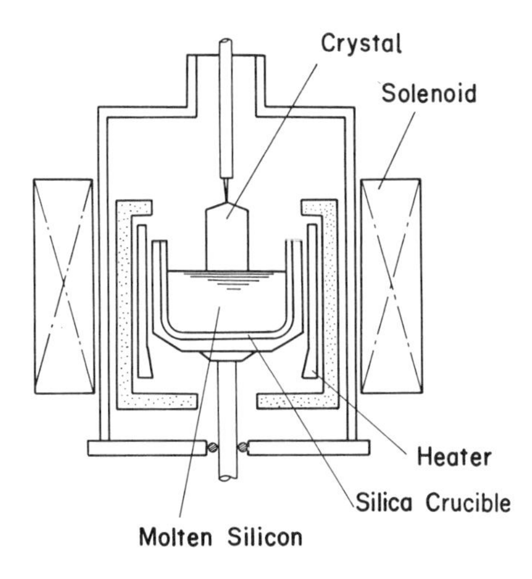

Fig. 10.3 Schematic diagram of the CZ-Si growth apparatus with vertical applied magnetic field.[5)]

magnets weigh as much as several tons, even in systems for growing 3-inch diameter crystals. This poses a practical problem that large-scale electromagnets are necessary to generate magnetic fields strong enough for crystal growth applications. A new approach to this problem has been proposed by Hoshikawa et al.[5)] In this new method, a vertical magnetic field is applied perpendicularly to the growth interface. They termed this the vertical magnetic field applied CZ (VMCZ) method. The proposed VMCZ method is based on the fact that, since convective fluid flows generated in the melt in a crucible must form a closed-loop, a component of the fluid flow invariably intersects with the magnetic field whether it is applied horizontally or vertically. As a result, the thermal convection should be suppressed by both of them. Langlois et al.[9)] confirmed on the basis of a theoretical analysis that thermal convection can be suppressed by a vertical magnetic field. The VMCZ apparatus[5)] is schematically shown in Fig. 10.3. In this case, a water-cooled coil is wound around the periphery of the 400 mm diameter chamber. The coil is very light, barely 80 kg, and so small that it appears to be a part of the chamber itself as shown in Fig. 10.3. With this coil, an axially-symmetric and fairly uniform magnetic field of up to 2000 Oe can be generated. Temperature fluctuations in the 3.5 kg of Si melt can be suppressed with a magnetic field of about 1000 Oe, and results similar to those obtained with a transverse magnetic field have been confirmed for a vertical magnetic field.

10.2.3. *Superconducting magnet*

Recently, superconducting magnets have been developed for applying magnetic fields horizontally[10)] and vertically.[11)] A schematic diagram of the magnetic field applied LEC (MLEC) apparatus for use in GaAs crystal growth reported by Terashima et al.[11)] is shown in Fig. 10.4. It consists of an in-house modified Melbourne high-pressure puller (Cambridge Instrument Co.) and a superconducting coil with a compact refrigerator system which is directly set to the cryostat. This apparatus can supply magnetic fields up to 5000 Oe.

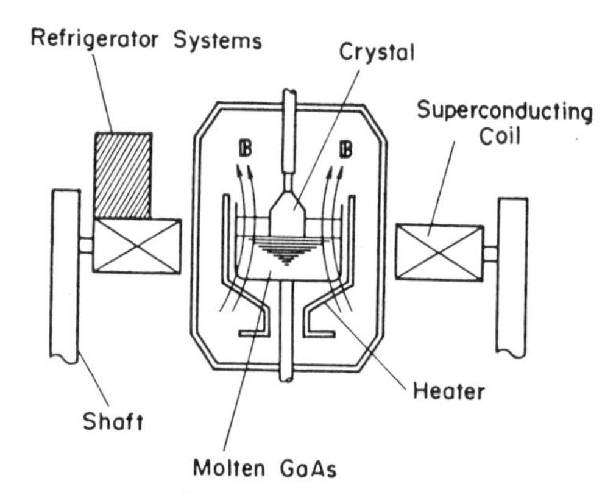

Fig. 10.4 Schematic diagram of the LEC growth apparatus with superconducting coil.[11]

10.2.4. *Effects of magnetic field on GaAs*

The magnetic field makes no essential difference to suppression of thermal convection between Si and GaAs. The effect of the vertical magnetic field strength on temperature fluctuations and temperature distributions in about 4 kg of molten GaAs in a 150 mm diameter silica crucible was obtained by Osaka et al.[12] as shown in Fig. 10.5. The temperature was measured at three points which were set at different distances from the center of the crucible. These results show that temperature fluctuations disappear above 900 Oe in the case of molten GaAs. Moreover, the temperature difference between the center and edge of the crucible increased as the magnetic field strength was increased. Osaka et al. concluded from these results that oscillational thermal convection corresponding to temperature fluctuations is easily suppressed; however, laminar thermal convection in the melt is not suppressed so easily but gradually decreases as the magnetic field strength increases.

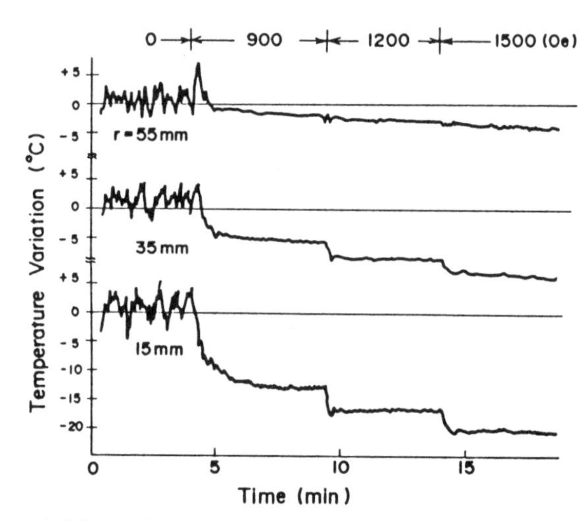

Fig. 10.5 Effect of a magnetic field on the melt temperature 5 mm below the melt surface. r indicates the distance from the center of the crucible.[12]

10.3. The growth of homogeneous crystals under application of a magnetic field

The principal subject of applying a magnetic field during the crystal growth process is to homogenize the distributions of impurities such as dopants by suppressing thermal convection in the melt. In this section, the results of research on homogeneous CZ-Si crystal growth using magnetic field application techniques are presented.

The results reported by Suzuki et al.[3] concerning the effects of magnetic field application on growth striations and variations in dopant concentration in CZ-Si crystals are shown in Fig. 10.6. An antimony-doped crystal with a ⟨100⟩ growth direction was

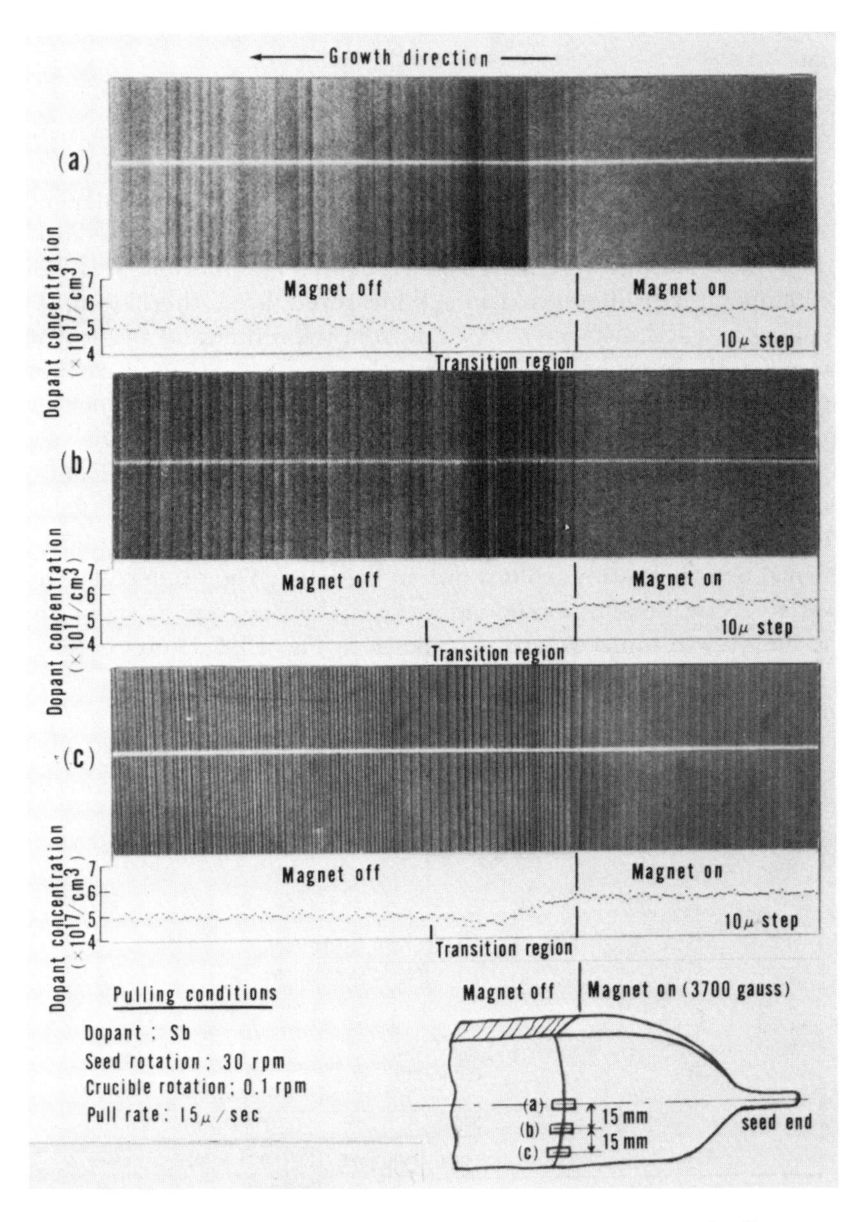

Fig. 10.6 Variations of growth striations and dopant concentration.[3]

grown in the presence of a 3700 Oe transverse magnetic field. Using substrates cut along the growth axis of the grown crystal, growth striations were revealed by Dash etching, and dopant variations across the striations were determined through spreading resistance (SR) measurements. The magnet-off regions in Fig. 10.6 show the complex variations of growth striations (irregular striations). In the magnet-on regions, where thermal convection was suppressed, the center of the crystal did not show striations. The dopant concentration had a relatively uniform distribution in the striation-free region of the crystal grown in a 3700 Oe magnetic field. On the other hand, the rotational striation patterns appeared and became more pronounced towards the crystal edge. Moreover, Suzuki et al.[3] pointed out that rotational striations varied depending on the crystal rotation conditions, and could be minimized by reducing the thermal asymmetry in the growth system.

Hirata et al.[13,14] studied the thermal asymmetry in the melt as the probable cause of rotational striations. Since they applied an axially-symmetric vertical magnetic field in their study, it is impossible to make a strict comparison between their study and that of Suzuki et al.[3] in which a transverse magnetic field was applied. The dependences of thermal asymmetry (regular temperature oscillation) and irregular temperature fluctuation on magnetic field strength are shown in Fig. 10.7. The thermal asymmetry is the peak-to-peak temperature difference ($\Delta T_{p\text{-}p}$) measured by a thermocouple rotating around the pulling axis at a distance $r = 45$ mm from the center and at $z = 10$ mm below the melt surface. On the other hand, the value of the irregular temperature fluctuation is the peak-to-peak temperature difference measured at a fixed point in the melt. It is clearly shown in Fig. 10.7 that the thermal asymmetry increases at 500 Oe but decreases above 1000 Oe to a level lower than that observed at 0 Oe. On the other hand, irregular temperature fluctuations in the melt steadily decrease as the magnetic field strength increases. The dependence of the periodic variation of the doped phosphorus concentration on the magnetic field strength is shown in Fig. 10.8. The results calculated on the basis of the variations in the diffusion boundary layer thickness and variations in thermal asymmetry at the growth interface are also shown in Fig. 10.8. Hirata et al. concluded from these results that a magnetic field of approximately 1000 Oe was suitable for growing homogeneous crystals without irregular and rotational striations.

On the basis of Hirata et al. studies,[13,14] Hoshikawa et al.[15] successfully obtained CZ-Si crystals with homogeneous dopant distribution. It was demonstrated in their report

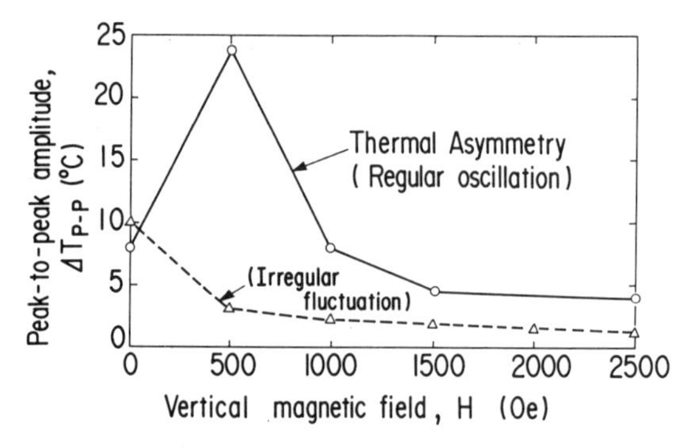

Fig. 10.7 Dependences of thermal asymmetry and irregular temperature fluctuation on vertical magnetic field strength.[14]

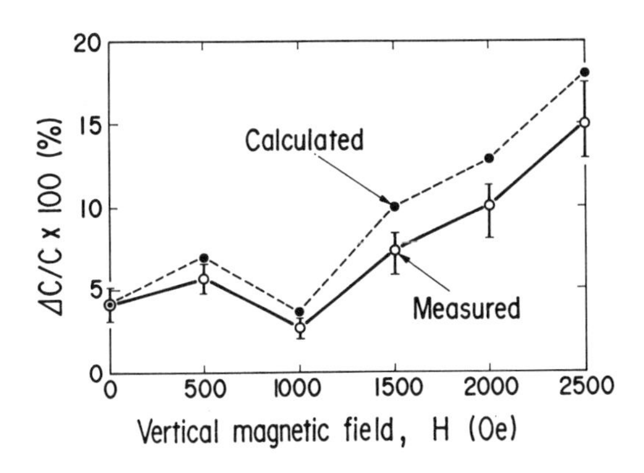

Fig. 10.8 Dependence of periodic phosphorus concentration variation ratio ($\Delta C/C$) on vertical magnetic field strength.[14]

Table 10.1 Growth conditions and measured results of resistivity variations for 6 crystals[15]

	Growth conditions			Resistivity variation*		
					Axial***	
Crystal	Magnetic field (Oe)	Crystal rotation rate (rpm)	Crucible**** rotation rate (rpm)	Radial** ΔR_r (%)	Center ΔR_c (%)	Periphery ΔR_p (%)
A	0	30	−5	5.1	10.5	8.0
B	1000	30	−5	18.2	8.1	3.9
C	1000	15	−5	4.2	3.6	6.3
D	1000	15	0	4.9	2.8	4.2
E	1000	15	5	4.4	3.8	6.2
F	1000	5	−5	21.3	4.0	19.4

* $\Delta R = 2(R_{max} - R_{min})/(R_{max} + R_{min})$
**Macroscopic radial variation measured by four-point probe method
***Microscopic axial variation measured by SR method
****Minus sign shows rotation direction opposite to crystal rotation direction

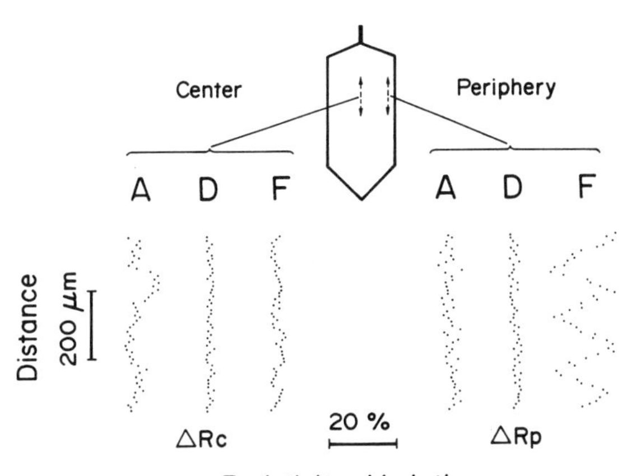

Fig. 10.9 Microscopic axial resistivity variations measured by SR method.[15]

that not only microscopic resistivity variations but also macroscopic radial resistivity distributions were eliminated. In order to accomplish this, the optimization of rotation conditions for both crystal and crucible was investigated. The growth conditions and the corresponding variations in resistivity for six phosphorus-doped crystals are summarized in Table 10.1. The microscopic axial resistivity variations for three representative crystals are shown in Fig. 10.9. On the basis of these results, a Si crystal was grown with a uniform dopant distribution resulting in less than a 5% macroscopic radial resistivity distribution and microscopic axial resistivity variation.[15]

A similar study on LEC-GaAs crystal was performed by Osaka et al.[12]

10.4. Applications of magnetic field-applied crystal growth technique

There are a number of significant results on magnetic field application other than those concerning the crystal growth of homogeneous dopant distribution just described. One result is the control of oxygen concentration in CZ-Si crystals.[2,3] Recently, studies into the effects of magnetic field application in LEC-GaAs growth on the native defect density[6] and the dislocation density[16] have also received a lot of attention. The results of these studies are presented in this section.

10.4.1. *Control of oxygen concentration in CZ-Si crystals*

Since the oxygen concentration in CZ-Si crystals has currently become the most important factor in determining the crystal quality, techniques for controlling the oxygen

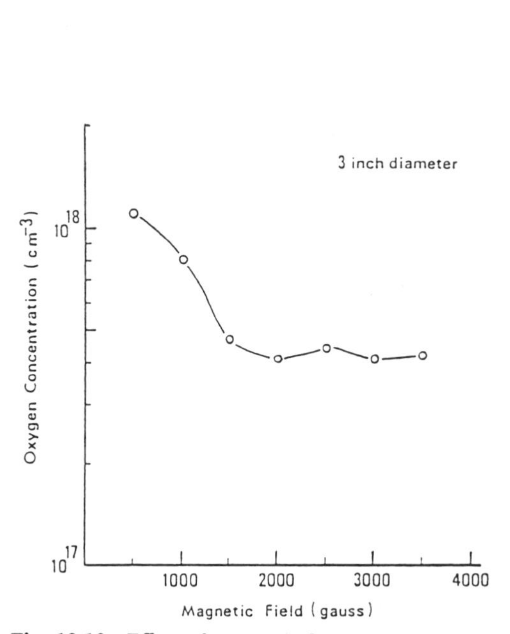

Fig. 10.10 Effect of magnetic field strength on the oxygen concentration in silicon crystal. Pulling conditions: crystal rotation 30 rpm, crucible rotation 0.1 rpm, pulling rate 15 μm/s.[3]

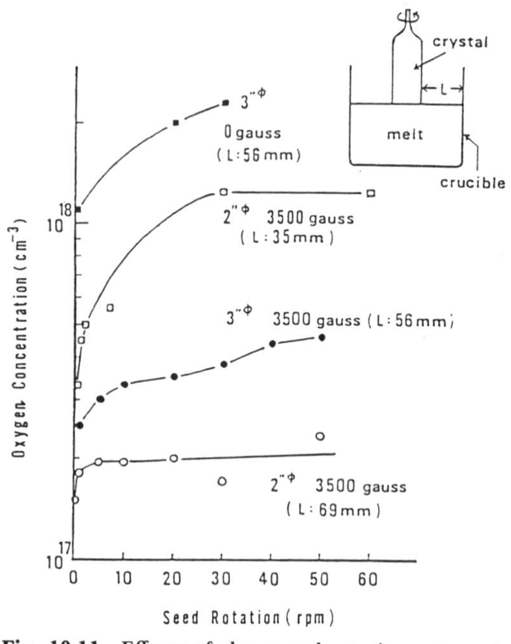

Fig. 10.11 Effects of the crystal rotation rate and distance (L) between the crystal edge and crucible wall on the oxygen concentration in silicon crystals. Pulling conditions: crucible rotation 0.1 rpm, pulling rate 15 μm/s.[3]

concentration during the growth process have aroused much interest. Hoshi et al.[2,3] demonstrated that the oxygen concentration was changed by the application of a magnetic field. Their experimental results for the effect of magnetic field strength on the concentration of oxygen atoms incorporated into a crystal are shown in Fig. 10.10. A substantial amount of oxygen is eliminated in the crystal grown in 1500 Oe. The reason why the oxygen concentration is controlled is as follows: the suppression of thermal convection in the melt causes a decrease in decomposed oxygen atoms moving from the silica crucible to the growth interface. As is shown in Fig. 10.11, the oxygen concentration in the crystals also depends on the crystal rotation rate and the ratio between the crystal and crucible diameters.

On the basis of the investigations of Hoshi et al., several crystal growers in Japan have controlled the oxygen concentration in a CZ-Si crystal 4–5 inches in diameter to a large extent by utilizing the magnetic field application technique.

10.4.2. *Control of the native defects in GaAs crystals*

Undoped semi-insulating GaAs crystals grown by the LEC process has been used as a substrate for GaAs-LSI. Holmes et al.[17] reported that they were able to grow an undoped semi-insulating crystal with high reproducibility from an in-situ synthesized GaAs melt using a pyrolithic boron nitride crucible by controlling the melt composition during growth. Terashima et al.[6] demonstrated that crystals grown by the above method lost their semi-insulating property, i.e. showed n-type low resistivity, as a result of magnetic field application during the growth process as shown in Fig. 10.12. They found that the density of deep level EL2 generally referred to as native defects was decreased by the application of a magnetic field. They also speculated that the observed reduction in native defect density may result from a reduction of temperature fluctuations in the melt by application of a magnetic field. On the other hand, Osaka et al.[18] did not find any evidence of deep level variance or decreased resistivity caused by magnetic field application.

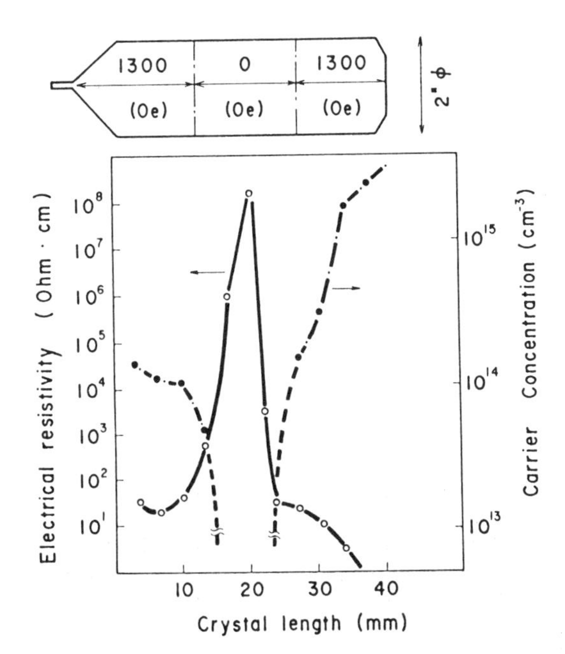

Fig. 10.12 Electrical resistivity (open circles) and carrier concentration (closed circles) along the pulling direction with and without the magnetic field.[6]

Although this problem has not yet been completely solved, it would be extremely useful to be able to control the density of native defects which play such an important role in the semi-insulating mechanism of GaAs crystals. It is expected that a solution will be found in the near future.

10.4.3. *Control of the dislocations in GaAs crystals*

A high dislocation density of 10^4–10^5 cm^{-2} exists in conventional LEC-GaAs crystals. Since these dislocations severely affect FET performance,[19] it has become important to either reduce them or distribute them uniformly. There is an interesting report by Fukuda et al.[16] concerning the relationship between application of a magnetic field and dislocations. The dislocation density distributions in wafers of MLEC crystals grown under high and low temperature gradients are shown in Fig. 10.13(a) and (b), respectively, and compared with distributions in wafers grown by the conventional LEC method. It is observed that the tendency for dislocations to be distributed in a W-shape was intensified and the density was reduced when the crystal was grown in a low temperature gradient. Although the mechanism of the effect of the magnetic field on the dislocations is not clear at present, it is of great interest if it should become possible to control the density and

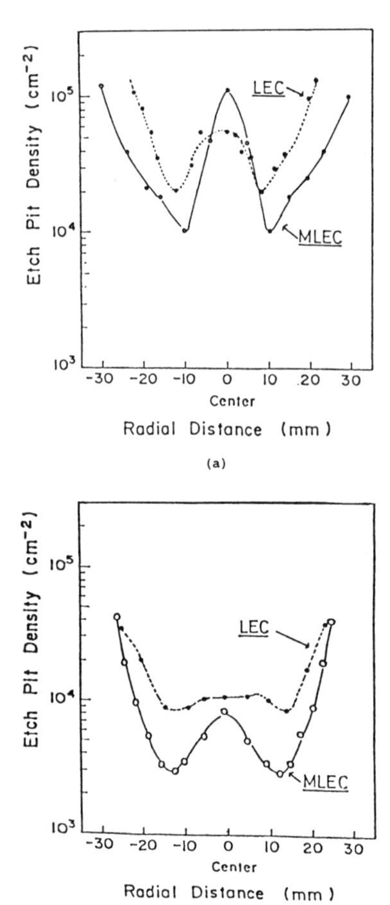

Fig. 10.13 Dislocation density (etch pit density) distribution on (100) wafer of MLEC crystals grown under a relatively high temperature gradient (a) and low temperature gradient (b), as compared with those grown by the LEC technique.[16]

distribution of dislocations through such factors as varying the native defects by application of a magnetic field.

10.5. Summary

Since Hoshi et al. reported the results of CZ-Si crystal growth in the presence of magnetic field in 1980, extensive studies have been carried out in Japan.

A method using a vertical applied magnetic field produced by a small coil was proposed and developed to solve the problem for which the previous method with a transverse applied magnetic field had required a large-scale electromagnet. Furthermore, a new apparatus to apply a magnetic field produced by a superconducting coil was also developed.

Regarding the goals of magnetic field application, i.e. the growth of homogeneous crystals, conditions for the growth of crystals with homogeneous dopant distribution have been realized by means of the following;

(1) decrease of irregular striations by suppression of thermal convection; and
(2) decrease of rotational striations by improvement of thermal symmetry.

By applying a magnetic field during the growth of CZ-Si crystals, it has become possible to a large extent to control the oxygen concentration. Crystals 4–5 inches in diameter in which the oxygen concentration has been controlled are now commercially produced in Japan.

Studies involving the effect of magnetic field application on the growth process of LEC-GaAs crystals are being carried out. There is a lot of data pertaining to the effects of a magnetic field on native defects and dislocations, and there are high expectations for future progress.

References

1) A. F. Witt, C. J. Herman, and H. C. Gatos: J. Mater. Sci., 5 (1970) 882.
2) K. Hoshi, T. Suzuki, Y. Okubo, and N. Isawa: 157th ECS Meeting Extended Abstracts (1980) p. 811.
3) T. Suzuki, N. Isawa, Y. Okubo, and K. Hoshi: Semiconductor Silicon 1981, Eds. H. R. Huff, R. J. Kreigler, and Y. Takeishi (The Electrochem. Soc., Pennington, 1981) p. 90.
4) K. M. Kim: J. Electrochem. Soc., 129 (1982) 427.
5) K. Hoshikawa: Japan. J. Appl. Phys., 21 (1982) L545.
6) K. Terashima, T. Katsumata, F. Orito, T. Kikuta, and T. Fukuda: Japan. J. Appl. Phys., 22 (1983) L325.
7) H. Kimura, M. F. Harvey, D. J. O'Connor, G. D. Robertson, and G. C. Valley: J. Cryst. Growth, 62 (1983) 523.
8) G. Fiegl: Solid State Technology (August, 1983) p. 121.
9) W. E. Langlois and K. J. Lee: IBM J. Res. Dev., 27 (1983) 281.
10) K. Terashima and T. Fukuda: J. Cryst. Growth, 63 (1983) 423.
11) K. Terashima, T. Katsumata, F. Orito, and T. Fukuda: Japan. J. Appl. Phys., 23 (1984) L302.
12) J. Osaka, H. Kohda, T. Kobayashi, and K. Hoshikawa: Japan. J. Appl. Phys., 23 (1984) L195.
13) H. Hirata and N. Inoue: Japan. J. Appl. Phys., 23 (1984) L527.
14) H. Hirata, K. Hoshikawa, and N. Inoue: J. Crystal Growth, 70 (1984) 330.
15) K. Hoshikawa, H. Kohda, and H. Hirata: Japan. J. Appl. Phys., 23 (1984) L37.
16) T. Fukuda, K. Terashima, T. Katsumata, F. Orito, and T. Kikuta: Extended Abstracts of 15th Conf. on Solid State Devices and Materials (Tokyo, 1983) p. 153.
17) D. E. Holmes, R. T. Chen, K. R. Elliot, and C. G. Kirkpatrick: Appl. Phys. Lett., 40 (1982) 46.
18) J. Osaka and K. Hoshikawa: Semi-insulating III–V Materials (Kah-nee-tu, 1984) Eds. D. C. Look and J. S. Blakmore (Shiva, Nantwich) p. 126.
19) S. Miyazawa, Y. Ishii, S. Ishida, and Y. Nanishi: Appl. Phys. Lett., 43 (1983) 853.

11 A SELF-ALIGNED GaAs MESFET WITH W–Al ALLOY GATE

Hiroshi NAKAMURA, Masanori TSUNOTANI, Yoshiaki SANO, Toshio NONAKA, Toshimasa ISHIDA, and Katsuzo KAMINISHI†

Abstract

A high-performance self-aligned GaAs MESFET using W–Al alloy as the gate metal is described. The W–Al alloy film is low-resistive and thermally stable up to 900°C. Using this gate metal, a reproducible and simple self-alignment FET fabrication process has been developed. FETs with high g_m (220 mS/mm) can be obtained, and this process is advantageous against short-channel effects. High-speed operation (21.8 ps/gate) was demonstrated using E/D gate ring-oscillators.

Keywords: GaAs MESFET, Self-alignment Process, Refractory Metal Gate, Short-channel Effect

11.1. Introduction

A GaAs MESFET is a very suitable device for high-speed digital integrated circuits. Especially, a normally-off MESFET (enhancement-type FET; E-FET) is a key device to realize low-power and high-speed LSIs. However, the effect of surface depletion between the source and gate regions strongly reduces the performance of E-FET. So, the source (drain) regions must be formed self-aligned to the gate electrode in order to obtain E-FETs with high g_m. Many kinds of self-alignment processes have been developed so far, including for example, a close-spaced technique,[1] a buried Pt gate process,[2] a gate pattern inversion process,[3,4] and a refractory metal gate process.[5,6]

Among these processes, the refractory metal gate process is very promising because of its simplicity. The refractory metal gate process resembles the poly-Si gate process in Si MOSFETs. The n^+ ions are ion-implanted using gate metal as a mask to form the source and drain regions, and ions are annealed at high temperature with the gate metal. In this process, the gate metal must maintain sufficient Schottky properties even after high-temperature annealing.

The refractory gate metal first reported by Yokoyama et al. was a Ti–W alloy.[5] Then, silicides such as Ti–W–Si,[6] Ta–Si,[7] and W–Si[8] were developed as better refractory gate materials. These materials can be formed by sputter deposition. Although these materials have fairly high thermal stabilities of Schottky contact, they have the disadvantages of high resistivity; the resistivities of silicides are about one order higher than those of pure metals. On the other hand, vacuum-evaporated pure W film has high thermal stability and low resistivity.[9] However, the evaporation technique is fairly difficult.

The W–Al alloy film, which we discuss in this chapter, has high thermal stability of Schottky contact up to 900°C and low resistivity, and can easily be deposited by

---

† Research Laboratory, Oki Electric Industry Co., Ltd., 550-5 Higashiasakawa, Hachioji, Tokyo 193.

JARECT Vol. 19, *Semiconductor Technologies* (1986), *J. Nishizawa (ed.)*
OHMSHA, LTD. and North-Holland

sputtering.[10] Because of its low resistivity, the gate metal thickness can be decreased. The use of thin gate metal has many advantages for processing.

In Section 2 of this chapter, the thermal stabilities of W–Al/GaAs and W/GaAs Schottky contacts are described. The FET fabrication process using W–Al gate self-alignment technique, FET properties and short-channel effects, and propagation delay time of E/D inverters are described in the following sections.

11.2. Thermal stability of Schottky contact

11.2.1. *Experimental*

The RF sputtering method was used to deposit W and W–Al alloy films. To obtain alloy films, W plates and Al wires were co-sputtered. The background pressure was less than 3×10^{-7} Torr, and the Ar pressure during deposition, substrate temperature and deposition rate were typically 2×10^{-2} Torr, 150°C, and 80 Å/min, respectively.

The concentration of Al in W can be controlled by the number of Al wires. As the result of a preliminary experiment, W–Al films of more than several at% Al had very high resistivity and could hardly be plasma-etched. Because the existence of a small amount of Al in the W film was essential, the Al concentration was kept at 1.0 at% in this experiment, and W–Al (1 at%) film and W film were compared. The Al concentration was measured by electron probe microanalysis (EPMA).

The experiment of the thermal stability of Schottky contacts were performed as follows:

(1) metals of 500 Å thickness are sputter deposited on n-GaAs ($n = 3 \times 10^{17}$ cm^{-3});
(2) metals are patterned to 160 μm diameter by CF_4/O_2 plasma;
(3) samples are annealed at 600–900°C for 20 min with CVD-SiO_2 cap;
(4) ohmic electrodes are formed and I–V characteristics are measured.

11.2.2. *Results and discussion*

The forward (log I)–V curves were almost linear over 3 or 4 orders. Using these curves, the barrier height and the ideality factor (n-value) were determined. Fig. 11.1 shows the dependence of barrier height (the effective Richardson constant is assumed to be 8.16 A/cm^2 K^2) and n-value on annealing temperature. The points at 450°C mean that the cap-SiO_2 was deposited at 450°C.

The n-value of a W/GaAs contact increases with the annealing temperature, whereas that of a W–Al/GaAs contact remains almost constant below 1.2. Barrier heights are lower in low annealing temperature than above 700°C, probably because of the damage during sputter deposition. The barrier height of the W–Al/GaAs contact is slightly higher than that of the W/GaAs. The difference is about 0.03 V at 800°C annealing.

Fig. 11.2 shows the dependence of the resistivity of the film on the annealing temperature, when the film thickness is 500 Å. The decrease in resistivity with annealing temperature is thought to be caused by the increase in grain size of the film. The resistivities of W and W–Al alloy after 800°C anealing were 40 $\mu\Omega$ cm and 60 $\mu\Omega$ cm, respectively. When the film thickness was 1000 Å, the resistivity of W–Al after annealing at 800°C was decreased to 40 $\mu\Omega$ cm because of the reduction of the thin-film effect. This

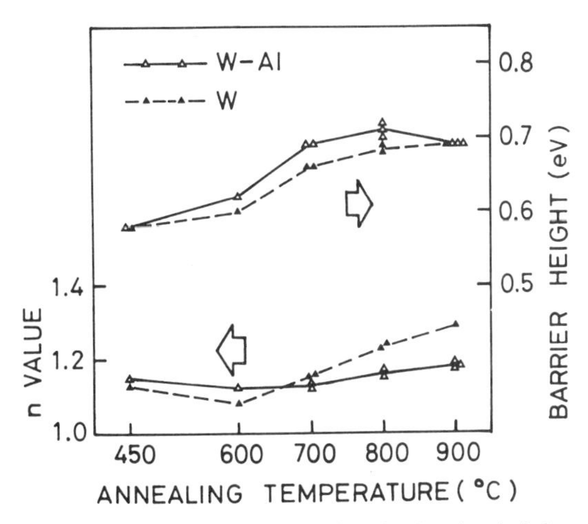

Fig. 11.1 Dependence of Schottky barrier height and n-value of W/GaAs and W–Al (1 at%)/GaAs contacts on annealing temperature.

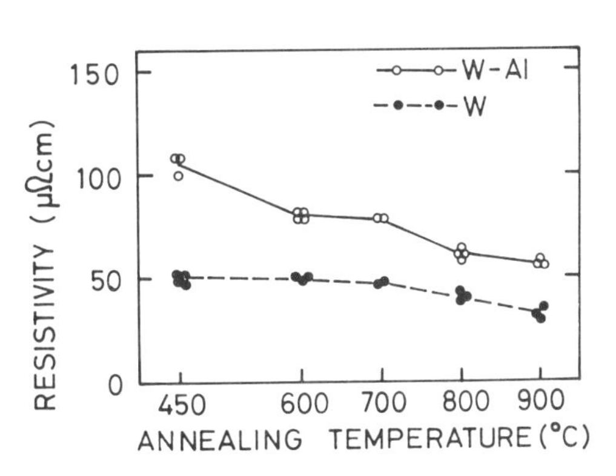

Fig. 11.2 Dependence of the resistivity of W and W–Al (1 at%) films (500 Å thickness) on annealing temperature.

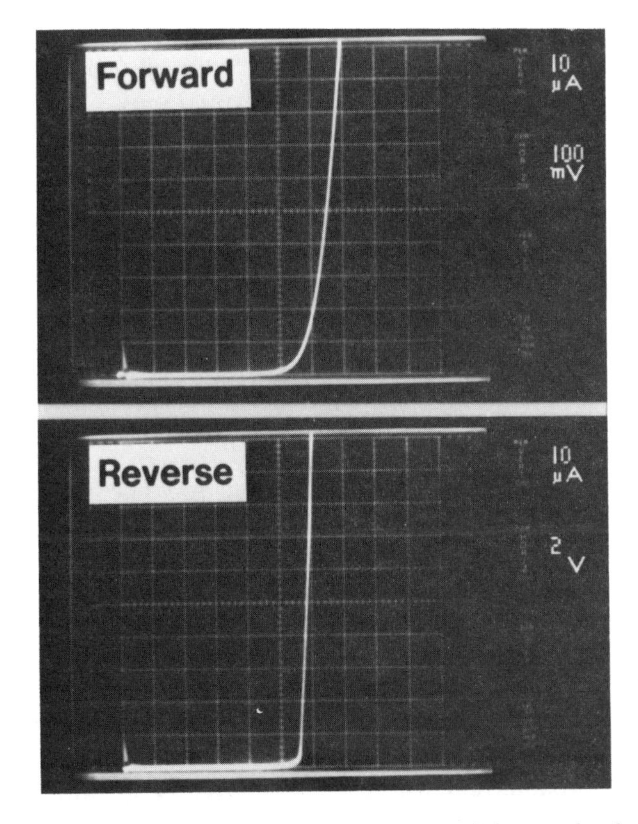

Fig. 11.3 Examples of forward and reverse gate–source I–V curves of the completed self-aligned W–Al gate ($1.0 \times 10\ \mu m^2$) MESFET.

value is about one order lower than the resistivity of silicides when the film thickness is same.

Fig. 11.3 shows examples of forward and reverse gate-source $I-V$ curves of the completed self-aligned W–Al gate FET. The gate area is $1.0 \times 10\ \mu m^2$. In addition to good forward curves, the reverse breakdown voltage is more than 10 V.

As mentioned above, the sputter-deposited W–Al (1 at%) alloy film has better thermal stability of Schottky contact than the sputter-deposited W film although the resistivity rises by about 50%. However, the mechanism of improvement in thermal stability is not clear. Preliminary SIMS (secondary ion mass spectroscopy) measurements show that the Al atoms do not diffuse into GaAs so much. The role of the Al atoms in the W/GaAs contact must be clarified through detailed experiments including many kinds of microanalysis.

11.3. Self-alignment process

11.3.1. *Gate mask metal*

Because the gate metal is low-resistive, the thickness of the gate metal can be decreased. To utilize thin gate metal in the self-alignment process, a gate mask metal layer was introduced. The role of gate mask metal is the mask of dry etching of gate metal and the mask of n^+ ion implantation. Before annealing, the gate mask metal is removed. So, the properties the gate mask metal must have are as follows:

(1) it must be easily deposited and patterned by lift-off technique;
(2) it must have large stopping power of implanted ions;
(3) it is not etched during dry etching of gate metal;
(4) it can be removed easily with no damage to GaAs and W–Al.

Typical metals suitable for gate mask metal are, therefore, Al, Cr and Ni. Among these, Ni has the largest stopping power of ions, but the adhesion between Ni and W–Al is poor. So, Ti/Ni two-layer metal was used as gate mask metal in this self-alignment process. The gate mask metal has also been used in the "T-gate" process reported in Refs. 11 and 12.

11.3.2. *Detailed process*

The typical FET fabrication process is as follows: (see Fig. 11.4)

(1) Channel implantation: ^{29}Si, 60 keV, $(1.2–1.5) \times 10^{12}\ cm^{-2}$ for E-FET and $(1.9–2.2) \times 10^{12}\ cm^{-2}$ for D-FET (dose depends on the substrate).
(2) Sputter deposition of W–Al film of 1000 Å thickness.
(3) Gate mask metal layer formation: Ti/Ni, total thickness of 2500–3000 Å, by lift-off technique (Fig. 11.4a).
(4) Reactive ion etching of W–Al film by CF_4/O_2 or SF_6 gas using a Ti/Ni layer as the etching mask. The gate metal is slightly side-etched.
(5) Selective n^+ implantation using a Ti/Ni layer as an implantation mask: ^{29}Si, 60–100 keV, $(1.5–2.0) \times 10^{13}\ cm^{-2}$. The effect of n^+ implantation energy on FET properties is shown in the next section (Fig. 11.4b).

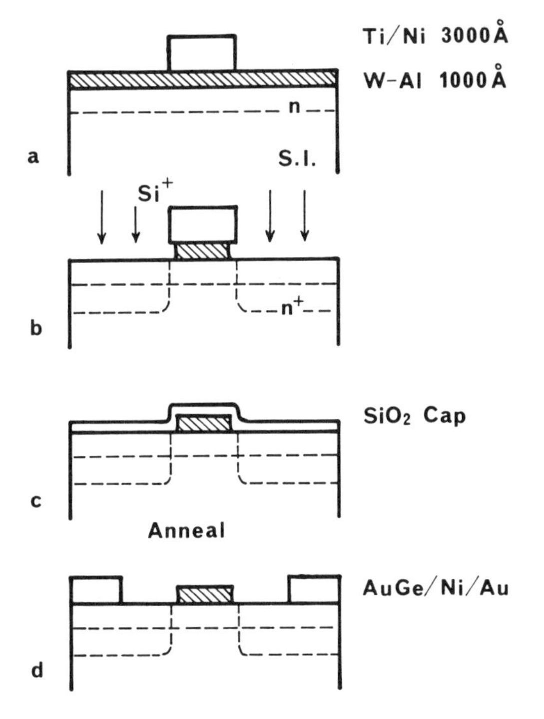

Fig. 11.4 Schematic cross-sectional diagram of the W–Al gate self-alignment process.

(6) Annealing at 800°C for 20 min (SiO_2 cap) after removing the Ti/Ni metal layer by HCl solution (Fig. 11.4c).
(7) Formation of Ohmic contacts: deposition of AuGe/Ni/Au metals and sintering (Fig. 11.4d).
(8) First level connecting line formation (Ti/Pt/Au).
(9) CVD of SiO_2 layer (3000 Å) and contact hole etching.
(10) Second level interconnection and the final passivation.

As the substrates, semi-insulating undoped LEC wafers were mainly used. Although the gate metal of more than 2000 Å thickness sometimes peeled off, that of 1000 Å thickness did not peel off. This fact shows that the use of thin gate metal is very advantageous in stress. Moreover, it is also advantageous in the reliability of interconnecting lines because the FET becomes more planar.

Because the gate mask metal is used, the n^+-implanted regions (source and drain regions) can be separated a little from the gate metal. This ability strongly improves the performance of the FET, because the n^+ depth can be decreased with no increase in parasitic capacitance.

11.4. FET properties and short-channel effects

11.4.1. *Typical FET properties*

Typical FET properties are shown in Fig. 11.5. The gate length (L_g) and gate width (W_g) are 1.0 μm and 10 μm, respectively. The threshold voltage (V_{th}) measured by

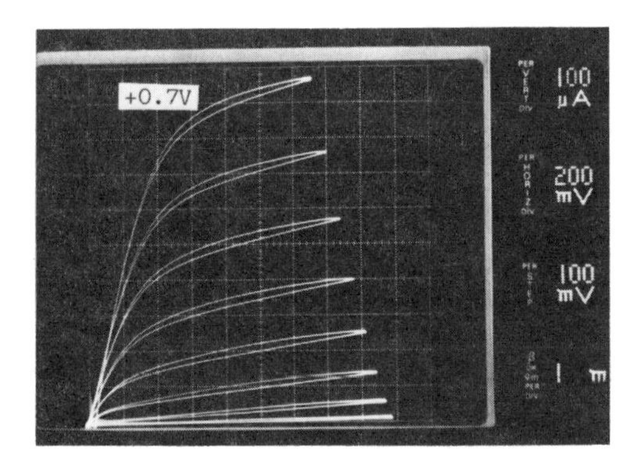

Fig. 11.5 Typical MESFET properties with 1.0 × 10 μm^2 gate.

$(I_d)^{1/2}-V_g$ relation is −0.02 V at $V_d = 1.0$ V. High transconductance of 220 mS/mm ($V_g = +0.6$ to +0.7 V, $V_d = 1.0$ V) and low on-resistance are obtained.

11.4.2. *The effect of gate direction*

The properties of self-aligned MESFETs are affected by various parameters: for example, gate direction, the depth of n^+ region and gate–n^+ spacing.

Lee et al. first reported[13] that V_{th} of a conventional GaAs MESFET was affected by the gate direction. In a refractory metal gate self-aligned FET, Yokoyama et al.[14] reported that there was much difference in short-channel effects between ⟨011⟩ and ⟨01$\bar{1}$⟩ FETs. The direction of ⟨011⟩ and ⟨01$\bar{1}$⟩ can be determined by chemical etching using Br–CH_3OH, as reported by Tarui et al.[15] It should be noted that the definition of the ⟨011⟩ direction in Refs. 13 and 14 is different from that in Ref. 15, and that the latter is a correct definition. It is shown in Fig. 11.6.

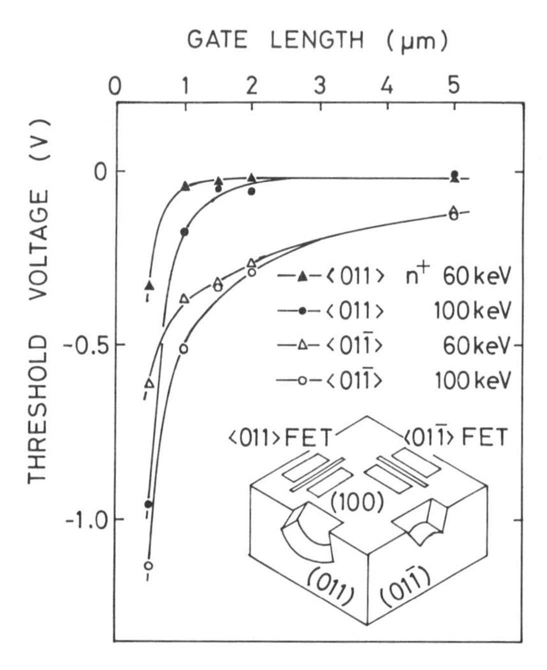

Fig. 11.6 Dependences of short-channel effects on gate direction and n^+ depth. The definition of the ⟨011⟩ and ⟨01$\bar{1}$⟩ directions is also shown in the figure.

Fig. 11.6 shows the dependences of short-channel effects on gate direction and n^+ depth. These four kinds of FET were fabricated on the same wafer and the conditions of ion implantation for the active layer were the same. The FETs of $\langle 01\bar{1}\rangle$ direction show lower V_{th} than those of $\langle 011\rangle$ direction. Moreover, V_{th} gradually shifts towards the negative from more than 5 μm gate length to short channel. The above phenomena are the same for all n^+ implantation energies between 60 keV and 100 keV.

11.4.3. *The effect of* n^+ *ion implantation energy*

Fig. 11.6 also reveals that FETs with high n^+ implantation energies show stronger short-channel effects than those with low n^+ implantation energies. More detailed results of the n^+ implantation dependence is shown in Fig. 11.7. The gate direction is fixed to the $\langle 011\rangle$ direction, and the n^+ ion implantation are with doses of $2\times 10^{13}\,\mathrm{cm}^{-2}$ at energies of 60, 100, 140 and 180 keV. The shift of V_{th} towards negative increases as the n^+ implantation energy increases. Moreover, the increase of drain conductance on the shortening of gate length is larger in the FET of high-energy n^+ implantation than that of low-energy n^+ implantation. These results are similar to the results in SAINT FETs.[16)]

The increase of n^+ ion implantation energy decreases the carrier concentration near the surface. Therefore, the parasitic gate–source and gate–drain capacitances decrease and the Schottky contact of the gate metal becomes stable even if the gate–source (drain) spacing (side etching of gate metal) is zero. However, FETs with such a condition

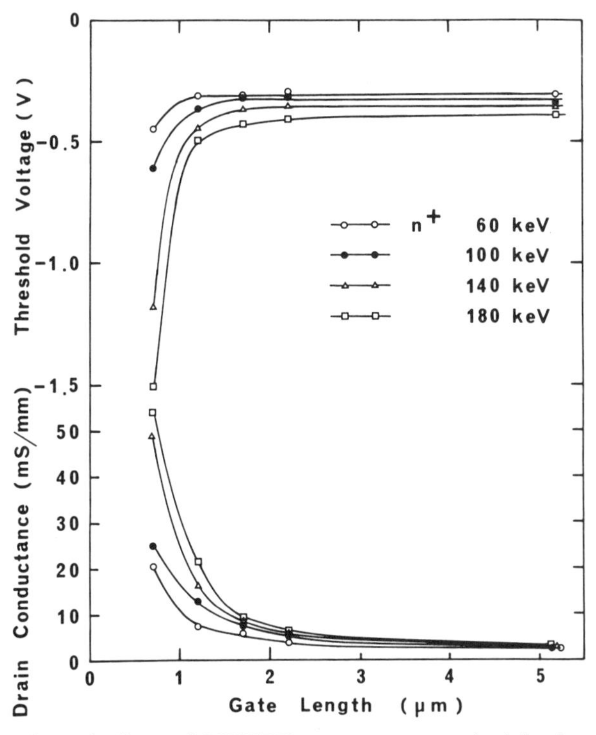

Fig. 11.7 Detailed short-channel effects of MESFET as a parameter of n^+ implantation energy. The gate direction is $\langle 011\rangle$.

inevitably show strong short-channel effects. In order to fabricate well-controlled FETs with 1.0 μm gate length, therefore, the n^+ implantation energy must be less than 100 keV. The slight side etching of gate metal is very effective to reduce the gate–source (drain) parasitic capacitance in the FET with low n^+ implantation energy. So, this process, in which the controlled slight side etching of gate metal is possible, has large advantages for fabricating logic gates with high switching speed using short-channel and low capacitance FETs.

11.4.4. *The effect of side etching of gate metal*

Fig. 11.8 shows the dependence of short-channel effects on side etching of the gate metal. The n^+ implantation energy was fixed as 140 keV, and the amount of side etching was varied from almost zero (less than 0.05 μm) to 0.15 μm. In this experiment, two wafers were used. So, the difference in V_{th} of a long-channel FET is not essential. As the gate length in Fig. 11.8, the gate length of the completed FET was used; so the n^+–n^+ spacing is about 0.3 μm larger in the FET of 0.15 μm etching than the other when the gate length of Fig. 11.8 is the same.

It seems that the curves of the FET of 0.15 μm etching are shifted by about 0.3 μm towards the left from those of the other. Therefore, the n^+–n^+ spacing is essential in short-channel effects. This result is similar to the result in SAINT FETs.[17)]

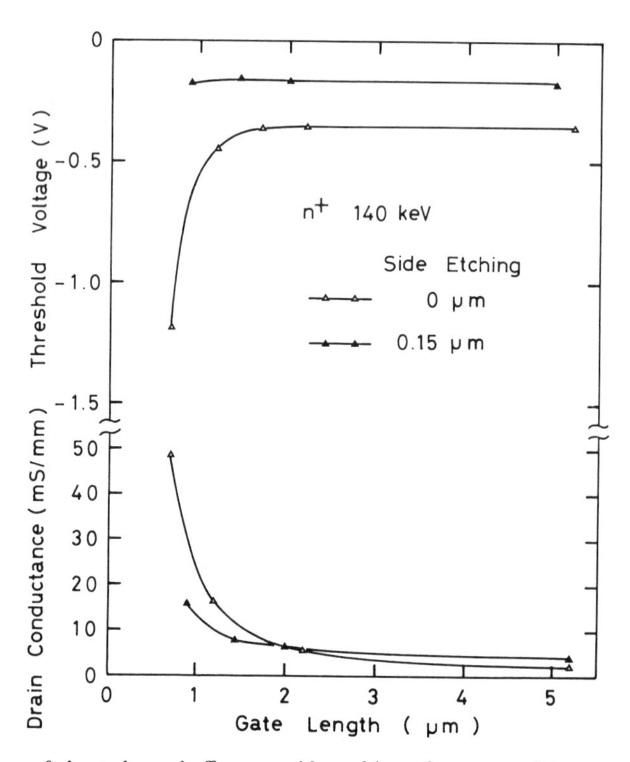

Fig. 11.8 Dependences of short-channel effects on side etching of gate metal (gate-source (drain) spacing).

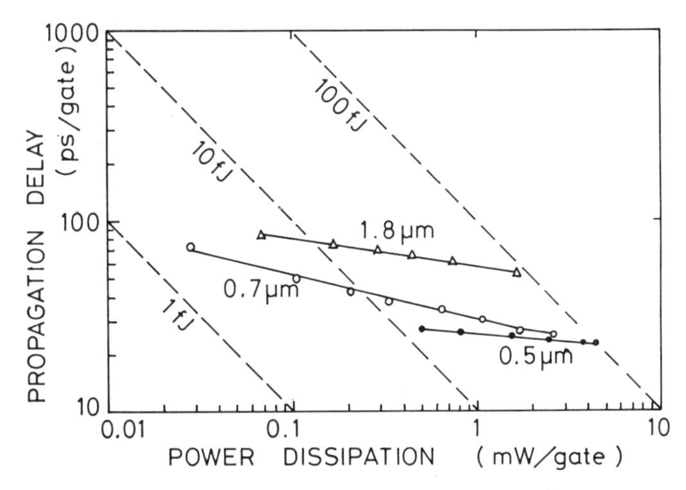

Fig. 11.9 Relations of propagation delay time and dissipated power per unit E/D gate as a parameter of gate length.

11.5. Propagation delay time of E/D gates

To demonstrate the high switching speed of the FET fabricated by this self-alignment process, various kinds of ring-oscillators were fabricated. The number of E/D stages is either 17 or 21. The gate width of E-FET is 10 μm or 20 μm. Fig. 11.9 shows examples of propagation delay time versus dissipated power relations as a parameter of gate length. The values of V_{th} of E-FET and D-FET are about +0.1 V and −0.5 V, respectively. Because the oscillation frequency of ring-oscillators are strongly affected by the static properties of the FET, parasitic capacitances and bias conditions, the delay times cannot be simply compared with each other. However, the fact that the delay time of about 20 ps/gate was obtained shows the effectiveness of this FET fabrication process. In a ring-oscillator of 0.5 μm gate length, the delay time of 26.9 ps/gate and 21.8 ps/gate were obtained when the power dissipation was 0.50 mW/gate and 5.3 mW/gate, respectively. For reference, the minimum delay time ever reported in a GaAs MESFET was 15.4 ps/gate at power dissipation of 5.4 mW/gate ($L_g = 0.6$ μm).[12]

11.6. Conclusions

A sputter-deposited W–Al (1 at%) alloy film was found to be a low-resistive and thermally stable Schottky gate metal for GaAs MESFET. This film can be deposited easily. Using this gate metal, a reproducible and simple self-alignment FET fabrication process was developed. FETs with high g_m can be obtained, and this process is advantageous against short-channel effects. High speed operation was demonstrated using E/D gate ring-oscillators. These results of our research show that this refractory metal gate self-alignment process has a large possibility of developing high-speed GaAs LSIs.

References

1) T. Furutsuka, T. Tsuji, F. Katano, M. Kanamori, A. Higashisaki, and Y. Takayama: Proc. 14th Conf. Solid State Devices (Tokyo, 1982); Japan. J. Appl. Phys., Suppl. 22-1 (1983) 335.

2) N. Toyoda, M. Mochizuki, T. Mizoguchi, R. Nii, and A. Hojo: Gallium Arsenide and Related Compounds, 1981 (Inst. Phys. Conf. Ser. No. 63, 1982) p. 521.
3) K. Yamasaki, K. Asai, and K. Kurumada: IEEE Trans. Electron Devices, ED-29 (1982) 1772.
4) T. Terada, Y. Kitaura, T. Mizoguchi, M. Mochizuki, N. Toyoda, and A. Hojo: GaAs IC Symp. Tech. Digest (1983) p. 138.
5) N. Yokoyama, T. Mimura, M. Fukuta, and H. Ishikawa: ISSCC Digest Tech. Papers (1981) p. 218.
6) N. Yokoyama, T. Ohnishi, K. Odani, H. Onodera, and M. Abe: IEEE Trans. Electron Devices, ED-29 (1982) 1541.
7) W. F. Tseng and A. Christou: Electron. Lett., (1983) 330.
8) N. Yokoyama, T. Ohnishi, H. Onodera, T. Shinoki, A. Shibatomi, and H. Ishikawa: IEEE Trans Solid-State Circuits, SC-18 (1983) 520.
9) K. Matsumoto, N. Hashizume, H. Tanoue, and T. Kanayama: Japan. J. Appl. Phys., 21 (1982) L393.
10) H. Nakamura, Y. Sano, T. Nonaka, T. Ishida, and K. Kaminishi: GaAs IC Symp. Tech. Digest (1983) 134.
11) H. M. Levy and R. E. Lee: IEEE Trans. Electron Devices Lett., EDL-4 (1983) 102.
12) R. A. Sadler and L. F. Eastman: IEEE Trans. Electron Devices Lett., EDL-4 (1983) 215.
13) C. P. Lee, R. Zucca, and B. M. Welch: Appl. Phys. Lett., 37 (1980) 311.
14) N. Yokoyama, H. Onodera, T. Ohnishi, and A. Shibatomi: Appl. Phys. Lett., 42 (1983) 270.
15) Y. Tarui, Y. Komiya, and Y. Harada: J. Electrochem. Soc., 118 (1971) 118.
16) N. Kato, K. Yamasaki, K. Asai, and K. Ohwada: IEEE Trans. Electron Devices, ED-30 (1983) 663.
17) N. Kato, Y. Matsuoka, K. Ohwada, and S. Moriya: IEEE Trans. Electron Devices Lett., EDL-4 (1983) 417.

12 ADVANCED GaAs IC TECHNOLOGY

Naoki YOKOYAMA†

Abstract

In GaAs MESFETs, negatively charged high-density surface states are localized on the GaAs surfaces exposed to the air or at dielectric/GaAs interfaces, and a surface depletion layer is formed in the channel layer in respect to the surface or interface. As a result, there has been a conspicuous increase in the source-series resistance in conventional GaAs MESFET structures, and this has led to deterioration in the device characteristics.

In this chapter, it is reported that tungsten silicide films form high-temperature-stable Schottky contacts on GaAs. Then by using these contacts to form the gates, self-aligned GaAs MESFETs, in which the effects of the surface depletion layer is greatly reduced, have been developed. Using the self-aligned GaAs MESFETs, the world's fastest GaAs 1 K and 4 K static RAMs have been successfully developed.

Keywords: GaAs, MESFET, Self-aligned, Tungsten Silicide, RAM

12.1. Introduction

The aim of reduction in size and increase in speed of computers spurs an effort in search of faster logic and memory. The superior electronic properties of GaAs, compared with Si, have made it of great interest for ultrahigh-speed logic and memory applications. Many ingenious device structures have been proposed and demonstrated for developing GaAs digital integrated circuits, based on the BFL (Buffered FET Logic)[1] and SDFL (Schottky Diode FET Logic)[2] circuits which comprise depletion GaAs MESFETs. However, these devices are not suitable for high-density GaAs LSI/VLSI applications because of their circuit complexity and high power dissipation. In contrast, DCFLs (Direct Coupled FET Logics)[3] using enhancement GaAs MESFETs are very promising candidates for such applications because of their simple circuitry and superiority in low power dissipation. The main problems with DCFL technology are the poor reproducibility and the high parasitic resistance of enhancement GaAs MESFETs due to conspicuous effects of surface depletion layer. This makes it very difficult to integrate GaAs MESFETs on a very large scale and realize high-speed operation.

One solution to these problems is to develop a novel self-aligned technology for GaAs MESFETs; this would be analogous to the well-known polysilicon gate self-aligned technique of Si LSI technology. Therefore, our basic approach was to identify a material analogous to polysilicon for use in Schottky gates on GaAs which is extremely stable at high temperatures.

† Fujitsu Laboratories Ltd., 10-1, Morinosako-Wakamiya, Atsugi 243-01.

JARECT Vol. 19, Semiconductor Technologies (1986), *J. Nishizawa (ed.)*
OHMSHA, LTD. and North-Holland

12.2. High-temperature-stable, tungsten silicide Schottky contacts to GaAs

Refractory metals form intermetallic compounds with Si which have stable metallic bonds. It was anticipated that refractory-metal silicides could replace metal layers as Schottky barrier contacts on GaAs and provide much improved high-temperature stability.[4–7] This section examines the thermal stability of tungsten silicide Schottky contacts to GaAs in terms of electric and crystallographical quantities.

WSi_x films were formed by using a multi-target magnetron co-sputtering apparatus[8] which has a perfectly oil-free vacuum system. Each cathode is connected to its own independently controllable power supply. The Si content of the film, determined by electron-probe microanalysis measurement, was controlled by varying the sputtering rate ratio of W and Si.

Fig. 12.1 shows Schottky barrier heights and standard deviations as functions of atomic ratio of Si to W after annealing at 800°C for 20 min. It has been found that these parameters depend on the Si content (x); at around $x = 0.60$, the barrier height reaches its maximum value and the standard deviation reaches its minimum value. As x increases beyond $x = 0.60$, the barrier height decreases and the standard deviation increases. At $x = 0.80$, characteristics similar to those of ohmic contacts were observed. As x decreases below $x = 0.60$, the barrier height decreases and the standard deviation increases. At $x = 0.50$, the film peeled off. Fig. 12.2 indicates the average ideality factor and standard deviation as functions of Si content. The ideality factor and standard deviation both reach their minimum, or best, value at around $x = 0.60$. Figs. 12.3 and 12.4 show a surface microphotograph of a tungsten-silicide film with $x = 0.40$ and a SEM cross-section of a film with $x = 0.80$, respectively. The thickness of these films is 800 nm. When $x = 0.40$, wrinkles appear as a result of compressive stress in the film. When $x = 0.80$, cracks occur due to tensile stress in the film. When $x = 0.60$, the film remains flat and lustrous. This implies that the film is stress-free at around $x = 0.60$. It appears that the Si content which provides the most stable Schottky diode characteristics coincides with that which minimizes stress in the tungsten silicide film.

Next, the crystalline structure of tungsten silicide films with $x = 0.57$, 0.64, and 0.70

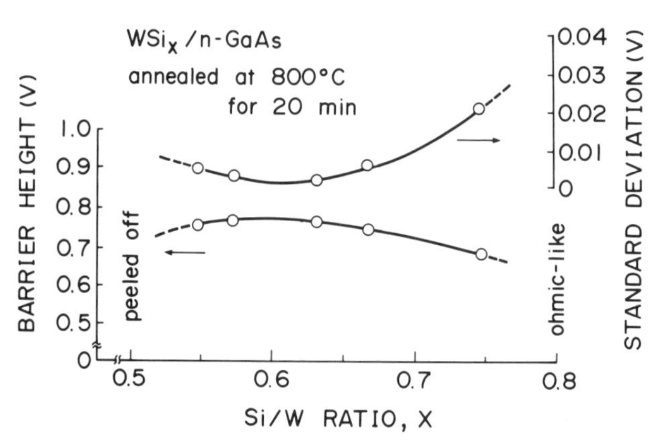

Fig. 12.1 Schottky barrier heights and standard deviations as functions of atomic ratio of Si to W after annealing at 800°C for 20 min.

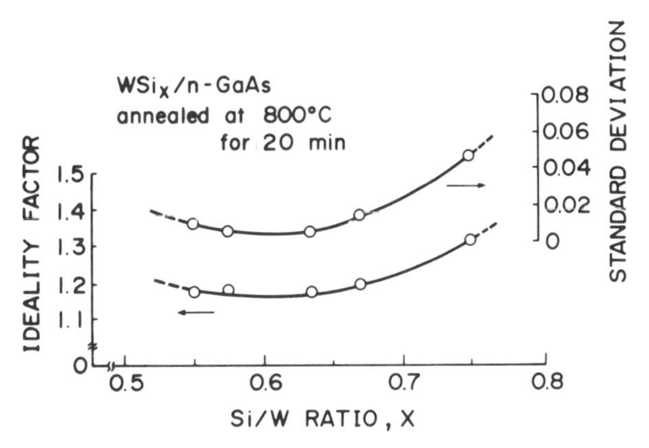

Fig. 12.2 Average ideality factors and standard deviations as functions of atomic ratio of Si to W after annealing at 800°C for 20 min.

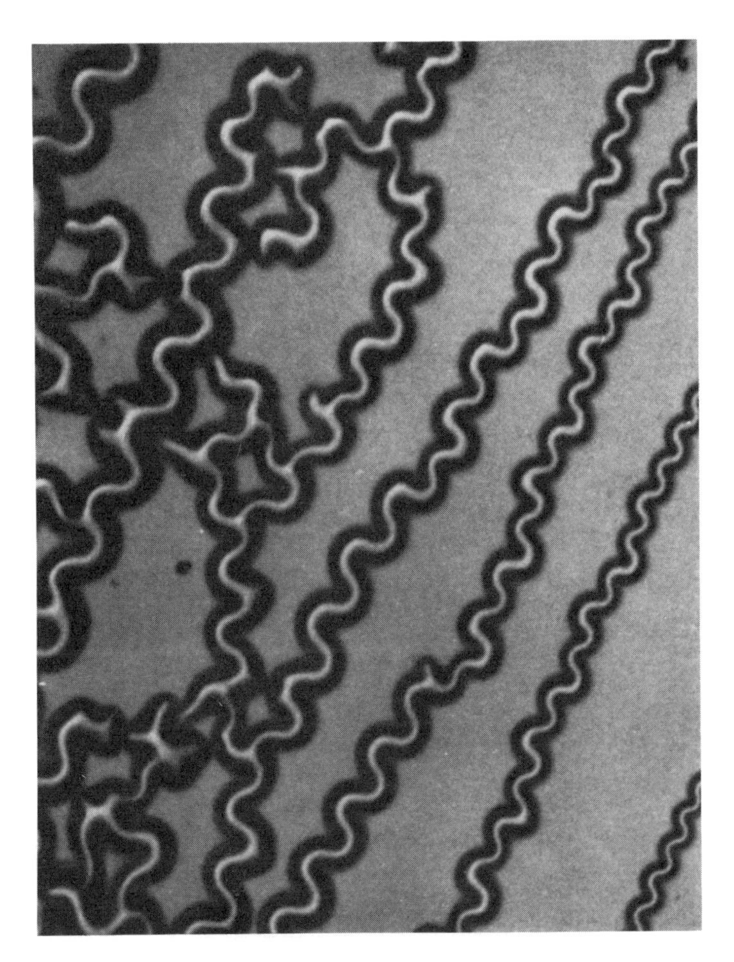

Fig. 12.3 Surface microphotograph of tungsten silicide film with $x = 0.40$.

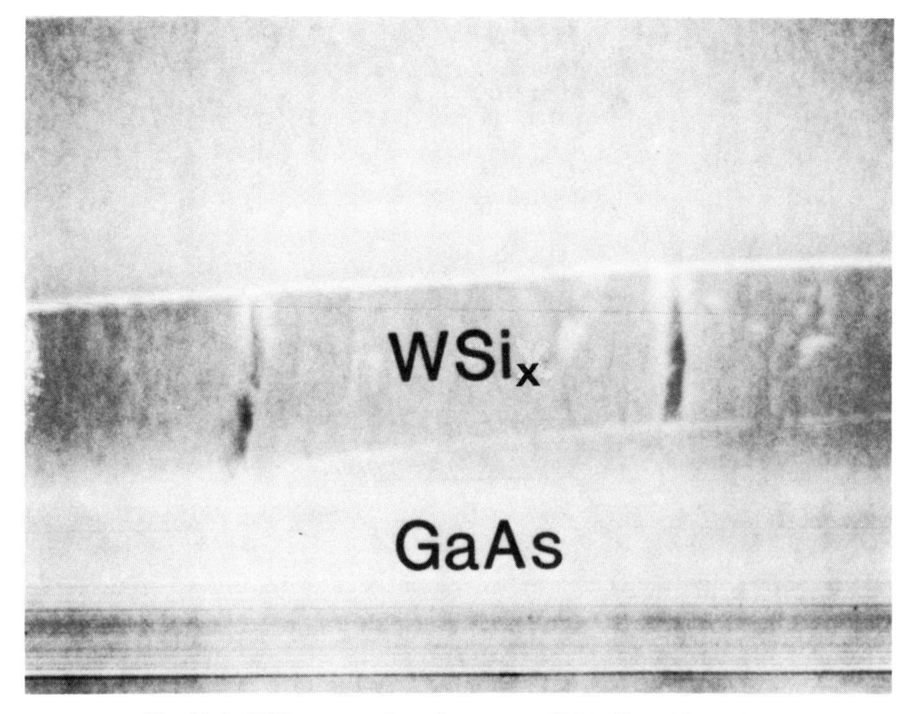

Fig. 12.4 SEM cross-section of tungsten silicide film with $x = 0.80$.

was investigated using X-ray diffraction measurements. The results are summarized in Table 12.1. These films are amorphous in the as-deposited state. At $x = 0.57$, the film remains amorphous after annealing up to 800°C. The film peeled off after annealing above 900°C. The films with $x = 0.64$ and 0.70 crystallized to the phase W_5Si_3 after annealing at temperatures above 700°C. Correlating the Schottky diode characteristics with results from X-ray diffraction analysis, it appears that Schottky diode characteristics are not affected by film crystallization. This is due to the fact that barrier height at $x = 0.57$ is nearly equal to that at $x = 0.64$.

Fig. 12.5 shows the Schottky barrier height and ideality factor of WSi_x/GaAs contacts, at $x = 0.64$, as a function of annealing temperature. Both barrier height and ideality factor remain constant and exhibit normal values for Schottky contacts on n-type GaAs after annealing at temperatures of up to 850°C. Fig. 12.6 shows the RBS spectrum for the

Table 12.1 Crystallographic properties of W–Si films.

x	as depo. 700 800 900 (°C)
0.57	amorphous / peeled
0.64	amorphous / β - phase
0.70	amorphous / β max - phase, WSi_2 was undetected

β phase: W_5Si_3

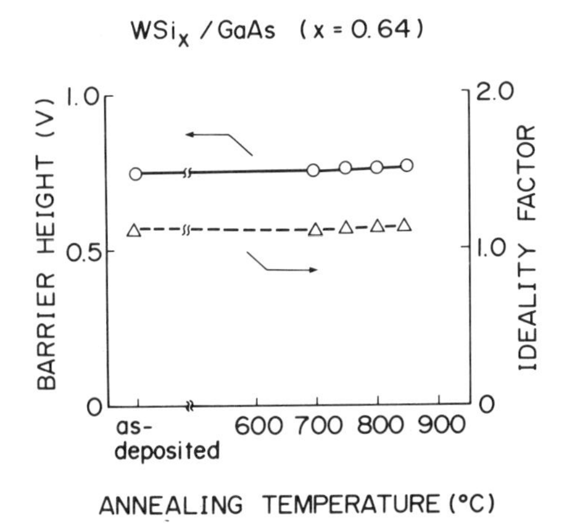

Fig. 12.5 Schottky barrier height and ideality factor of W–Si contacts on GaAs as functions of annealing temperature.

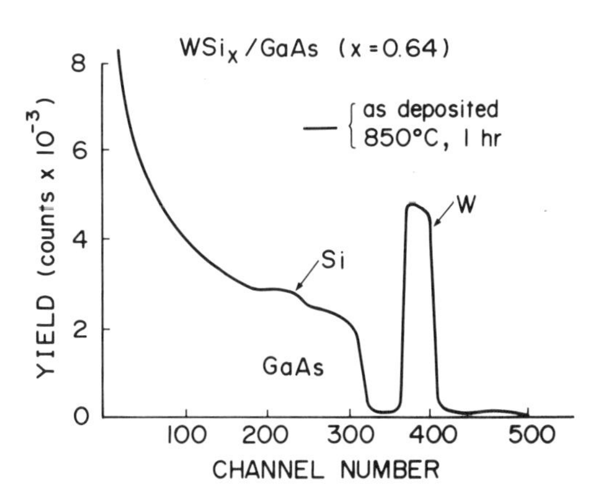

Fig. 12.6 RBS spectrum of a W–Si/GaAs contact before and after annealing at 850°C for 1 h.

$WSi_{0.64}$/GaAs system in both the as-deposited state and after annealing at 850°C for 1 h. Clearly, there is no evidence of crystallographical reactions between the tungsten silicide films and GaAs.

The ability to generate a fine line pattern with a good rectangular cross-section is one of the most important factors governing the usefulness of gate metallization in self-aligned MESFETs, since the gate material must act as a complete ion-implantation mask. Fig. 12.7 shows the cross-section of tungsten silicide films after fine-line pattern engraving. The etching was performed with a parallel-plate reactive-ion etching technique using a CF_4 and O_2 gas mixture. The etching parameters were power of 75 W, self-bias voltage of 200 V, and gas pressure of 0.6 Pa for both films. As shown in the figure, the tungsten silicide has a good rectangular cross-section.

Fig. 12.7 Cross-section of tungsten silicide film after fine-line engraving.

From these, it is found that tungsten silicide films can be used to form high-temperature stable Schottky contacts on GaAs for gate materials.

12.3. Self-aligned technology for GaAs MESFETs using tungsten silicide Schottky contacts

Use of high-temperature-stable Schottky contacts, such as tungsten silicides, has made it possible to develop self-aligned source/drain planar GaAs MESFETs.[9] Fig. 12.8 shows the major stages in the self-aligned MESFET fabrication process developed by the author. First, an active layer of n-type GaAs is formed using a conventional selective ion-implantation technique. Second, tungsten silicide gates are formed. Third, a high-dosage Si^+ implantation is made with the gate acting as an implantation mask. Fourth, annealing is carried out at 800°C for 10 min with a SiO_2 encapsulation film to activate dopants and to form the self-aligned n^+ regions. Fabrication is completed by ohmic metallization with AuGe/Au. It should be noted here that the gate electrical characteristics are not changed during this annealing process since the tungsten silicide film is used for the gate electrode.

The Schottky-gate reverse breakdown voltage in self-aligned MESFETs depends on the donor density profiles of the n^+ regions in direct contact with the gate electrodes. Therefore, high-dosage implantation is made with a dosage of 1.7×10^{13} cm^{-2} at as much as 175 keV so as to maintain a reverse breakdown voltage of at least 6 V and a peak carrier density of 1×10^{18} cm^{-3}.

Fig. 12.9 compares the schematic cross-section of a self-aligned MESFET with that of a conventional MESFET.[9] In the conventional MESFET, it is not possible to avoid extension of the surface depletion layer due to traps localized at the surface of the GaAs; the source series resistance r_s is very high because of the thinness of the undepleted n-type layer. High source series resistance and fluctuations in the charge state of these traps adversely affect the performance and producibility of MESFETs. In the self-aligned MESFETs, however, the self-aligned n^+ regions are expected to prevent extension of the surface depletion layer, so that the undepleted n^+ layer considerably reduces the parasitic

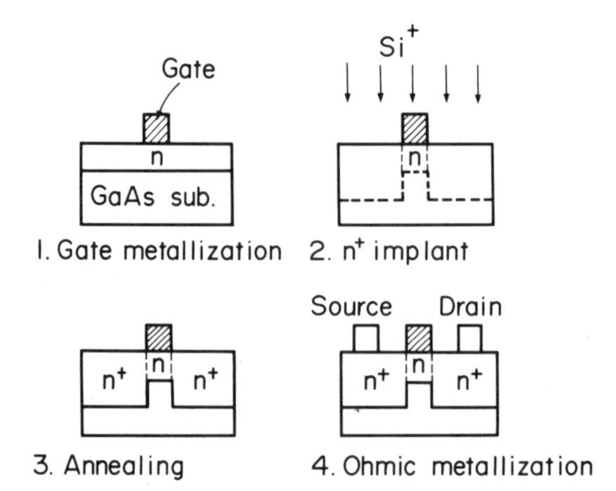

Fig. 12.8 Major stages in the self-aligned GaAs MESFET fabrication process.

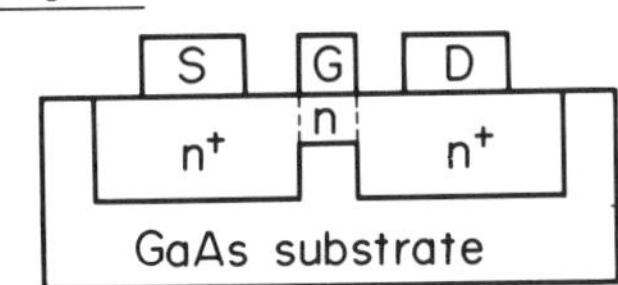

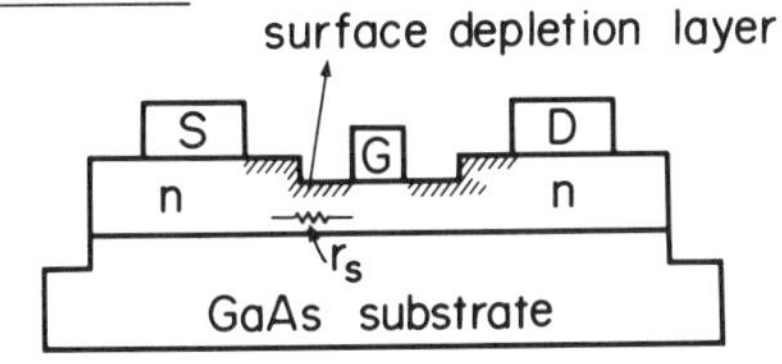

Fig. 12.9 Schematic cross-sections of a self-aligned MESFET and a conventional MESFET.

resistance. Also, the self-aligned MESFET allows a higher integration density because of its full planar structure and lack of requirement for accurate gate alignment.

Fig. 12.10 shows the current–voltage characteristics of a self-aligned enhancement GaAs MESFET with a gate 2.0 μm long by 10 μm wide, and compares it with that of a conventional enhancement MESFET with a gate 1.2 μm long by 50 μm wide. Selective Si^+ implantation for the self-aligned MESFET n-channel was made into a horizontal-bridgeman Cr-doped GaAs substrate at 59 keV with a dosage of 1.10×10^{12} cm^{-2}. The maximum transconductance of the self-aligned MESFET is 1.27 mmho with a threshold voltage of 0.145 V. Estimated source series resistance is 75 Ω. The maximum trans-

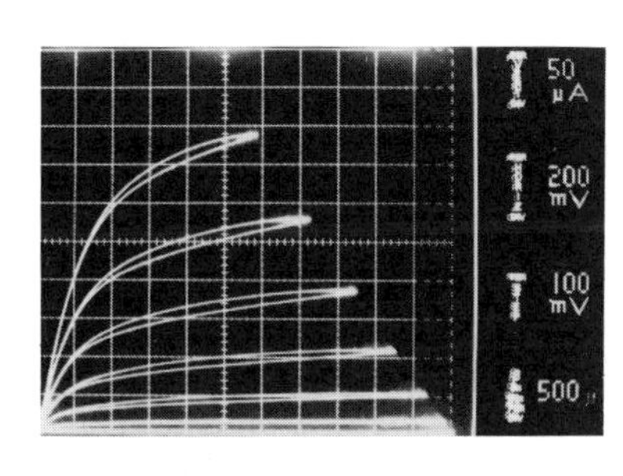

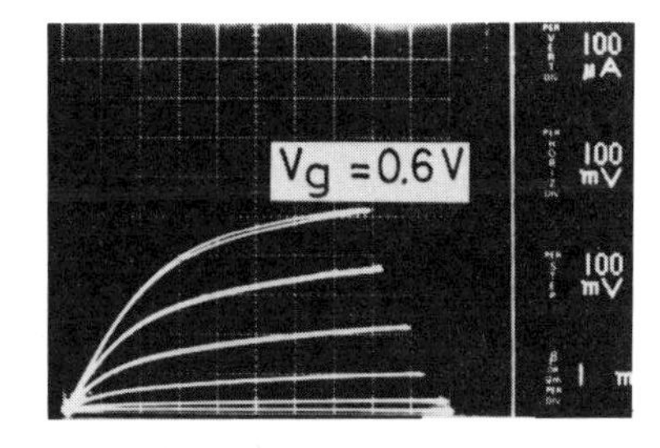

Fig. 12.10 *I–V* characteristics of a self-aligned and conventional enhancement GaAs MESFETs.

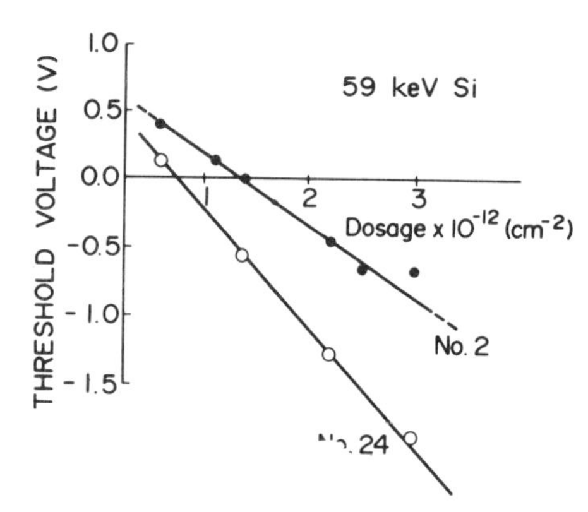

Fig. 12.11 The threshold voltage of self-aligned MESFETs as a function of dosage with the implantation energy fixed at 59 keV.

conductance of the conventional FET is 1.5 mmho with a threshold voltage of 0.1 V and an estimated source series resistance of 75 Ω. The n-channel for the conventional MESFET was grown by VPE with a carrier concentration of $1\times10^{17}\,cm^{-3}$. When mathematically normalized to a 1 mm wide gate, transconductance of the self-aligned MESFET is about four times greater and source-series resistance is five times smaller than that of the conventional FET. Reduction in parasitic resistance certainly contributes to the enhancement of transconductance, while a reduction of gate length due to lateral spread of implanted Si ions (discussed in Chapter 6) is another factor that enhances transconductance.

Fig. 12.11 shows the threshold voltage of 2 μm gate self-aligned MESFETs as a function of dosage with the implantation energy fixed at 59 keV; where 25 wafers were cut from an HB grown Cr-doped GaAs ingot and two of them (No. 2 and No. 24) were used for the measurements. The No. 2 wafer was cut from the tail region and No. 24 wafer was cut from the seed region. It was found that the tail region wafer exhibits more positive threshold voltage than the seed side wafer. The author believes that this is due to the increasing compensation by Cr, of which the content increases in the tail region as compared with the seed region.[10] It should be noted here that there is an approximately linear relationship between threshold voltage and dosage for both wafers. Using these relationships and considering the location in the ingot, any MESFET threshold voltage in the range from 0.40 to −0.65 V can be obtained using other wafers from the same ingot. These indicate that the self-aligned process itself is highly reliable and stable. Furthermore, the standard deviation of threshold voltage in an area of $30\times30\,mm^2$ was only 0.022 V, with an average threshold voltage of 0.145 V. The small standard deviation can be routinely achieved using the tungsten silicide gate self-aligned technology.

12.4. GaAs static RAMs using the tungsten silicide gate self-aligned technology

GaAs 1 K and 4 K static RAMs[6,7,11] were fabricated using the tungsten silicide gate self-aligned technology with full-ion implantation into 2 inch LEC GaAs substrates. Table 12.2 summarizes the ion-implantation conditions for enhancement and depletion FETs, and averaged threshold voltages with their standard deviations across a wafer. The n^+ regions were made by implanting Si^+ with a dosage of $1.7\times10^{13}\,cm^{-2}$ at 175 keV. As shown here, the threshold voltages of enhancement and depletion FETs were 0.11 V and

Table 12.2 Ion-implantation conditions and threshold voltages for enhancement and depletion GaAs MESFETs.

	Dosage (cm^{-2})	Energy (keV)	V_{th} (V)	σ (mV)
E-FET	0.9×10^{12}	59	0.11	40
D-FET	2.0×10^{12}	59	-0.71	110
n^+ region	1.7×10^{13}	175		

−0.71 V, respectively. Note that the standard deviation of enhancement FET is only 40 mV across a wafer. Fig. 12.12 shows a typical $I-V$ characteristic of the 1.5 μm gate enhancement self-aligned GaAs MESFET. It should be noted that the transconductance is high at 160 mS per 1 mm wide gate with a threshold voltage of 0.11 V.

Fig. 12.13 shows the top view of a basic memory cell of the 1 K and 4 K GaAs static RAMs. The basic memory cell consists of an E/D FET cross-coupled flip-flop circuit with two transfer enhancement FETs. The gates for the switching FETs are 1.5 μm long and 15 μm wide. The gates for the depletion load FETs are 4 μm long and 6 μm wide. The cell, 47×32 μm^2, is the smallest GaAs memory cell reported to date. The small cell was realized using a minimum design rule of 1.5 μm. Fig. 12.14 is a microphotograph of a completed GaAs 1 K static RAM. The chip is 3.4×2.6 mm^2 in size. Fig. 12.15 is a microphotograph of a completed GaAs 4 K static RAM. Over 26,000 FETs are integrated on a chip 4.6×3.4 mm^2. This is the most complex and densest GaAs IC reported to date.

Fig. 12.16 shows a circuit diagram of a GaAs 4 K static RAM. Enhancement/depletion FET DCFL NOR gates are used for x and y decoders. The x decoder selects a desired word line through two stage word amplifiers. Bit lines, selected by the y decoder, are connected

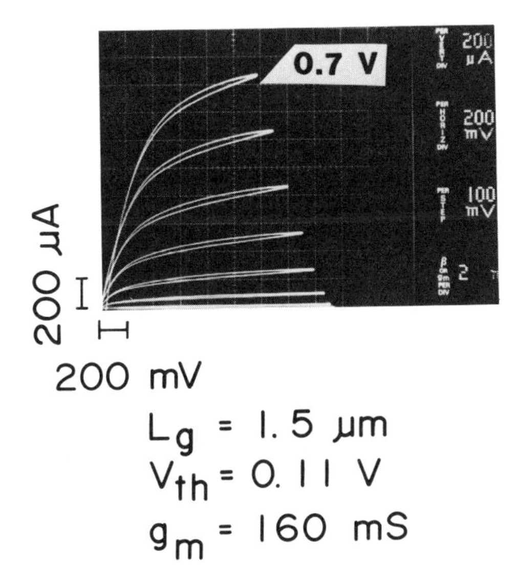

Fig. 12.12 Typical drain-current–voltage characteristics of a 1.5 μm gate self-aligned GaAs MESFET.

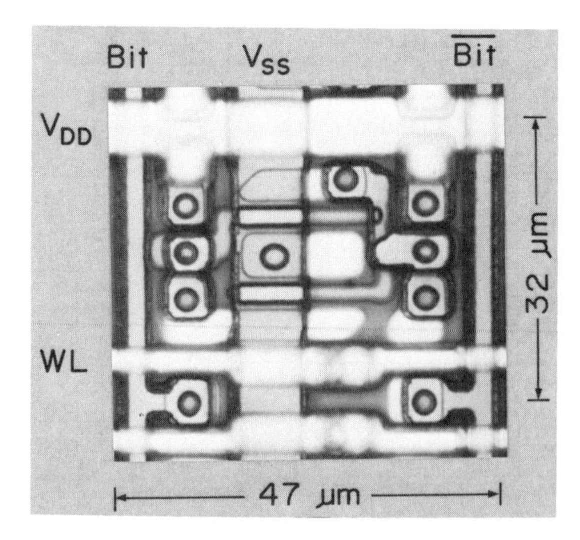

Fig. 12.13 Top view of the basic memory cell.

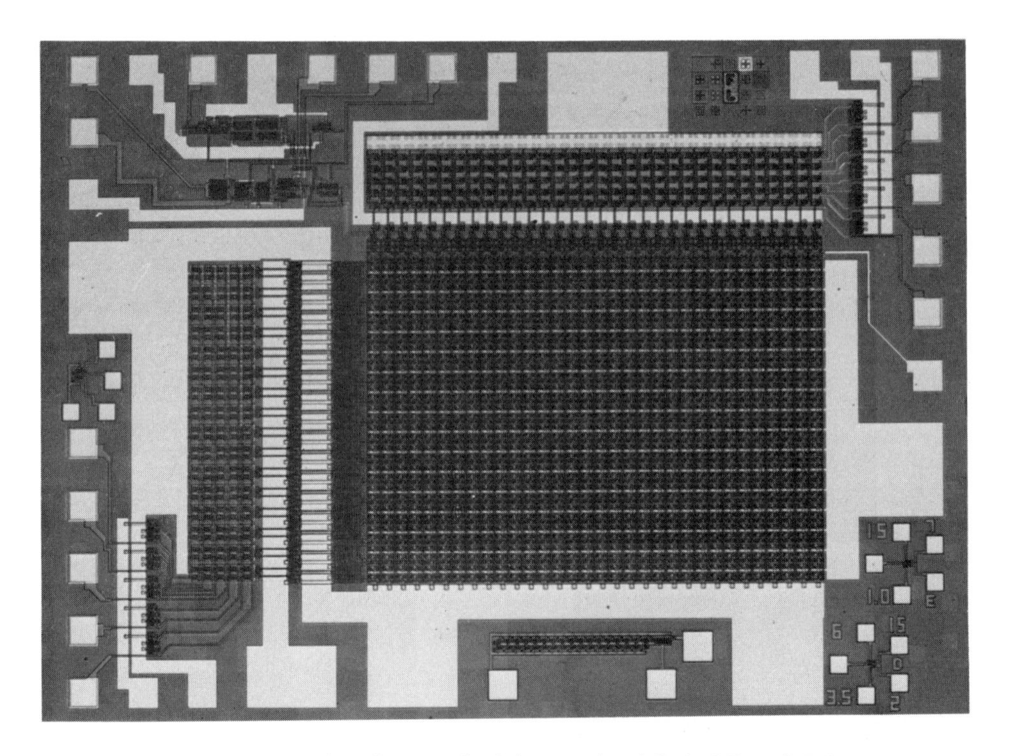

Fig. 12.14 Microphotograph of the completed GaAs 1 K static RAM.

to an input or output buffer. The output buffer is comprised of a differential sense amplifier, an E/D inverter, and a push-pull circuit. Depletion FETs were used to pull up bit lines and thereby to reduce their charging time. The gates of depletion load FETs for peripheral circuits are 2 μm long to reduce deviations in driving current. Word amplifiers use very wide (180 μm) gate driver FETs to reduce word line transition times.

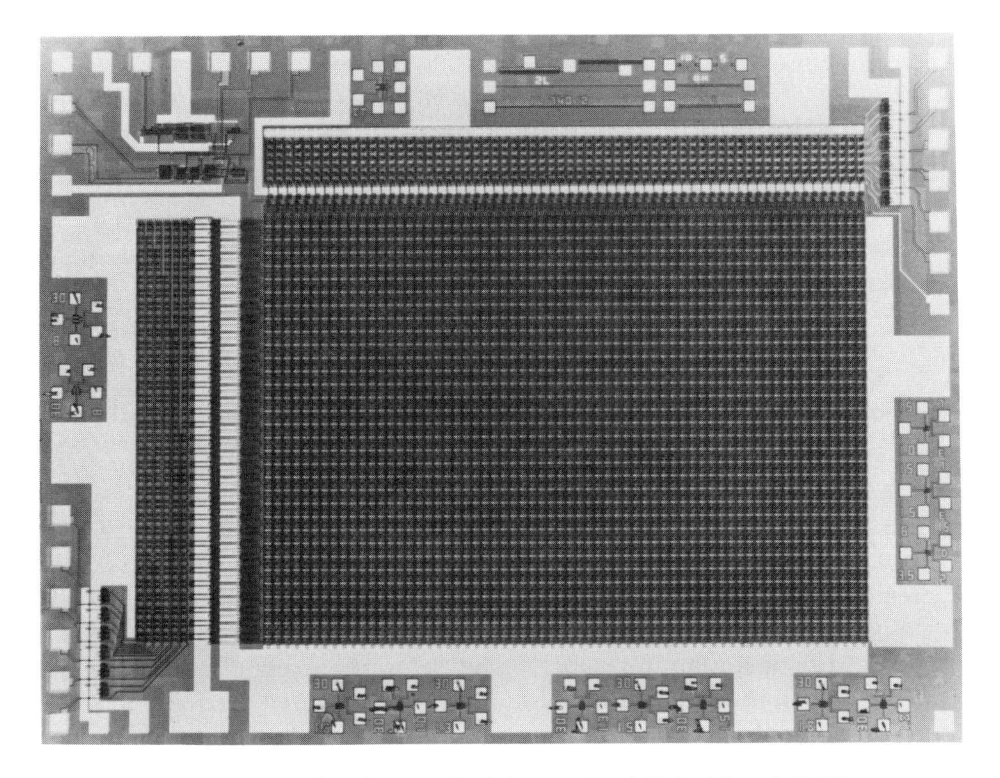

Fig. 12.15 Microphotograph of the completed GaAs 4 K static RAM.

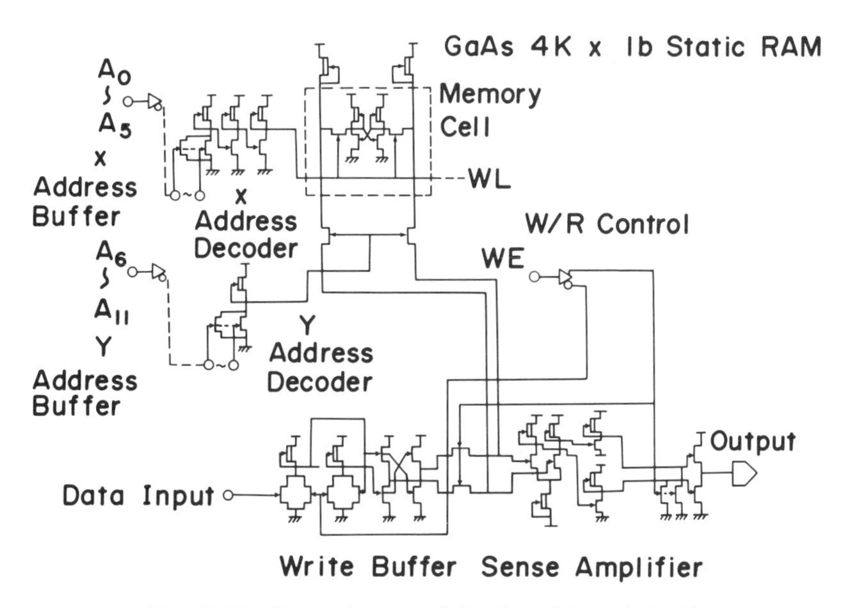

Fig. 12.16 Circuit diagram of the GaAs 4 K static RAM.

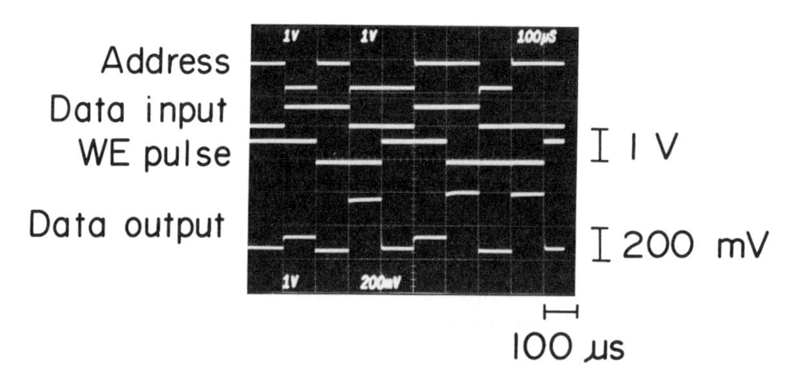

Fig. 12.17 Oscillograph of read/write operations for the 4 K RAM observed at supply voltages of 0.8 V for memory cell array and decoder circuits. The supply voltages for data output circuits are 1.0 and −0.8 V.

Fig. 12.17 is a typical oscillograph of read/write operations for a 4 K RAM. The supply voltage was 0.8 V for memory cell array and decoder circuits. The supply voltages for data output circuits were 1.0 and −0.8 V. From top to bottom, these traces are an address-input, data-input, write-enable pulses, and data output. These traces indicate that the basic read and write operations for data "1" and "0" were successfully performed.

Next, the minimum write-enable pulse width was measured for the 4 K RAM. Fig. 12.18 is an oscillograph of data-input pulse, write-enable pulses, and data output at a minimum write-enable pulse width. These traces show that the minimum write-enable pulse width was less than 2.5 ns. The minimum write-enable pulse width of the 1 K RAM was less than 1.5 ns.

The x-address access time was measured using a probe card modified for high-speed testing in a 50 Ω measurement system. Fig. 12.19 shows an oscillograph of x-address input and output waveforms at a typical address access time. The oscillograph shows that the typical address access time was 3 ns. The power dissipated by the memory cell array and the peripheral circuits was 450 and 250 mW, respectively, a total power dissipation of 700 mW. Fig. 12.20 shows that a minimum address access time of the 4 K RAM was 2.7 ns. The address access time is favorably compared with that of the 4 K RAM,[12)] fabricated using a self-aligned implantation for n^+-layer technology (SAINT).[13)]

Fig. 12.21 shows the simulated and measured access times of 4 K RAM as functions of

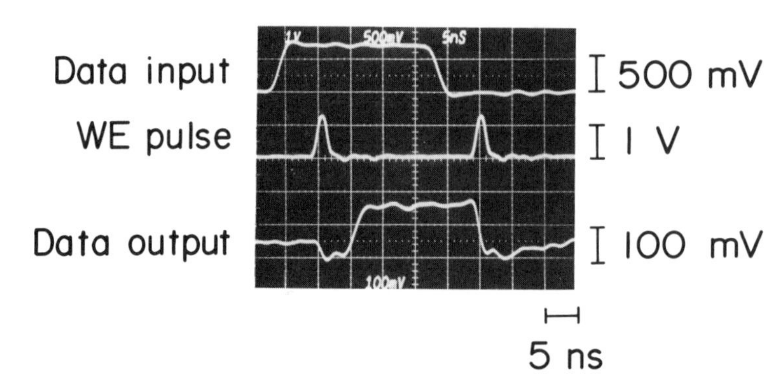

Fig. 12.18 Oscillograph observed at minimum write-enable pulse width for the 4 K RAM.

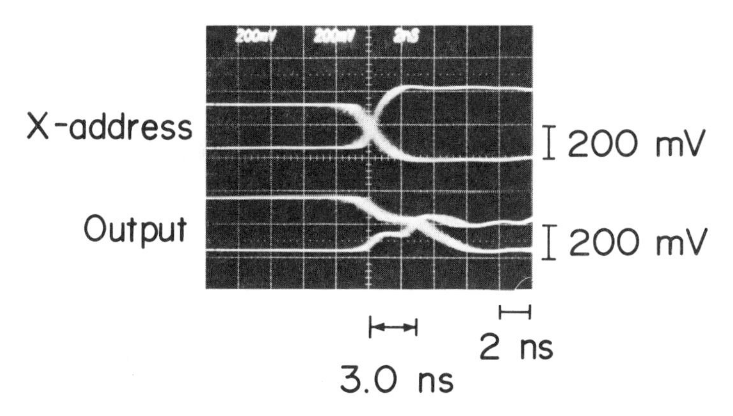

Fig. 12.19 Oscillograph of typical address access time for the 4 K RAM.

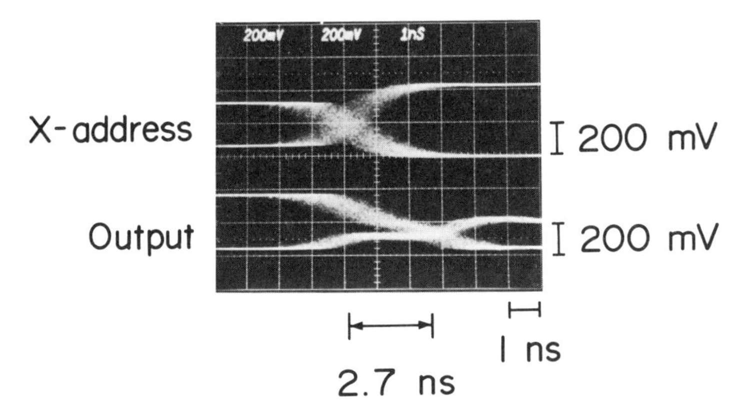

Fig. 12.20 Oscillograph of minimum address access time for the 4 K RAM.

threshold voltages of enhancement and depletion FETs. The circle plot shows the measured performance, 3 ns, described above. The "star" and "X" in the figure are plots for 4 K RAMs, fabricated with different pairs of threshold voltages by changing the ion-implantation conditions. The address access time of the "star" was 6 ns. For the "X", read/write operations were not observed. From these, it is found that the measured performance agreed comparatively well with the simulated performance.

Fig. 12.22 compares the access time and power performance of the GaAs 4 K static RAM with those of Si 4 K bipolar and NMOS RAMs recently reported. It is found that the access time of the GaAs 4 K × 1b RAM superior to that of 4 K × 1b NMOS[14)] and bipolar RAM,[15)] and is comparable with that of high-speed 1 K × 4b or 256 × 16b bipolar RAMs,[16,17)] even though the GaAs RAM has greatly reduced power dissipation.

Fig. 12.23 shows an oscillograph of *x* address and output waveforms at a minimum address access time for the GaAs 1 K static RAM. The oscillograph shows that the minimum address access time of the 1 K RAM was 1.0 ns, which was achieved with 300 mW power dissipation. The address access time is the fastest of any memory devices. Fig. 12.24 compares access time and power performance of GaAs and Si 1 K static RAMs. This figure clearly shows the excellence of GaAs memory circuits in high-speed and low-power operations.

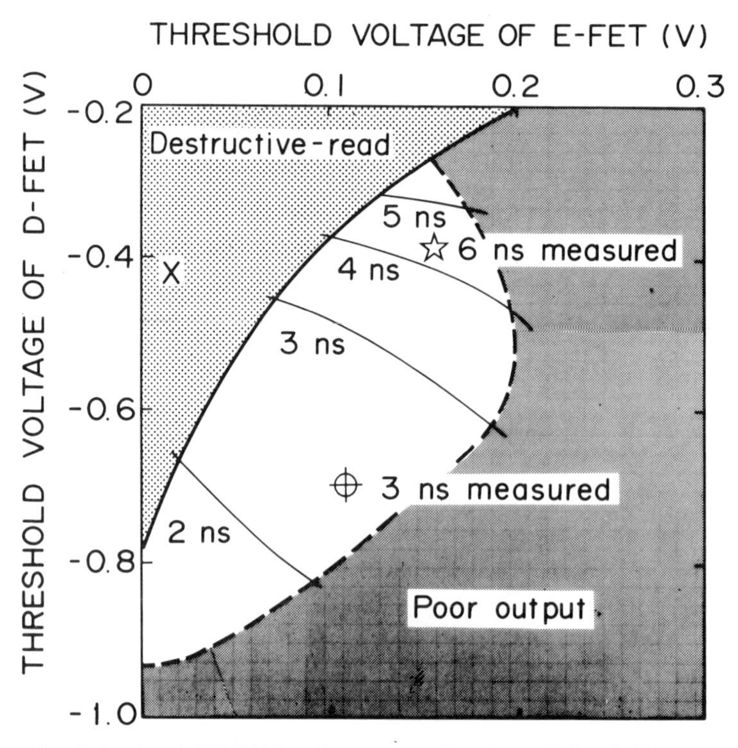

Fig. 12.21 Measured and simulated 4 K RAM performance as functions of threshold voltage of enhancement and depletion FETs.

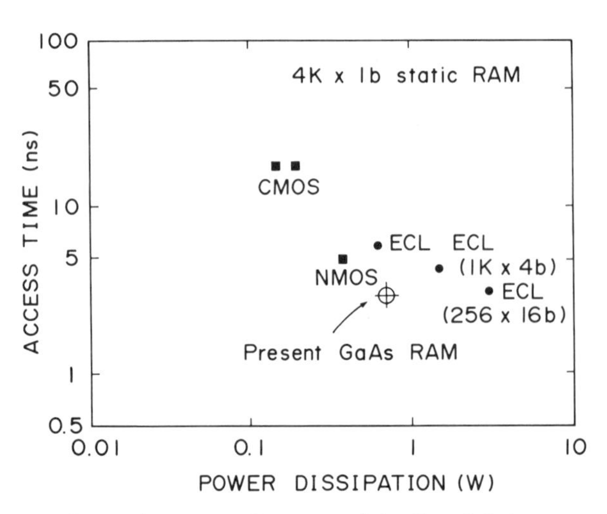

Fig. 12.22 Map of access time and power performance of the Si and GaAs 4 K static RAM reported to date.

Fig. 12.25 shows present and projected timing budgets for 4 K RAM. For the projected performance, it was assumed that the 4 K RAM was fabricated with 1 μm gate self-aligned GaAs MESFETs and 1.5 μm rule in wiring metallizations. In the present RAM, delay of address decoding and set-up of word lines is estimated to be 0.73 ns, the set-up time of bit lines 0.52 ns, and set-up time of bit bus lines and read-sense response time is 1.05 ns. Thus, the total read-response time is 2.30 ns. In the projected RAM, the

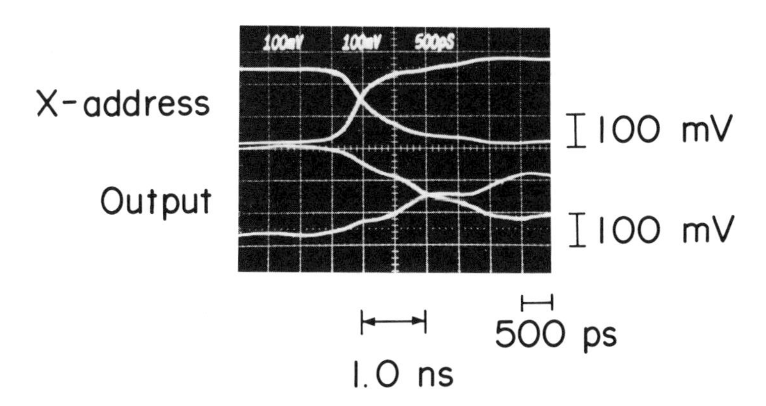

Fig. 12.23 Oscillograph of minimum address access time for GaAs 1 K static RAM.

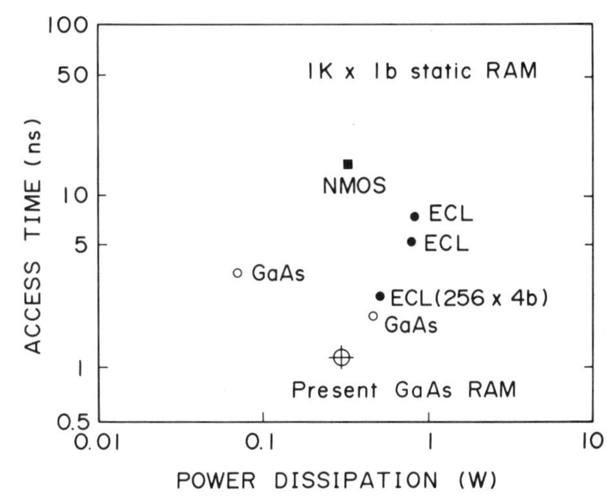

Fig. 12.24 Map of access time and power performance of the Si and GaAs 1 K static RAM reported to date.

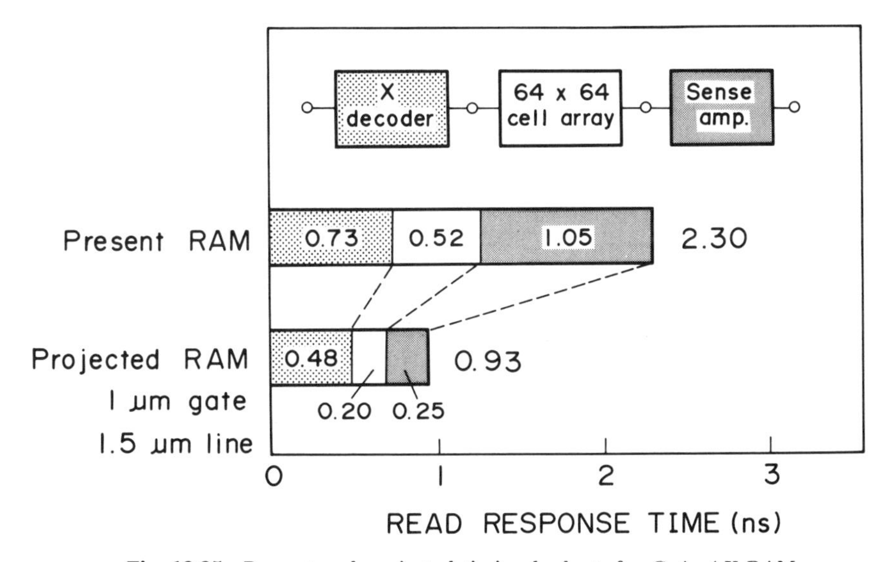

Fig. 12.25 Present and projected timing budgets for GaAs 4 K RAM.

delay in address decoding and set-up of word lines will be reduced to 0.48 ns, the set-up time of bit lines 0.20 ns, and read-sense response time 0.25 ns. Thus, a total read-sense response time of 0.93 ns will be achieved. Power dissiptation is estimated to be 900 mW.

12.5. Conclusions

It was found that tungsten silicides form extremely high-temperature-stable Schottky contacts on GaAs. Using the tungsten silicide film to form the gates, self-aligned GaAs MESFETs, in which the effects of the surface depletion layer is greatly reduced, have been developed. The self-aligned GaAs MESFET also allows a higher integration density because of its full planar structure and lack of requirement for accurate gate alignment. The world's fastest GaAs 1 K and 4 K static RAMs have been successfully developed using the self-aligned GaAs MESFETs.

Acknowledgements

The author is indebted to Messrs. Toyokazu Ohnishi, Hiroyuki Onodera, Haruo Kawata, Shoichi Suzuki for their cooperation in developing self-aligned technology, to Messrs. Touru Shinoki, Hiroaki Ohnishi, Yoshiroh Nakayama, and Akira Taguchi for their helpful discussions and support in developing high-speed GaAs static RAMs. He wishes to extend his thanks to Drs. Hidetoshi Nishi, Akihiro Shibatomi, Masaaki Kobayashi, Osamu Ryuzan, and Takahiko Misugi for their many suggestions and for their support during the performance of this work.

References

1) R. L. Van Tuyl and C. A. Liechti: IEEE J. Solid State Circuits, SC-9 (1974) 269.
2) R. C. Eden, B. M. Welch, and R. Zucca: IEEE J. Solid State Circuits, SC-13 (1978) 419.
3) H. Ishikawa, H. Kusakawa, K. Suyama, and M. Fukuta: ISSCC Digest of Technical Papers (1977) p. 200.
4) N. Yokoyama, T. Ohnishi, K. Odani, and M. Abe: Digest of IEDM (1981) p. 80.
5) N. Yokoyama, T. Ohnishi, K. Odani, H. Onodera, and M. Abe: IEEE Trans. Electron Devices, ED-29 (1982) 1541.
6) N. Yokoyama, T. Ohnishi, H. Onodera, T. Shinoki, A. Shibatomi, and H. Ishikawa, ISSCC Digest of Technical Papers (1983) p. 44.
7) N. Yokoyama, T. Ohnishi, H. Onodera, T. Shinoki, A. Shibatomi, and H. Ishikawa, IEEE J. Solid State Circuits, SC-18 (1983) 520.
8) T. Ohnishi, N. Yokoyama, H. Onodera, S. Suzuki, and A. Shibatomi: Appl. Phys. Lett., 43 (1983) 600.
9) N. Yokoyama, T. Mimura, M. Fukuta, and H. Ishikawa: ISSCC Digest of Technical Papers (1981) p. 218.
10) H. Nishi, S. Okamura, T. Inada, and H. Hashimoto: Proc. 12th Conf. Solid State Devices (Tokyo, 1980); Japan. J. Appl. Phys., 20, Suppl. 20–1 (1981).
11) N. Yokoyama, H. Onodera, T. Shinoki, H. Ohnishi, H. Nishi, and A. Shibatomi: ISSCC Digest of Technical Papers (1984) p. 44.
12) H. Hirayama, M. Ino, Y. Matsuoka, and M. Suzuki: ISSCC Digest of Technical Papers (1984) p. 46.
13) Y. Yamasaki, K. Asai, and K. Kurmada: IEEE Trans. Electron Devices, ED-29 (1982) 1772.
14) K. J. O'Connor: ISSCC Digest of Technical Papers (1983) p. 104.
15) M. Inadachi, N. Homma, K. Yamaguchi, T. Ikeda, and H. Higuchi: ISSCC Digest of Technical Papers (1979) p. 108.
16) J. Nokubo, T. Tamura, M. Nakame, H. Shiraki, and Y. Ikushima: ISSCC Digest of Technical Papers (1983) p. 112.
17) K. Ooami, M. Tanaka, Y. Sugo, R. Abe, and T. Takada: ISSCC Digest of Technical Papers (1983) p. 114.

13 EPITAXIAL GROWTH OF GROUP-IIa FLUORIDES FOR SEMICONDUCTOR/INSULATOR LAYERED STRUCTURES

Hiroshi ISHIWARA and Tanemasa ASANO†

Abstract

Heteroepitaxial growth of group-IIa fluorides (CaF_2, SrF_2, BaF_2 and their mixed crystals) on Si substrates and growth of semiconductors (Si and Ge) on the fluoride/Si structures are reviewed. Fluoride and semiconductor films were prepared by vacuum deposition onto heated substrates (room temperature to 800°C). From investigations of growth conditions and structures of pure fluoride films on (111) and (100) oriented Si, it has been found that single crystalline films grow only when the matching of lattice parameters is approximately satisfied. In the epitaxial growth experiments of Si films, it has been found that the interface between deposited Si and underlying CaF_2 is successfully stabilized by depositing thin Si layers at room temperature prior to deposition of Si at high temperatures. Finally, the lattice-matched epitaxial growth of Ge films onto mixed-fluoride/Si structures is described.

Keywords: SOI, Heteroepitaxy, Fluorides, Si, Ge

13.1. Introduction

Semiconductor-on-insulator (SOI) structures stacked on Si substrates are of great interest in fabrication of such three-dimensional devices as high-speed and high-density integrated circuits, optoelectronic devices and intelligent sensors. Several methods, such as the beam recrystallization of semiconductor films on amorphous insulators,[1] the lateral seeding epitaxial growth of semiconductor films in the solid phase,[2–4] and the hetero-epitaxial growth of insulator and semiconductor films,[5–7] have been proposed in order to form the semiconductor/insulator layered structures. Among these approaches, the heteroepitaxial growth is highly attractive because it possesses the possibility to grow layered structures composed not only of elemental semiconductors but also of compound semiconductors.

For the heteroepitaxial growth of insulator and semiconductor films, group-IIa fluorides, such as CaF_2, SrF_2 and BaF_2, have the following advantages over other insulators, such as Al_2O_3 and $MgO{\cdot}Al_2O_3$, in addition to the prerequisite fact that they are good insulators at room temperature:

(1) The fluorides crystallize in the cubic fluorite structure which is closely related to the diamond, zincblende, and NaCl structures.

† Graduate School of Science and Engineering, Tokyo Institute of Technology, Nagatsuda, Midoriku, Yokohama 227.

JARECT Vol. 19, *Semiconductor Technologies* (1986), *J. Nishizawa (ed.)*
OHMSHA, LTD. and North-Holland

(2) The fluorides form mixed crystals[8] whose lattice constants vary continuously from 0.546 nm of CaF_2 to 0.620 nm of BaF_2. Thus the pure or mixed fluorides can be closely or exactly lattice-matched to the most interesting semiconductors such as Si, Ge, GaInAsP and PbSnTe (see Fig. 13.1).

(3) Since fluorides sublime as molecules,[9] stoichiometric fluoride films can easily be obtained by vacuum evaporation.

In Tables 13.1 and 13.2, the fluoride/semiconductor and semiconductor/fluoride/semiconductor systems for which epitaxial growth has been obtained are listed.

In this chapter, we describe growth conditions and structures of pure or mixed

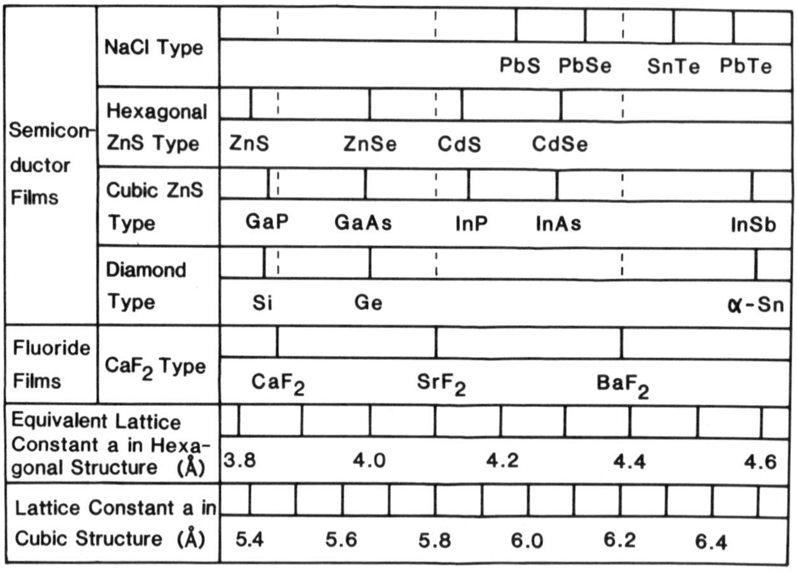

Semiconductor Films
NaCl Type: PbS PbSe SnTe PbTe
Hexagonal ZnS Type: ZnS ZnSe CdS CdSe
Cubic ZnS Type: GaP GaAs InP InAs InSb
Diamond Type: Si Ge α-Sn
Fluoride Films
CaF_2 Type: CaF_2 SrF_2 BaF_2
Equivalent Lattice Constant a in Hexagonal Structure (Å): 3.8 4.0 4.2 4.4 4.6
Lattice Constant a in Cubic Structure (Å): 5.4 5.6 5.8 6.0 6.2 6.4

Fig. 13.1 Lattice constants of fluorides and typical semiconductors. [after H. Ishiwara et al.[22,23]]

Table 13.1 Epitaxial fluoride/semiconductor systems

Fluoride	Semiconductor	Orientation	References
CaF_2	Si	(100), (110), (111)	5, 10–15
	InP	(100)	9, 16, 17, 21
	GaAs	(100)	26
SrF_2	Si	(111)	12, 13
	InP	(100)	17
	GaAs	(100)	25
BaF_2	Si	(111)	12, 13
	Ge	(111)	18–20
	InP	(100), (111)	9, 15–19
	CdTe	(100)	9
$(Ca, Sr)F_2$	Si	(111)	22, 23
	GaAs	(100)	26
$(Ca, Ba)F_2$	InP	(100)	16
$(Sr, Ba)F_2$	InP	(100)	17, 21, 24

Table 13.2 Epitaxial semiconductor/fluoride/semiconductor systems

Semiconductor (epi.)	Fluoride (epi.)	Semiconductor (sub.)	Orientation	References
Si	CaF_2	Si	(100), (111)	5, 10–12, 27, 28
Ge	CaF_2	Si	(100), (111)	29
	SrF_2	Si	(111)	22, 23
	$(Ca, Sr)F_2$	Si	(111)	22, 23
InP	CaF_2	InP	(100)	21
	$(Sr, Ba)F_2$	InP	(100)	21
GaAs	SrF_2	GaAs	(100)	25

fluoride films on Si(111) and (100) substrates and those of Si or Ge films on the fluoride/Si structures. The experimental procedure is described in Section 13.2. The growth conditions of pure fluoride films and effects of lattice mismatch on the structures of the fluoride films are discussed in Section 13.3. In Section 13.4, the growth of Si films on CaF_2/Si structures is investigated. It is found that the interface between deposited Si and underlying CaF_2 is successfully stabilized by depositing thin Si layers at room temperature prior to deposition of Si at high temperatures. Finally, in Section 13.5, the growth of mixed fluoride films on Si substrates and Ge films on the mixed fluoride films is investigated.

13.2. Growth and characterization techniques

The growth of fluoride, Si and Ge films was performed by vacuum evaporation in a 60 l/s ion-pumped vacuum system having a base pressure of 1×10^{-6} Pa. (111) and (100) oriented single-crystal Si wafers were chemically cleaned in $NH_4OH:H_2O_2:H_2O$ and $HCl:H_2O_2:H_2O$ solutions, and dipped in HF acid. Prior to deposition of films, the substrate was heated to 900°C for 10 min in order to evaporate contaminants from the surface. Fluoride grains with 99.9–99.99% purity were evaporated from a resistively heated crucible. The deposition rate of fluoride films was 1–2 nm/s. During the growth of fluoride films, the Si substrate was kept at temperatures from room temperature (RT) to 800°C. For the formation of top Si or Ge films, polycrystalline Si or Ge were evaporated from an electron beam gun. The deposition rates of Si and Ge films were 0.1 nm/s and 0.5 nm/s, respectively.

The composition ratio of the fluoride films and the crystalline quality of both fluoride and semiconductor films were measured by Rutherford backscattering and channeling measurements with 1–2 MeV $^4He^+$ ions. The dose of the $^4He^+$ probe beam was typically 1 μC with a beam diameter of 1 mm. It has been confirmed that effects of the probe beam itself on the composition and crystalline quality of the films are not detectable at doses up to 10 μC. In addition, we have carried out investigations of ion irradiation damage in epitaxial CaF_2 films. Results suggested that fluoride films are less sensitive to ion bombardment than Si.[30] The crystalline quality was also examined by X-ray diffraction analysis with a Cu target and transmission electron microscopy (TEM). The surface morphology of the films was observed with a Nomarski interference microscope.

13.3. Pure-Fluoride/Si structures[5,10–13)]

13.3.1. *Growth conditions*

Fig. 13.2 shows random and ⟨111⟩ aligned channeling spectra for CaF_2, SrF_2 and BaF_2 films grown on Si(111) substrates at 600°C. The composition ratio of each fluoride film, which is calculated from the ratio of signal heights between Ca, Sr or Ba atoms and F atoms in the random spectra, is stoichiometric and uniform. No interdiffusion between the films and substrates can be observed. The aligned spectra show the orientated overgrowth of the fluoride films. The channeling minimum yield χ_{min}, which is defined as the ratio of the aligned yield to the random yield of the film, is less than 5% for each fluoride film, indicating that these fluoride films are of good crystalline quality.

Fig. 13.3(a) shows variations of the χ_{min} of CaF_2, SrF_2 and BaF_2 films grown on Si(111) with substrate temperature. The composition ratio of the fluoride films was stoichiometric and independent of the substrate temperature. The films deposited at RT show no channeling effects. The oriented overgrowth is observed for each fluoride at temperatures higher than 400°C. The crystalline quality of CaF_2 films is improved with increase of the substrate temperature, and the χ_{min} of CaF_2 films grown at 600–800°C are less than 5%. SrF_2 films show similar dependence of χ_{min} to CaF_2 films. But the values of χ_{min} are slightly higher than those of CaF_2 films at each substrate temperature, which is probably attributed to the larger lattice mismatch with Si. The values of χ_{min} of BaF_2 films grown at 400–600°C are lower than those of CaF_2 and SrF_2 films in spite of the larger lattice mismatch between BaF_2 and Si. At temperatures higher than 700°C, however, the crystalline quality of BaF_2 films degrades.

Stoichiometric fluoride films were also formed on Si(100) substrates at RT to 800°C, but the dependence of the crystalline quality of the fluoride films on the substrate temperature was rather complex. Fig. 13.3(b) shows variations of the χ_{min} of CaF_2, SrF_2 and BaF_2 films formed on Si(100) substrates with the substrate temperature. CaF_2 films grown at 500–600°C are excellent in crystalline quality, showing values of χ_{min} less than 5%. The increase of the substrate temperature degrades the crystalline quality of CaF_2 films and no orientated overgrowth is observable at 800°C. For SrF_2 films, the orientated overgrowth is observable at temperatures around 500°C, but the crystalline quality of SrF_2 films is much worse than that of the other fluoride films. BaF_2 films grown at 500–600°C show values of χ_{min} less than 6%, but the quality of BaF_2 films becomes poor at 700–800°C like the BaF_2 films grown on Si(111) substrates.

13.3.2. *Structures*

X-ray diffraction analysis of CaF_2, SrF_2 and BaF_2 films on Si(111) has shown that the (111) face of the fluoride films is parallel to the (111) face of the substrate. However, ion channeling measurements at various directions have revealed two distinct crystal orientations of the epitaxial fluoride films; that is, they have orientations either identical to those of the Si substrate (type A), or rotated through 180° about the surface normal ⟨111⟩ axis of the substrate (type B).[31)] Fig. 13.4 shows channeling spectra taken in the ⟨111⟩, ⟨114⟩ and ⟨110⟩ directions of the substrate for a CaF_2 film grown at 800°C. The values of χ_{min} of the film along the ⟨111⟩, ⟨114⟩ and ⟨110⟩ axes of the substrate are 3.5, 2.9 and 19%,

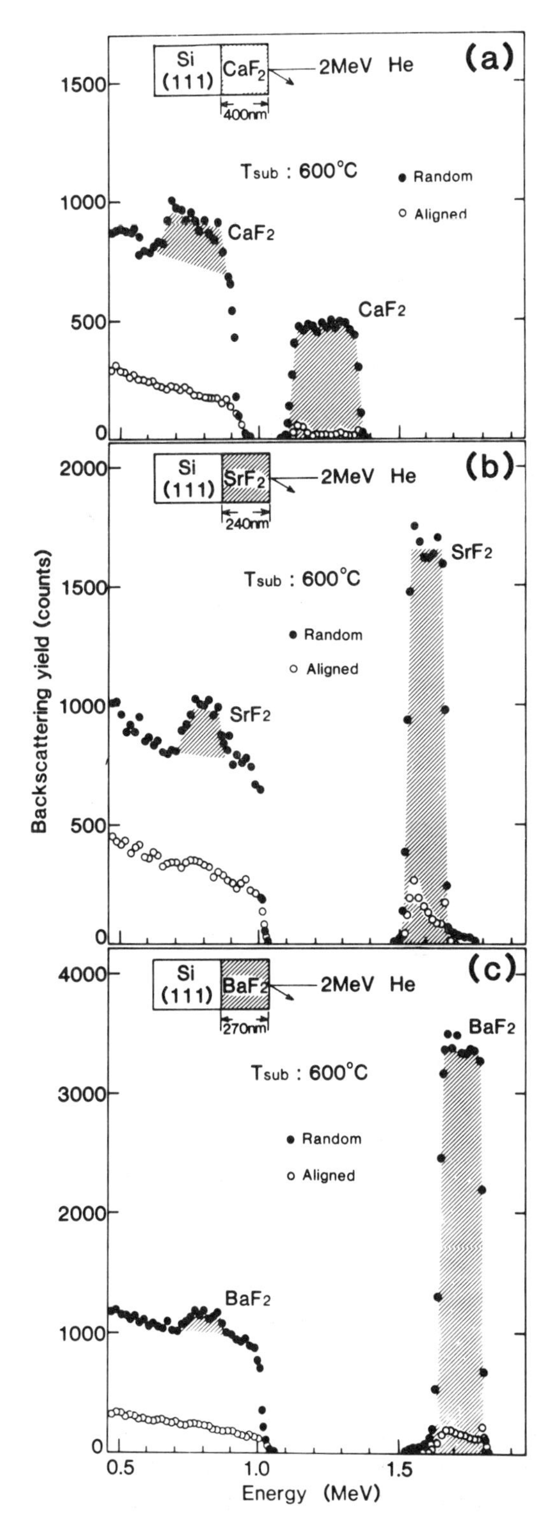

Fig. 13.2 Random and ⟨111⟩ aligned channeling spectra for CaF_2(a), SrF_2(b) and BaF_2(c) films grown on Si(111) at 600°C. [after T. Asano et al.[13)]]

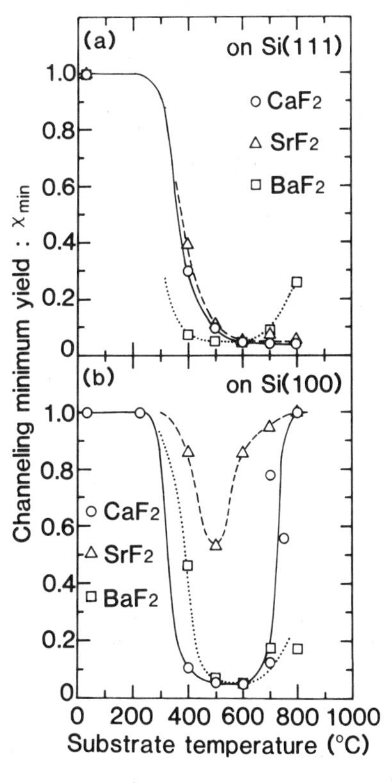

Fig. 13.3 Dependence on the substrate temperature of the channeling minimum yield χ_{min} for CaF_2, SrF_2 and BaF_2 films grown on Si(111)(a) and Si(100)(b). χ_{min} were measured with 2 MeV $^4He^+$ ions. Films were 200–400 nm in thickness. [after T. Asano et al.[13)]]

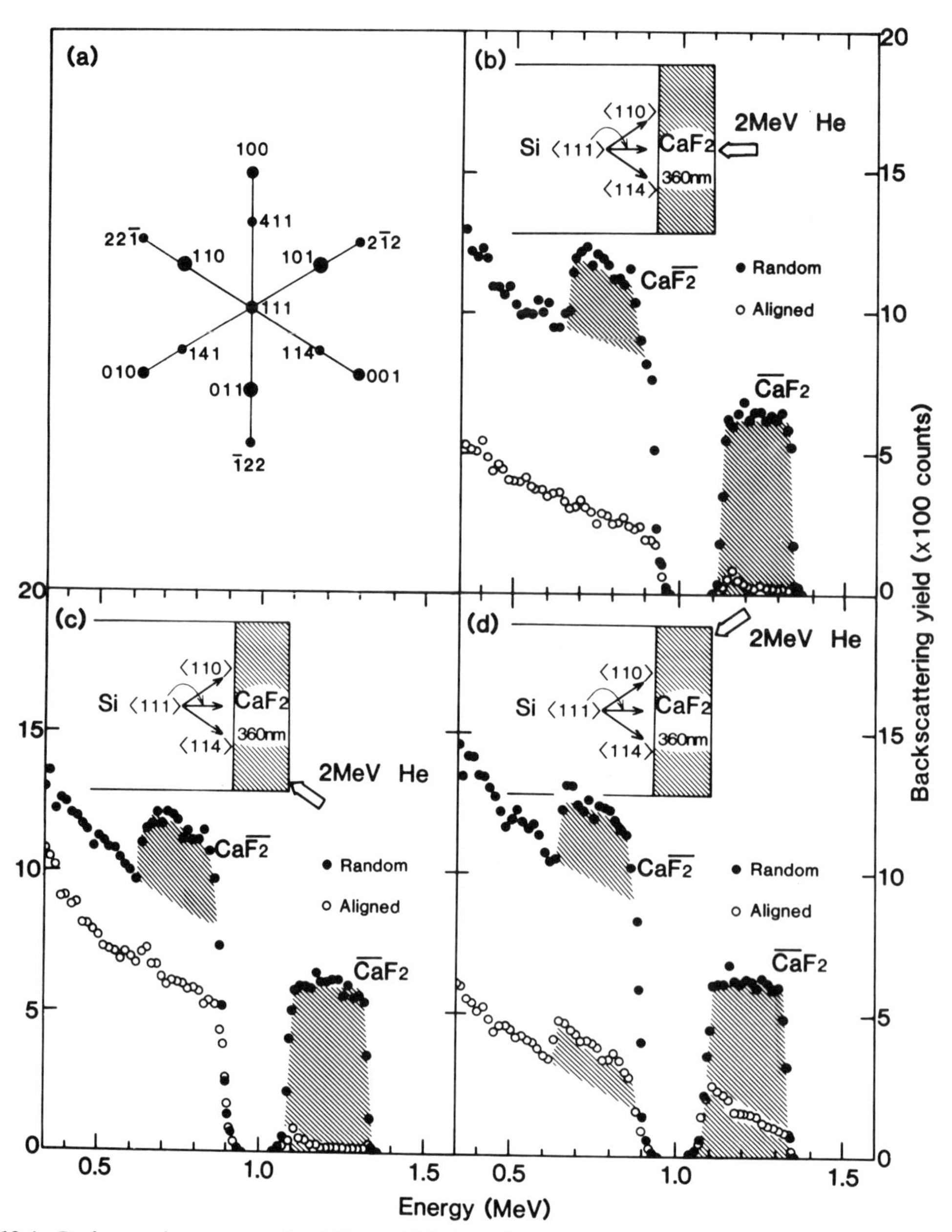

Fig. 13.4 Backscattering spectra of a 360 nm thick CaF_2 film on Si(111), measured in various directions. (a) Standard (111) projection of a cubic crystal. (b) Random and channeling spectra taken in the ⟨111⟩ direction of the substrate. (c), (d) Random and channeling spectra measured in the ⟨114⟩ and ⟨110⟩ directions of the substrate. [after T. Asano et al.[12]]

respectively. This relation between the values of the χ_{min} is contrary to that expected from the fluorite lattice or the diamond lattice. (In these lattices, we expect the ⟨110⟩ axis to have a much lower channeling minimum yield than the ⟨114⟩ axis.) This result clearly demonstrates that the vast majority of the CaF_2 film is of type B orientation. TEM analysis has shown that CaF_2 films grown on Si(111) at 600–800°C are single crystal having crystal orientations of type B.

This discontinuous epitaxial relation is not peculiar to CaF_2, but is observed in other fluoride films. In Fig. 13.5, the channeling minimum yields along the ⟨114⟩, ⟨111⟩ and ⟨110⟩ directions of the substrate are shown for SrF_2 and BaF_2 films grown on Si(111) at 600°C. The variation of χ_{min} of the SrF_2 film is similar to that of the CaF_2 film, so we can say that the growth of SrF_2 is also dominated by the type B epitaxial relation. In the case of BaF_2, however, the values of χ_{min} along both the ⟨114⟩ and ⟨110⟩ axes are larger than that along ⟨111⟩. This result indicates that the BaF_2 film is composed of both types A and B. TEM analysis has shown that the SrF_2 and BaF_2 films are composed of both type A and type B crystallites. The total volumes of the type A crystallites are a few per cent and approximately 25% for the SrF_2 and BaF_2 films, respectively. The origin of the type A and type B growth is not well understood at present. However, results for a $Ca_{0.44}Sr_{0.56}F_2$ mixed crystal film on Si(111)[12] (also plotted in Fig. 13.5) and for varying substrate temperature[13] suggest that the type A and type B growth is affected by local stress near the interface due to lattice mismatch or due to incorporation of foreign atoms with a different atomic radius.

X-ray diffraction and TEM analyses have shown that CaF_2 films grown on Si(100) at 500–600°C are single-crystal.[13] However, it has been found that SrF_2 and BaF_2 films grown on Si(100) contain (111) oriented crystallites. Fig. 13.6 shows X-ray diffraction spectra for SrF_2 and BaF_2 films grown on Si(100) at 500°C. The spectrum of the SrF_2 film shows the presence of (111) oriented crystallites in addition to (100) oriented crystallites. In the spectrum of the BaF_2 film, signals only from (111) oriented BaF_2 can be observed in spite of the use of the (100) oriented substrate. It has been demonstrated that (111) oriented BaF_2 films can be grown even on amorphous SiO_2.[13] Therefore, the appearance of (111) oriented cyrstallites in SrF_2 and BaF_2 films on Si(100) is considered to be mainly due to the preferential growth of these fluorides along the ⟨111⟩ axis. However, channeling measurements for BaF_2/Si(100) along various directions of the substrate have suggested that the alignment of each crystallite is greatly enhanced by the atomic ordering of the Si(100) substrate.[13]

The above experimental results on the comparative study of the growth of pure

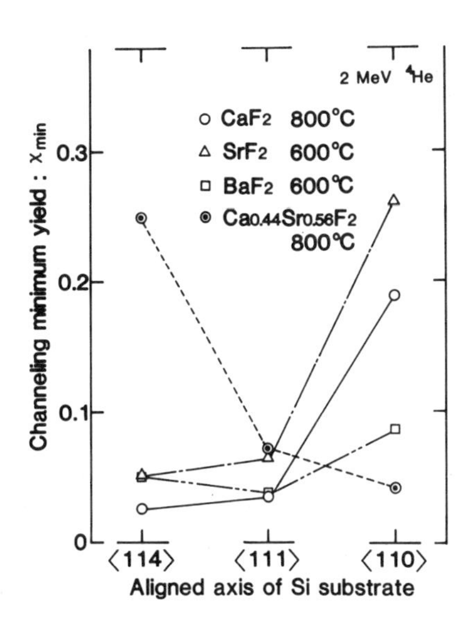

Fig. 13.5 Channeling minimum yield values of χ_{min} along ⟨114⟩, ⟨111⟩ and ⟨110⟩ directions of the Si(111) substrate for CaF_2, SrF_2, BaF_2 and (Ca, Sr)F_2 films. The films were 240–360 nm in thickness. [after T. Asano et al.[12]]

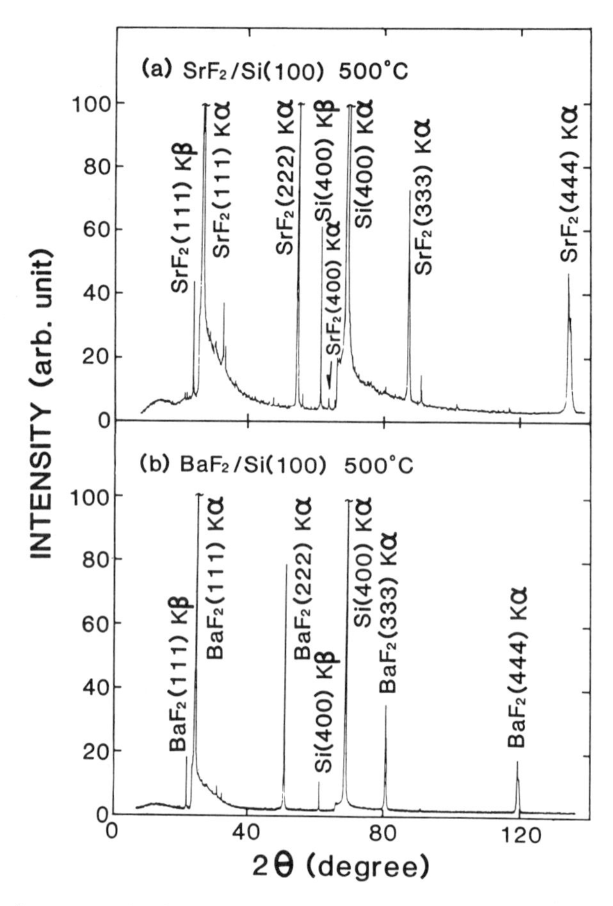

Fig. 13.6 X-ray diffraction spectra for SrF_2(a) and BaF_2(b) films grown on Si(100) at 500°C, showing diffracted signals from (111) oriented crystallites. Thicknesses of SrF_2 and BaF_2 films were 230 nm and 240 nm, respectively. [after T. Asano et al.[13)]]

fluoride films on Si(111) and Si(100) clearly demonstrate that the fluoride films are more likely to grow on Si(111) than Si(100) in the substrate temperature range of 400–800°C. The preferential (111) growth of the fluorides leads the fluoride films to the nearly perfect epitaxial growth on Si(111) even if the lattice mismatch is as high as 14% (BaF_2). It also leads SrF_2 (mismatch = 6.8%) and BaF_2 films to the (111) oriented growth even on Si(100) substrates. But, CaF_2 which has the smallest lattice mismatch (0.6%) grows as single crystal on Si(100).

13.4. Si/CaF_2/Si structures[5,10–12,28,33)]

13.4.1. *Direct deposition of Si onto heated substrates*

In order to form Si/CaF_2/Si structures, CaF_2 films were first deposited onto Si(111) at 800°C and Si(100) at 600°C. Si films were then deposited onto the CaF_2/Si structures without breaking the vacuum at temperatures from 400 to 800°C. Fig. 13.7 shows the dependence of χ_{min} of the top Si films on CaF_2/Si(111) and CaF_2/Si(100) structures on the

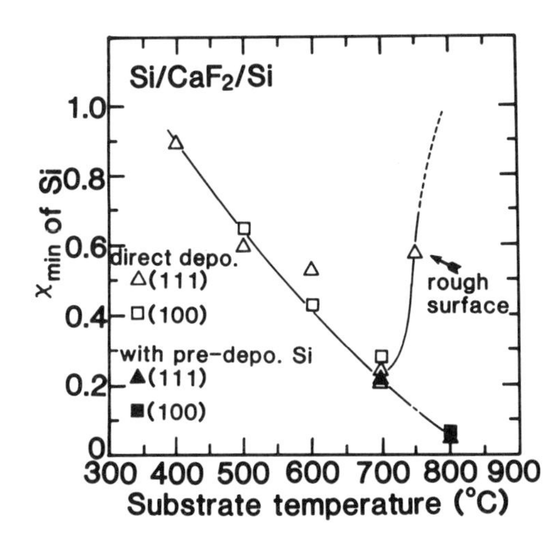

Fig. 13.7 Dependence on the substrate temperature of the channeling minimum yield χ_{min} for Si films grown on CaF_2/Si(111) and CaF_2/Si(100) structures. Values of χ_{min} were measured with 1 MeV $^4He^+$ ions. Si films were about 400 nm in thickness. [after H. Ishiwara et al.[33]]

substrate temperature. The epitaxial growth becomes evident at 500°C, and the crystalline quality of Si films is improved with increase of the substrate temperature. No significant difference between the Si films on (111) and (100) oriented substrates can be observed. The surface of the Si films was smooth at temperatures up to 700°C. Above 750°C, however, the surface of the Si films became rough, which might result from reactions between deposited Si and underlying CaF_2. According to the occurrence of the surface reaction, the crystalline quality of the Si films became poor at 750°C as shown by the open triangle in the figure.

13.4.2. *Thin Si predeposition technique*

In order to prevent the surface reaction between deposited Si and underlying CaF_2 during the growth of Si at high temperatures, we have introduced a thin Si layer deposited in-situ onto the CaF_2 surface prior to deposition of a thick Si film at elevated temperatures. A similar technique has been successfully applied to the chemical vapor deposition of Si onto sapphire substrates.[32] The preparation procedure in our method is as follows. After the CaF_2 deposition, the temperature of the sample was decreased to RT and a thin Si layer was deposited on top of the CaF_2/Si structure. The thickness of the thin Si layer was varied from 4 to 32 nm. Then the temperature of the substrate was elevated to 800°C, and Si deposition was continued. The final thickness of the Si film was about 400 nm.

Fig. 13.8 shows backscattering random and aligned spectra of Si films grown on the CaF_2/Si(111) and CaF_2/Si(100) structures. In these samples, 4 nm thick Si layers were deposited at RT prior to deposition of 400 nm thick Si films at 800°C. Despite the fact that the substrate temperature is high enough for the reaction to take place if the CaF_2 surface is not covered with predeposited Si, the random spectra in Fig. 13.8 show the formation of uniform 400 nm thick Si films. The surface of these samples appeared smooth through examination by an optical interference microscope. We can see from the aligned spectra in Fig. 13.8 and from electron diffraction analysis that Si films grow epitaxially on both (111) and (100) oriented substrates. The channeling minimum yield χ_{min} of the Si film on the (111) substrate is about 21% near the interface, and it reduces to about 7% near the

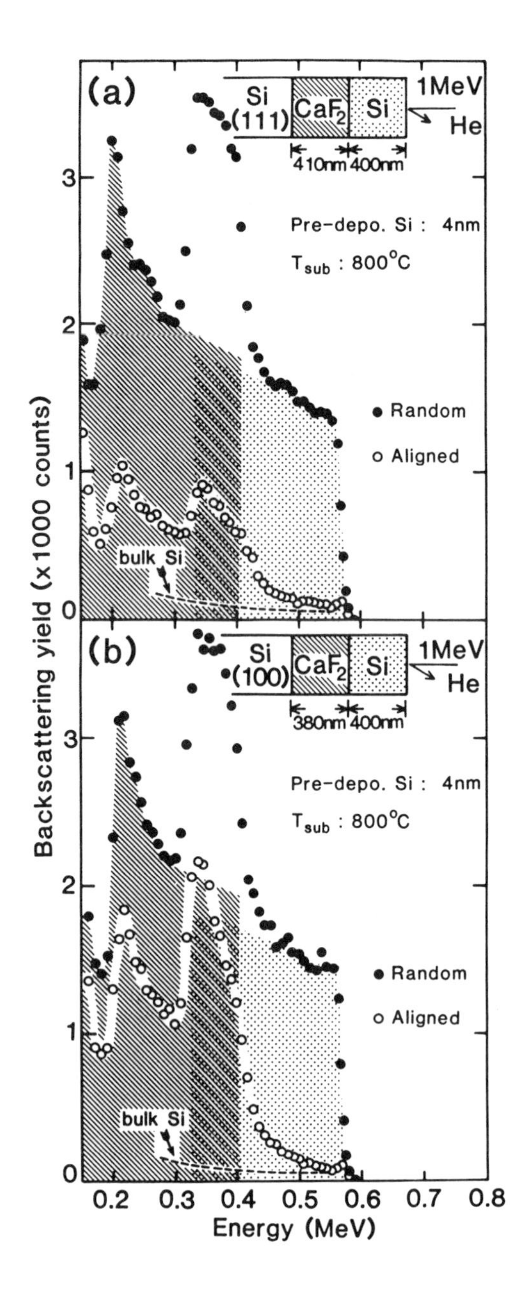

Fig. 13.8 Random and aligned backscattering spectra of 400 nm thick Si films grown at 800°C on CaF_2/Si(111)(a) and CaF_2/Si(100)(b) structures predeposited with 4 nm thick Si layers at room temperature. These spectra were taken with 1 Mev $^4He^+$ ions. [after T. Asano et al.[28)]]

surface. In the case of the Si film on the (100) substrate, the value of χ_{min} near the interface is about 67%, but it is markedly reduced toward the surface, and the value of χ_{min} near the surface is about 6%.

Fig. 13.9 shows the variation of the near surface χ_{min} of approximately 400 nm thick Si films on (111) and (100) substrates with thickness of the predeposited Si layer. No particular difference in these plots can be observed between (111) and (100) substrates. The values of χ_{min} of Si films on both (111) and (100) substrates are less than 8% when the predeposited Si layer is about 10 nm or less in thickness. However, the increase of thickness of the predeposited Si layer results in poor crystalline quality of Si films, and no

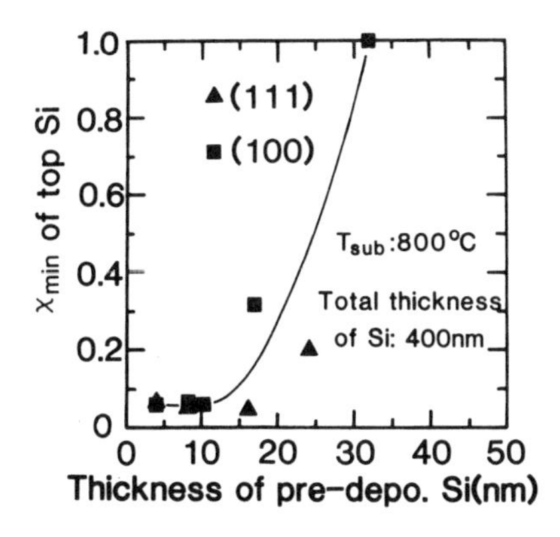

Fig. 13.9 Variation of the near surface χ_{min} of 400 nm thick Si films grown on CaF_2/Si structures with thickness of predeposited Si layers. [after T. Asano et al.[28]]

epitaxial growth is observed on Si(100) substrates when the thickness of the predeposited Si layer is more than 30 nm.

In order to investigate the effects of the predeposited Si layer on the epitaxial growth of Si at lower temperatures, a 5 nm thick Si layer was predeposited onto the CaF_2/Si(111) structure, then a 400 nm thick Si film was deposited at 700°C. The resultant crystalline quality of the Si film was very similar to that of a Si film grown directly on the CaF_2/Si(111) structure at 700°C. These results were also plotted in Fig. 13.7. From these results, we can say that, by the use of the thin (≤ 10 nm) Si predeposition technique, the reaction between underlying CaF_2 and deposited Si at high temperatures can be prevented without destroying the epitaxial growth. A possible model of the epitaxial growth mechanism in this technique has been proposed based on investigations of the early stage of the growth.[28]

13.5. Mixed-fluoride/Si and Ge/(Ca, Sr)F_2/Si structures[22,23]

13.5.1. *Mixed-fluoride/Si*

Fig. 13.10 shows backscattering random and aligned spectra of a $Ca_{0.44}Sr_{0.56}F_2$ film grown on a Si(111) substrate at 700°C. The random spectrum shows that the in-depth profile of the composition ratio is uniform. The aligned spectrum shows that the mixed fluoride film grows epitaxially. The channeling minimum yield χ_{min} of the film is about 7% for both Ca and Sr portions. Fig. 13.11 shows the dependence of χ_{min} of $Ca_{0.44}Sr_{0.56}F_2$ films on the substrate temperature during deposition. (Ca, Sr)F_2 films of good crystalline quality can be grown at 600–800°C which is similar to the temperature range for pure fluoride films to grow in good crystalline quality. It has been confirmed by X-ray diffraction analysis that the (111) face of the fluoride film is parallel to the (111) face of the substrate and that the film is not composed of unmixed CaF_2 and SrF_2 crystallite but a mixed crystal.

The epitaxial growth of (Ca, Sr)F_2 films with other composition ratios was also observed. Fig. 13.12 shows the variation of the lattice constant of epitaxial (Ca, Sr)F_2 films on Si(111) with the composition ratio of the films. The composition ratio was determined from backscattering measurements. The lattice constant of the epitaxial films varies linearly with the composition ratio, that is the epitaxial mixed fluoride films satisfy Vegard's law as observed in bulk crystals.[8]

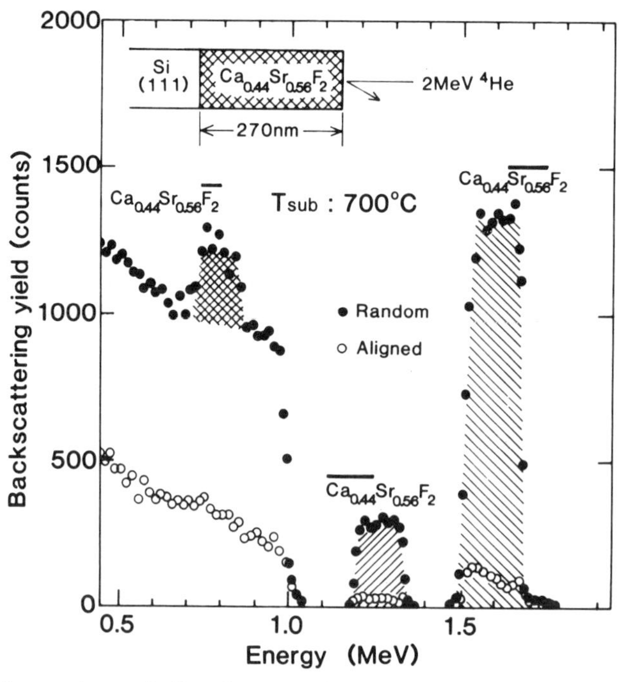

Fig. 3.10 Backscattering random and aligned spectra of a $Ca_{0.44}Sr_{0.56}F_2$ film grown on Si(111) at 700°C. [after H. Ishiwara et al.[22]]

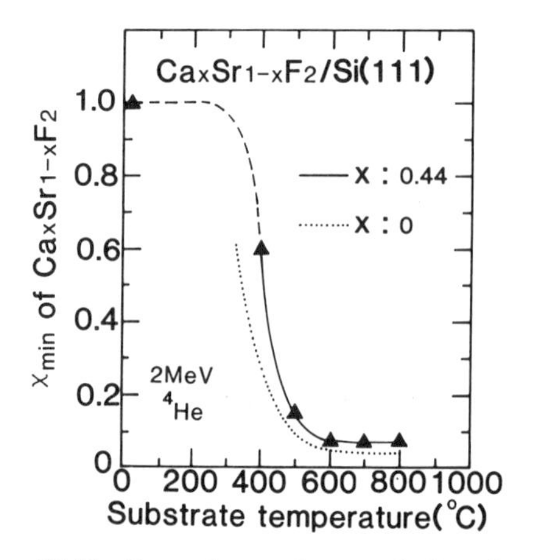

Fig. 13.11 Dependence of χ_{min} of $Ca_{0.44}Sr_{0.56}F_2$ films on the substrate temperature. The films were 200–320 nm in thickness. [after H. Ishiwara et al.[22]]

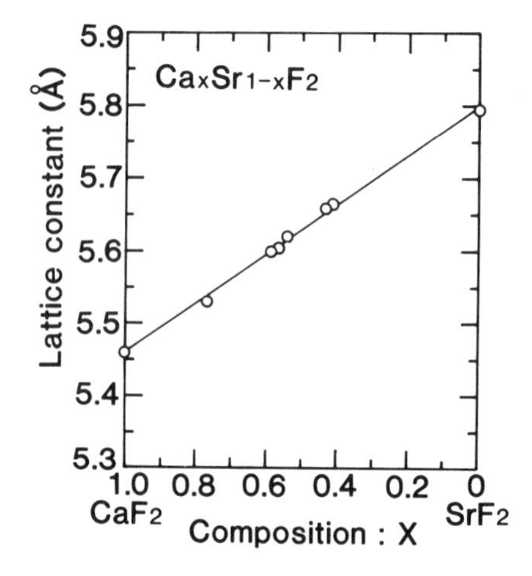

Fig. 13.12 Variation of the lattice constant of epitaxial (Ca, Sr)F_2 films with the composition ratio. [after H. Ishiwara et al.[22]]

13.5.2. *Ge/(Ca, Sr)F_2/Si structures*

Finally, the epitaxial growth of Ge films onto lattice-matched epitaxial (Ca, Sr)F_2 films is demonstrated. Fig. 13.13 shows the backscattering spectra of a Ge film grown on a $Ca_{0.44}Sr_{0.56}F_2$/Si(111) structure at 500°C. The lattice constant of the underlying (Ca, Sr)F_2 film determined by X-ray diffraction was 0.566 nm, which matched to the lattice constant

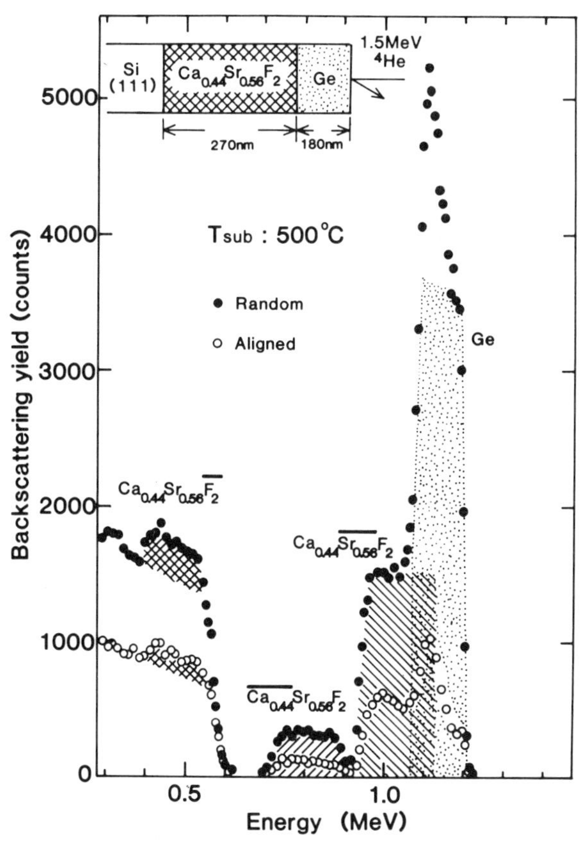

Fig. 13.13 Backscattering spectra of a Ge/$Ca_{0.44}Sr_{0.56}F_2$/Si(111) structure. The Ge film was grown at 500°C. [after H. Ishiwara et al.[22,23]]

of Ge within experimental error. The aligned spectrum in the figure shows the epitaxial growth of the Ge film. The value of χ_{min} of the Ge film is 9% near the surface and 14% near the interface. The crystalline quality of the Ge film is better than that for Ge films grown on CaF_2/Si(111) and SrF_2/Si(111) structures at the same substrate temperature.[29]

13.6. Summary

The epitaxial growth of group-IIa fluorides on Si substrates and the growth of Si and Ge films on epitaxial fluoride films have been investigated. The films were prepared by vacuum deposition. The results of this work are summarized as follows:

(1) Single-crystal CaF_2 films can be grown on both (111) and (100) oriented Si substrates at temperatures of 600–800°C and 500–600°C, respectively.
(2) Single-crystal CaF_2 films grown on Si(111) have orientations rotated through 180° about the surface normal ⟨111⟩ axis of the substrate.
(3) SrF_2 and BaF_2 films of good crystalline quality can be grown on Si(111) substrate at temperatures around 600°C. But they are composed of two types of crystallites, i.e. one has orientations identical to those of the substrate and the other has orientations rotated through 180° about the normal to the substrate surface.

(4) SrF_2 and BaF_2 films grown on Si(100) at 400–800°C contain (111) oriented crystallites. In an extreme case, completely (111) oriented BaF_2 films were obtained on Si(100).
(5) Si films grow epitaxially on CaF_2/Si structures at 500–800°C. At high growth temperatures (about 800°C), thin($\leq$ 10 nm) Si layers predeposited at room temperature on the CaF_2 surface are useful to prevent the reaction between deposited Si and underlying CaF_2 without destroying the epitaxial growth.
(6) Mixed fluoride (Ca, Sr)F_2 films of good crystalline quality can be grown on Si(111) at 600–800°C. The lattice constant of the epitaxial fluoride films can be controlled by changing the composition ratio of the fluoride films.
(7) Epitaxial Ge films can be grown on (Ca, Sr)F_2/Si structures at about 500°C. Ge films of better crystalline quality can be obtained when the lattice matching between Ge and underlying (Ca, Sr)F_2 is satisfied.

From these results, we conclude that semiconductor/insulator layered structures formed by low-temperature epitaxial growth of group-IIa fluorides and semiconductors such as Si, Ge, GaInAsP and other compounds are promising in fabrication of future electronic and optical devices.

Acknowledgements

The authors are grateful for a useful discussion with Professor Seijiro Furukawa. A part of this work was carried out with Kouzo Orihara (Nippon Electric Co.), Noriyuki Kaifu (Cannon Inc.) and Seigo Kanemaru (Tokyo Institute of Technology). The authors wish to gratefully acknowledge their cooperation.

References

1) B. R. Appleton and G. K. Celler (Eds.): Laser and Electron-Beam Interactions with Solids (North-Holland, Amsterdam, 1982).
2) Y. Ohmura, Y. Matsushita, and M. Kashiwagi: Japan. J. Appl. Phys., 21 (1982) L152.
3) Y. Kunii, M. Tabe, and K. Kajiyama: J. Appl. Phys., 54 (1983) 2847.
4) H. Ishiwara, H. Yamamoto, S. Furukawa, M. Tamura, and T. Tokuyama: Appl. Phys. Lett., 43 (1983) 1028.
5) H. Ishiwara and T. Asano: Appl. Phys. Lett., 40 (1982) 66.
6) M. Ihara, Y. Arimoto, M. Jifuku, T. Kimura, S. Kodama, H. Yamawaki, and T. Yamaoka: J. Electrochem. Soc., 129 (1982) 2569.
7) K. Nonaka, C. K. Kim, and K. Shono: J. Crystal Growth, 50 (1980) 549.
8) E. G. Chernevskaya and G. V. Ananeva: Sov. Phys. Solid State, 8 (1966) 169.
9) R. F. C. Farrow, P. W. Sullivan, G. M. Williams, G. R. Jones, and D. C. Cameron: J. Vac. Sci. Technol., 19 (1981) 415.
10) T. Asano and H. Ishiwara: Proc. 13th Conf. Solid State Devices (Tokyo, 1981); Japan. J. Appl. Phys., 21 (1982) Suppl. 21-1, p. 187.
11) T. Asano and H. Ishiwara: Thin Solid Films, 93 (1982) 143.
12) T. Asano and H. Ishiwara: Appl. Phys. Lett., 42 (1983) 517.
13) T. Asano, H. Ishiwara, and N. Kaifu: Japan. J. Appl. Phys., 22 (1983) 1474.
14) T. R. Harrison, P. M. Mankiewich, and A. H. Dayem: Appl. Phys. Lett., 41 (1982) 1102.
15) P. W. Sullivan, T. I. Cox, R. F. C. Farrow, G. R. Jones, D. B. Gasson, and C. S. Smith: J. Vac. Sci. Technol., 20 (1982) 731.
16) P. W. Sullivan, R. F. C. Farrow, and G. R. Jones: J. Crystal Growth, 60 (1982) 403.
17) C. W. Tu, T. T. Sheng, M. H. Read, A. R. Schlier, J. G. Johnson, W. D. Johnstron, Jr., and W. A. Bonner: J. Electrochem. Soc., 130 (1983) 2081.
18) J. M. Phillips, L. C. Feldman, J. M. Gibson, and M. L. McDonald: J. Vac. Sci. Technol., B1 (1983) 246.

19) J. M. Gibson and J. M. Phillips: Appl. Phys. Lett., 43 (1983) 828.
20) J. M. Phillips, L. C. Feldman, J. M. Gibson, and M. L. McDonald: Thin Solid Films, 107 (1983) 217.
21) C. W. Tu, S. R. Forrest, and W. D. Johnstron, Jr.: Appl. Phys. Lett., 43 (1983) 569.
22) H. Ishiwara and T. Asano: Proc. 14th Conf. Solid State Devices (Tokyo, 1982); Japan. J. Appl. Phys., 22 (1983) Suppl. 22-1, p. 201.
23) H. Ishiwara, T. Asano, and S. Furukawa: J. Vac. Sci. Technol., B1 (1983) 266.
24) C. W. Tu, T. T. Sheng, A. T. Macrander, J. M. Phillips, and H. J. Guggenheim: J. Vac. Sci. Technol., B2 (1984) 24.
25) P. W. Sullivan: Appl. Phys. Lett., 44 (1984) 190.
26) S. Siskos, C. Fontaine, and A. Munoz-Yague: Insulating Films on Semiconductors, Eds. J. F. Verweij and R. R. Wolters (North-Holland, Amsterdam, 1983) p. 71.
27) T. Asano, H. Ishiwara, K. Orihara, and S. Furukawa: Japan. J. Appl. Phys., 22 (1983) L118.
28) T. Asano and H. Ishiwara: J. Appl. Phys., (1984) 3566.
29) T. Asano and H. Ishiwara: Japan. J. Appl. Phys., 21 (1982) L630.
30) H. Ishiwara, K. Orihara, and T. Asano: Japan. J. Appl. Phys., 22 (1983) L458.
31) R. T. Tung, J. M. Poate, J. C. Bean, J. M. Gibson, and D. C. Jacobson: Thin Solid Films, 93 (1982) 77.
32) M. Ishida, H. Ohyama, S. Sasaki, Y. Yasuda, T. Nishinaga, and T. Nakamura: Japan. J. Appl. Phys., 20 (1981) L541.
33) H. Ishiwara and T. Asano: Thin Films and Interfaces, Eds. J. Baglin, D. Campbell, and W. K. Chu (North-Holland, Amsterdam, 1984) p. 393.

14 INFRARED RAPID THERMAL ANNEALING FOR ION-IMPLANTED GaAs

Hideaki KOHZU, Masaaki KUZUHARA and Yoichiro TAKAYAMA†

Abstract

A rapid thermal process, utilizing infrared radiation from halogen lamps, is used to post-anneal ion-implanted GaAs. The annealing conditions for implantation of Si in GaAs are discussed from the viewpoint of applying this technique to GaAs MESFET fabrication. Also, the properties of S and Mg implantation in GaAs followed by infrared rapid thermal annealing are studied and compared with the results after conventional furnace annealing. High electrical activation and minimized implant diffusion for both low- and high-dose implantation are the principal features of this technique. The fabricated MESFET shows much higher transconductance without any anomalous characteristics, indicating that this technique is a promising alternative to conventional furnace annealing.

Keywords: GaAs, Ion-Implantation, Annealing

14.1. Introduction

In recent years, rapid thermal annealing techniques have been extensively studied for ion-implanted III–V compound semiconductors, such as GaAs and InP, to accomplish activation of implanted dopants in the solid phase. These techniques involve annealing periods of 1–10 s, which is much shorter than that for conventional furnace annealing, resulting in minimized implant diffusion during annealing. These features are listed in Table 14.1.

Various heating sources have so far been explored for rapid thermal annealing, including incoherent light sources,[1–9] graphite strip heaters,[10,11] and multiply scanned electron beams.[12,13]

The most simple and inexpensive method is the use of graphite strip heaters. Their advantages include their facility of securing uniformity because they are intrinsically area-heating sources. However, the graphite heater technique has the possibility of contamination from the graphite heater itself. Furthermore, the annealing chamber must be evacuated to obtain a strictly oxygen-free ambient. In the case of electron-beam annealing, it is possible to anneal the implanted region selectively or separately. However, it must be carried out in a vacuum. Therefore, annealing under ambient gases, such as an As pressure controlled atmosphere, cannot be used in this annealing method.

Compared with these two heating sources, incoherent light sources have some inherent advantages as follows: (1) no evacuation system is needed; (2) implanted wafers are thermally isolated in the annealing chamber; (3) the emitted light power can be

† Microelectronics Research Laboratories, NEC Corporation, 4-1-1, Miyazaki, Miyamae-ku, Kawasaki 213.

JARECT Vol. 19, Semiconductor Technologies (1986), *J. Nishizawa (ed.)*
OHMSHA, LTD. and North-Holland

Table 14.1 Comparison of features between IRTA and furnace annealing.

Item	IRTA	Furnace annealing
Annealing time	1–10 s	600–1800 s
Annealing temperature	900–1100°C	800–900°C
Capless annealing	Yes	No (Yes, under As pressure)
Operating time	2–4 min	40–60 min
Implant diffusion	Negligible	Large

collected by reflectors; (4) large annealing systems can be constructed by increasing the number and/or length of the light sources.

Several researchers have reported some preliminary results on the rapid thermal annealing for implanted GaAs, including its application to device fabrication. Kohzu et al.[8] first applied rapid thermal annealing to GaAs MESFET fabrication. The Si-implanted active layers, activated by 950°C, 2 s rapid annealing, lead to a high transconductance of 113 mS/mm due to a realization of steep carrier concentration profile. Asbeck et al.[11] used a graphite strip heater to anneal a Be-implanted p-type region in MBE-grown GaAlAs/GaAs heterojunction bipolar transistors. They obtained 67% electrical activation for $1 \times 10^{15}\,cm^{-2}$ Be implantation in GaAs and reported a transistor current gain of 200 from 925°C, 10 s rapid annealing.

This chapter describes rapid thermal annealing using halogen lamps, called infrared rapid thermal annealing (IRTA), for Si-implanted GaAs for fabricating GaAs MESFETs. The conditions for capless annealing and uniformity of the activated layer are discussed. Also, the annealing behavior of S and Mg implantation in GaAs for the formation of n^+ and p^+ layers is discussed. Finally, the application of IRTA to GaAs MESFETs is presented with the good results obtained.

14.2. IRTA apparatus

A cross-sectional view of the IRTA apparatus, with the upper and lower lamp units, is shown in Fig. 14.1. These units, being connected with eight and nine halogen lamps, respectively, are facing each other at alternate interleaved positions to improve the

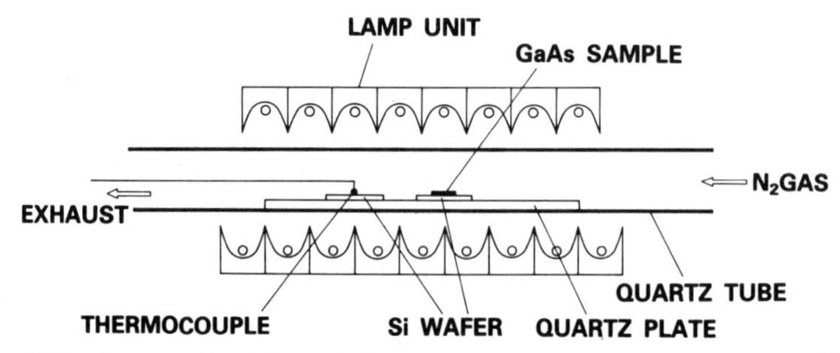

Fig. 14.1 Cross-sectional view of the IRTA apparatus. [after H. Kohzu et al.[8]]

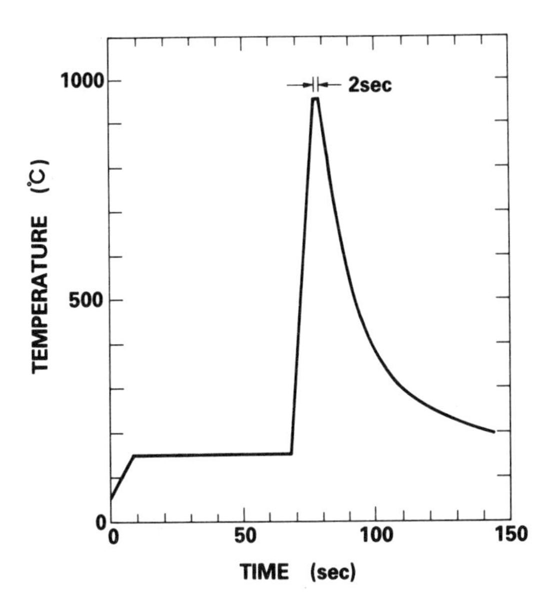

Fig. 14.2 Measured temperature cycle for IRTA [after H. Kohzu et al.[8]]

temperature uniformity of the samples. Each halogen lamp of 127 mm length has a parabolic Au-coated reflector to produce parallel reflected rays as well as to raise the irradiation efficiency. A maximum electric power of 1.2 kW can be applied to each lamp.

A rectangular quartz tube is put between the lamp units and is filled with flowing nitrogen gas during annealing. An implanted GaAs sample is placed on the sample holder in the quartz tube, so that the implanted surface faces the holder.

The sample temperature, which is monitored by a C–A thermocouple on another Si wafer, is controlled by a PID controller. An example of a measured temperature cycle is shown in Fig. 14.2. Preheating the GaAs sample at 150°C for 1 min is performed before annealing, to make the initial sample temperature identical from run to run. The heating rate can be changed up to 200°C/s. However, it is fixed at 100°C/s in this experiment. The maximum maintained temperature and its duration are defined as the annealing temperature and time, respectively. The reproducibility of the annealing temperature is within ±10°C for 950°C rapid annealing. After annealing, the sample cools at an uncontrolled rate, and its cooling rate down to 600°C is more than 100°C/s.

14.3. Conditions for capless annealing

One of the advantages of the IRTA process is its capability of capless annealing, due to its short annealing time. The surface dissociation conditions for GaAs samples followed by capless IRTA with two types of sample holder, i.e. GaAs wafer (broken line) and Si wafer (solid line), are shown in Fig. 14.3. Under the annealing conditions at the upper portion of each line, thermal pits were observed on the GaAs surface through a microscope. As shown in Fig. 14.3, the GaAs wafer is superior to the Si wafer as a sample holder to suppress the surface dissociation during IRTA. However, the Si wafer sample holder is also acceptable for actual annealing conditions, such as 950°C for 2 s. Therefore, the Si wafer was generally used as a sample holder in this study.

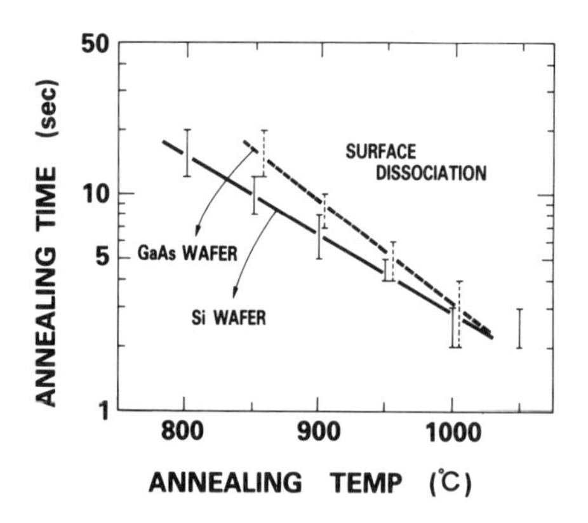

Fig. 14.3 Annealing conditions for capless IRTA. [after H. Kohzu et al.[8]]

14.4. Uniformity of the activated layer

In order to apply the IRTA technique to fabrication of devices, such as GaAs ICs, uniformly activated layers are essential. The temperature uniformity in this type of annealing apparatus is mainly determined by the spacing between the lamp units and the sample, because each lamp is long and narrow. Fig. 14.4 shows the pinch-off voltage uniformity evaluated by *C–V* methods on half of a horizontal Cr-doped GaAs wafer grown by the Bridgman method. The wafer was implanted with 100 keV Si to a dose of 5×10^{12} cm^{-2} and subsequently annealed at 950°C for 2 s under 70 mm spacing between the sample and the lamp units. The average and standard deviation of the pinch-off voltage for 1027 diodes are −3.2 V and 177 mV, respectively. On the other hand, the same values for IRTA under 55 mV spacing are −3.82 V and 798 mV, respectively. The uniformity shown in Fig. 14.4 represents almost the same result as those after furnace annealing with Si_3N_4 encapsulant.

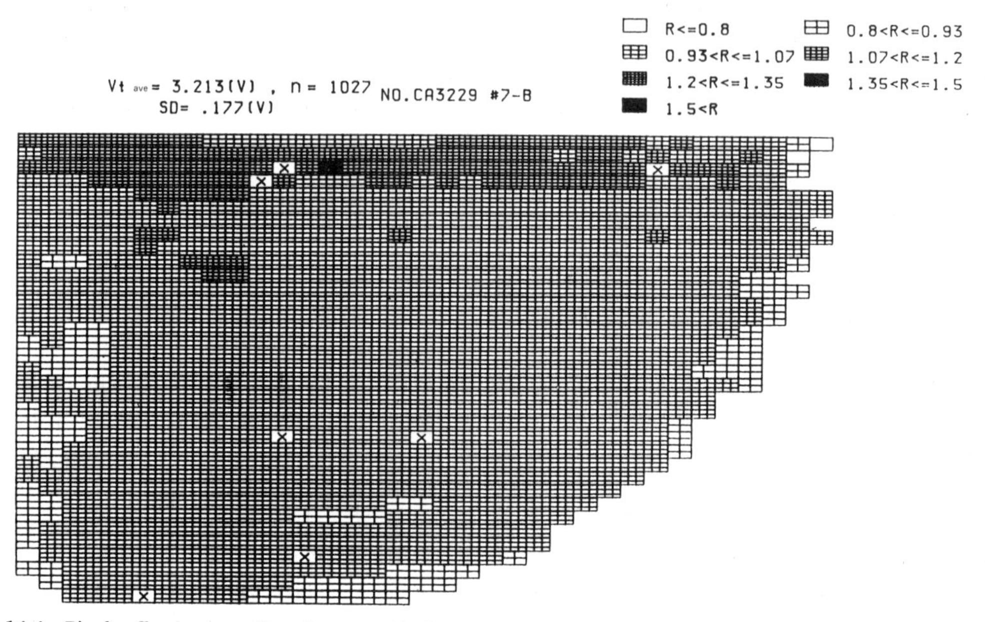

Fig. 14.4 Pinch-off voltage uniformity over a half of HB GaAs wafer. *R* is the ratio of the measured pinch-off voltage to the averaged pinch-off voltage.

14.5. Implantation of Si in GaAs

14.5.1. *Low-dose implantation of Si*

Fig. 14.5 shows the dependence of electrical activation on annealing temperature for 150 keV Si implantation in GaAs with doses of 3×10^{12} to 3×10^{13} cm^{-2}. The annealing time was fixed at 2 s for each annealing temperature from 750 to 1100°C. The range of annealing temperature to achieve high electrical activation is 900 to 1000°C. For annealing above 1000°C, the electrical activation begins to decrease. However, the similar experiment for implantation of S in GaAs showed an increase in electrical activation above 1000°C, indicating that the decrease in electrical activation for Si implantation above 1000°C can be ascribed to an increase in the compensation ratio, because of the amphoteric behavior of Si implants in GaAs.

The dependence of electrical activation on annealing time at annealing temperatures of 850 and 950°C is shown in Fig. 14.6, where Si was implanted at 100 keV to a dose of 5×10^{12} cm^{-2}. The maximum electrical activation is obtained for 2 s at 950°C and 10 s at 850°C, respectively. Since these annealing times are almost the maximum times allowable for which no thermal pits are observed on the annealed surfaces, as shown in Fig. 14.3, the decrease in electrical activation shown in Fig. 14.6 is considered to be due to the surface dissociation of GaAs.

Fig. 14.7 shows the carrier concentration profile for 100 keV Si implantation in GaAs with a dose of 5×10^{12} cm^{-2} followed by IRTA for 2 s at 950°C, together with the result after Si_3N_4-capped furnace annealing for 15 min at 850°C. A marked difference in carrier

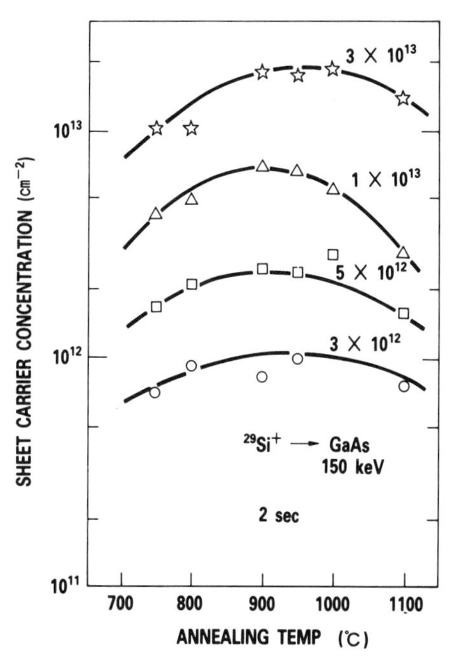

Fig. 14.5 Dependence of sheet carrier concentration on annealing temperature. [after H. Kohzu et al.[8)]]

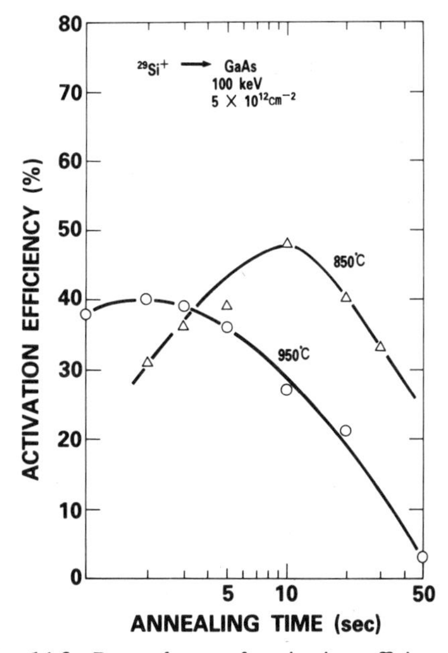

Fig. 14.6 Dependence of activation efficiency on annealing time. [after H. Kohzu et al.[8)]]

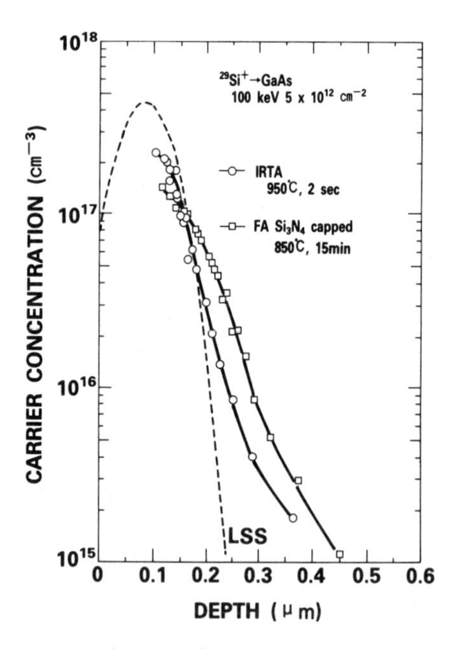

Fig. 14.7 Carrier concentration profiles for low-dose implantation of Si in GaAs, followed by IRTA and furnace annealing. [after H. Kohzu et al.[8)]]

concentration profile is observed between these two annealing methods. The IRTA-made layer has a higher peak carrier concentration and steeper carrier concentration profile than that of the layer made by furnace annealing. In view of the carrier concentration profile desired for the fabrication of high-performance GaAs MESFETs, the active layer made by IRTA is superior to that made by furnace annealing.

14.5.2. *High-dose implantation of Si*

Carrier concentration profiles for high-dose Si implantation (5×10^{13} cm^{-2}, 150 keV) activated by capless IRTA at 950°C for 4s are shown in Fig. 14.8, together with that

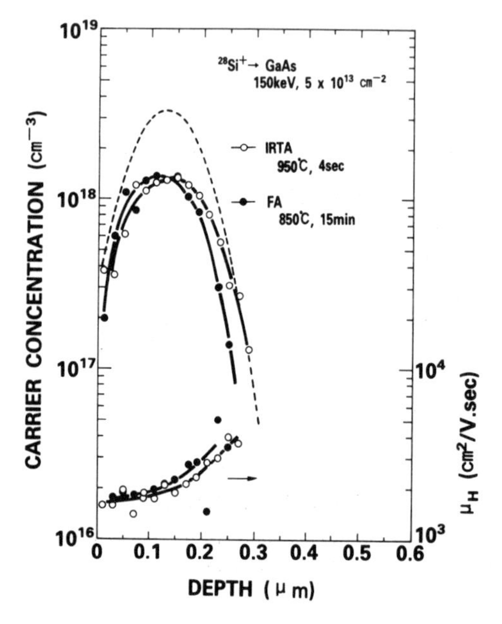

Fig. 14.8 Carrier concentration profiles for high-dose implantation of Si in GaAs, followed by IRTA and furnace annealing. [after H. Kohzu et al.[8)]]

Table 14.2 Results of Hall measurements for capped IRTA (7×10^{13} cm^{-2}, 150 keV).

Anneal temperature (°C)	Encapsulant	n_s (10^{13} cm^{-2})	μ_s (cm^2/V s)	R_s ($\Omega/\square$)
1000	SiO_2	2.86	2123	103
1000	Si_3N_4	1.95	2339	137
1000	SiO_xN_y	4.19	2074	71.8
1100	SiO_2	1.41	2224	200
1100	Si_3N_4	3.28	1982	96.1
1100	SiO_xN_y	5.40	2022	57.2

activated by Si_3N_4-capped furnace annealing. Carrier concentration and mobility profiles for the IRTA sample are almost the same as those for the furnace-annealed sample, when the accuracy of the technique for removal of successive layers is taken into account. The maximum peak carrier concentrations were 1.3×10^{18} cm^{-3} for both annealing methods.

In order to achieve much higher carrier concentration for high-dose Si implantation in GaAs, the capless IRTA conditions obtainable such as 950°C, 4s, are not sufficient, although a comparable electrical activation with that after furnace annealing can be obtained by capless IRTA. Therefore, IRTA with encapsulating films was examined to prevent the surface dissociation of GaAs during high-temperature IRTA. Three types of encapsulant, namely SiO_2, Si_3N_4, and SiO_xN_y with refractive index of 1.75, were used to activate 7×10^{13} cm^{-2}, 150 keV Si implantation in GaAs. The thickness of each encapsulant was fixed about 1000 Å. The results obtained after IRTA for 4s at temperatures from 1000 to 1100°C are listed in Table 14.2. It should be noted that a marked improvement in electrical activation is observed for SiO_xN_y-capped IRTA, as has also been observed in the standard furnace annealing.[17] This effect can be ascribed to the generation of a certain concentration of Ga vacancies in the GaAs substrate due to outdiffusion of Ga, which facilitates the incorporation of Si dopants into Ga lattice sites. A peak carrier concen-

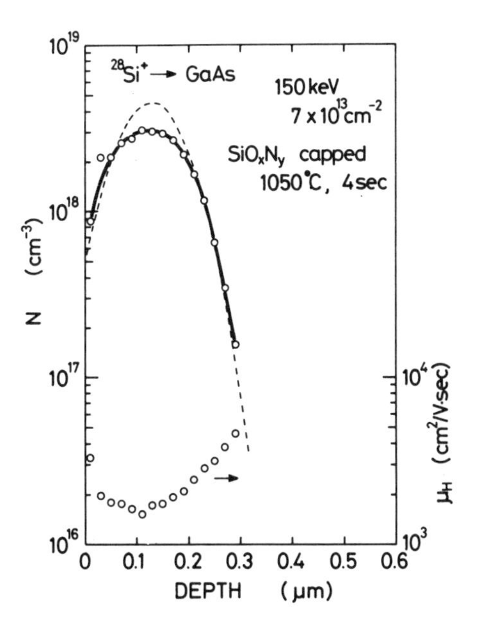

Fig. 14.9 Carrier concentration profile for high-dose implantation of Si in GaAs, followed by SiO_xN_y ($n = 1.75$)-capped IRTA at 1050°C for 4s.

tration of $3 \times 10^{18}\,cm^{-3}$ is achieved by SiO_xN_y-capped IRTA at 1050°C, with negligible dopant diffusion (see Fig. 14.9).

14.6. Implantation of S in GaAs

14.6.1. *Electrical activation*

Sulphur is a well-known n-type dopant in GaAs. However, the electrical properties for implantation of S in GaAs have been reported to involve a significant redistribution of S during the standard furnace annealing.[14)] In this case, if the electrical activation of S implanted in GaAs is mainly determined by the annealing temperature, much higher electrical activation with minimized dopant diffusion would be realized by IRTA.

Fig. 14.10 shows the sheet carrier concentrations obtained as a function of ion dose from 1×10^{13} to $1 \times 10^{14}\,cm^{-2}$ for 150 keV S implantation activated by capless IRTA at 950 and 1000°C for 2 s, together with those after furnace annealing. The samples after IRTA show higher electrical activation than those furnace-annealed at 850°C for 15 min. In particular, more than 50% electrical activation is constantly attained after 1000°C IRTA for the dose range investigated. This indicates that the annealing temperature, not the annealing time, mainly determines the electrical activation of implantation of S in GaAs.

14.6.2. *Depth profiles*

Fig. 14.11 shows depth profiles of the carrier concentrations and Hall mobilities for 150 keV S implants with a dose of $3 \times 10^{13}\,cm^{-2}$, followed by IRTA for 2 s at 750, 900, and 1000°C. Since no activation is observed near the surface for the sample annealed at 750°C, electrical activation seems to occur initially from the deeper side of the implanted layer. On the other hand, the electrical activation near the surface begins to occur with

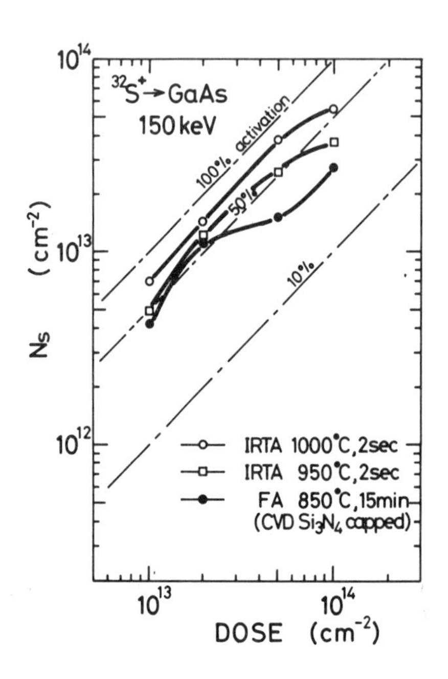

Fig. 14.10 Dependence of sheet carrier concentration on ion does for 150 keV implantation of S in GaAs. [after M. Kuzuhara et al.[4)]]

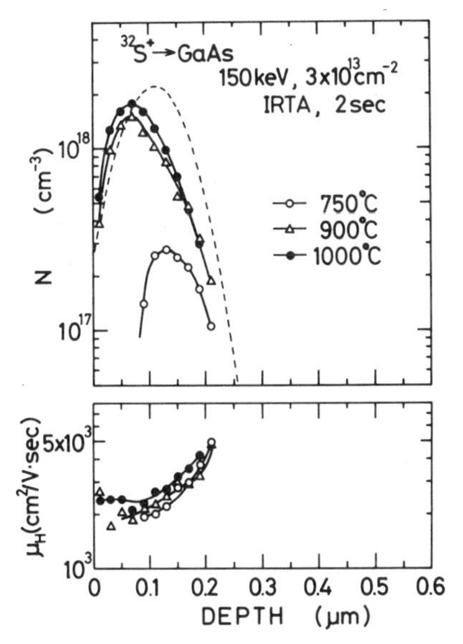

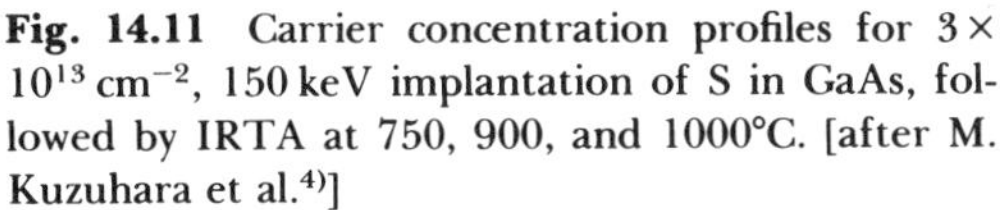

Fig. 14.11 Carrier concentration profiles for $3 \times 10^{13}\ \mathrm{cm}^{-2}$, 150 keV implantation of S in GaAs, followed by IRTA at 750, 900, and 1000°C. [after M. Kuzuhara et al.[4]]

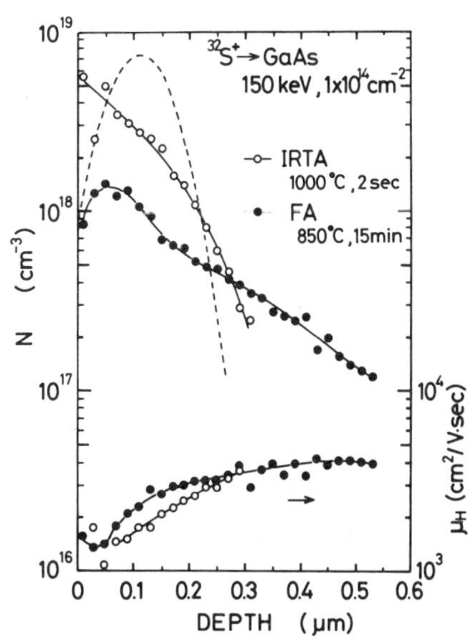

Fig. 14.12 Carrier concentration profiles for $1 \times 10^{14}\ \mathrm{cm}^{-2}$, 150 keV implantation of S in GaAs, followed by IRTA and furnace annealing. [after M. Kuzuhara et al.[4]]

increasing annealing temperature. No significant diffusion of S is observed for this dose level and a maximum carrier concentration of $2 \times 10^{18}\ \mathrm{cm}^{-3}$ is attained for the sample after IRTA at 1000°C.

The results for a dose of $1 \times 10^{14}\ \mathrm{cm}^{-2}$, followed by IRTA and furnace annealing, are shown in Fig. 14.12. Near a depth of 0.3 μm, an intersection in carrier concentration profile is seen between the two annealing methods, indicating the suppression of S diffusion for IRTA. The maximum carrier concentration of $5 \times 10^{18}\ \mathrm{cm}^{-3}$ is achieved after IRTA at 1000°C.

For implantation of S in GaAs, the resultant carrier concentration profile is closely related to the implantation damage which depends on the ion dose. If the recovery speed of the implantation damage precedes the dopant diffusion, the resultant carrier profile will follow a Gaussian distribution (LSS profile) as shown in Fig. 14.11. However, as the amount of implantation damage becomes too significant to be annealed instantly, the implanted S will undergo an anomalous diffusion enhanced by the implantation induced damage, as shown in Fig. 14.12.

14.6.3. *Ohmic contact formation*

In order to investigate the effect of high electrical activation for implantation of S on an ohmic contact formation, contact resistances were measured for the alloyed Ni/AuGe electrodes formed on 150 keV S-implanted GaAs with various ion doses. Annealing was performed caplessly at 980°C for 2 s. Contact resistances were evaluated from the resistance measurements between electrodes with different spacings. The results are listed in Table 14.3 for ion doses from 3×10^{13} to $3 \times 10^{14}\ \mathrm{cm}^{-2}$. The contact resistance decreases with increasing ion dose up to $1 \times 10^{14}\ \mathrm{cm}^{-2}$, where the lowest value of 0.15 Ωmm is

Table 14.3 Measured contact resistances for S-implanted GaAs.

Dose (10^{13} cm^{-2})	Energy (keV)	R_c (Ω mm)	R_s ($\Omega/\square$)
3	150	0.20	91
5	150	0.16	67
10	150	0.15	61
30	150	0.26	63

obtained. This result is consistent with the dose dependence of sheet carrier concentration shown in Fig. 14.10. The increase in contact resistance for a dose of 3×10^{14} cm^{-2} can be ascribed to the decrease in electrical activation. For such high-dose S implants, higher-temperature annealing would result in much lower contact resistance.

14.7. Implantation of Mg in GaAs

The characteristics of furnace-annealed Mg-implanted GaAs have been reported to have a greatly broadened carrier concentration profile due to a thermal diffusion.[15)] Here, the application of IRTA to implantation of Mg in GaAs is shown.

Fig. 14.13 shows the carrier concentration profiles for 80 keV Mg implants in GaAs with a dose of 2×10^{14} cm^{-2} activated by 2 s capless IRTA at 700, 800, and 900°C, respectively. The depth of the peak carrier concentration for each sample is slightly shallower than that of the LSS profile. In addition, the second peak in the carrier profile is seen for the sample annealed at 900°C. Such a dual-peak phenomenon was also reported for high-dose Zn implants in GaAs.[16)] Presumably, it can be explained by the accumulation of Mg into point defects produced by high-dose implantation. However, the diffusion of implanted Mg after IRTA is relatively small and a maximum hole concentration of

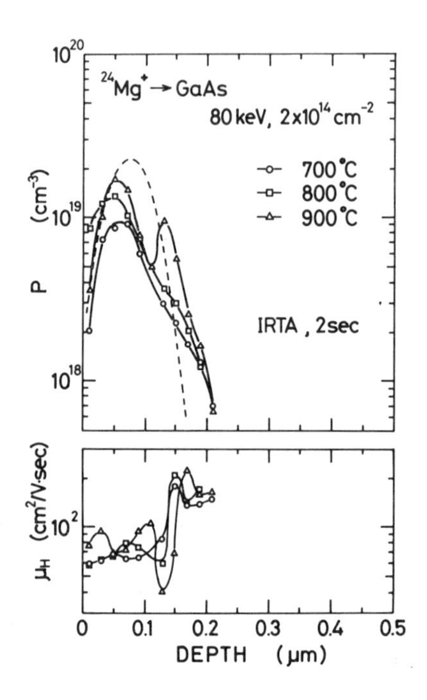

Fig. 14.13 Carrier concentration profiles for 2×10^{14} cm^{-2}, 80 keV implantation of Mg in GaAs, followed by IRTA at 700, 800, and 900°C.

Table 14.4 Results of Hall measurements for Mg-implanted GaAs.

Dose (10^{13} cm^{-2})	Energy (keV)	Anneal temperature (°C)	n_s (10^{13} cm^{-2})	μ_s (cm^2/V s)	R_s ($\Omega/\square$)
20	80	700	8.10	89	868
20	80	800	11.5	90	603
20	80	900	13.1	102	468

1.8×10^{19} cm^{-3} is achieved for the sample annealed at 900°C. The results of Hall measurements are summarized in Table 14.4.

14.8. Application of IRTA to GaAs MESFETs

The IRTA technique was applied to the fabrication of planar GaAs MESFETs with 1 μm gate length, 300 μm gate width and 3 μm source–drain spacing. The active layers

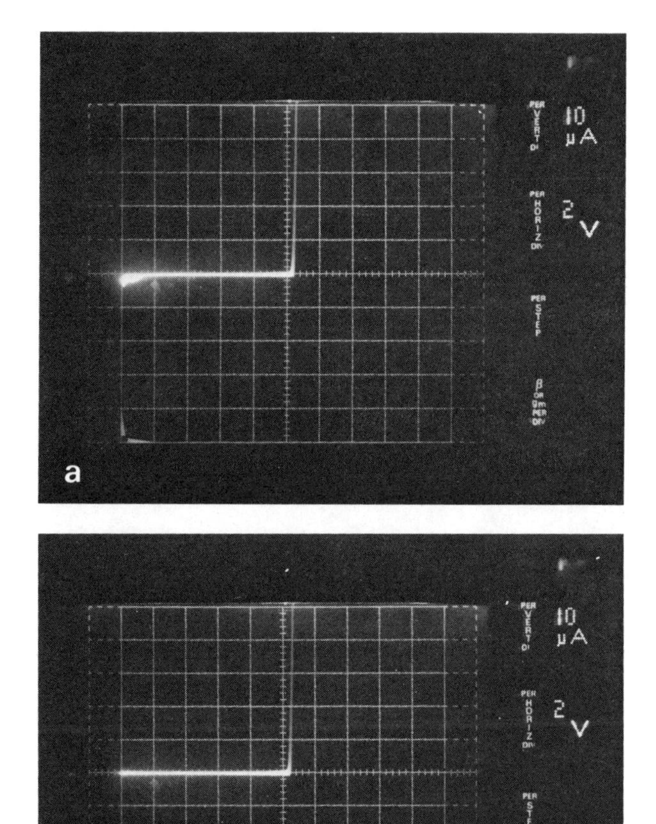

Fig. 14.14 Gate reverse *I–V* characteristics for GaAs MESFETs fabricated by (a) IRTA and (b) furnace annealing. [after H. Kohzu et al.[8)]]

Table 14.5 Electrical characteristics for GaAs MESFETs. [after H. Kohzu et al.[8)]]

Item	MESFET		Conditions
	IRTA	Furnace anneal	
Drain saturation current (mA)	64.0	65.5	$V_{ds} = 3$ V, $V_{gs} = 0$ V
Transconductance (mS)	34	24	$V_{ds} = 3$ V, $V_{gs} = 0$ V
Threshold voltage (V)	2.7	3.6	$V_{ds} = 3$ V, $I_{ds} = 100\ \mu$A
Gate leakage current (nA)	27	26	$V_{ds} = 0$ V, $V_{gs} = -1$ V
Gate breakdown voltage (V)	10–12	14–18	$V_{ds} = 0$ V, $I_{gs} = 10\ \mu$A

were formed by Si implantation at 100 keV to a dose of 5×10^{12} cm^{-2}, followed by IRTA at 950°C for 2s or by Si_3N_4-capped furnace annealing at 850°C for 15 min. MESFETs were fabricated using the self-aligned metallization process. After each device region was isolated by mesa etching, an Al gate was evaporated and the gate patterns were formed by the side-etching technique.[18)] Then, source and drain ohmic contacts were made by evaporation of Ni/AuGe, followed by alloying at 420°C.

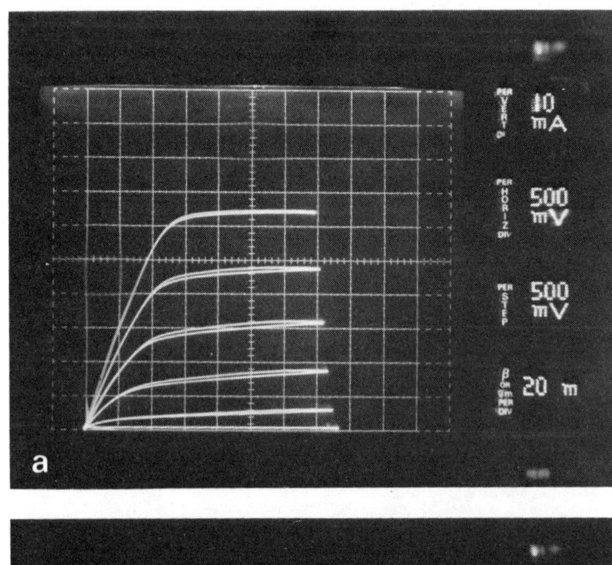

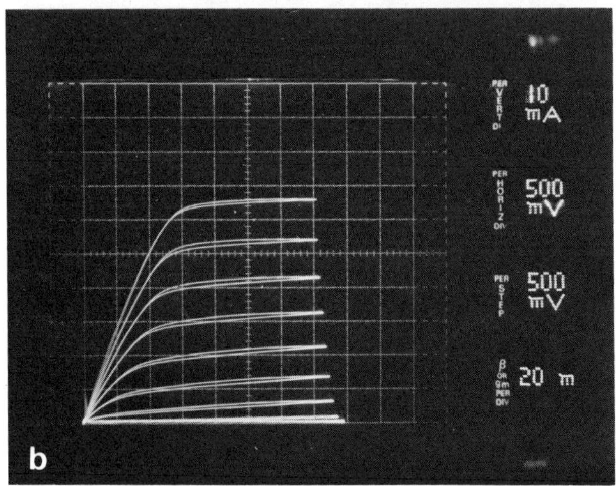

Fig. 14.15 Drain characteristics for GaAs MESFETs fabricated by (a) IRTA and (b) furnace annealing. [after H. Kohzu et al.[8)]]

The electrical characteristics for both MESFETs are shown in Figs. 14.14 and 14.15, and are summarized in Table 14.5. The gate breakdown voltage of the FET made by IRTA is a little lower than that of the FET made by furnace annealing. This can be ascribed to the higher carrier concentration of the active layer, as shown in Fig. 14.7. However, there is no difference in the gate reverse leakage currents between the two FETs, indicating that the capless IRTA in a N_2 atmosphere does not affect GaAs surface conditions on the device level.

The MESFET on the active layer made by IRTA shows 30–40% higher transconductance and lower threshold voltage than that of the device made by furnace annealing. This higher transconductance is due to a higher peak carrier concentration with a steeper carrier concentration profile for the active layer made by IRTA, indicating the superiority of the IRTA technique in the application of device fabrication.

14.9. Conclusion

Recent activities of the rapid thermal annealing technique for ion-implanted GaAs are described. The intrinsic features of high electrical activation and minimized implant diffusion are confirmed for various implants, such as Si, S, and Mg. The highest carrier concentrations achieved are $3 \times 10^{18}\,cm^{-3}$ for Si implants, $5 \times 10^{18}\,cm^{-3}$ for S, and $1.8 \times 10^{19}\,cm^{-3}$ for Mg. The GaAs MESFET fabricated using this technique showed much higher transconductance with no anomalous characteristics, indicating that this technique is promising for high-performance GaAs MESFET fabrication.

A recent publication has reported transconductance as high as 330 mS/mm for a TiW self-aligned gate MESFET annealed by a rapid thermal method.[19] Also, more recently, the application of rapid thermal annealing to a GaAlAs/GaAs two-dimensional electron gas structure (HEMT structure) was reported. Its initial high electron mobility is unchanged by rapid annealing up to 900°C.[20] These results indicate that the rapid thermal process can establish a new area of application not only as an alternative to conventional techniques but also as a unique process technology in the future.

Acknowledgements

The authors would like to thank Toshiharu Ozawa, Yuji Tanaka, and Mari Ohyagi for their technical assistances. They also thank Drs. Kazuo Ayaki, Hihehiko Katoh, and Hiroki Muta for their continuous support and encouragement during this work.

References

1) M. Arai, K. Nishiyama, and N. Watanabe: Japan. J. Appl. Phys., 20 (1981) L124.
2) D. E. Davies, P. J. McNally, and J. P. Lorenzo: IEEE Electron Device Lett., EDL-3 (1982) 102.
3) M. Kuzuhara, H. Kohzu, and Y. Takayama: Appl. Phys. Lett., 41 (1982) 755.
4) M. Kuzuhara, H. Kohzu, and Y. Takayama: J. Appl. Phys., 54 (1983) 3121.
5) K. Tabatabaie-Alavi, A. N. M. Masum Choudhury, C. G. Fonstad, and J. C. Gelpey: Appl. Phys. Lett., 43 (1983) 505.
6) K. Tabatabaie-Alavi, A. N. M. Masum Choudhury, H. Kanbe, C. G. Fonstad, and J. C. Gelpey: Appl. Phys. Lett., 43 (1983) 647.
7) K. Ito, M. Yoshida, M. Otsubo, and T. Murotani: Japan. J. Appl. Phys., 22 (1983) L299.
8) H. Kohzu, M. Kuzuhara, and Y. Takayama: J. Appl. Phys., 54 (1983) 4998.

9) A. N. M. Masum Choudhury, K. Tabatabaie-Alavi, C. G. Fonstad, and J. C. Gelpey: Appl. Phys. Lett., 43 (1983) 381.
10) R. L. Chapman, J. C. C. Fan, J. P. Donnelly, and B-Y. Tsaur: Appl. Phys. Lett., 40 (1982) 805.
11) P. M. Asbeck, D. L. Miller, E. J. Babcock, and C. G. Kirkpatrick: IEEE Electron Device Lett., EDL-4 (1983) 81.
12) N. J. Shah, H. Ahmed, I. R. Sanders, and J. F. Singleton: Electron Lett., 16 (1980) 433.
13) N. J. Shah, H. Ahmed, and P. A. Leigh: Appl. Phys. Lett., 39 (1981) 322.
14) R. Kwor, Y. K. Yeo, and Y. S. Park: J. Appl. Phys., 53 (1982) 4786.
15) B. D. Choe, Y. K. Yeo, and Y. S. Park: J. Appl. Phys., 51 (1980) 4742.
16) J. Kasahara, H. Sakurai, Y. Kato, and N. Watanabe: Japan. J. Appl. Phys., 21 (1982) L103.
17) M. Kuzuhara, and H. Kohzu: Appl. Phys. Lett., 44 (1984) 527.
18) T. Furutsuka, T. Tsuji, F. Katano, and A. Higashisaka: Electron. Lett., 17 (1981) 944.
19) T. Furutsuka, F. Katano, M. Ishikawa, T. Nozaki, T. Tsuji, and A. Higashisaka: 1983 Nat. Conv. Rec. on Sci. Tech., IECE Japan 166. [in Japanese]
20) S. Tatsuta, T. Inada, S. Okamura, and S. Hiyamizu: Japan. J. Appl. Phys., 23 (1984) L147.

15 THERMAL OXIDATION OF SiC AND ELECTRICAL PROPERTIES OF Al–SiO_2–SiC MOS DIODES

Hiroyuki MATSUNAMI†

Abstract

Thermal oxidation of SiC at 850–1100°C in wet oxygen has been carried out and the structure of the oxide layer has been studied by Auger electron spectroscopy and ellipsometry. The relation between the oxide thickness and the oxidation time is well explained by the model used for thermal oxidation of Si. The oxidation mechanism is discussed using the temperature dependence of the oxidation rate constants. The homogeneous oxide layer confirmed to be SiO_2 has a narrow transition region at the oxide–SiC interface. The resistivity and the breakdown field of the oxide are $2 \times 10^{12}\ \Omega\,cm$ and 2×10^6 V/cm, respectively. C–V characteristics of Al–SiO_2–SiC MOS diodes have been measured in the frequency range between 10 Hz and 1 MHz. Under illumination, three regions of accumulation, depletion and inversion are clearly observed. In the dark, inversion does not occur owing to the very small number of minority carriers because of the large band-gap of SiC. The minimum surface density of states is around $2 \times 10^{12}\ cm^{-2}\,eV^{-1}$.

Keywords: Thermal Oxidation, SiC, MOS Diodes, Inversion

15.1. Introduction

Silicon carbide (SiC) is an attractive semiconductor with remarkable chemical inertness and stability even at several hundreds of degrees Celsius.[1] Because of its wide band-gap, high electric breakdown field[2] and high saturated electron drift velocity,[3] SiC is expected to be a promising material for high-temperature, high-power and high-frequency electronic devices as well as optoelectronic devices in the visible wavelength region.[1] Devices such as rectifiers, particle detectors, tunnel diodes and light-emitting diodes based on p–n junctions have been fabricated.[1] p–n junctions have been prepared by impurity diffusion,[1] chemical vapor deposition,[4–8] liquid-phase epitaxy[7,9–14] and ion implantation.[15]

In semiconductor device technology, oxide layers play an important role either as a gate insulator or as a masking material for photolithography. Surfaces of SiC can be thermally oxidized using dry or wet oxygen (O_2) gas at around 1000°C.[16–20] The oxide film of SiC is silicon dioxide (SiO_2) grown by the inward diffusion of O_2 and the outward diffusion of carbon monoxide (CO).[16,17] The oxide layers have been used for identification of the surface polarity,[12,13,19] masking in selective gas etching[2,3,21] and passivation of p–n junctions. Mesa-type and planar-type light-emitting diodes[12] and transistors[21] have been

† Department of Electrical Engineering, Kyoto University, Yoshidahonmachi, Sakyo, Kyoto, 606.

JARECT Vol. 19, Semiconductor Technologies (1986), *J. Nishizawa (ed.)*
OHMSHA, LTD. and North-Holland

fabricated by photolithographic techniques using thermal oxides. However, there have been few researches on the structures and properties of the thermal oxides of SiC.[16,20,22]

Studies of the electrical properties of oxide–semiconductor interfaces thermally grown on SiC are of great interest, since device applications will be further developed if SiC MOS (metal–oxide–semiconductor) structures with good electrical characteristics become available. A brief report by Kee et al.[22] stated the interface characteristics of SiC MOS diodes mainly using Auger electron analysis for CVD (chemical vapor deposition) grown 3C–SiC. They claimed that *C–V* (capacitance–voltage) measurements showed no capacitance change for bias voltages upto ±40 V for frequencies of 10 kHz, 100 kHz and 1 MHz even after various heat treatments of the MOS diodes. The author's group investigated the structure of thermally-grown SiO_2 layers on SiC[20] and studied the electrical characteristics of the MOS diodes.[23,24]

In this chapter, thermal oxidation of SiC in a way similar to that used for thermal oxidation of Si[25] is described. The structure of the thermally-grown SiO_2 on SiC was analyzed using Auger electron spectroscopy and ellipsometry. Electrical characteristics, especially *C–V* characteristics with and without illumination, are discussed.

15.2. Thermal oxidation

Thermal oxidation of SiC has been studied from early days to understand the oxidation behavior of SiC heating elements and refractories in industrial atmospheres.[16,17] Although most of the present knowledge of oxides of SiC is derived from the standpoint of material protection, the previous works give a useful background for the investigation of an insulating oxide for electronic devices. The remarkable difference in the oxidation rates depending on the surface polarity of SiC has been recognized.[18,19]

15.2.1. *Oxidation procedure*

Thermal oxidation was carried out at 850–1100°C for 0.1–10 h in a wet oxygen flow using a horizontal quartz tube. The flow rate of oxygen by bubbling through deionized water at 950°C was fixed at 100 cm^3/min. Crystals of n-type 6 H–SiC prepared by the Lely method were used for oxidation. The surface polarity of the plate-shaped crystals with the (0001) Si and $(000\bar{1})$ C faces were identified by etching in hydrogen (H_2) gas or in molten potassium hydroxide (KOH).[8,13] The SiC crystals were cleaned by dipping them in hydrofluoric acid (HF) for several minutes to remove any oxides initially present and rinsed in deionized water. The crystals thus cleaned were inserted keeping the [0001] faces vertical with a small quartz slab into the furnace.

The furnace was heated up to the oxidation temperature before insertion of the crystals. The oxidation temperature was calibrated using the temperature difference between the positions at the crystals and a thermocouple measured in advance. After oxidation, the crystals were rapidly moved to the cool zone of the quartz tube.

15.2.2. *Oxide thickness versus oxidation time*

Fig. 15.1 shows the oxidation time dependence of the thickness of the oxide layer grown at 1000°C on the (0001) Si and the $(000\bar{1})$ C faces. The results for the (111) face of Si

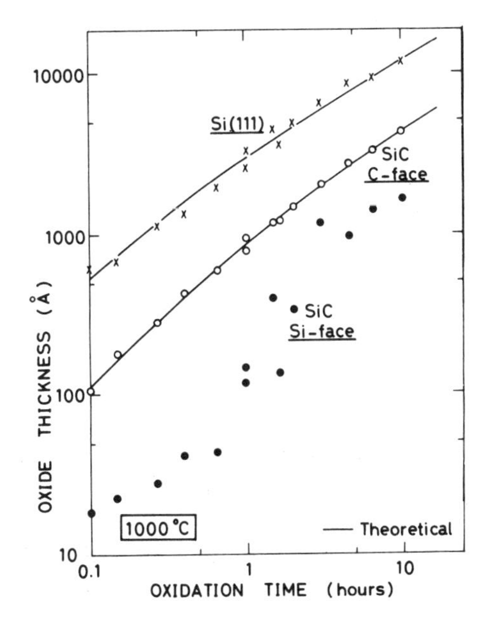

Fig. 15.1 Oxidation time dependence of the thickness of the oxides grown at 1000°C on the (0001) Si and the (000$\bar{1}$) C faces of SiC together with those on the (111) face of Si. Solid lines are theoretical curves.[24]

obtained in the same oxidation runs are also shown in Fig. 15.1 for comparison. The thickness of the oxide was calculated together with the refractive index using the results of measurement by an ellipsometer with a He–Ne laser (632.8 nm).

The oxide layer on the (000$\bar{1}$) C face was thicker than that on the (0001) Si face for any oxidation times between 0.1 and 10 h. The oxide layers on both faces were considerably thinner than that of the oxide layer on the (111) face of Si. This tendency in the thickness difference among the three faces was observed for all of the oxides grown at 850–1100°C.

The thickness x_0 of the oxide on the (000$\bar{1}$) C face of SiC is found to be related to the oxidation time t by the well-known expression[25]

$$x_0^2 + Ax_0 = B(t + \tau), \tag{15.1}$$

where A, B and τ are constants depending on the oxidation conditions. The solid lines for the (000$\bar{1}$) C face of SiC and the (111) face of Si in Fig. 15.1 are the theoretical curves calculated using the constants A, B and τ determined from the experimental results, which will be described in detail in section 15.2.4.

The results for the (0001) Si face of SiC are rather scattered as shown in Fig. 15.1 and cannot be represented by Eq. (15.1).

15.2.3. *Structure of oxides*

The refractive indices of the oxide layers of thickness 500–1500 Å grown on the (000$\bar{1}$) C face at oxidation temperatures between 850 and 1100°C are shown in Fig. 15.2. The average values of the measured refractive indices for each oxidation temperature are indicated by circular points. The value of 1.463 ± 0.003 is very close to the value of 1.457 for fused quartz (SiO_2)[26] at the wavelength of 632.8 nm shown with the broken line in Fig. 15.2. From this result, all the oxide layers grown at 850–1100°C can be considered to be SiO_2. The deviations from the average values shown with error bars are ±0.01 or less for

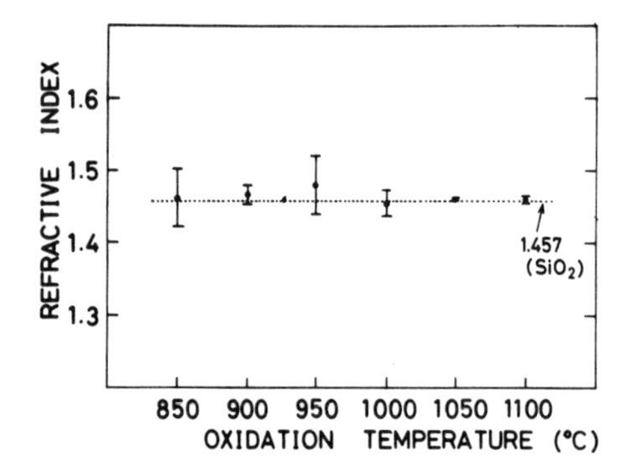

Fig. 15.2 Refractive index of the oxide layers on the $(000\bar{1})$ C face of SiC at different oxidation temperatures.[24]

the oxide layers grown above 1000°C. Significant deviations are found for the oxide layers grown below 1000°C, which may reflect some change in the chemical structure of the oxides.

The refractive indices of the oxide layers on the (0001) Si face also showed similar values to that of SiO_2. However, the measured values are rather scattered compared with the results for the oxide layers on the $(000\bar{1})$ C face.

Both the oxide layers on the (0001) Si and the $(000\bar{1})$ C faces of SiC are found by Auger electron spectroscopy[20] to have the structure of SiO_2 and contain no residual carbon atoms. An example of the Auger electron spectra for the $(000\bar{1})$ C face of SiC is shown in Fig. 15.3. The upper spectrum is from an SiC crystal, and the lower from a thermally-grown oxide layer on the $(000\bar{1})$ C face. For the primary electron beam, an accelerating voltage of 5 kV and a beam current of several tenths of μA are employed. The beam scanned the sample surface over a region of $50 \times 50\ \mu m^2$ to avoid an electron-irradiation effect.[27] One can recognize that the spectral shape of the Si_{LVV} peak for the oxide is different from that of SiC, and that the Si_{KLL} peak for the oxide is located at about 9 eV lower than that of SiC. It should be noted that the C_{KLL} peak cannot be

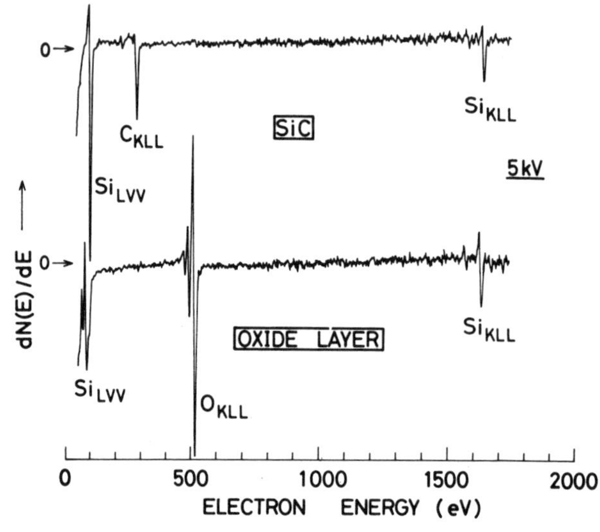

Fig. 15.3 Auger electron spectra of SiC and the oxide layer on the $(000\bar{1})$ C face.[20]

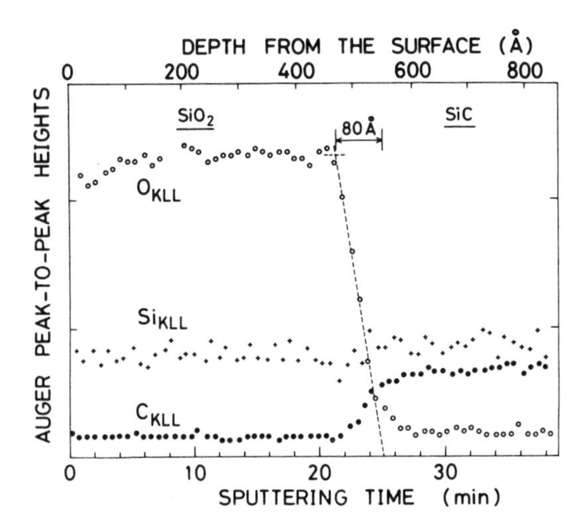

Fig. 15.4 Auger depth profile of the oxide layer on the $(000\bar{1})$ C face of SiC.[24]

observed in the Auger spectra for the oxide layer in Fig. 15.3. The same spectrum as the lower spectrum is obtained from the thermally-grown oxide layer on Si, which is a well-known spectrum of SiO_2.[27]

Fig. 15.4 shows an Auger depth profile of a typical oxide layer of thickness 503 Å on the $(000\bar{1})$ C face of SiC grown by a 45 min oxidation at 1000°C. The depth profile was obtained automatically by measuring the peak-to-peak heights of the differentiated Auger peaks (dN/dE) combined with simultaneous sputtering using an argon (Ar) ion beam accelerated at 3 kV. The signal levels of C_{KLL} in the oxide region and O_{KLL} in the SiC region in Fig. 15.4 are the noise level. The oxide layer seems to be homogeneous SiO_2 with a narrow transition region at the oxide–SiC interface. The width of the transition region is estimated to be 80 Å as shown in Fig. 15.4. The value is derived from extrapolating the linear change of the O_{KLL} profile to both the maximum and zero values and by using the sputtering etch rate (22 Å/min) for the region from the oxide to the interface. Taking into account the escape depth of Auger electrons and the knock-on effects of sputtering Ar ions,[27,28] the actual transition region width is assumed to be 73 Å. About the same value was obtained for the thermal oxide layer on Si simultaneously oxidized with SiC. Much narrower widths have been reported for the thin thermal oxide layers on Si using dry O_2.[27,28] The transition region broadening may be caused by the present oxidation process (wet oxidation) or by the depth-resolution limit of the present Auger analysis.

Similar depth profiles are obtained for thin oxide layers on the (0001) Si face of SiC with a remarkably thick transition region width above 150 Å.[29] The broad transition region may cause the scattered experimental data of the thickness and the refractive index of the oxide layers on the (0001) Si face as previously described.

15.2.4. *Analysis of oxidation mechanism*

Eq. (15.1) is rewritten as

$$x_0 = B(t + \tau)/x_0 - A, \tag{15.2}$$

which indicates that x_0 versus $(t + \tau)/x_0$ has a linear relation. The constant τ can be estimated to be zero for all the samples oxidized at 850–1100°C from the x_0 versus t curve, which is the same as for the thermal oxide of Si grown with wet oxygen.[25] All the experimental data in Fig. 15.1 are plotted for x_0 versus $(t+\tau)/x_0$ as in Fig. 15.5. Each straight line in Fig. 15.5 is obtained using the least-squares method. The intercept on the vertical axis and the slope give the constants A and B, respectively. Using the values of A, B and τ thus obtained, theoretical curves expressed by Eq. (15.1) are calculated and shown as solid lines in Fig. 15.1 for both the oxides on the $(000\bar{1})$ C face of SiC and the (111) face of Si.

The relation expressed by Eq. (15.1) is based on the diffusion process through the oxide and the reaction processes at the oxide surface and the oxide–semiconductor interface, which is widely applied for the thermal oxidation of Si.[25] The value of B (the parabolic rate constant) is proportional to the diffusion coefficient, while the value of B/A (the linear rate constant) is proportional to the slower reaction rate at the surface or at the interface. The rate constants B and B/A for wet oxidations of Si are found to have exponential temperature dependences with the activation energies of 16.3 and 45.3 kcal/mol, respectively.[25] The former value agrees with the activation energy of 18.3 kcal/mol for diffusion of the water (H_2O) molecule in SiO_2 and the latter value is similar to the energy of 42.2 kcal/mol required for the breaking of a Si–Si bond.[25]

From the temperature dependences of the constants B and B/A for wet oxidation of the $(000\bar{1})$ C face shown in Fig. 15.6, the activation energies of 48 and 26 kcal/mol respectively, are obtained. Although the oxidizing species H_2O diffuses through the oxide of SiC, the value of 48 kcal/mol is very large compared with 18.3 kcal/mol for the diffusion of H_2O in SiO_2. In the case of dry oxidation of SiC, oxygen gas diffuses through the oxide to the oxide–SiC interface and form SiO_2 and CO.[16,17] The by-product CO gas diffuses out through the oxide to the surface. The activation energy of 47 kcal/mol is reported for dry oxidation of SiC.[18] Based on these results, the activation energy associated with the constant B may be concerned with the diffusion of by-product CO and not of the oxidizing species H_2O

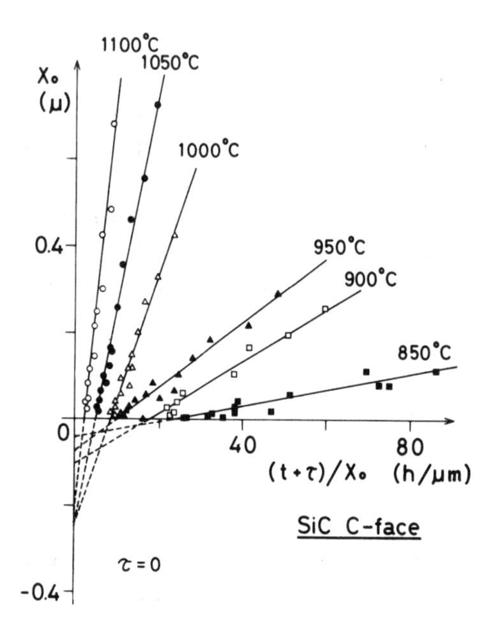

Fig. 15.5 The relation of x_0 versus $(t+\tau)/x_0$ for evaluation of the rate constants of the oxide on the $(000\bar{1})$ C face of SiC.[24]

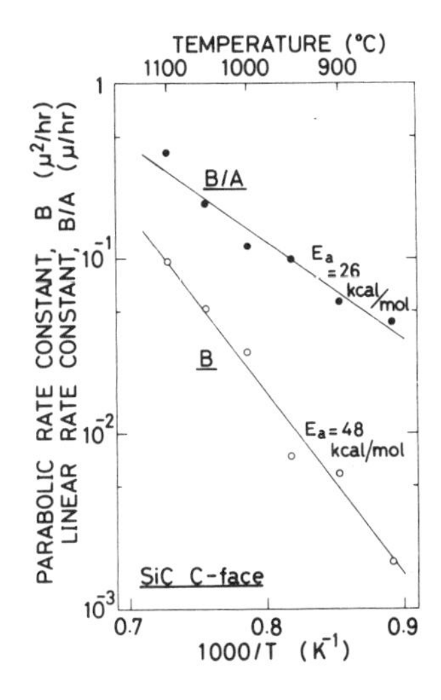

Fig. 15.6 Temperature dependences of the parabolic rate constant B and the linear rate constant B/A for the oxide on the $(000\bar{1})$ C face of SiC.[24)]

The activation energy of 26 kcal/mol for the constant B/A also does not agree with 69.3 kcal/mol[30)] required for breaking of a Si–C bond. Moreover, the oxidation rate differs between the $(000\bar{1})$ C and (0001) Si faces as shown in Fig. 15.1. Therefore, the reaction which controls the oxidation rate at the oxide–SiC interface is determined by other factors than the breaking of a Si–C bond. In order to reveal the thermal oxidation mechanism of SiC, more detailed data for both wet and dry oxidation are required together with the more precise analysis of the chemical structures at the oxide–SiC interface.

15.3. Electrical properties of SiC MOS diodes

15.3.1. *Samples*

Thermal wet oxidation was carried out at 1000°C using n-type 6 H–SiC single crystals prepared by liquid-phase epitaxial growth with a dipping technique.[10)] The ionized donor concentration of the epitaxial layer was 5×10^{17} cm^{-3} determined from the C–V measurements using a Au–SiC Schottky diode. The oxygen gas flowed at a rate of 200 cm^3/min by bubbling through deionized water at 85–90°C. After annealing at 1000°C for 30–60 min in an Ar flow of 200 cm^3/min, the oxide on the back side of the wafer was removed and an ohmic contact of nickel (Ni) was evaporated on SiC. The aluminum (Al) gate electrode with a diameter of 0.5 mm was evaporated on the oxide layer in a vacuum of 6.5×10^{-4} Pa. The resistivity and the breakdown field of the oxide layers are typically 2×10^{12} Ω cm and 2×10^6 V/cm, respectively.

15.3.2. *C-V characteristics*

Fig. 15.7 shows C–V curves measured at 1 MHz with the sweep rate of 0.02 V/s in the dark and under illumination with a 100 W tungsten lamp at room temperature for several samples with different oxide thicknesses. The C–V characteristics have no frequency dispersion in the frequency range between 10 Hz and 1 MHz. Under illumination, the C–V curves show the three distinct regions of accumulation (> +5 V), depletion (0 to

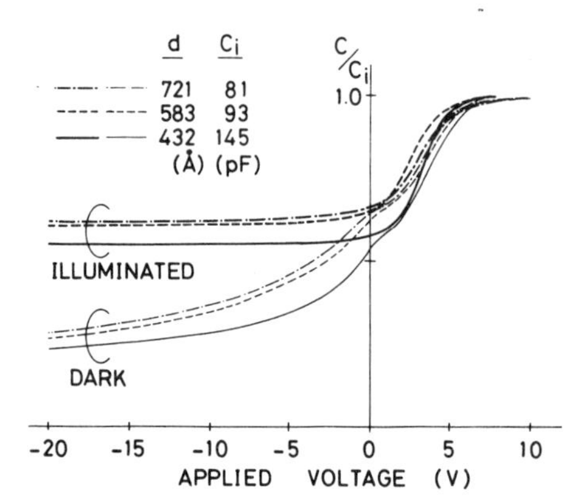

Fig. 15.7 *C–V* characteristics of Al-SiO_2-SiC MOS diodes with different oxide thicknesses. The measurement was carried out at 1 MHz in the dark and under illumination with the sweep rate of 0.02 V/s.[24)]

+5 V) and inversion (<0 V), which look like the standard MOS high-frequency *C–V* curves. The insulator capacitance C_i derived from the accumulation region is shown in Fig. 15.7 together with the oxide thickness *d*. The capacitances in Fig. 15.7 are normalized using the values of C_i for each sample. In the dark condition, the *C–V* curves in the accumulation and the depletion regions are almost the same as those under illumination. As the negative bias increases, however, the capacitance continues to decrease even at the bias voltages corresponding to the inversion condition, which indicates the deep depletion characteristics.[31)]

For the slow sweep rate (0.02 V/s), the *C–V* curves show a slight hysteresis both in the dark and under illumination. Even when the sweep rate increases up to 2 V/s, almost the same characteristics are obtained in the dark. However, the characteristics under illumination strongly depend on the sweep rate as shown in Fig. 15.8. For the low sweep rate of 0.1 V/s, the same characteristics as in Fig. 15.7 are obtained without significant hysteresis. For the high sweep rates (0.5 and 2 V/s), the capacitance in the inversion region decreases to the value in the dark condition with decreasing bias voltage from 0 to −10 V, like the deep depletion characteristics. As the bias voltage is reversed from −10 to 0 V, the capacitance recovers the value obtained for the low sweep rates as shown in Fig. 15.8. Thus the illumination effect observed in the inversion region seems to have some response time.

The hysteresis loop for the high sweep rates under illumination is explained as follows. For the sufficiently low sweep rate, the increase in the number of the minority

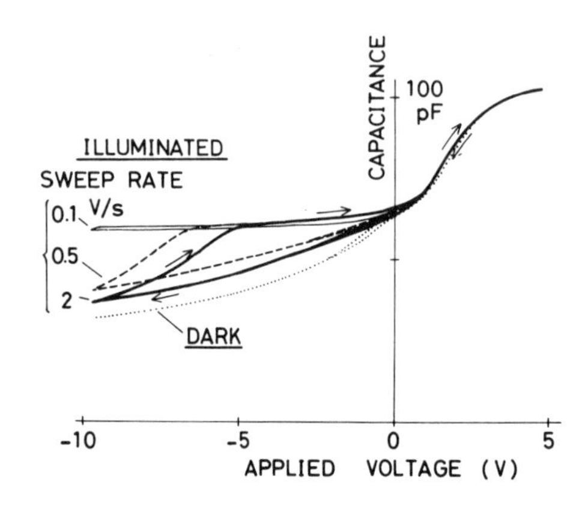

Fig. 15.8 Dependence of *C–V* characteristics on the sweep rate of bias measured at 1 MHz under illumination.[24)]

carriers can follow the increase in charges on the Al gate metal with increasing negative bias. Thus, the capacitance gives the equilibrium value. However, for the high sweep rate, the lack of minority carriers occcurs at the SiC surface due to the limited generation rate. This causes an increase in the width of the depletion layer and thus a decrease in the capacitance by an amount depending on the sweep rate, which results in the deep depletion characteristics. As the bias is reversed towards 0 V, the number of accumulated carriers is sufficiently large to follow the voltage change, and the capacitance recovers the equilibrium value. A similar hysteresis loop is reported for Si MIS diodes at high sweep rates at room temperature even without illumination.[32] However, in the case of SiC the hysteresis is not observed without illumination owing to the very small number of minority carriers.

Under weak illumination, similar characteristics are also obtained, though a much longer response time is necessary to recover the capacitance. The recovered capacitance does not decrease even for several hours after the illumination is stopped, which shows that there is no leakage of the accumulated minority carriers near the surface through the oxide or other passages.

MOS diodes without thermal annealing after oxidation show inferior *C–V* characteristics both in the dark and under illumination.

15.3.3. *Analysis of C–V curves*

In Fig. 15.9 are shown experimental and theoretical *C–V* curves for typical Al–SiO_2–SiC MOS diodes with the oxide thickness of 432 Å.[24] Experimental *C–V* curves under illumination give high-frequency characteristics even at 10 Hz. The experimental flat-band voltage (indicated by FB in Fig. 15.9) is approximately +4 V larger than the theoretical value. The negative effective surface-charge density in the oxide producing the above-described flat-band shift is estimated to be $2\text{–}5 \times 10^{12}\,\text{cm}^{-2}$ for the sample shown in Fig. 15.9. The minimum surface density of states in the band-gap is approximately $2 \times 10^{12}\,\text{cm}^{-2}\,\text{eV}^{-1}$ from calculation with the Terman method. Both values seem to be independent of the oxide thickness.

The experimental *C–V* curve in the dark can be regarded as a non-equilibrium case,[31] because both the minority carrier concentration* and the thermal generation rate are

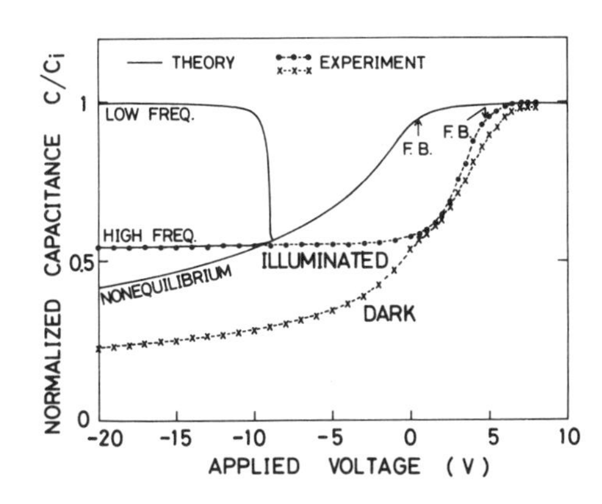

Fig. 15.9 Experimental and theoretical *C–V* characteristics for typical MOS diodes.[24]

* The intrinsic carrier concentration is as small as $3 \times 10^{-6}\,\text{cm}^{-3}$ at 300 K.

extremely small in 6 H–SiC due to the large band-gap of 3.0 eV. Similar non-equilibrium characteristics are reported for Si MIS diodes at low temperatures, where the number of the minority carriers is quite small.[32)]

15.4. Conclusion

Thermally-grown oxide layers on the $(000\bar{1})$ C face of 6 H–SiC at 850–1100°C in wet oxygen were studied by Auger electron analysis and ellipsometry, and the following results were obtained:

(1) The relation between the oxide thickness and the oxidation time can be expressed by the well-known model for the thermal oxidation of Si.
(2) The parabolic and linear rate constants for oxidation have exponential temperature dependences with the activation energies of 48 and 26 kcal/mol, respectively.
(3) The oxidation mechanism is tentatively explained by the inward diffusion of H_2O through the oxide, the reaction with SiC at the interface and the outward diffusion of by-product CO.
(4) The oxide layer is found to be homogeneous SiO_2 with a narrow transition region less than 80 Å at the oxide–SiC interface.

C–V characteristics of Al–SiO_2–SiC MOS diodes using the thermal oxide of LPE-grown 6 H–SiC have been measured in the frequency range between 10 Hz and 1 MHz in the dark and under illumination, and the following results were obtained:

(5) The resistivity and the breakdown field of the oxide layer are typically 2×10^{12} Ω cm and 2×10^{6} V/cm, respectively.
(6) The *C–V* characteristics at the frequencies of 10 Hz and 1 MHz are almost the same without any frequency dispersion both in the dark and under illumination.
(7) Under illumination, the *C–V* curves give the standard MOS high-frequency characteristics and clearly show the distinct regions of accumulation, depletion and inversion.
(8) In the dark condition, inversion does not occur, showing the deep depletion characteristics due to the very small number of minority carriers because of the large band-gap of SiC.
(9) A considerable hysteresis loop is observed in the inversion region under illumination for the high sweep rates of bias, which is probably caused by the limited generation rate of minority carriers.
(10) The minimum surface density of states in the band-gap is roughly 2×10^{12} cm^{-2} eV^{-1}. The negative effective surface-charge density in the oxide is estimated to be 2–5×10^{12} cm^{-2}, which gives the flat-band shift.

Acknowledgements

The author would like to express many thanks to Dr. Akira Suzuki (now at Sharp Co. Ltd.) who initiated the work on oxidation of SiC in Kyoto University. He also shows his gratitude to Messrs. Hisashi Ashida, Nobuyuki Furui and Kazunobu Mameno for their contributions to the experiments.

References

1) R. B. Campbell and H. C. Chang: *Semiconductors and Semimetals*, Vol. 7, Eds. R. K. Willardson and A. C. Beer (Academic Press, New York, 1971) p. 625.
2) W. v. Münch and I. Pfaffeneder: J. Appl. Phys., 48 (1977) 4831.
3) W. v. Münch and E. Pettenpaul: J. Appl. Phys., 48 (1977) 4823.
4) S. Minagawa and H. C. Gatos: Japan. J. Appl. Phys., 10 (1971) 1680.
5) G. Gramberg and M. Königer: Solid State Electron., 15 (1972) 285.
6) W. v. Münch and I. Pfaffeneder: Thin Solid Films, 31 (1976) 39.
7) W. v. Münch, W. Kürzinger, and I. Pfaffeneder: Solid State Electron., 19 (1976) 871.
8) S. Nishino, A. Ibaraki, H. Matsunami, and T. Tanaka: Japan. J. Appl. Phys., 19 (1980) L353.
9) R. W. Brander and R. P. Sutton: J. Phys., D2 (1969) 309.
10) A. Suzuki, M. Ikeda, N. Nagao, H. Matsunami, and T. Tanaka: J. Appl. Phys., 47 (1976) 4546.
11) H. Matsunami, M. Ikeda, A. Suzuki, and T. Tanaka: IEEE Trans. Electron. Devices, ED-24 (1977) 958.
12) W. v. Münch and W. Kürzinger: Solid State Electron., 21 (1978) 1129.
13) M. Ikeda, T. Hayakawa, S. Yamagiwa, H. Matsunami, and T. Tanaka: J. Appl. Phys., 50 (1979) 8215.
14) G. Ziegler and D. Thesis: IEEE Trans. Electron Devices, ED-28 (1981) 425.
15) E. V. Kalinina, N. K. Profof'eva, A. V. Suvorov, G. K. Kholuyanov, and V. E. Chelnokov: Sov. Phys. Semicond., 12 (1978) 1372.
16) E. A. Gulbransen, K. F. Andrew, and F. A. Brassart: J. Electrochem. Soc., 113 (1966) 1311.
17) E. Fitzer and R. Ebi: Silicon Carbide – 1973, Eds. R. C. Marshall, J. W. Faust, Jr. and C. E. Ryan: (Univ. of South Carolina Press, Columbia, 1974) p. 320.
18) R. C. A. Harris and R. L. Call: Silicon Carbide – 1973, Eds. R. C. Marshall, J. W. Faust, Jr. and C. E. Ryan (Univ. of South Carolina Press, Columbia, 1974) p. 329.
19) W. v. Münch and I. Pfaffeneder: J. Electrochem. Soc., 122 (1975) 642.
20) A. Suzuki, H. Matsunami and T. Tanaka: J. Electrochem. Soc., 125 (1978) 1896.
21) W. v. Münch and P. Hoeck: Solid State Electron., 21 (1978) 479.
22) R. W. Kee, K. M. Geib, C. W. Wilmsen, and D. K. Ferry: J. Vac. Sci. Technol., 15 (1978) 1520.
23) A. Suzuki, K. Mameno, N. Furui, and H. Matsunami: Appl. Phys. Lett., 39 (1981) 89.
24) A. Suzuki, H. Ashida, N. Furui, K. Mameno, and H. Matsunami: Japan. J. Appl. Phys., 21 (1982) 579.
25) B. E. Deal and A. S. Grove: J. Appl. Phys., 36 (1965) 3770.
26) W. S. Rodney and R. J. Spindler: J. Opt. Soc. Am., 44 (1954) 677.
27) J. S. Johannessen and W. E. Spicer: J. Appl. Phys., 47 (1976) 3028.
28) C. R. Helms and W. E. Spicer: Solid State Commun., 25 (1978) 673.
29) A. Suzuki, H. Matsunami and T. Tanaka: J. Vac. Soc. Japan (Shinku), 22 (1979) 49. [in Japanese]
30) L. Pauling: The Nature of the Chemical Bond and the Structure of Molecules and Crystals, 3rd edn. (Cornell Univ. Press, Ithaca, NY, 1960) p. 85.
31) S. M. Sze: Physics of Semiconductor Devices (Wiley, New York, 1969) p. 425.
32) L. S. Wei and J. G. Simmons: Solid State Electron., 17 (1974) 1021.

Note added in proof

After the submission of the manuscript, MOS characteristics using thermal oxides on chemical vapor deposited 3C-SiC were reported by:

K. Shibahara, S. Nishino, and H. Matsunami: Jpn. J. Appl. Phys., 23 (1984) L862.

16 SCREEN PRINTED CdS/CdTe SOLAR CELLS

Hitoshi MATSUMOTO and Seiji IKEGAMI†

Abstract

Thin-film CdS/CdTe solar cells are prepared on a borosilicate glass substrate by successively repeating screen printing and heating (sintering) in a belt furnace of each paste of CdS, Cd + Te, C, Ag + In and Ag. In a small cell of area 0.78 cm^2, the intrinsic conversion efficiency of 12.8% is obtained; this value is the highest yet achieved in thin-film-type solar cells. On a large glass substrate of 30 × 30 cm^2, 36 unit cells connected in series are constructed by this printing technique, and a 4.8 W single substrate module is obtained under 100 mW/cm^2 illumination. Under rooftop conditions, no change in output power was observed in the encapsulated solar cells over 400 days. Thus, the screen printed CdS/CdTe solar cells can be expected to be low-cost, highly efficient and stable solar cells.

Keywords: Solar Cell, II–VI Compound Semiconductor, CdS Sintered Film, CdTe Sintered Film, Screen Printing

16.1. Introduction

Heterojunction solar cells with a wide-bandgap window and a narrow-bandgap absorber have been studied in order to develop efficient, stable and low-cost cells.

CdS, having an energy bandgap of 2.42 eV, is one of the most promising materials for the wide-bandgap window, because a low-resistivity CdS film can easily be obtained by various film deposition methods.[1] CdTe is suitable as a narrow-bandgap absorber, because it has a direct bandgap of 1.44 eV, which is optimum for solar energy conversion.[2] Therefore, CdS/CdTe heterojunction offers the possibility of forming an efficient solar cell.

Fahrenbruch et al. theoretically estimated the solar conversion efficiency for a CdS/CdTe heterojunction, and obtained a value of 17%.[3] Yamaguchi et al. made a solar cell with an intrinsic efficiency (η_i) of 11.7% by growing a CdS epitaxial layer on a p-CdTe crystal plate, and clarified the usefulness and stability of this junction.[4] Bonnet et al., Uda et al. and Tyan et al. made thin-film-type CdS/CdTe solar cells by using different thin-film techniques, and obtained cells with η_i of 5.4, 8.7 and 10.5%, respectively.[5–7]

In order to obtain low-cost solar cells, we must use as little material as possible, that is, the cell must be of thin-film type. Secondly, we must use processes suitable for mass production. One of the production methods which can satisfy both these requirements is the screen printing methods.[8] Recently, CdS/CdTe solar cells produced by this screen printing method have made remarkable advances, and a cell with η_i of 12.8% was obtained,[9] which is the highest value obtained in thin-film-type cells.

---

† Solar Battery Research and Development Group, Carbon Products Division, Matsushita Battery Industrial Co., Ltd., Moriguchi, Osaka 570.

JARECT Vol. 19, Semiconductor Technologies (1986), *J. Nishizawa (ed.)*
OHMSHA, LTD. and North-Holland

In this chapter, we report the preparation methods, electrical and photovoltaic properties and stability of screen printed CdS/CdTe solar cells.

16.2. Preparation and properties

16.2.1. *Structure of a cell*

Fig. 16.1(a) and (b) respectively show the cross-section and the CdTe pattern of a CdS/CdTe cell made on a $5 \times 10\,cm^2$ glass substrate.[10] The cell is manufactured on a

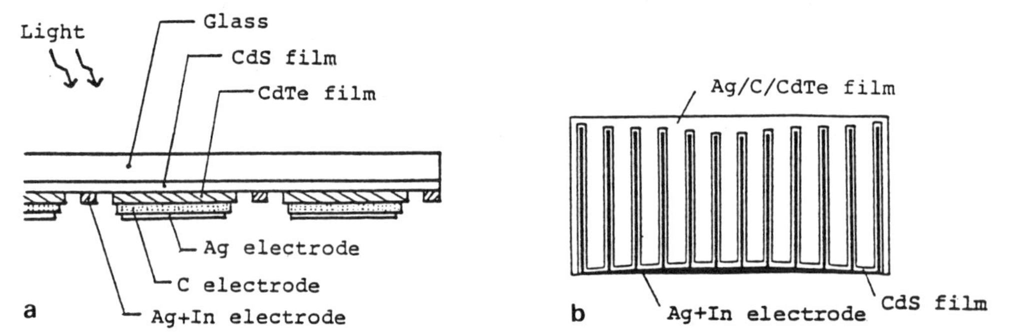

Fig. 16.1 (a) Cross-section of CdS/CdTe solar cell. (b) Printed pattern of CdS/CdTe solar cell. [after H. Matsumoto et al.[10]]

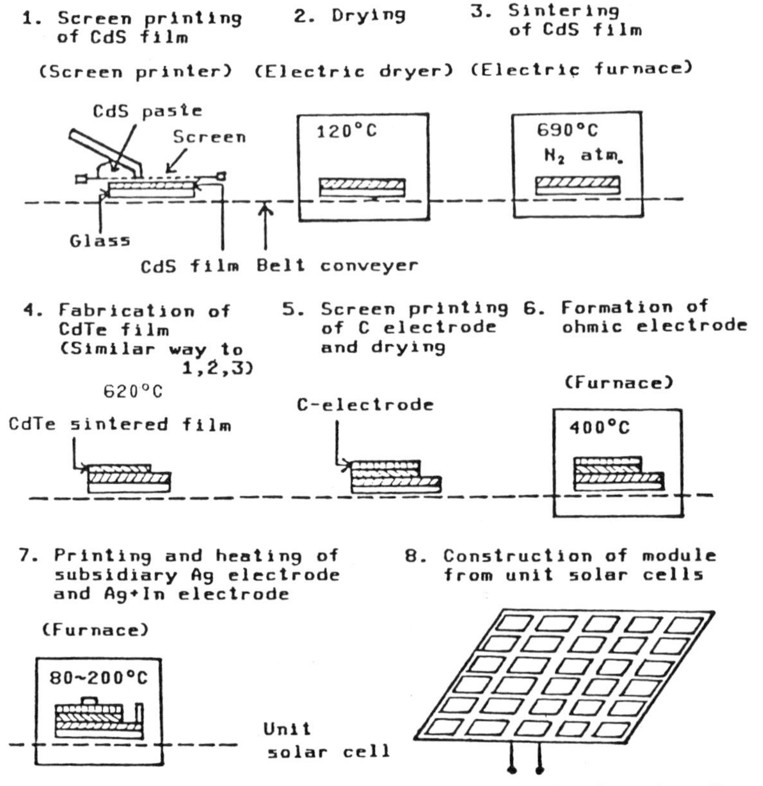

Fig. 16.2 Manufacturing processes of screen printed CdS/CdTe solar cell.

borosilicate glass substrate by successively repeating screen printing and heating in a belt furnace of each paste of CdS, Cd + Te, C, Ag + In and Ag as shown in Fig. 16.2.[11]

16.2.2. *CdS film*

As the starting material for the CdS film, commercially-available CdS powder (5N) was used. This CdS powder has hexagonal grains of sizes ranging from 1 to 2 μm. CdS paste for screen printing was obtained by mixing 100 g of CdS powder and 10 g of $CdCl_2$ powder with an appropriate amount of propylene glycol.

The CdS paste was screen printed on a borosilicate glass and dried at 120°C for 1 h. The printed CdS film was placed in an alumina case with a cover which had many small holes, and sintered in a nitrogen atmosphere at 690°C for 90 min. $CdCl_2$ acts as a flux to promote CdS particle fusion and granule regrowth, and evaporates through the holes of the cover during sintering. The CdS film prepared in this way had a thickness of about 30 μm.

The residual amount of $CdCl_2$ in the CdS sintered film gave a large influence on the resistivity of the CdS film and the properties of the cell.[12] Fig. 16.3 shows the relation between the resistivity and the residual amount of Cl ions in as-sintered CdS films. The CdS sintered film containing a large amount of Cl ions had a large electrical resistivity and did not give a high-efficiency cell.

The area of the holes in an alumina cover influences the grain size and the residual amount of Cl ions of a CdS sintered film.[13] The grain size of the film decreased with increasing hole area. The rate of evaporation of $CdCl_2$ through the hole increased with increasing hole area, and resulted in a decrease in grain size. A CdS film sintered with a cover without holes had a high resistivity of 500 Ωcm, a large grain size of about 50 μm and a large amount of residual Cl ions, owing to the very small evaporation of $CdCl_2$.

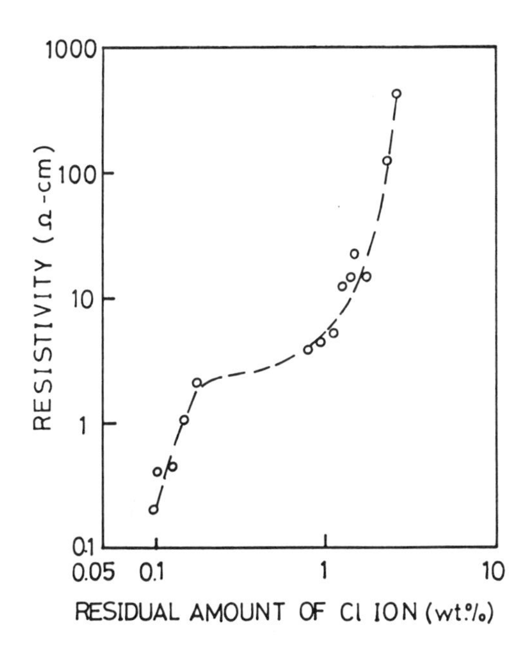

Fig. 16.3 Relation between resistivity and residual Cl ions of as-sintered CdS film. [after H. Uda et al.[13]]

The CdS film which gave a high-efficiency cell had Cl ions less than 0.2 wt%, grain size of 20–30 μm, resistivity of less than 0.5 Ωcm, electron concentration of about 10^{18} cm^{-3} and electron mobility of about 14 cm^2/V s.[13] Since the main role of the CdS film is to serve as a window material, it is necessary to increase the conductivity as well as the transparency of the film.

16.2.3. *CdTe film*

As the starting material for the CdTe film, Cd powder and Te powder were used. A nearly equimolar ratio of Cd powder (5N) and Te powder (5N) was mixed with distilled water and crushed down to a grain size of about 0.5 μm in an agate ball mill and dried. Paste for screen printing was obtained by mixing 100 g of the crushed Cd + Te mixture and 0.5 g of $CdCl_2$ powder with an appropriate amount of propylene glycol. The paste was screen printed on a CdS sintered film and dried. The printed Cd + Te film was placed in the same alumina case as was described in Section 16.2.2, and sintered in a nitrogen atmosphere at 620°C for 1 h. During sintering, Cd and Te react to form CdTe sintered film.

Fig. 16.4 shows the spectral responses of short circuit current in typical cells prepared at different sintering temperatures of the CdTe film.[11] The response was normalized to a constant incident energy. Cells A, B and C were prepared at 620, 660 and 700°C, respectively. The spectral response of cell A begins to increase from about 510 nm which nearly corresponds to the fundamental absorption edge of CdS and falls off rapidly at a

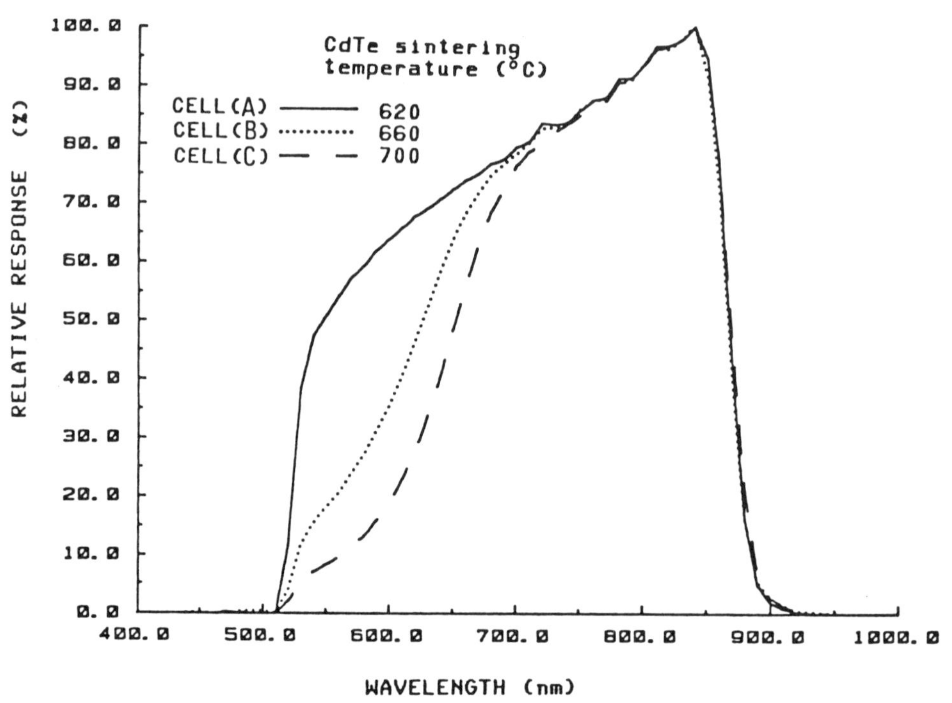

Fig. 16.4 Spectral responses of the cells fabricated at different CdTe sintering temperatures. [after H. Uda et al.[11]]

wavelength of about 860 nm which nearly corresponds to the fundamental absorption edge of CdTe. Cells B and C show a poor response in the short-wavelength region.

The X-ray microanalysis (XMA) of the elements S and Te along the cross-section near the CdS/CdTe interface showed that the transition region of cell C was wider than that of cell A. With a higher temperature of CdTe sintering, the interfacial reaction between S and Te become more active, forming a thick CdS_xTe_{1-x} layer.

The formation of a thick CdS_xTe_{1-x} layer may be responsible for the poor spectral response in the short-wavelength region of cells B and C. Because the thick CdS_xTe_{1-x} layer absorbs short-wavelength light before the light arrives at the depletion layer, which is considered to be formed at the CdTe region due to the low carrier concentration of CdTe film.

16.2.4. *Carbon electrode*

In order to give the photovoltaic property in the CdS/CdTe film, it is necessary to apply a suitable electrode material containing an acceptor impurity on the CdTe film, and diffuse the impurity towards the CdS/CdTe interface by a heat treatment, so as to form a p(CdTe)–n(CdS) junction. As this electrode, a carbon electrode was used.

A carbon electrode was formed on the CdTe film by screen printing carbon paste containing 50–100 ppm Cu and heating at 400°C for 30 min in a nitrogen atmosphere containing 1–1.5 mol% O_2.[14] The solar cell of highest efficiency was obtained by heating the carbon electrode in an atmosphere containing 1.0–1.5 mol% O_2 at 400°C for 30 min. Heat treatment at the same temperature in O_2 concentrations higher or lower than 1.0–1.5 mol% O_2 resulted in an increase in the series resistance R_s because of the increase in contact resistivity between the CdTe and the carbon electrode. The effects of oxygen in the heating atmosphere on efficiency of the solar cell is not clear. Two explanations, however, might be possible. In the first, Tyan and Perez-Albuerne[7] reported that a small amount of oxygen introduced during deposition of the CdS and CdTe layers enhanced the p-type character of the CdTe film and ensured the shallow-junction behavior of the device. An appropriate amount of oxygen in the heating atmosphere is probably essential in order to obtain the highest-efficiency CdS/CdTe solar cell. Oxygen, however, is not known to be a p-type dopant in CdTe. The second, a decomposition of resin in the carbon electrode, may be associated with the oxygen concentration in the furnace. The sheet resistance of carbon electrodes heated at 400°C for 30 min in different atmospheres between 1.0 mol% and 2.5 mol% O_2 showed nearly the same value of about 9 $\Omega/\square$. The heat treatment at the same temperature in atmospheres with 0.1 mol% O_2 and 6 mol% O_2 gave the sheet resistivity of about 10 and 12 $\Omega/\square$, respectively. Adhesion between the carbon electrode and the CdTe layer may also be dependent on the amount of residual resin in the carbon electrode. The carbon electrode heated in the 6 mol% O_2 atmosphere resulted in a poor adhesion to the CdTe layer and high R_s. A carbon electrode which has a low sheet resistivity and gives a good contact to CdTe is essential in the fabrication of high-efficiency solar cells.

The addition of Cu improved η_i as shown in Fig. 16.5. This increase resulted from increases in V_{oc} and the fill factor (FF) and decreases in R_s and the diode factor (n). The decreases of R_s and n lead to the increase in FF. The reverse saturation current density (J_0) was reduced from $\sim 10^{-7}$ A/cm^2 to $\sim 10^{-8}$ A/cm^2 by the addition of Cu. R_s and n were

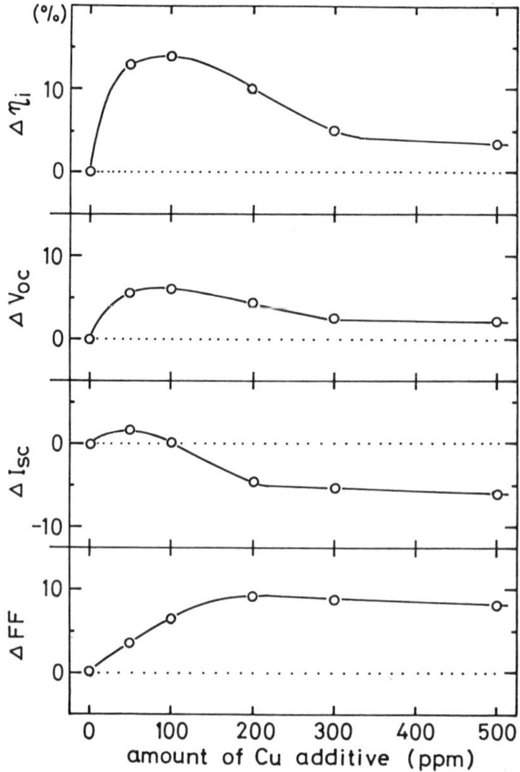

Fig. 16.5 Effects of addition of Cu on characteristics of solar cell heated at 400°C for 30 m. [after K. Kuribayashi et al.[14]]

decreased from 10–12 Ωcm^2 to ~10 Ωcm^2 and 2.8–3.3 to 2.0–2.4 respectively, with the addition of Cu up to approximately 100 ppm.

Addition of an appropriate amount of Cu in the carbon paste and heat treatment under optimum conditions might make the CdTe layer p^+-type and lower the contact resistance of CdTe/C electrode through diffusion of an optimum amount of Cu into CdTe.

16.2.5. *Ag + In and Ag electrode*

The electrode for CdS has been formed, so far, by the vacuum evaporation of In, Ga or Al, by soldering of In, by electrolytic deposition of In, by painting with In or In–Ga metal alloy, or by other methods. For the continuous production of solar cells, it is desirable to develop a screen-printable electrode material which forms an ohmic contact to an n-type CdS film.

The Ag + In electrode was developed as a screen-printable electrode material which formed an ohmic contact to an n-type CdS film.[15] Ag + In paste for screen printing was obtained by mixing a conventional Ag paint (about 67 wt%) with In powder (about 33 wt%). The paste was screen printed on the remaining CdS film (Fig. 16.1) and heated at 180°C for 1 h in a nitrogen atmosphere. Heat treatment might make the CdS layer n^+-type and lower the contact resistivity of CdS/Ag + In electrode through diffusion of In into CdS.

Finally, the Ag paste, which was of conventional type, was screen printed on the carbon layer as a subsidiary electrode, and heated at 150°C for 30 min.

16.3. Effect of CdTe width

In the print pattern of Fig. 16.1(b), a change in the width of the CdTe pattern causes a change in the practical conversion efficiency η_P (for total substrate area) through changes in the series resistance of the CdS layer and in the ratio of the active area to the substrate area.

In order to determine the optimum width of the CdTe pattern to obtain the maximum value of η_p, cells with different widths of CdTe ranging from 2 to 8 mm were made on a 5×10 cm^2 substrate, as shown in Fig. 16.6.[16] The maximum η_p was obtained at the CdTe width of 5 mm from the CdS film around 100 $\Omega/\Box$. The value of FF increased with decrease in x as expected, and the maximum value of FF was about 0.6.

The maximum η_i was obtained at the minimum CdTe width of 2 mm as expected. Fig. 16.7 shows the photovoltaic characteristics of the cell of the maximum η_i at AM 1.5, 100 mW/cm^2. Since the carbon paste is printed inside 0.15 mm from the edge of CdTe, the active area of this cell is 1.7×46 mm^2 (0.78 cm^2) corresponding to the area of the carbon electrode. The value of η_i for the active area was 12.8%,[9] which is the highest obtained in thin-film-type solar cells.

The forward V–I characteristics of the 12.8% efficiency cell could be expressed by the equation

$$I = I_0 \exp(qV/nkT).$$

The values of the diode quality factor n and reverse saturation current I_0 were 2.2 and 2.3×10^{-8} A respectively. The deviation from an exponential relation at high currents in the forward-bias region indicated the series resistance was 3.8 Ω, while the series resistance

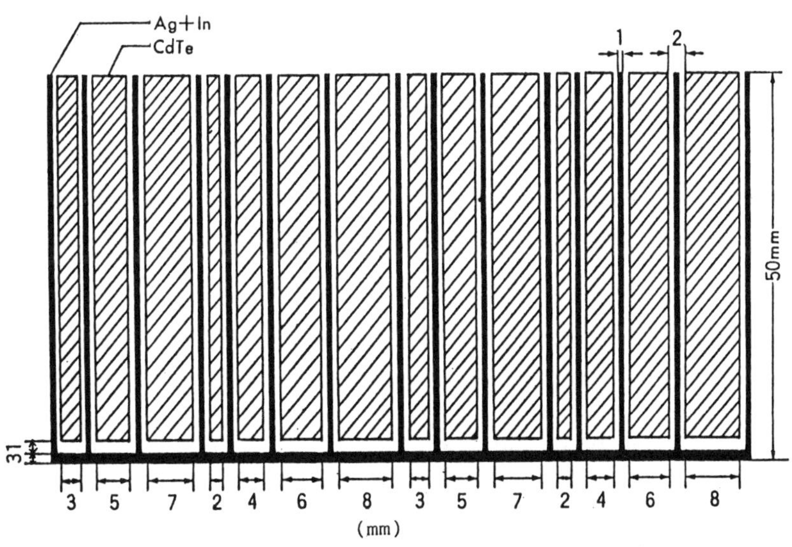

Fig. 16.6 CdTe pattern with different width and Ag + In pattern on 5×10 cm^2 substrate. [after S. Ikegami et al.[8]]

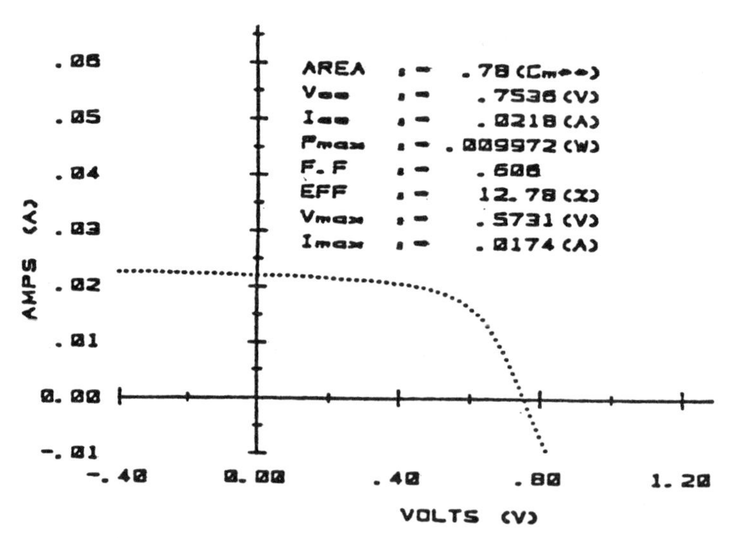

Fig. 16.7 *V–I* curve of the highest-efficiency solar cell under AM 1.5, 100 mW/cm² solar simulator.

which arises from the sheet resistance of the CdS film (100 $\Omega/\square$) was calculated as 0.80 Ω by the method described by Handy[17)] and Sahai.[18)]

The difference between the experimental (3.8 Ω) and the calculated (0.8 Ω) values of R_s is due to factors other than the resistance of the CdS film. Analysis of the other factors which increase the value of R_s will be useful in the realization of a higher-efficiency cell.

16.4. Large-area substrate cells

In studies of screen printed CdS/CdTe solar cells, a substrate of 4×4 cm² (0.8 mm thickness) was initially used.[19)] A 5×10 cm² substrate (1.2 mm thickness) which was used at the next step was a fundamental size: the comb-type print pattern as shown in Fig. 16.1(b) was established by this size of substrate.[16)]

As the third step, a 10×10 cm² substrate was used:[11)] this substrate contains just two 5×10 cm² unit cells connected in series. By using 10×10 cm² substrates (1.2 mm thickness) fundamental manufacturing techniques were established.

A further advance in manufacturing techniques was made by using a 30×30 cm² substrate (3 mm thickness). The first 30×30 cm² single substrate module consisted of 4×7 unit cells; these unit cells are connected in series simultaneously at the final Ag printing process as shown in Fig. 16.8. The p–n junction was not formed at the periphery of the substrate, because the periphery was covered with an aluminum frame in module encapsulation. This resulted in a decrease in the active area. The module efficiency (η_m) of the first module was 4.7%, though η_i was as high as 8.5%.[20)]

The second 30×30 cm² single-substrate module consisted of 4×9 unit cells connected in series. The active area of the second module was increased by decreasing the areas of the periphery and Ag + In electrodes. η_m of the second module was 5.4%, and η_i was 8.7%.[21)] This module produced 4.8 W under 100 mW/cm² illumination. The photovoltaic characteristics are shown in Table 16.1 together with those of the highest efficiency cell and 10×10 cm² substrate cell.

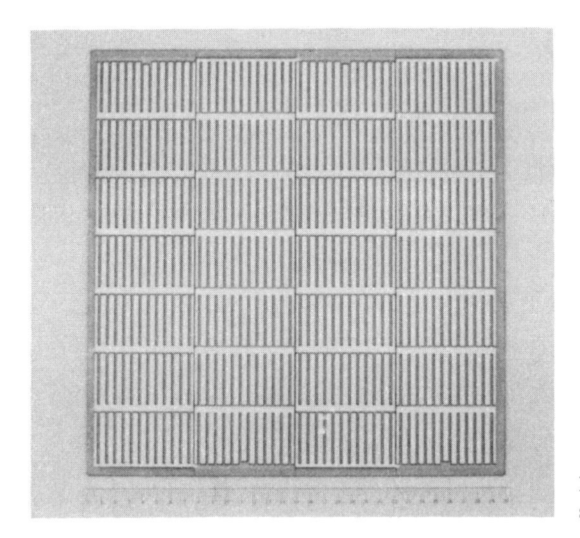

Fig. 16.8 Photograph of $30 \times 30\,cm^2$ single substrate module.

Table 16.1 Photovoltaic characteristics of screen printed CdS/CdTe solar cells (AM1.5, $100\,mW/cm^2$)

Cell	Small area cell	Middle substrate cell	Large substrate cell
Size (cm^2)	4.6×0.17	10×10	30×30
Area (cm^2)	0.78	100	900
Active area (cm^2)	0.78	64	559
V_{oc} (V)	0.75	1.54	27.5
I_{sc} (A)	0.022	0.79	0.35
FF	0.61	0.54	0.51
P_{max} (W)	0.01	0.65	4.84
V_{max} (V)	0.57	1.04	18.0
I_{max} (A)	0.17	0.63	0.27
η_p (%)	—	6.53	5.4
η_i (%)	12.8	10.21	8.7
No. of cell	1	2	36

η_i: intrinsic efficiency for active area.
η_p: practical efficiency for substrate area.

16.5. Modules

Large modules consisting of 4×12 pieces of $10 \times 10\,cm^2$ substrate cells (totally 96 unit cells) were constructed. Under an AM 1.5, $100\,mW/cm^2$ solar simulator, photovoltaic characteristics of the module were as follows:[11)]

$$V_{oc} = 68.5\ V, \quad I_{sc} = 0.725\ A, \quad FF = 0.485$$

$$P_{max} = 24.1\ W, \quad \eta_m = 5.02\%\ \text{(for total substrate area).}$$

16.6. Temperature coefficient

It is well known that the output of solar cells is decreased with increasing temperature. Solar cells for terrestrial use sometimes reach 40–70°C under sunlight. Therefore, it is desirable for the temperature coefficient of solar cells to be small.

The temperature coefficient of screen printed CdS/CdTe solar cells was measured from 25°C to 70°C. The values obtained were as follows:[22]

V_{oc} − 0.2 to −0.3%/°C, I_{sc} 0 to +0.04%/°C

P_{max} −0.2 to −0.3%/°C.

The temperature coefficient of output (P_{max}) is smaller than that of Si crystal solar cells, so the screen printed CdS/CdTe cells are advantageous for terrestrial use.

16.7. Reliability

Some II–VI compound solar cells, for example, $Cu_{2-x}S$/CdS and $Cu_{2-x}Te$/CdTe solar cells show a considerable degradation in efficiency. Therefore, stability is another important factor, especially in II–VI compound solar cells. The degradation in these two cells is caused mainly by ionic conduction of Cu ions in $Cu_{2-x}S$ or $Cu_{2-x}Te$. It may be expected that CdS/CdTe solar cells will exhibit high stability because they contain no Cu compounds.

16.7.1. *Outdoor tests of encapsulated cells*

Outdoor tests were conducted on encapsulated $10 \times 10\,cm^2$ substrate cells.[23] Tests were made on more than 25 cells, under open-circuit, short-circuit and maximum power load conditions. The cells were sometimes removed from the life test station, and their output characteristics were measured under the AM 1.5, $100\,mW/cm^2$ solar simulator at

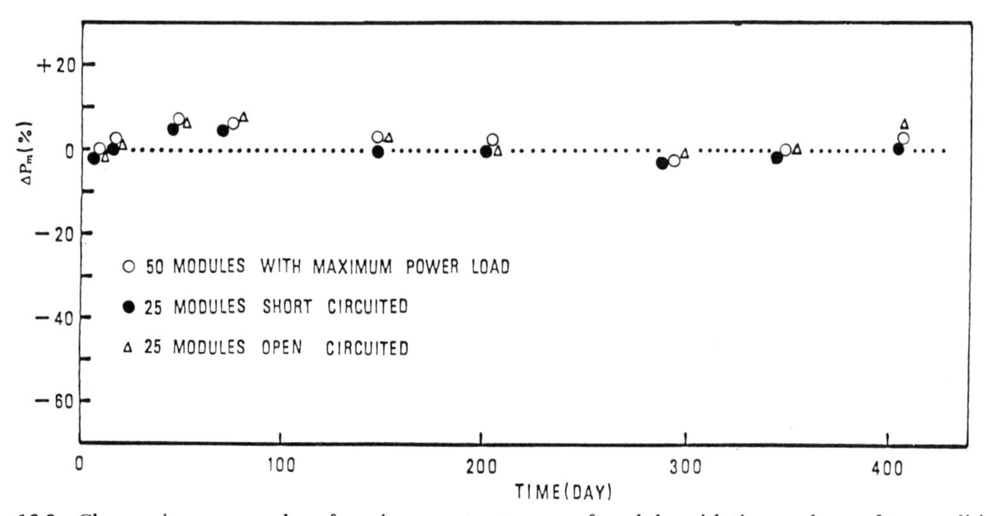

Fig. 16.9 Changes in average value of maximum output power of modules with time under rooftop conditions.

room temperature. The time dependences of the average value of the maximum output power (P_m) are shown in Fig. 16.9. No change was observed in the average maximum output power for 400 days under these three conditions. Thus, it was clarified that the screen printed CdS/CdTe solar cells are stable under rooftop conditions and are reliable for practical applications.

16.7.2. *Accelerated tests of unencapsulated cells*

The accelerated tests were conducted on unencapsulated cells in order to study rapidly the degradation mechanism. Continuous illumination tests (40°C, 80 mW/cm^2 sunlight) were performed under open-circuit and short-circuit conditions. These tests caused no degradation of cells after 90 days.

The unencapsulated cells were exposed to elevated temperatures ranging from 50 to 130°C over a period of more than 70 days. The cells exposed to 50 and 70°C exhibited excellent stability in efficiency over 72 days. The cells exposed to 100 and 130°C for 72 days, however, showed a decrease in efficiency by about 15 and 35%, respectively, which was caused by the decrease in fill factor due to an increase in the series resistance of the cell.

The unencapsulated cells exposed to 95% relative humidity for 60 days also showed a decrease in efficiency by about 8 and 35%, respectively, at 40 and 65°C. The decrease in efficiency in this test was also due to an increase in the series resistance.

These increases in series resistance of the cell result from an increase in contact resistance between CdTe and carbon in the high-temperature and high-humidity test, and from an increase in resistance of the carbon layer in the high-humidity test. It was found that the increase in contact resistance was caused by a decrease in adhesion between CdTe and carbon. Also, it found was that the increase in carbon resistance was due to a swelling of the carbon layer with water absorption. Improved cells are now being made to endure these accelerated tests by improving the preparation conditions of the carbon electrode.

16.8. Conclusion

Thin-film CdS/CdTe solar cells are prepared on a borosilicate glass substrate by successively repeating screen printing and heating (sintering) of each paste of CdS, Cd + Te, C, Ag + In and Ag. These cells can be expected to be low-cost, highly efficient and stable solar cells. The main conclusions are as follows:

(1) The CdS sintered film containing a large amount of $CdCl_2$ has a large electrical resistivity and does not give a high-efficiency cell.

(2) The photovoltaic characteristics of a cell are influenced by the CdTe sintering temperature. The cell sintered at a higher temperature shows a poor spectral response in the short-wavelength region.

(3) A low contact resistance between the carbon electrode and CdTe, and consequently a higher-efficiency solar cell are obtained by heating the carbon electrode containing 50–100 ppm Cu at 400°C for 30 min in an atmosphere including 1.0–1.5 mol%O_2.

(4) A cell of 12.8% efficiency was obtained for an active area of 0.78 cm^2. This is the highest value obtained for any thin-film-type cells.

(5) A 4.8 W single substrate module was obtained on a $30 \times 30\ cm^2$ glass substrate under $100\ mW/cm^2$ illumination by screen printing method.
(6) The temperature coefficient of the output of a cell is between -0.2 and $-0.3\%/°C$. This value is smaller than that of Si crystal solar cells.
(7) Encapsulated cells show excellent stability for about 400 days under rooftop conditions.

Acknowledgements

The authors thank Dr. Yoshio Iida, Vice-President of Matsushita Electronic Component Co., Mr. Masaya Nakajima, Director of the Laboratory and Mr. Kensuke Kuchiba for their encouragement. They also acknowledge Dr. Hiroshi Uda, Mr. Yasumasa Komatsu, Dr. Akihiko Nakano and Dr. Kiyoshi Kuribayashi for their co-works. This work was supported in part by the Agency of Industrial Science and Technology under the Sunshine Project.

References

1) S. Ikegami: Energy and Resources, 3 (1982) 34.
2) J. J. Loferski: J. Appl. Phys., 27 (1956) 777.
3) A. L. Fahrenbruch, V. Vasilchenko, F. Buch, K. Mitchell, and R. H. Bube: Appl. Phys. Lett., 25 (1974) 605.
4) K. Yamaguchi, N. Nakayama, H. Matsumoto, and S. Ikegami: Japan. J. Appl. Phys., 16 (1977) 1203.
5) D. Bonnett and H. Rabenhorst: Proc. 9th IEEE Photovoltaic Specialists Conf., May 1972 (IEEE, New York, 1972) p. 129.
6) H. Uda, H. Taniguchi, M. Yoshida, and T. Yamashita: Japan. J. Appl. Phys., 17 (1978) 585.
7) Y. S. Tyan and E. A. Perez-Albuerne: Proc. 16th IEEE Photovoltaic Specialists Conf., Sept. 1982 (IEEE, New York, 1982) p. 794.
8) S. Ikegami, H. Matsumoto, H. Uda, Y. Komatsu, A. Nakano, and K. Kuribayashi: Proc. 5th EC Photovoltaic Solar Energy Conf., Oct. 1983 (Reidel, Dordrecht, The Netherlands) p. 740.
9) H. Matsumoto, K. Kuribayashi, H. Uda, Y. Komatsu, A. Nakano, and S. Ikegami: Solar Cells, 11 (1984) 367.
10) H. Matsumoto, A. Nakano, Y. Komatsu, H. Uda, K. Kuribayashi, and S. Ikegami: Japan. J. Appl. Phys., 22 (1983) 269.
11) H. Uda, H. Matsumoto, Y. Komatsu, A. Nakano, and S. Ikegami: Proc. 16th IEEE Photovoltaic Specialists Conf., Sep. 1982, (IEEE, New York, 1982) p. 801.
12) H. Matsumoto, A. Nakano, H. Uda, and S. Ikegami: Japan. J. Appl. Phys., 21 (1982) 800.
13) H. Uda, H. Matsumoto, K. Kuribayashi, Y. Komatsu, A. Nakano, and S. Ikegami: Japan. J. Appl. Phys., 22 (1983) 1832.
14) K. Kuribayashi, H. Matsumoto, H. Uda, Y. Komatsu, A. Nakano, and S. Ikegami: Japan. J. Appl. Phys., 22 (1983) 1828.
15) A. Nakano, H. Matsumoto, N. Nakayama, and S. Ikegami: Proc. 1st Photovoltaic Science and Engineering Conf. in Japan (Tokyo, 1979); Japan. J. Appl. Phys. 19 (1980) Suppl. 19-2, 157.
16) H. Matsumoto, H. Uda, Y. Komatsu, A. Nakano, and S. Ikegami: Proc. 3rd Photovoltaic Science and Engineering Conf. in Japan, (1982); Japan. J. Appl. Phys., 21 (1982) Suppl. 21-2, 103.
17) R. J. Handy: Solid State Electron., 10 (1967) 765.
18) R. Sahai and A. G. Milnes: Solid State Electron., 13 (1970) 1289.
19) N. Nakayama, H. Matsumoto, A. Nakano, S. Ikegami, H. Uda, and T. Yamashita: Japan. J. Appl. Phys., 19 (1980) 703.
20) H. Matsumoto, K. Kuribayashi, Y. Komatsu, A. Nakano, H. Uda, and S. Ikegami: Japan. J. Appl. Phys., 22 (1983) 891.
21) H. Matsumoto, A. Nakano, Y. Komatsu, H. Uda, and S. Ikegami: Technical Digest of Int. PVSEC-1 (Kobe, Japan, 1984) p. 393.
22) H. Matsumoto, H. Uda, Y. Komatsu, A. Nakano, S. Ikegami, and K. Kuribayashi: 44th Autumn Meeting of the Japan Society of Applied Physics (Sendai, Sept. 1983).
23) H. Uda, A. Nakano, K. Kuribayashi, Y. Komatsu, H. Matsumoto, and S. Ikegami: Japan. J. Appl. Phys., 22 (1983) 1822.

17 ROOM-TEMPERATURE-OPERATED HIGH-SPEED INFRARED PHOTOSENSOR

Masanori OKUYAMA and Yoshihiro HAMAKAWA†

Abstract

Various types of pyroelectric sensors which can be operated at room temperature have been developed. The infrared-sensitive parts consist of $PbTiO_3$ ferroelectric thin films deposited by RF sputtering. Typical voltage responsivity and detectivity of the film deposited on Pt-coated mica are 330 V/W and 1.5×10^8 cm $Hz^{1/2}$/W with 1 Hz bandwidth at 20 Hz, respectively. A fast response with a rise time of 1.3 μs has been obtained by measuring the output under CO_2 laser pulse irradiation. The film fabricated on a Si wafer can be combined directly with active devices such as FET(IR-OPFET) and bipolar transistor. Sensitivity of the Si monolithic sensor has been increased considerably by reducing the heat capacity of the Si substrate beneath the sensitive area with anisotropic etching. Linear array sensors having 16 elements have been developed on some types of substrate and their basic characteristics have been investigated for improving resolution. Infrared imaging has been also tried by using the linear array sensor.

Keywords: Infrared Sensor, Pyroelectric Sensor, $PbTiO_3$, IR-OPFET, Linear Array Sensor, Infrared Imaging

17.1 Introduction

In recent years, much attention has been paid to room-temperature-operated infrared sensors as there is a considerable potential need for applications such as remote sensing, biomedical thermography, gas detection and alarms. A pyroelectric infrared sensor shows some advantages compared with the other types of infrared sensors such as photo-quantum-effect sensors operated at very low temperature and Golay cells with large response time,[1,2] because the pyroelectric can be operated even at room temperature, has little wavelength dependence of the response over a wide infrared range and has a fast response.[3] In particular, $PbTiO_3$ shows excellent pyroelectric characteristics because of its high pyroelectric coefficient and high Curie temperature.[4,5] However, conventional pyroelectric sensors are made of bulk materials such as $PbTiO_3$ ceramics or $LiTaO_3$ single crystals,[5,6] and pyroelectric vidicon has also been developed already. Then there are such problems as connection to processing circuit, compactness and operating voltage. Using thin films of the pyroelectric material can solve these problems considerably because the film can easily be connected to the peripheral device by depositing it on the circuit and can be operated at a low voltage. In the last ten years, various kinds of ferroelectric thin films showing good pyroelectric characteristics have been prepared by several methods.[7-11] We have also succeeded in making $PbTiO_3$, PLT and PLZT thin films by RF sputtering or

† Department of Electrical Engineering, Faculty of Engineering Science, Osaka University, Toyonaka, Osaka 560.

JARECT Vol. 19, Semiconductor Technologies (1986), *J. Nishizawa (ed.)*
OHMSHA, LTD. and North-Holland

chemical vapor deposition.[12–16] Moreover their device applications for infrared sensing have also been developed. This chapter reviews the preparation method and some properties of the thin film, infrared-sensitive IR-OPFET (Infrared–Optical FET) having a $PbTiO_3$ thin-film gate on a Si MOSFET,[17,18] a highly sensitive infrared sensor fabricated on a Si membrane of low heat capacity,[19] a linear array sensor and infrared imaging by the array sensor.[20]

17.2. Preparation of $PbTiO_3$ thin films and their basic properties

17.2.1. *Deposition of the films and their dielectric properties*

Thin films of ferroelectric $PbTiO_3$ were deposited on a platinum foil, platinum-coated mica and Si wafer by RF sputtering. The target used was a powder mixture of $PbTiO_3$ made from a sintered mixture of Pb_3O_4 and TiO_2. Excess Pb_3O_4 of 10 wt% was also added to the target to compensate for the Pb deficiency in the deposited film. The target pressed on a quartz plate was sputtered at substrate temperatures of 300–550°C in an atmosphere of 90% argon and 10% oxygen gas. The deposition rate was about 30 Å/min. The detailed deposition conditions are given in Table 17.1.

The crystalline structures of the films were investigated by X-ray diffraction analysis and were strongly dependent on the deposition conditions such as the substrate temperature, the input power and the film thickness and heat treatment after the sputtering. The films deposited below 400°C show amorphous or pyrochlore structures with only paraelectric properties. However in the films deposited at 400–450°C, the structure is a mixture of pyrochlore and perovskite structures, and in the films deposited above 450°C it is completely perovskite with ferroelectric properties. Fig. 17.1 shows the substrate–temperature dependence of: (a) dielectric constant ϵ and loss tangent tan δ measured at 100 kHz; (b) remanent polarization P_r and coersive force E_c under applied field of 420 kV/cm; and (c) grain size S_g estimated by SEM observation of the film of 2.4 μm thickness deposited on the platinum foil. The dielectric constant increases gradually with the substrate temperature and for the film deposited at 500–550°C it is about 200 which is similar to that of its ceramic. This increase is attributed to crystallographic improvement at high temperature, and the accompanying increase of the grain size. The dielectric constant falls abruptly above about 550°C because the Pb having high vapor pressure tends to leave the deposited film at the high temperatures and the composition of the film shifts considerably from the stoichiometric ratio of $PbTiO_3$. Fig. 17.2 shows dependence on the film thickness of the dielectric constant and the remanent polarization measured at an applied electric field of 420 kV/cm. The dielectric constant increases with the thickness

Table 17.1 Sputtering conditions

Target diameter	80 mm
Target–substrate distance	35–40 mm
RF voltage	1.6 kV
RF input power	150 W
Sputtering gas	Ar(90%) + O_2(10%)
Gas pressure	27 Pa
Substrate temperature	350–550°C

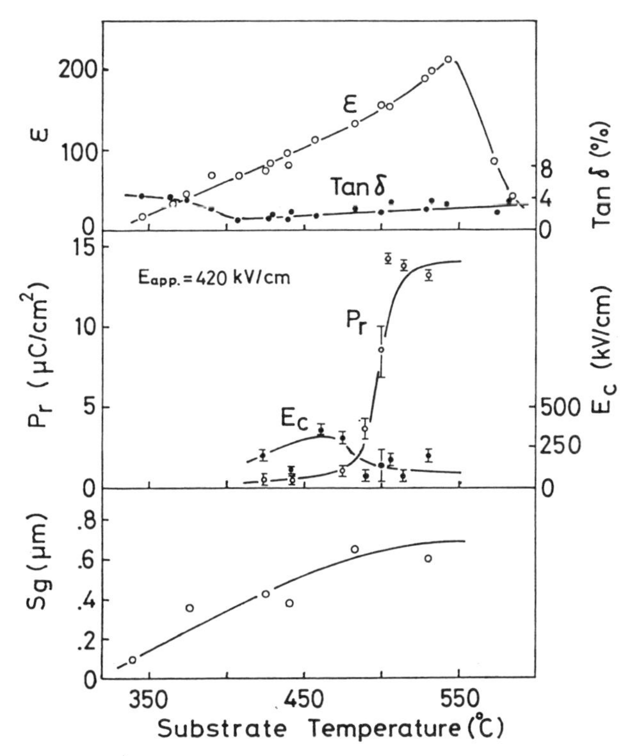

Fig. 17.1 Substrate temperature dependence of (a) dielectric constant ϵ and loss tangent tan δ, (b) remanent polarization P_r and coersive force E_c and (c) grain size S_g. [after M. Okuyama et al.[15]]

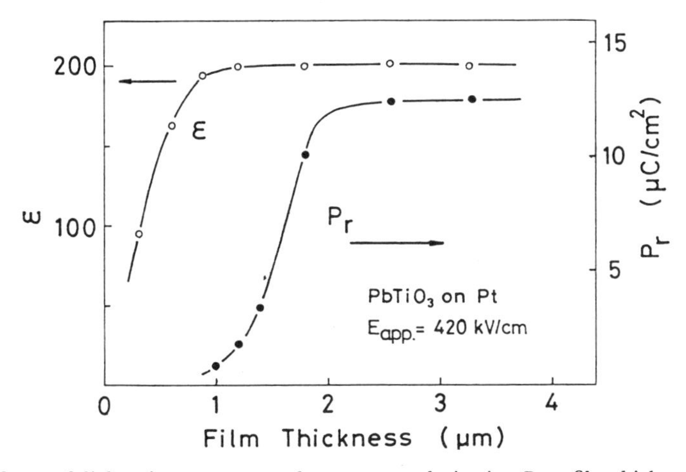

Fig. 17.2 Dependence of dielectric constant ϵ and remanent polarization P_r on film thickness. [after M. Okuyama et al.[17]]

above about 2000 Å and saturates to about 200 above about 2 μm. The remanent polarization increases up to about 12 $\mu C/cm^2$ above about 2 μm. It is therefore considered that good ferroelectric properties can be obtained in the films of thickness above about 2 μm. The maximum remanent polarization obtained was 27 $\mu C/cm^2$. Similar crystalline and dielectric properties have been obtained in the films deposited on the platinum-coated mica and silicon.

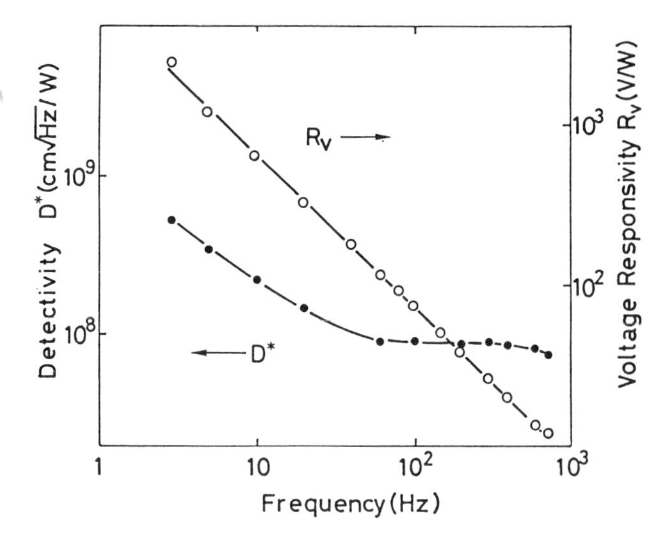

Fig. 17.3 Chopping frequency dependence of voltage responsivity R_v and detectivity D^* of $PbTiO_3$. [after M. Okuyama et al.[17)]]

17.2.2. *Pyroelectric properties*

Pyroelectric measurements have been made on the film deposited near 500°C on the Pt-coated mica. A black electrode was deposited on the sample as an infrared-absorbing layer after formation of the aluminum electrode of $0.36\,mm^2$ area. The sample was mounted on a TO-5 stem by supporting both ends of the mica to float the sensitive area in air and the film was poled by applying 20 V at 200°C for 20 min. Infrared light from a lamp was chopped and applied to the black electrode. The pyroelectric voltage of the film was amplified through a high-input-impedance FET by a lock-in amplifier. Fig. 17.3 shows the frequency dependences of voltage responsivity R_v and detectivity D^*. Typical R_v and D^* are 330 V/W and $1.5 \times 10^8\,cm\,Hz^{1/2}/W$ with a bandwidth of 1 Hz at a chopping frequency of 20 Hz. R_v is inversely proportional to the frequency as shown in the pyroelectric theory.[2)] D^* is also inversely proportional to the square root of the frequency, but becomes flat at higher frequencies. These pyroelectric characteristics are comparable with those of the $PbTiO_3$ ceramic. The fast response was measured by using a digital memory (IWATSU DM-701) when a CO_2 laser pulse was applied to the sample. The rise time of the response was about 1.3 μs and was fairly fast compared with the other types of thermal infrared sensors.

17.3. Infrared optical FET(IR-OPFET)

17.3.1. *Device structure and operational principles*

By combining these prominent pyroelectric effects with sophisticated Si IC technology, a Si monolithic infrared sensor can be fabricated. We have proposed a new type of infrared sensing FET with the $PbTiO_3$ pyroelectric thin film gate and call this device the IR-OPFET (Infrared-Optical FET). The device structure of the IR-OPFET is shown in

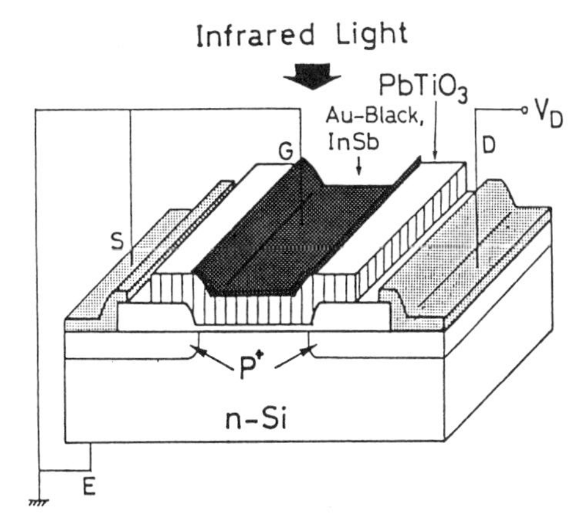

Fig. 17.4 Device structure of IR-OPFET. [after M. Okuyama et al.[17)]]

Fig. 17.4. This structure is almost the same as the non-volatile memory FET with the ferroelectric film gate (MFSFET), which was reported by some workers,[21–24)] except for its having an infrared-light-absorbing gate electrode. The change in electric polarization of the pyroelectric film induced by the temperature change of the gate electrode modulates the surface potential of the semiconductor over the channel. This change in surface potential induces a change in the drain current. Consequently infrared light is detected as a voltage change in a load resistance. The IR-OPFET has the function of amplification of the pyroelectric signal of the film and low output impedance, and so is very useful for practical applications.

17.3.2. *Device fabrication*

An epitaxial layer of n-Si (3.5×10^{14} cm^{-3}) grown on a low-resistive wafer was used as a substrate for fabrication of the MOSFET. Typical sizes of channel length and width are 10 and 460 μm, respectively. SiO_2 film of 300–500 Å thickness was thermally grown over the channel in order to avoid semiconductor–dielectric interface damage induced by the sputtering of the $PbTiO_3$ and prevent carriers from being injected through the ferroelectric layer into the Si wafer. The $PbTiO_3$ film of 2.1 μm thickness was grown on this MOSFET with RF sputtering at a substrate temperature of 500°C, and InSb or Au black was formed on the $PbTiO_3$ as an infrared-light-absorbing electrode. After the poling of the $PbTiO_3$ film at 200°C for 20 min at 8 V, the FET worked in depletion mode.

17.3.3. *Infrared response*

The infrared response of the IR-OPFET was measured in the similar way as the film's with the conditions of short-circuit of the gate electrode and the substrate. The drain current decreased under infrared irradiation because the dielectric polarization of the $PbTiO_3$ film is reduced by the slight temperature increase due to the infrared irradiation and the surface potential supported by the polarization is decreased. On the other hand, if visible or near-infrared light with photon energy above the Si band-gap energy is applied to the IR-OPFET, the current is increased by photo-carrier generation. Fig. 17.5 shows the chopping frequency dependence of R_v and D^* of the FET. The infrared input power P_{in} is 2.58 μW, the load resistance is 30 kΩ and drain voltage is 8 V. R_v is almost inversely

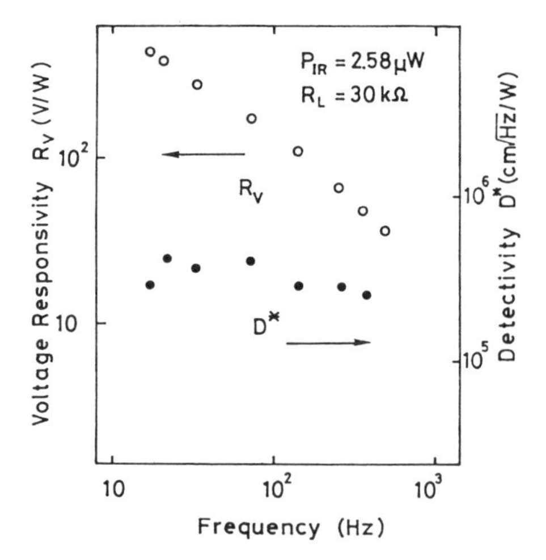

Fig. 17.5 Dependence on chopping frequency of voltage responsivity R_v and detectivity of IR-OPFET. [after M. Okuyama et al.[17)]]

proportional to the frequency as well as the dependence of the film. Typical data obtained were $R_v = 390$ V/W, NEP $= 2 \times 10^{-8}$ W/Hz$^{1/2}$ and $D^* = 3.5 \times 10^5$ cm Hz$^{1/2}$/W with a bandwidth of 1 Hz at the chopping frequency of 20 Hz. R_v is comparable with that of the film, but D^* is three orders of magnitude smaller than that of the film. The small value of D^* is attributed to the fact that the noise generated in the MOSFET is much larger than that of a conventional FET which was used in the pyroelectric measurement of the film, because the sputtering of the $PbTiO_3$ film on the SiO_2 layer damaged the Si-SiO_2 interface[25)] and so this damage enlarged the noise. However, the damage can be avoided by inserting Pt layer between the SiO_2 and the $PbTiO_3$ films.[25)]

The response of the IR-OPFET under irradiation by a fast CO_2 laser pulse was also measured. The light pulse was monitored by a Ge; Au photoconductor. A typical oscillograph trace of the response is shown in Fig. 17.6. The upper trace of the Ge; Au detector indicates the intensity transient of the pulse because of its much faster response.

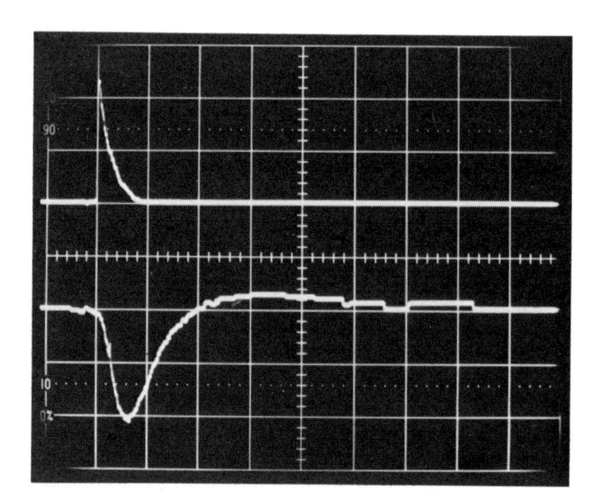

Fig. 17.6 Response of IR-OPFET under CO_2 laser pulse. Upper trace: Ge:Au. Lower trace: IR-OPFET. Horizontal: 10 μs/div. [after M. Okuyama et al.[17)]]

The lower trace shows the transient signal of the IR-OPFET with the opposite polarity. The rise time is about 3.5 μs, which is considered to be attributable to the circuit delay of the FET or thermal delay. As the rise time of the $PbTiO_3$ film on the Pt-coated mica under the same pulse irradiation was 1.3 μs, a faster response of the IR-OPFET is expected by improvements of the FET characteristics and the infrared-absorbing electrode.

17.3.4. *Analysis of the characteristics*

The infrared irradiation supplies the charge to the capacitances of the MOS and the $PbTiO_3$, C_{os} and C_f, by the pyroelectric effect. The voltage induced in the SiO_2 and the substrate of the channel region, V_i can be expressed as follows:

$$V_i = \frac{\eta p A I_\omega}{(C_{os} + C_f) H\omega} \tag{17.1}$$

where H is the heat capacity, η is the emissivity of the electrode, p is the pyroelectric coefficient of $PbTiO_3$, A is the channel area and I_ω is the incident power with angular frequency ω. V_i is amplified by the FET and the output voltage V_o is induced in the load resistance R_L. When the chopping frequency is much larger than the inverse of the thermal and the electrical time constant, V_o is expressed as follows:

$$V_o = -\frac{Z}{L} \mu_p C_i \frac{\eta p I_\omega V_D R_L}{(C_{os} + C_f) H\omega} \tag{17.2}$$

where Z and L are the channel width and length, μ_p is the hole mobility, C_i is the oxide capacitance and V_D is the drain voltage. The value of V_o at 20 Hz is 1.0 mV and the transconductance is 400 $\mu\Omega^{-1}$. Then the voltage responsivity of the $PbTiO_3$ film of the FET is estimated to be 32 V/W which is one tenth as large as that of the film deposited on the mica. This is attributed to the four following reasons: (1) the heat capacity of the device is very large as the thermal energy absorbed by the electrode escapes to the Si substrate; (2) the pyroelectric property of the film on amorphous SiO_2 is a little worse than that of the film on the Pt-coated mica; (3) the gate electrode has smaller infrared absorption; (4) the induced pyroelectric charge is shared between the $PbTiO_3$, the channel and the other parasitic capacitances, and so the dielectric charge induced at the Si–SiO_2 interface on the current channel is reduced. Some improvements of the sensitivity or the detectivity are proposed, namely: the reduction of the heat capacity, the use of a highly infrared-absorbing electrode, and the increase of the oxide capacitance on the current channel.

17.4. Improvement of responsivity in Si monolithic sensor

17.4.1. *Thermal analysis of infrared response of Si monolithic sensor*

In the IR-OPFET, as the pyroelectric $PbTiO_3$ film is attached directly to the Si substrate, the thermal energy absorbed in the black electrode conducts through the $PbTiO_3$ and escapes to the substrate. This thermal loss on the pyroelectric response can be analyzed by solving the heat conduction equation. Fig. 17.7 shows the calculated frequency dependence of the current responsivity, R_i, as a function of the thicknesses of the $PbTiO_3$

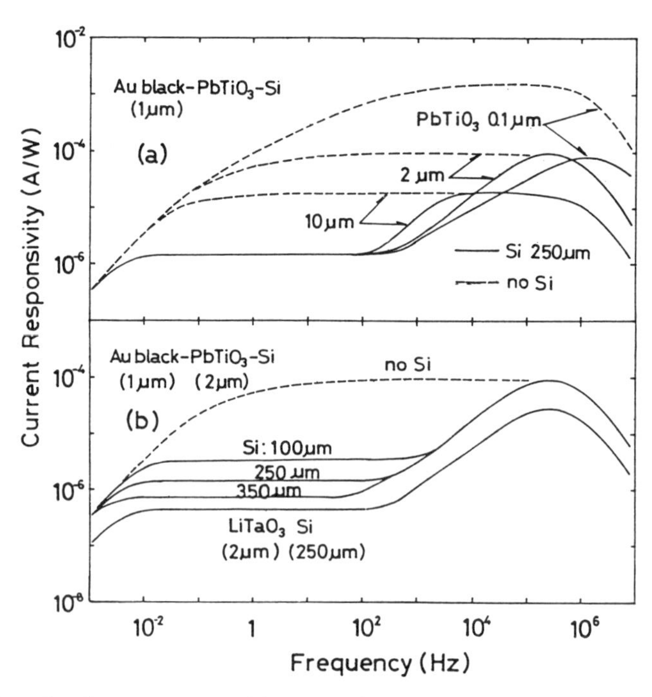

Fig. 17.7 Calculated results of current responsivity versus frequency of gold-black–$PbTiO_3$–Si system [after M. Okuyama et al.[17]]

and the Si.[17,26] The thickness of the black electrode is 1 μm. The voltage responsivity R_v in the usual frequency range is proportional to the current responsivity divided by the frequency because of the small electrical and thermal time constants. Solid lines show the responsivities of gold black–$PbTiO_3$–Si tandems and broken lines show those of gold black–$PbTiO_3$. R_i decreases with the frequency in the low-frequency region (for example, below 10^{-2} Hz for the sample $PbTiO_3$ 2 μm thick and Si 250 μm thick) as the heat escapes to air. R_i flattens in the intermediate region (10^{-2}–10^2 Hz) as the heat spreads in the whole sample. In the high-frequency region (10^2–10^5 Hz), R_i increases with the frequency as the heat flow into the thick Si decreases gradually. At higher frequencies (above 10^5 Hz), R_i decreases as the frequency increases as little heat can be transferred to the pyroelectric $PbTiO_3$. In Fig. 17.7(b), the dependence on the Si thickness of R_i in the sample of $PbTiO_3$ of constant thickness and R_i of $LiTaO_3$ are also shown. The plateau part decreases with the Si thickness as the heat capacity becomes large.

17.4.2. *Device structure of sensor having substrate of low heat capacity*

From the responsivity analysis of the previous subsection, an attempt to raise the output signal has been made by reducing the heat capacity of the substrate. Fig. 17.8 shows an overview of one element of the SiO_2–Si substrate of the improved device. Si beneath the sensitive area was etched off from narrow SiO_2 holes using a preferential etchant of ethylenediamine, pyrocatechol and water mixture. The narrow holes are oriented to the ⟨100⟩ direction and its width is about 25 μm. Square hollows with the diagonal line of the hole are formed by preferential etching as the etching rate of the ⟨010⟩

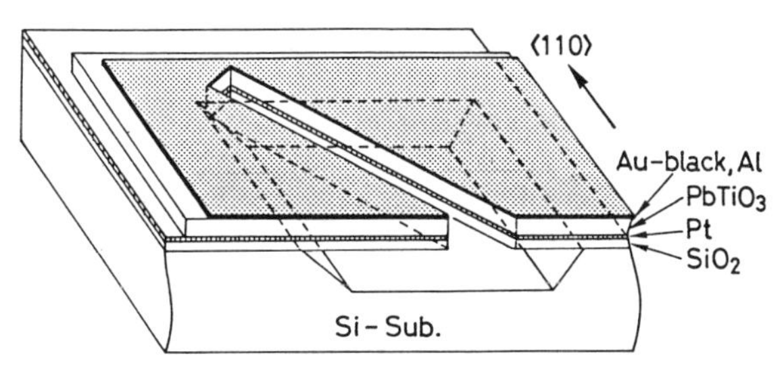

Fig. 17.8 Structure of sensitive area with floating SiO_2 membrane. [after M. Okuyama et al.[19)]]

and ⟨001⟩ directions of the (100) wafer is much higher than that of the other faces. Eventually thin SiO_2 (5000Å) membranes bridging Si plateaux are floated over the Si substrate. The side length of the square membrane is typically 200–250 μm and the area of the membrane is about one half of the whole area. A $PbTiO_3$ thin film of 2.1 μm thickness was deposited on Pt- and SiO_2-coated n-Si wafer with many membranes by RF sputtering, and Al and gold black layers were formed on the $PbTiO_3$ film. Samples with various sensitive areas were fabricated and a typical sensitive area was 0.037 cm^2. The pyroelectric current of the $PbTiO_3$ film was amplified by a bipolar transistor fabricated on the same Si wafer and infrared light can be detected as a collector current.

17.4.3. *Infrared response*

Fig. 17.9 shows the frequency dependence of the pyroelectric current of the $PbTiO_3$ thin films fabricated on Si with and without the SiO_2 membranes. The current of the sample with the membranes is larger by one order of magnitude than that of the film attached directly to the SiO_2–Si substrate. The frequency dependence of the sample without the membrane is similar to R_i drawn in Fig. 17.7 and is well explained by the vertical thermal conduction to the Si wafer. The current of the sample with the membranes also increases with frequency in the low-frequency region, and can be explained by horizontal thermal conduction. The infrared response of the device with a bipolar transistor was measured in the same way as that of the film. The output voltage of the bipolar transistor increases with frequency in the low-frequency region and saturates in the high-frequency region as well as in Fig. 17.7. This output voltage is larger than that of the IR-OPFET in the high-frequency region as mentioned in the previous section. The impulse response of the device was measured using a CO_2 laser and the rise time was estimated to be about 2.5 μs.

In order to investigate the frequency dependence of the output in the low-frequency region, the horizontal heat flow along the Si should be analyzed. It is assumed for simplicity that heat flows in the square side in the x and y directions. The solution of the heat conduction equation can be obtained by Fourier's method under sinusoidal infrared irradiation ($Q \sin \omega t$). The pyroelectric current density J_p is calculated as a summation of the whole current in the membrane,

$$J_{\mathrm{P}} = \frac{1}{a^2}\int_0^a \int_0^a p \frac{\partial u}{\partial t}\, \mathrm{d}x\, \mathrm{d}y$$

$$= \frac{64pQ}{\pi^4} \sum_{m,n} \frac{\omega(\omega \sin \omega t - \alpha_{mn} \cos \omega t)}{(2m-1)^2(2n-1)^2(\alpha_{mn} + \omega^2)}, \tag{17.3}$$

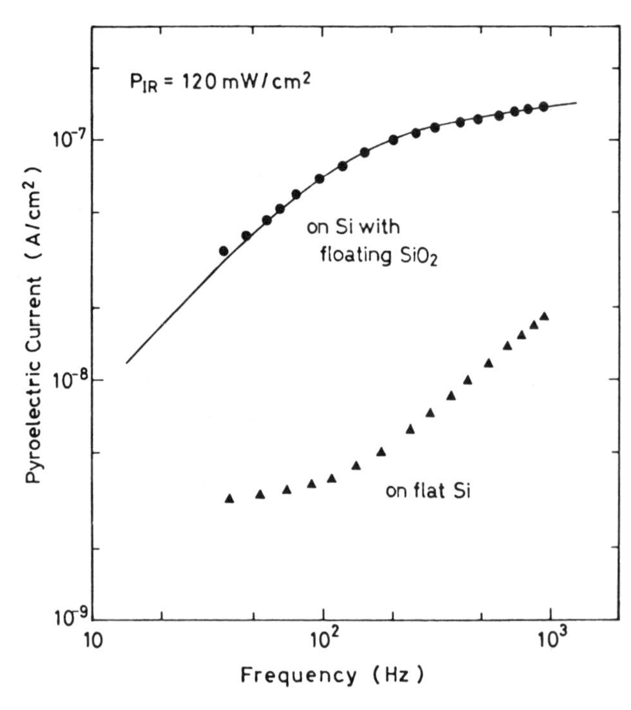

Fig. 17.9 Frequency dependence of pyroelectric current of the sensor with and without floating sensitive area. The solid line shows the theoretical result obtained by solving horizontal heat conduction. [after M. Okuyama et al.[19)]]

where a is the side length of the square membrane, u is the temperature of the membrane, $\alpha_{mn} = \kappa^2\pi^2((2m-1)^2+(2n-1)^2)/a^2+b^2$ and b^2 is a constant proportional to the heat loss to the ambient atmosphere. The calculated results of Eq. (17.3) is shown with a solid line in Fig. 17.9. The thickness of Au black, $PbTiO_3$, Pt and SiO_2 were 3, 2, 0.16 and 0.5 μm, respectively, and the side length of the membrane was 250 μm in the calculation. The calculated results agree well with the experimental data. In the high-frequency region, as the temperature follows the illumination in a chopping cycle, the pyroelectric current is constant; however, in the low-frequency region, the temperature distribution approaches a steady state and so the current decreases with decrease of the frequency. The heat conduction of the Pt layer affects the pyroelectric current most among the four layers because of its large thermal conductivity. In order to get a better frequency dependence even in the low-frequency region, the horizontal heat conduction must be suppressed by improving the sample structure. The considered improvements are: (1) the thickness of each layer, especially that of the Pt, should be small; and (2) the side length of the membrane should be made as large as possible, keeping its mechanical strength.

17.5. Linear array sensor and infrared imaging

17.5.1. *Structure of linear array and basic characteristics*

A linear array sensor consisting of a thin $PbTiO_3$ film has been fabricated for infrared imaging. Substrates are a mica sheet of 20–50 μm thickness, a Si bridge and a Si cantilever.

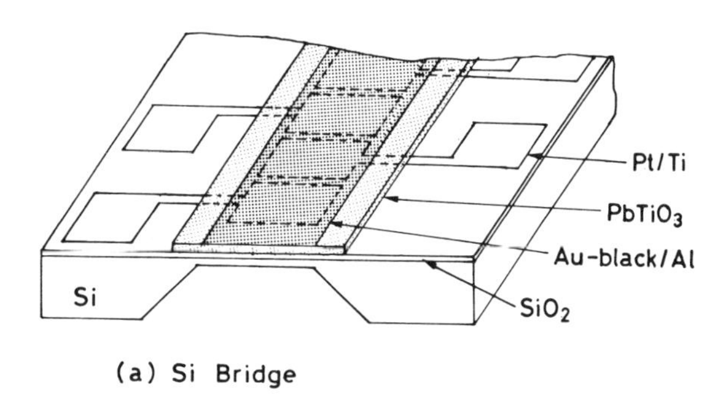

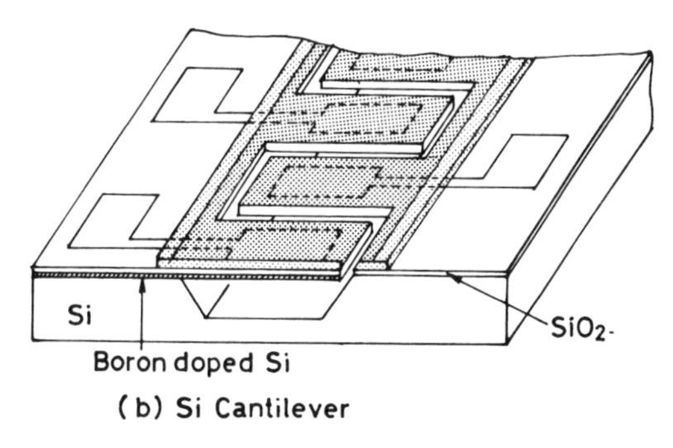

Fig. 17.10 Device structure of infrared sensor array with 16 elements fabricated on (a) the Si bridge and (b) the Si cantilever. [after M. Okuyama et al.[27)]]

Fig. 17.10 shows the sample structures of the array sensors on the Si bridge and Si cantilever. Platinum electrodes of 16 elements were formed on the substrate, the $PbTiO_3$ film was deposited by RF sputtering, and then Al and gold-black layers were formed as the infrared-absorbing electrodes.

In order to investigate the resolution of the array sensor, the output has been measured when a focused micro-spot of a He–Ne laser beam is scanned on the sensor array. Fig. 17.11 shows the dependence on the spot position of the normalized output of the array on (a) the mica, (b) the Si bridge and (c) the Si cantilever as a parameter of the chopping frequency. The response of the array on the mica does not change even if the frequency is changed from 200 to 950 Hz. On the other hand, the response of that on the Si bridge becomes broad as the frequency decreases because the horizontal heat conduction along the Si is much larger than that of the mica. However that of the Si cantilever is similar to that of the mica because the horizontal conduction along the Si is prevented by the air gap between the cantilevers.

The modulation transfer function (MTF) characterizing resolution of the sensor can be observed from the measurement of the micro-spot irradiation. The MTF can be expressed as $(S_w - S_b)/(S_w + S_b)$, where S_w is the integral of the response when the spot is scanned on the measured electrode, and S_b is the integral when the spot is scanned on the neighboring electrodes. The MTF of the sensor on the cantilever is about 0.96 at a spatial frequency of 20 lp/cm and a chopping frequency of 400 Hz. This value is considerably larger than that of the bridge, 0.05 at a spatial frequency of 27 lp/cm, and almost equal to that of $PbTiO_3$. This means that the resolution of the cantilever is superior to that of the bridge.

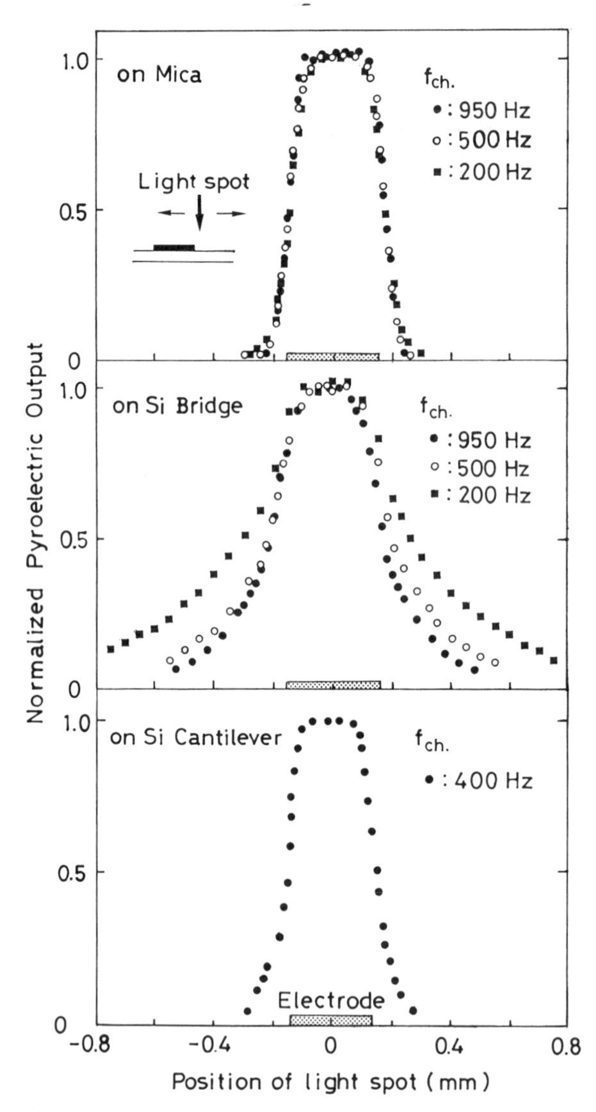

Fig. 17.11 The normalized output of the array fabricated on mica, Si bridge and Si cantilever, when a micro-spot of laser beam is scanned on the sensor. [after M. Okuyama et al.[27)]]

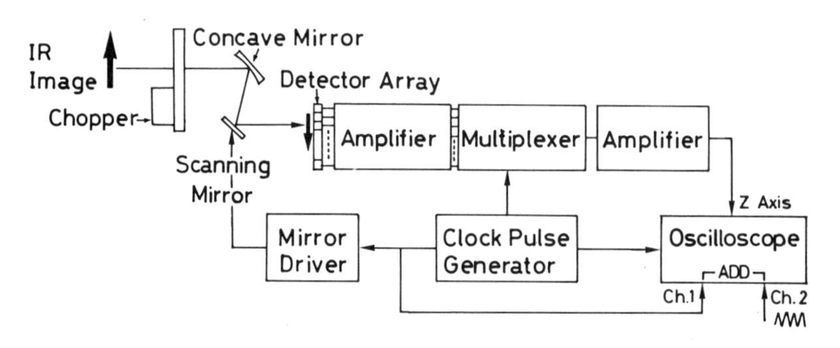

Fig. 17.12 Schematic diagram of infrared imaging system. [after M. Okuyama et al.[27)]]

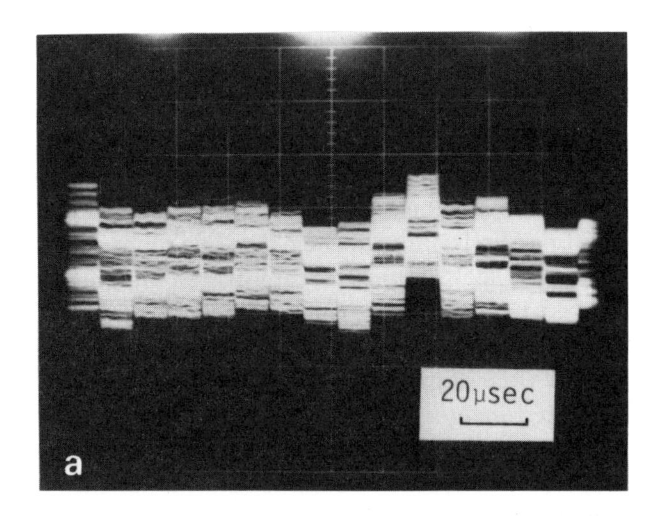

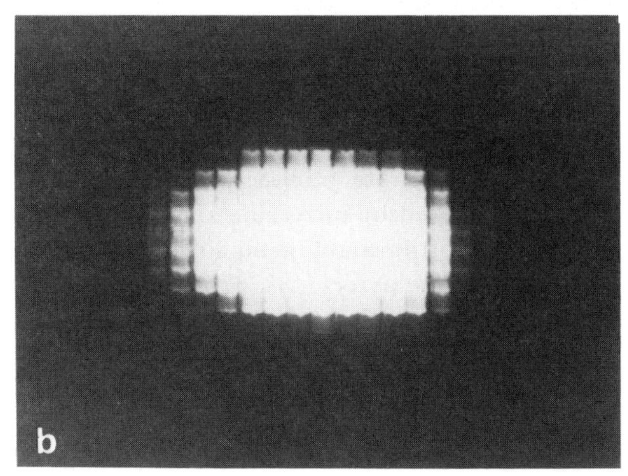

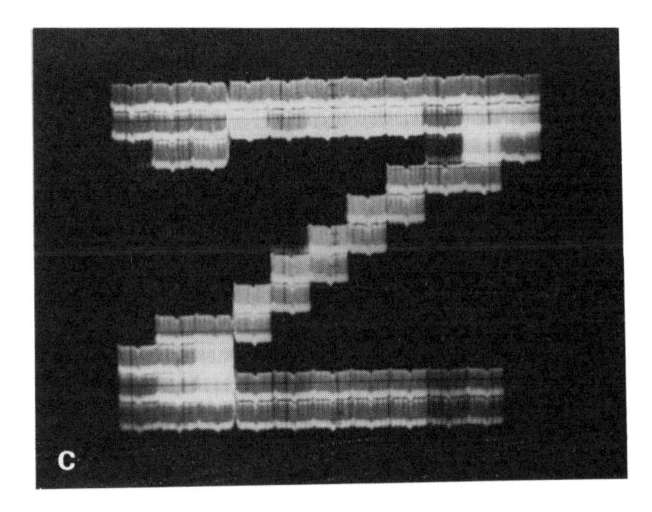

Fig. 17.13 Examples of (a) a serial pulse train of the array outputs, (b) an image of weak CO_2 laser spot, and (c) an image of an image of the letter "Z".

17.5.2. *Infrared imaging by the linear array sensor*

There are two ways to process the output of the sensor for the imaging. In the first method, the output of the sensor is transformed by a multiplexer to a serial pulse train, amplified by a current amplifier and sampled and held. In the other method, the output of the sensor is firstly amplified by a voltage amplifier and then is switched by a multiplexer. In either method, the pulse train is put into the *z* input of an oscilloscope to modulate its brightness. Fig. 17.12 shows a block diagram of an infrared imaging system using voltage amplification. The infrared light emitted from an object comes through a chopper to vertical and concave mirrors, and is focused on the sensor array. By using this signal processing system, a trial method of getting a simple image has been carried out using the film on the mica and infrared images of alphabetical letters have been obtained. Typical width of clock pulse and frequency of mirror scanning were about 10 μs and 30 Hz, respectively. Fig. 17.13 shows examples of a serial pulse train of the outputs, an image of weak CO_2-laser spot and an image of the letter "Z".

17.6. Conclusion

$PbTiO_3$ ferroelectric thin films have been prepared on Pt sheet, Si wafer and Pt-coated mica by RF sputtering at the substrate temperatures of 300–550°C. The dielectric properties of the films were improved by increasing the substrate temperature and the film thickness, and good characteristics could be obtained in the films of thickness more than 2 μm deposited near 500°C. The pyroelectric properties of the films are fairly good compared with those of the ceramics. Typical voltage responsivity and detectivity obtained are 330 V/W and 1.5×10^8 cm $Hz^{1/2}$/W with a bandwidth of 1 Hz at 20 Hz, respectively. The fast response was measured under irradiation of a CO_2 laser pulse and its rise time was about 1.3 μs.

An IR-OPFET having the structure of a thin $PbTiO_3$ film gate on a Si-FET has been fabricated. Typical voltage responsivity and detectivity of the IR-OPFET are 390 V/W and 3.5×10^5 cm $Hz^{1/2}$/W with a bandwidth of 1 Hz at 20 Hz, respectively. The rise time of the response measured under the CO_2 laser pulse is about 3.5 μs. The current responsivity of the thin $PbTiO_3$ film on Si has been increased by reducing the heat capacity of the sensitive area with anisotropy etching of Si, and is one order of magnitude larger than that of the film attached directly to Si.

Linear array sensors having 16 elements have been developed by the $PbTiO_3$ film on Pt-coated mica and Si wafer, and their basic characteristics have been investigated with a view to improving resolution. Infrared imaging has also been tried by using the linear array and simple infrared images have been obtained.

Acknowledgements

The authors would like to thank Drs. Kikuo Wakino and Katsuhiko Tanaka of Murata Mfg. Co. for supplying the target materials of the $PbTiO_3$ sputtering and Prof. Kazuhiko Miyazaki and Dr. Naoyuki Yamabayashi for fast response measurements using a CO_2 laser. The authors would also like to thank Prof. Taichi Nakagawa of Setunan University for helpful discussion, and Dr. Yasushi Matsui, Messrs Chitose Sada, Hiroyuki Nakano,

Hiroyuki Seto, Shunichi Mashima, Motohiro Kojima, Kohzo Ohtani and Toshiyuki Ueda for their technical assistance and helpful discussions.

References

1) A. J. Steckel (Ed.): IEEE Trans. Electron Devices, ED-27 (1980) 1.
2) E. H. Putley: Semiconductors and Semimetals, Vol. 56, Eds. R. K. Willardson and A. C. Beer (Academic Press, New York and London, 1970).
3) S. C. Stotlar, E. J. Mcllan, A. J. Gibbs, and J. Webb: SPIE, 246 (1980) 10.
4) H. Beerman: Infrared Phys., 15 (1975) 225.
5) E. Yamaka, T. Hayashi and M. Matsumoto: Infrared Phys., 11 (1971) 247.
6) C. B. Roundy: Appl. Optics, 18 (1979) 943.
7) A. Okada: J. Appl. Phys., 48 (1977) 2905.
8) R. N. Castellano and L. G. Feinstein: J. Appl. Phys., 50 (1979) 4406.
9) M. Oikawa and H. Toda: Appl. Phys. Lett., 29 (1976) 491.
10) K. Tanaka, Y. Higuma, K. Tyokoyama, T. Nakagawa, and Y. Hamakawa: Japan. J. Appl. Phys., 15 (1976) 1381.
11) M. Ishida, H. Matsunami, and T. Tanaka: J. Appl. Phys., 48 (1977) 951.
12) T. Nakagawa, Y. Matsui, T. Usuki, M. Okuyama, and Y. Hamakawa: Japan. J. Appl. Phys., 15 (1979) 897.
13) M. Okuyama, Y. Matsui, H. Nakano, T. Nakagawa, and Y. Hamakawa: Japan. J. Appl. Phys., 18 (1979) 1633.
14) M. Okuyama, T. Usuki, Y. Hamakawa, and T. Nakagawa: Appl. Phys., 21 (1980) 339.
15) M. Okuyama, Y. Matsui, H. Nakano, H. Seto, and Y. Hamakawa: Paper of Technical Group, TGED80-28, TGCPM80-20, IECE Japan (May 1980). [in Japanese]
16) T. Nakagawa, J. Yamaguchi, M. Okuyama, and Y. Hamakawa: Japan. J. Appl. Phys., 21 (1982) L655.
17) M. Okuyama, Y. Matsui H. Seto and Y. Hamakawa: Japan. J. Appl. Phys., Suppl. 20-1 (1981) 315.
18) M. Okuyama, Y. Matsui, H. Nakano, and Y. Hamakawa: Ferroelectrics, 33 (1981) 235.
19) M. Okuyama, H. Seto, M. Kojima, Y. Matsui, and Y. Hamakawa: Japan. J. Appl. Phys. Suppl., 21-1 (1982) 225.
20) M. Okuyama, H. Seto, M. Kojima, Y. Matsui, and Y. Hamakawa: Japan. J. Appl. Phys. Suppl., 22-1 (1983) 465.
21) S. Y. Wu: Ferroelectrics, 11 (1976) 379.
22) K. Sugibuchi, Y. Kurogi, and N. Endo: J. Appl. Phys., 40 (1975) 2871.
23) Y. Hamakawa, Y. Matsui, Y. Higuma, and Y. Hamakawa: Technical Digest of 1977 IEEE IEDM (1977) p. 294.
24) Y. Matsui, H. Nakano, Y. Higuma, M. Okuyama, Y. Hamakawa, and T. Nakagawa: FMA-2 (FMA Office, Kyoto, 1979) p. 239.
25) Y. Matsui, M. Okuyama, M. Noda, and Y. Hamakawa: Appl. Phys., A28 (1982) 161.
26) S. Y. Wu: IEEE Electron Devices, ED-27 (1980) 88.
27) M. Okuyama, K. Ohtani, T. Ueda, and Y. Hamakawa: Int. J. Infrared and Millimeter Waves; 6 (1985) 71.

18 HIGH-BRIGHTNESS LOW-THRESHOLD-VOLTAGE THIN-FILM AC ELECTROLUMINESCENT DEVICES AND THEIR TUNABLE MULTI-COLORING

Yoshihiro HAMAKAWA, Yoshiro OISHI and Takatoshi KATO†

Abstract

A series of experimental studies on low-threshold-voltage thin-film electroluminescent (EL) devices and their multi-coloring have been made. The EL cells developed here are mostly AC types having the structure of a ZnS: REF_3 layer sandwiched between ferroelectric thin films. Firstly, a series of technical data on the reduction of threshold voltage has been demonstrated together with a key technology of high dielectric thin film growth. Secondly, the results of R&D efforts to develop multi-coloring EL cells are summarized, and various colors including red, orange, yellow, green, blue and white light-emitting devices have been realized. Finally, on the basis of these technologies, a new device having a function of tunable color EL devices which could be able to change emission spectra by controlling applied voltages are developed. Newly developed dual coloring and full coloring EL devices might make a unique contribution to the active type display technology field.

Keywords: Electroluminescence, Display Panel, Tunable Color EL, Optoelectronics, Semiconductor

18.1. Introduction

Electroluminescence (EL) was first discovered by Destriau in 1937 in the form of ZnS: Cu powders dispersed in oil.[1] After this report, the first work concerning a thin-film EL device was made by Halstead in 1954.[2] Since these reports, many investigators had made efforts to improve the performances of EL devices. EL devices have a number of attractive advantages such as low power dissipation, the possibility of forming a large-area flat-type display, and being solid-state devices. In spite of these advantages, EL devices are not yet considered to be practical display devices because of their low efficiency and their performance lifetime. A practical-level thin-film EL device composed of a Mn-doped ZnS thin film sandwiched between two Y_2O_3 insulating layers, the so-called doubly insulating layer structure, was developed by the Sharp group in 1977.[3] The device developed has a high brightness level of more than 1000 ft-L with a very good stability in operating for more than 10,000 h. However, there still remain at least two serious problems: it has a high operating voltage of more than 200 V_{rms}, and gives monochromatic light of Mn orange. With the recent progress in computer technology, there exists a tremendous potential demand for color imaging display processing. To respond to

† Faculty of Engineering Science, Osaka University, Toyonaka, Osaka 560.

JARECT Vol. 19, Semiconductor Technologies (1986), *J. Nishizawa (ed.)*
OHMSHA, LTD. and North-Holland

this requirement, we have conducted a systematic investigation to improve the performance, particularly the reduction of threshold voltage down to IC drive, and multi-coloring. The results of these R&D efforts are summarized as follows:

(a) Low-threshold-voltage devices:
  (a1) $V_{th} = 50$–60 V for AC EL
  (a2) $V_{th} = 20$–50 V for DC EL.

(b) Multi-coloring:
  (b1) red ZnS:SmF_3
  (b2) orange ZnS:Mn, ZnS:NdF_3
  (b3) yellow ZnS:DyF_3
  (b4) green ZnS:TbF_3, ZnS:HoF_3, ZnS:ErF_3
  (b5) blue ZnS:TmF_3
  (b6) white ZnS:PrF_3.

(c) Tunable-color EL devices:

  (c1) Two stacked EL cells doped with different luminescent centers, light emission spectra from red to orange, yellow, yellow-green and green can be achieved with a spatial address having a brightness of more than about 100 ft-L;
  (c2) Three tandem EL cells with the three primary colors, a tunable full-color device has been developed.

With these successful results, the EL devices developed could be used not only for the flat panel color display devices such as computer output display, automobile console panel, flat-panel color TV, but also for a new function of the multi-indication systems in information display engineering.

18.2. Low-threshold-voltage EL devices

A high driving voltage in the EL device severely limits the use of commercially available IC as a driving circuit and leads to high cost and complexity of the driving circuit. Moreover, the application of such a high driving voltage to the device prevents not only the application fields but also device reliability. For these reasons, a decrease of threshold voltage less than 50 V in the EL device would be of prime importance. To fulfil this requirement, several approaches for lowering the threshold voltage have been attempted so far, for example the use of low-bandgap host materials such as ZnSe,[4,5] low-energy Mn ion implantation into ZnS,[6] and thin active and insulating layers etc.[7–9]

Recently, our group has developed a bright green-emitting DC EL device with a threshold voltage of 20–30 V in the system of Al–ZnS:TbF_3–ZnSe–ITO, as shown in Fig. 18.1.[10] The ZnSe thin film was utilized as a primer layer which improves the crystal quality of ZnS:TbF_3 by a kind of epitaxial growth. This type of device could lead to a remarkable reduction of operation voltage and to improved efficiency. The device emits 400 fL around 60 V in the DC excitation mode.

We have also conducted an intensive program of work to develop a low-threshold-voltage thin-film EL device with high brightness and long life.[11–13] Thin-film EL devices

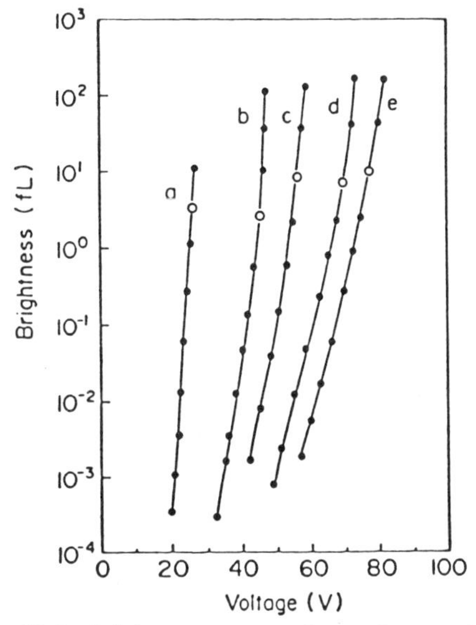

Fig. 18.1 Brightness versus voltage characteristics of Al–ZnS:TbF_3–ZnSe–ITO heteroface EL devices with the active layer thickness of (a) = 0.03, (b) = 0.19, (c) = 0.25, (d) = 0.33, (e) = 0.43 μm.

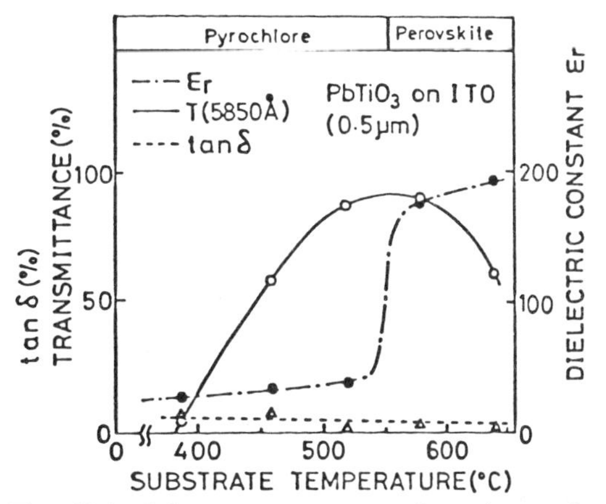

Fig. 18.2 Substrate temperature dependence of RF-sputtered $PbTiO_3$ thin film on the dielectric constant, transmittance and tan δ.

with a high dielectric constant film as an insulating layer have been fabricated by replacing the conventional Y_2O_3 insulating layer by a ferroelectric insulating layer such as $PbTiO_3$ and PLT which have a dielectric constant of more than 160. They were successfully operated at about four times lower threshold voltage and gave a similar level of brightness as those of conventional EL devices.

A lanthanum-modified lead zirconate titanate solid solution system (PLZT) is a well known ferroelectric material used for various electro-optic and photochromic applications. Recently, several attempts have been made to prepare a PLZT thin film by RF sputtering[14–16] and electron beam evaporation.[17] $PbTiO_3$ and PLT thin films have been prepared for the purpose of applying an insulating layer in the thin-film EL devices. $PbTiO_3$ and PLT were chosen because they have a good optical transparency, high dielectric constant and relatively high breakdown strength, which are required for the insulating layer used in the EL structure.[18–19]

Fig. 18.2 shows the dependence on the substrate temperature of the dielectric constant, transmittance and tan δ of the $PbTiO_3$ thin film of thickness 500 nm deposited on an ITO-coated quartz substrate. The transmittance was measured at 585 nm which is the peak wavelength of the emission from the ZnS:Mn EL device. Typical sputtering conditions are listed in Table 18.1. As can be seen from Fig. 18.2, a good transparency with a transmittance above 80% has been observed in the films deposited at the substrate temperature range between 500 and 600°C. The dielectric constant changes drastically at the substrate temperature of 550°C: below 550°C, it was measured to be 30–40, while above 550°C, a high dielectric constant of 160–190 has been consistently obtained. Almost constant values of tan δ of 3–5% have been obtained.

The change in dielectric constant is caused by a change in the crystal structure of the

Table 18.1 Sputtering conditions of $PbTiO_3$ thin film

RF voltage	1.6 kV
RF input power	150 W
Target–substrate distance	40 mm
Target diameter	80 mm
Sputtering gas	Ar(90%) + O_2(10%)
Gas pressure	1×10^{-1} Torr
Sputtering rate	40 Å/min.

deposited film. Fig. 18.3 shows the X-ray diffraction patterns of $PbTiO_3$ thin films deposited at 520°C (a) and 640°C (b). A clear difference in the X-ray diffraction pattern is observed, that is the peaks in Figs. 18.2(a) and 18.2(b) are characteristic of the pyrochlore and perovskite structures, respectively.[20]

The characteristics of the $PbTiO_3$ thin film changed with the film thickness. Fig. 18.4 shows the thickness dependence of the dielectric constant, transmittance and tan δ. The transmittance strongly decreases when the film thickness exceeds 600 nm, which is perhaps due to the scattering by the largely grown $PbTiO_3$ grain. The largest dielectric constant was obtained for the film 500 nm in thickness. Above 500 nm, the dielectric constant decreases monotonically with the film thickness. From the characteristics described above, the film of thickness 500 nm deposited at the substrate temperature of 580°C is the most suitable as the insulating layer of the EL devices.

A schematic illustration of the EL structure with a $PbTiO_3$ thin film as the insulating layer is shown in Fig. 18.5. A very thin MgO layer less than 50 nm was used as a substrate for the deposition of the second $PbTiO_3$ layer, because $PbTiO_3$ thin film grown directly on

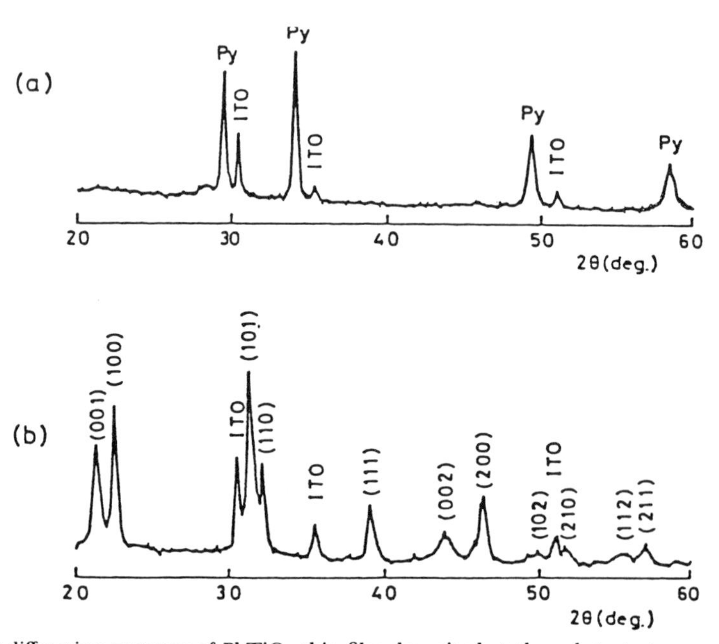

Fig. 18.3 X-ray diffraction patterns of $PbTiO_3$ thin film deposited at the substrate temperatures of 520°C (a) and 650°C (b).

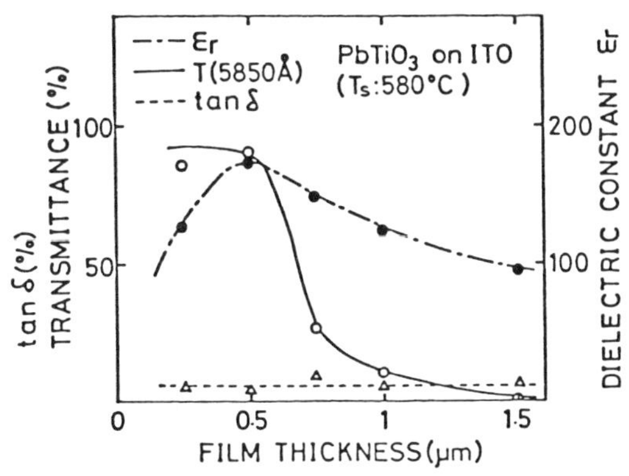

Fig. 18.4 Thickness dependence of $PbTiO_3$ thin film on the dielectric constant, transmittance and tan δ.

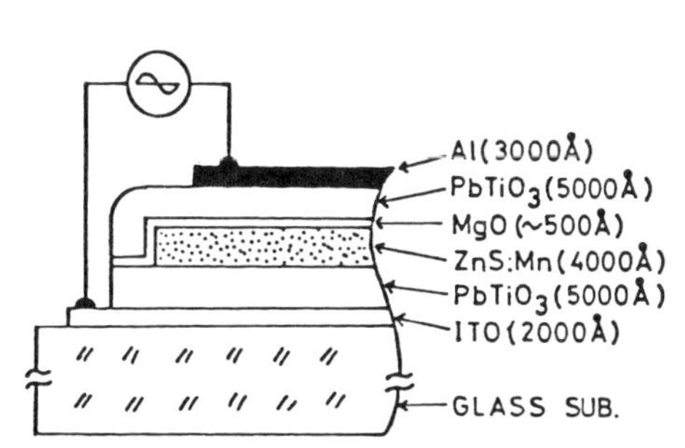

Fig. 18.5 Schematic illustration of low-threshold-voltage thin-film EL device.

the ZnS:Mn layer usually tends to peel. By evaporating the MgO layer on the ZnS:Mn film, an adherent and transparent $PbTiO_3$ thin film can be successfully deposited. Both the ZnS:Mn and MgO films were deposited by an electron beam evaporation method. The ZnS:Mn thin film was subjected to a heat treatment at 550°C for 1 h after deposition to improve the optical and electrical properties of the ZnS:Mn layer. The experimental results made on another ferroelectric thin-film material PLT shows similar characterisitics to the $PbTiO_3$ thin film.

Fig. 18.6 shows the brightness versus applied voltage (B–V) characteristics of the EL device together with that of a conventional EL device using Y_2O_3 as the insulating layers. The threshold voltage of the device is around 50 V, which is about four times lower than that of the conventional EL device of around 200 V. A typical brightness of more than

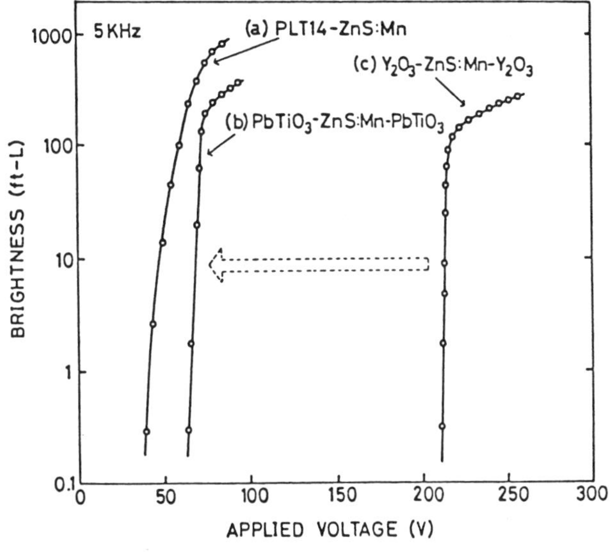

Fig. 18.6 Typical B–V characteristics of the developed EL device. (a) ITO–PLT14–ZnS:Mn–Al, (b) ITO–$PbTiO_3$–ZnS:Mn–$PbTiO_3$–Al, and (c) the conventional EL device ITO–Y_2O_3–ZnS:Mn–Y_2O_3–Al.

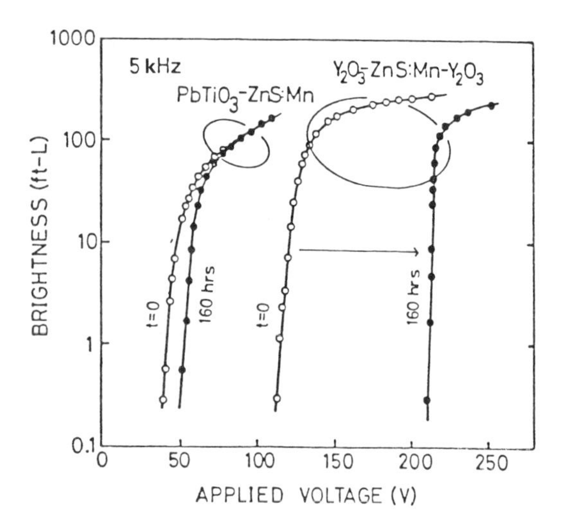

Fig. 18.7 Comparison of *B–V* characteristics between the developed EL device and conventional EL device.

800 ft-L has been obtained at the driving voltage of 80 V in the two-layered device of ITO–PLT14–ZnS: Mn–Al structure.

The EL device is very stable to the aging treatment and shows no degradation as reported elsewhere.[3,21] The *B–V* characteristics of the newly developed EL device measured both before and after aging treatment for 160 h are shown in Fig. 18.7. The *B–V* characteristics of the conventional EL device using Y_2O_3 insulating layers largely shifts to the higher voltage region with a difference of 50–100 V during the initial 160 h operation.

In case of the conventional EL device using Y_2O_3 as the insulating layer, the applied voltage required to obtain a brightness of 10 ft-L gradually increases with increasing aging time of even more than 150 h. On the other hand, in case of the device using ferroelectric insulating layer, the stabilization of the *B–V* characteristics is completed during the very short aging time of a few hours. In addition to this, difference of applied voltage measured before and after the aging is only 1.8 V. These characteristics such as very small difference of operation voltage and fast stabilization are remarkable merits of the EL device utilizing ferroelectric thin film as the insulating layer, since no aging treatment is required for the practical use of this device.

In addition, another type of low-threshold-voltage EL device has been reported by RCA[22] and Tottori University groups.[23] They have developed stable DC EL devices by replacing the insulating layer with a current-limiting cermet film or ZnSe film.

18.3. Multi-coloring efforts

Recently, high brightness and long life have been attained successfully in a ZnS: Mn double insulated AC thin-film EL device and mass production has been initiated.[24] However, the only emission color having a practically acceptable brightness level is orange which is attained in the ZnS: Mn system. As regards multi-coloring of EL devices, much R & D effort has been made because of the wide range of potential applications. In particular, it is necessary to produce light of the three primary colors, namely red, green and blue, with a higher brightness level to realize a full-color display. Many kinds of luminescent centers, for example transition metal ions and rare-earth fluoride molecules,

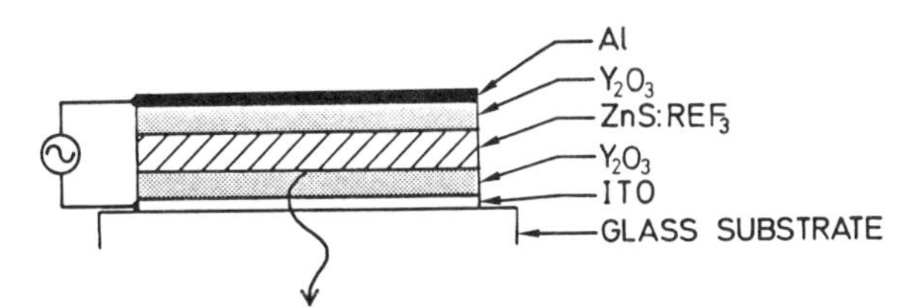

Fig. 18.8 Schematic illustration of multi-color-emitting EL device.

have been investigated elsewhere.[25–29] Among these rare-earth fluoride molecules have at least two advantages for practical applications.[30] The first is their comparatively high conversion efficiency due to the large ionization cross-section of molecular centers. The second is their capacity to give a wide variation of emission colors by selecting appropriate molecular centers.

Recently we have succeeded in realizing a practically available green emission EL device with a ZnS:TbF_3 thin-film having the structure In_2O_3–Y_2O_3–ZnS:TbF_3–Y_2O_3–Al.[31] A schematic illustration of the device structure is shown in Fig. 18.8. A high-brightness green-light emission EL has been obtained by using a semiconductive Y_2O_3 layer instead of an insulating layer. It has been shown that a new carrier generation mechanism relating to the thermal process exists in the introduced semiconductive Y_2O_3 layer. The thermal process is due to a Frenkel–Poole emission of electrons from donor levels to the conduction band of Y_2O_3. An efficient injection of such thermally generated carriers from Y_2O_3 to ZnS contributes to the high brightness.

Fig. 18.9 shows a typical emission spectrum measured by an OMA (Optical Multi-channel Analyzer) together with the spectrum of a commercially available GaP:N light-emitting diode. The spectrum consists of four emission lines in the visible region. As reported by Chase and Kahng, the main peak of green emission at 545 nm corresponds to $^5D_4$–$^7F_5$ transitions arising from the $(4f)^8$ electron configuration of a Tb^{3+} ion.[25,30] The three other peaks can be attributed to the $^5D_4$–$^7F_4$, $^5D_4$–$^7F_3$ and $^5D_4$–$^7F_6$ transitions,

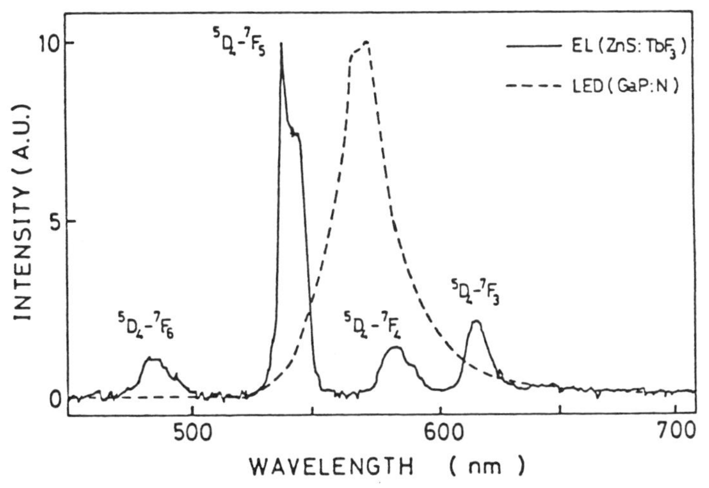

Fig. 18.9 Emission spectrum of ZnS:TbF_3 thin-film EL device. The dashed line denotes the spectrum of commercially available GaP:N LED for comparison.

respectively. As can be seen in this figure, the main peak at 545 nm is about 20 nm shorter than that of the GaP:N LED at 565 nm, and so the emission light looks like a purer green compared with that of GaP:N LED.

Based upon the results of the optimization for TbF_3, our colleagues Suyama et al. have tried to make multi-coloring of EL cells doped with various rare-earth fluorides,[32] namely PrF_3, NdF_3, SmF_3, EuF_3, TbF_3, DyF_3, HoF_3, ErF_3, TmF_3 and YbF_3, which give various emission colors ranging from blue to red in the order given above. No emission of rare-earth ions have been obtained in the visible region with the devices doped with EuF_3 and YbF_3. Fig. 18.10 shows the emission spectra of the devices doped with various rare-earth fluoride molecules. The emission colors obtained are red(SmF_3), orange(NdF_3), yellow(DyF_3), green(TbF_3, ErF_3 and HoF_3), blue(TmF_3) and white(PrF_3). These spectra are almost the same as those already reported.[25] The new result is obtained that the present HoF_3-doped device emits pure green light, which is different from the results reported by Chase et al. The difference may arise from the difference of the crystal field around the Ho^{3+} ions. This is probably due to the different device fabrication process between the present work and the previous work.

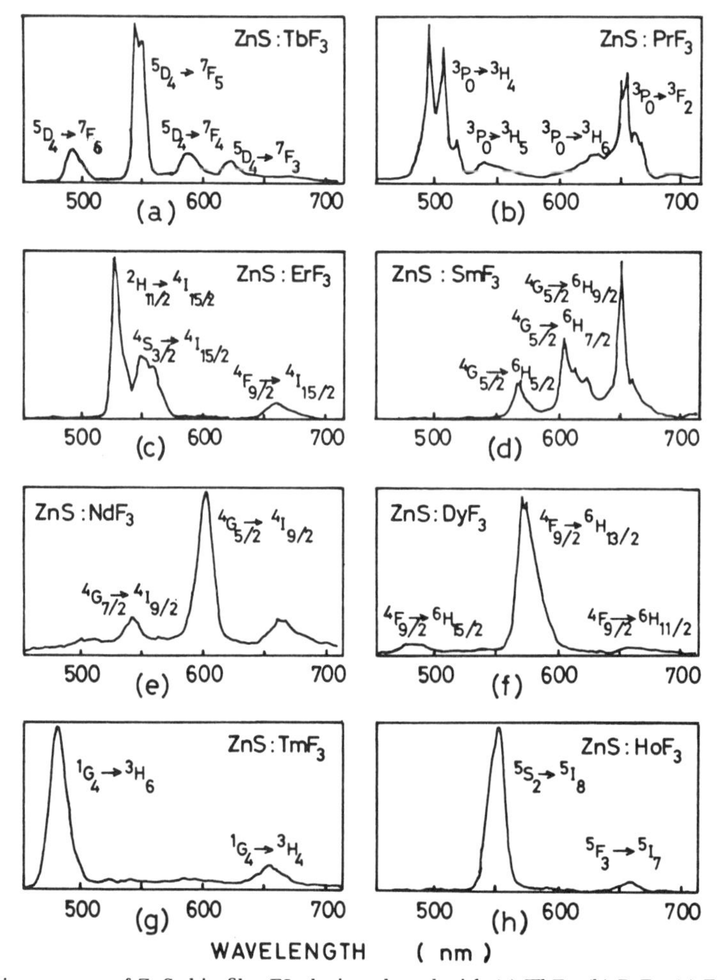

Fig. 18.10 Emission spectra of ZnS thin-film EL devices doped with (a) TbF_3, (b) PrF_3, (c) ErF_3, (d) SmF_3, (e) NdF_3, (f) DyF_3, (g) TmF_3 and (h) HoF_3.

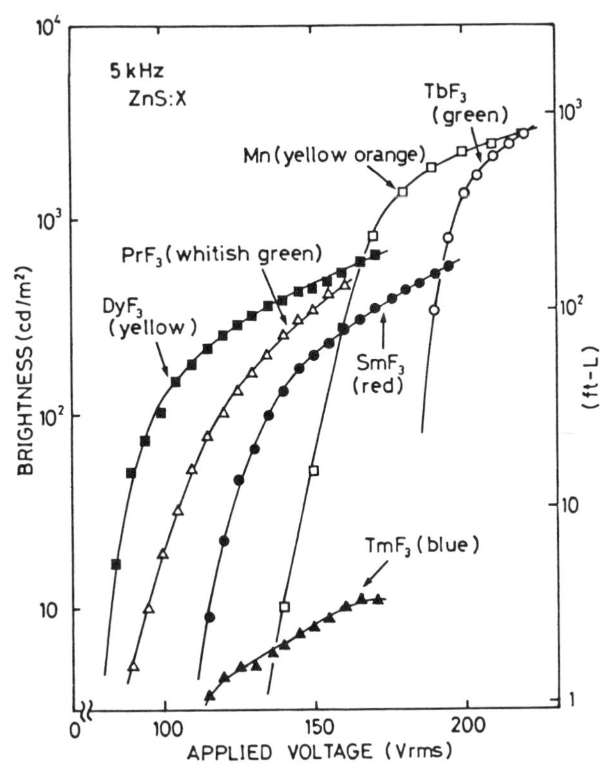

Fig. 18.11 Typical *B–V* characteristics for various rare-earth fluoride- and Mn-doped ZnS thin-film EL device under 5 kHz sinusoidal voltage excitation.

Table 18.2 The presently obtained brightness level of a thin-film EL device emitting various colors, together with their dopant materials

Color	Dopant	Brightness
green	TbF_3	780 ft-L
red	SmF_3	200 ft-L
yellow	DyF_3	190 ft-L
white	PrF_3	120 ft-L
blue	TmF_3	3 ft-L

Fig. 18.11 shows typical brightness versus applied voltage characteristics obtained in these multi-coloring EL studies and Table 18.2 shows the presently obtained brightness level of these devices. The brightness of the ZnS:TbF_3 green-light-emitting device was 780 ft-L under 5 kHz sinusoidal voltage excitation. The SmF_3-, DyF_3- and PrF_3-doped devices provide relatively high brightness of 170, 190 and 120 ft-L, respectively, but the brightness of the TmF_3-doped devices is about two orders of magnitude lower than that of the others. Further study is required for new bright blue-emitting EL devices.

18.4. Tunable EL device and full-color EL device

In the view of the recently postulated multi-indication system by Hamakawa, a tunable-color display might be quite useful for a wide variety of application fields. To realize this kind of display with EL, we have developed a new type of tunable-color-emitting EL device with two stacked thin-film EL cells involving different luminescent centers,[33,34] utilizing some unique properties: (a) all films in these AC thin-film EL cells are transparent; (b) the horizontal separation between two stacked cells does not matter, since these films are thin; (c) various colors can be realized in one address.

Fig. 18.12 shows a schematic illustration of the cross-section structure of tunable-color EL devices. Two unit thin-film AC EL cells are stacked vertically on a glass substrate. In this case, $ZnS:TbF_3$ and $ZnS:SmF_3$ are used as emission layers partly because they emit two of the three primary colors, and because their brightness is relatively high. The driving voltage can be applied to the middle ITO as the common electrode.

Various emission spectra for several pairs of applied voltages are summarized in Fig. 18.13. The voltages applied to the green cell (V_g) and to the red cell (V_r) are shown in the figure. When V_g is 113 V and V_r is 110 V, which is lower than the threshold voltage of the red cell, only green emission can be observed, as is shown in Fig. 18.13(a). Conversely, when V_g is 90 V and V_r is 135 V, then only the red emission is observed since V_g is lower than the threshold voltage of the green cell, as can be seen in Fig. 18.13(e). Furthermore, when both V_g and V_r become higher than the threshold voltages of each cell, mixed color emission was observed. In the case of (b), the green emission intensity is greater than the red one, and then the emission color looks yellow-green. In the case of (c), the emission color looks yellow. In the case of (d), the red emission is dominant, and the mixed color

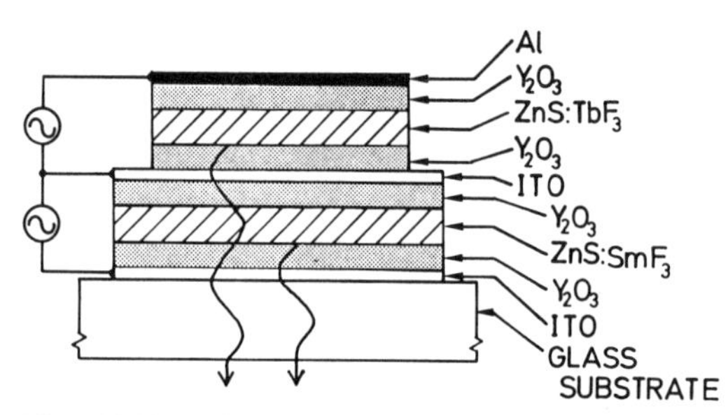

Fig. 18.12 Schematic illustration of the thin-film EL device consisting of two stacked cells with green and red emission.

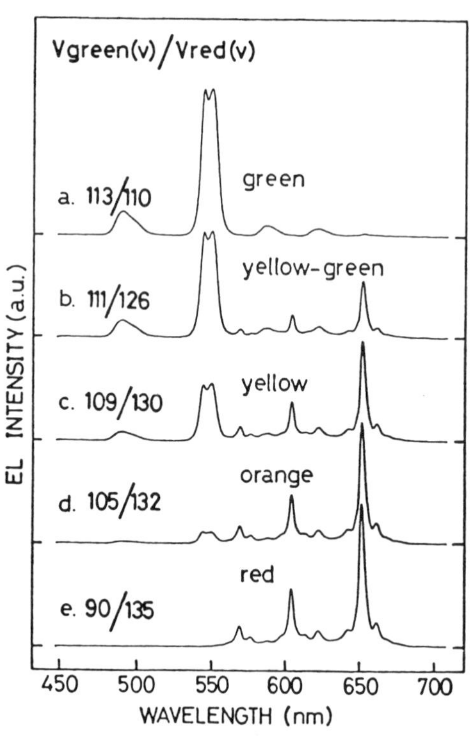

Fig. 18.13 The emission spectra of the tunable-color EL device with the driving voltage applied across each cell as a parameter. Any color between green and red can be obtained by changing the driving voltage of each cell about 30 volts.

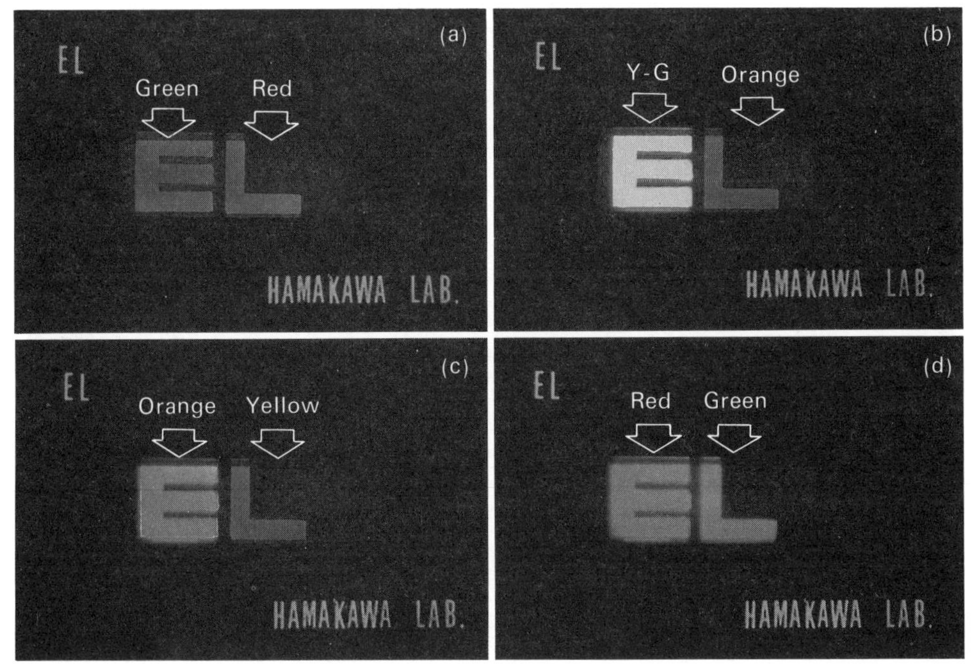

Fig. 18.14 Photograph of the tunable-color EL device. Emission color spectra can be changed continuously from red to green by controlling color regulation voltage (0–20 V_{rms}) superposed with sustaining voltage (100 V_{rms}) as shown in (a)–(d).

looks orange. Consequently, any colors between green and red would be produced by controlling the color regulation voltage (0–30 V_{rms}) superposed with a sustaining voltage of 100 V_{rms}. Stable variable color emission of about 100 ft-L was observed with one address of this tunable-color EL device. Examples of the display are shown in Fig. 18.14. The emission color varies continuously from red to green.

Moreover, any emission color can be achieved using the device consisting of cells for the three primary colors which are stacked vertically. The device structure is illustrated in Fig. 18.15. Fig. 18.16 shows the emission spectrum of the device when it emits the white light. The emission spectrum consists of three elements, that is the emission spectra of ZnS:SmF_3(red), ZnS:TbF_3(green) and ZnS:TmF_3(blue), as shown in Fig. 18.10. Fig. 18.17

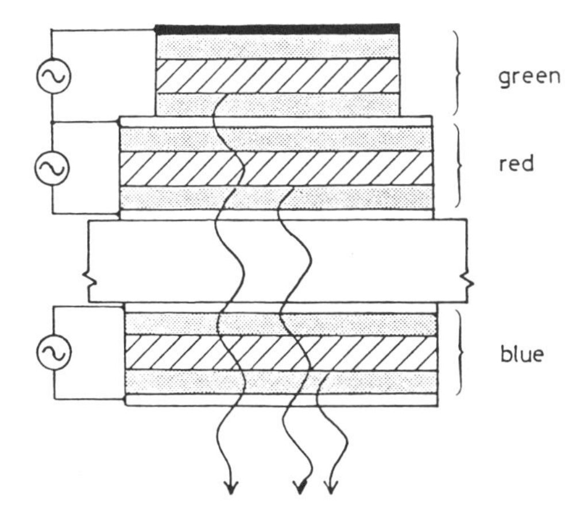

Fig. 18.15 Schematic illustration of the full-color EL device.

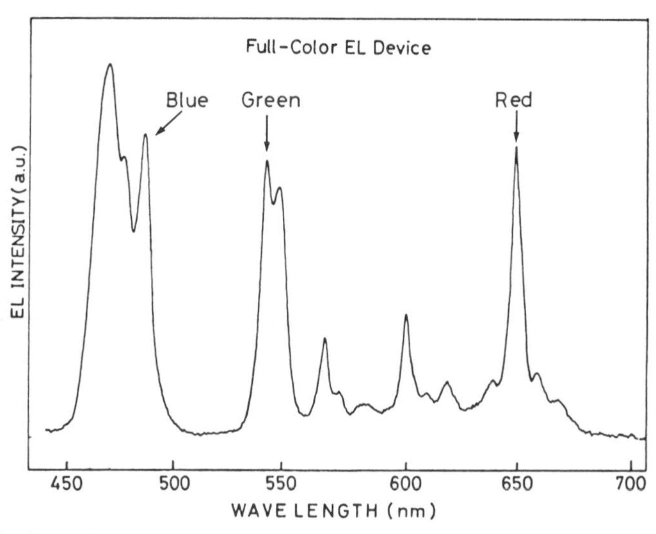

Fig. 18.16 Emission spectra of the full-color EL device when it emits white light.

shows a photograph of the device and Fig. 18.18 shows a chromatic diagram. The closed circles in the figure denote the points of three colors used in this full-color EL device. As a reference, the colors used in the CRT of a color TV set are shown in the figure with open circles. The brightness of each element color can be controlled by changing the applied voltage and/or width of the applied pulse voltage. By regulating this color control signal, any color of the shaded region can be observed with one spatial address point of this type of device. This function is entirely new and has never previously been attained, which means that one spatial address point shows multi-function by color changing. For a simple example, a traffic signal giving a red–yellow–green display could be managed with only one full-color display device. With this technical innovation, a wide variety of new application fields would be expected in the information sciences.

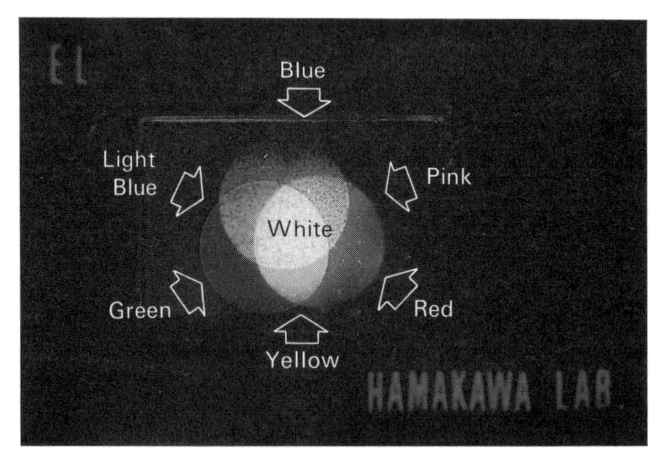

Fig. 18.17 Desk pattern of color mixing for the full-color EL device. With this device any color emission can be displayed by controlling the color regulation voltage.

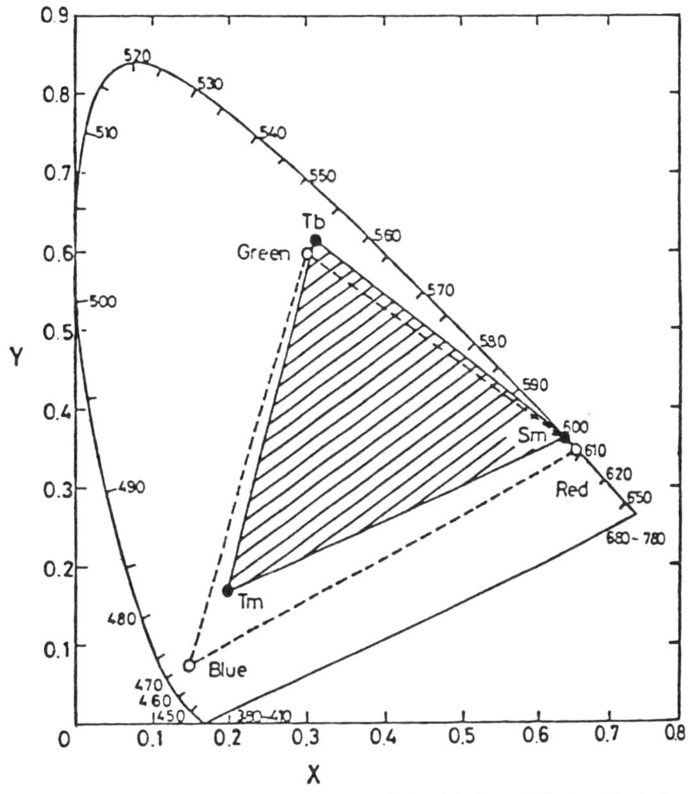

Fig. 18.18 Chromaticity diagram showing emitted color of the fabricated device. Emission can be varied within the shaded region. The open circles indicate the three primary colors used in the CRT of a color TV set.

18.5. Conclusion

A thin-film EL device utilizing $PbTiO_3$ as the insulating layer has been developed, and its threshold voltage could be reduced to four times lower than that of the conventional EL device having Y_2O_3 insulating layers. Multi-coloring including red, orange, yellow green, blue and white has been realized by changing the luminescent centers in the ZnS layer. Based upon this technology, tunable-color EL devices emitting any color from green to red with a brightness of 100 ft-L have been developed in doubly stacked cells which emit green (TbF_3) and red (SmF_3) light. The color emitted can easily be controlled by changing each cell. Moreover full-color devices have been also fabricated in cells stacked triply on both sides of a grass substrate and can present any color by controlling three applied voltages. The emitted color covers a wide range because the primary colors are very pure ones which are comparable with those of the CRT of a conventional color TV set. The characteristics obtained with these EL devices has been improved very much, but are not yet sufficient to apply to actual color display panel at the present stage. In order to obtain high-brightness full-color EL, optimization of fabrication conditions and a full understanding of the physical mechanism of the light emission are required. As thin-film EL devices have prominent advantages such as emission of bright light, pure color, compactness and high resolution, the EL device is most attractive as a display panel and realizing of a flat color TV set using the EL device is sincerely desired in the near future.

Acknowledgements

The authors wish to thank Prof. Taneo Nishino and Dr. Masanori Okuyama for their encouragements and valuable discussions. We also would like to thank Dr. Hideomi Ohnishi, Dr. Kenji Okamoto, Mr. Yasuhiro Nasu, Mr. Takahiro Suyama and Mr. Tetsuya Kitagawa for their helpful discussions and technical assistance.

References

1) G. Destriau: J. Chem. Phys., 33 (1936) 620.
2) R. E. Halstead and R. L. Koller: Phys. Rev., 93 (1954) 394.
3) T. Inoguchi and S. Mito: in Electroluminescence, Ed. J. I. Pankove (Springer, New York, 1977) p. 197.
4) Jagdeep Shar and A. E. Digiovanni: Appl. Phys. Lett., 33 (1978) 995.
5) R. Mack, J. Vonkalben, G. O. Müller, W. Gericke, and G. V. Reinspergen: J. Appl. Phys., 54 (1983) 4657.
6) T. Takagi, I. Yamada, A. Sasaki, and T. Ishibashi: IEEE Trans. Electron. Devices, ED-20 (1973) 110.
7) H. Ohnishi and Y. Hamakawa: Japan. J. Appl. Phys., 17 (1978) 1225.
8) Y. Fujita, J. Kuwata, M. Nishikawa, T. Tohda, T. Matsuoka, A. Abe, and T. Nitta: Japan Display '83 Digest (1983) p. 76.
9) T. Mishima, M. Konagai, and K. Takahashi: IEEE Trans. Electron. Devices, ED-30 (1983) 282.
10) H. Ohnishi, H. Yoshino, N. Sakuma, and Y. Hamakawa: Japan Display '83 Digest (1983) p. 88.
11) K. Okamoto, Y. Nasu, K. Okuyama, and Y. Hamakawa: Japan. J. Appl. Phys., 20, Suppl. 20-1 (1981) 215.
12) K. Okamoto, Y. Nasu, and Y. Hamakawa: Conf. Record of 1980 Biennial Display Research Conf. (1980) p. 143.
13) K. Okamoto, Y. Nasu, and Y. Hamakawa: IEEE Trans. Electron. Devices, ED-28 (1981) 698.
14) M. Ishida, H. Matsunami, and T. Tanaka: J. Appl. Phys., 48 (1977) 951.
15) K. Tanaka, Y. Higuma, K. Yokoyama, T. Nakagawa, and Y. Hamakawa: Japan. J. Appl. Phys., 15 (1976) 491.
16) A. Okada: J. Appl. Phys. 48 (1977) 2905.
17) M. Oikawa and K. Toda: Appl. Phys. Lett., 29 (1976) 491.
18) M. Okuyama, Y. Matsui, H. Nakano, and Y. Hamakawa: Proc. 8th Int. Vacuum Congress, Vol. 1 (1980) p. 503.
19) M. Okuyama, Y. Matsui, H. Nakano, T. Nakagawa, and Y. Hamakawa: Appl. Phys., 21 (1980) 339.
20) M. Okuyama, Y. Matsui, H. Nakano, T. Nakagawa, and Y. Hamakawa: Japan. J. Appl. Phys., 18 (1979) 1163.
21) K. O. Fugate: IEEE Trans. Electron Devices, ED-24 (1977) 909.
22) J. J. Hanak: Proc. 6th Int. Vacuum Congr.; Japan. J. Appl. Phys., Suppl. 2 (1974), 809.
23) H. Sasakura: IEEE Trans. Electron Devices, ED-25 (1978) 1170.
24) M. Takeda, Y. Kanatani, H. Kishishita, A. Fujimori, and K. Okano, SID Int. Symp. Digest (1981) p. 29.
25) E. W. Chase, R. T. Happlewhite, D. C. Krupka, and D. Kahng: J. Appl. Phys., 40 (1969) 2512.
26) T. Suyama, K. Okamoto, and Y. Hamakawa: Appl. Phys. Lett., 41 (1982) 462.
27) E. C. Freeman, D. H. Baird, and J. R. Weaver: Japan Display '83 Digest (1983) p. 592.
28) H. Kobayashi, S. Tanaka, T. Kunou, M. Shiiki, and H. Sasakura: Japan Display '83 Digest (1983) p. 592.
29) T. Tokita, Y. Fujita, T. Matsuoka, A. Abe, and T. Nitta: Japan Soc. Prog. Sci. Soc. 125 (Conversion between light and electricity) 108th Meeting 418 (1984) p. 18 [in Japanese]
30) D. Kahng: Appl. Phys. Lett., 13 (1968) 210.
31) K. Okamoto and Y. Hamakawa: Appl. Phys. Lett., 35 (1979) 508.
32) T. Suyama, N. Sawara, K. Okamoto, and Y. Hamakawa: Proc. 13th Conf. Solid State Device (Tokyo, 1981); Japan. J. Appl. Phys., 21, Suppl. 21-1, (1982) 383.
33) Y. Oishi, T. Kato, and Y. Hamakawa: Extended Abstract of the 15th Conf. on Solid State Devices and Materials (Tokyo, 1983) p. 353.
34) Y. Oishi, T. Kato, and Y. Hamakawa: Japan Display '83 Digest (1983) p. 570.

19 CRYSTAL GROWTH AND PROPERTIES OF GALLIUM NITRIDE AND ITS BLUE LIGHT-EMITTING DIODE

Isamu AKASAKI*, Hiroshi AMANO*, Nobuhiko SAWAKI*,
Masafumi HASHIMOTO†, Yoshimasa OHKI‡ and Yukio TOYODA‡

Abstract

Single crystals of GaN are grown by HVPE and MOVPE techniques on sapphire (0001) substrates. Structural defects characteristic of GaN are found in the former; and they are intentionally formed and applied to an n^+ contact. This enables both types of electrode to be made on one surface of a chip and a new structure of m–i–n diode to be developed. A practical blue LED with luminous intensity of 10 mcd at a forward current of 10 mA and an operating voltage of 7.5 V typically have been realized for the first time. The effects of growth conditions on the properties of the crystal and LED characteristics are reported.

By the MOVPE technique, in which TMG reacts with NH_3 in a slow stream of N_2, the crystal morphology, thickness uniformity and luminescence properties are much improved. Cracks and some deep level lattice defects are also markedly reduced.

Keywords: GaN, Crystal Growth, Photoluminescence, Blue LED

19.1. Introduction

Gallium nitride (GaN) is eminently suitable for use in blue light-emitting devices, as well as ZnS and ZnSe, because it has a direct energy band-gap of 3.39 eV[1,2] and has been expected to show high luminescence efficiency.[2] In contrast to other III–V compounds, however, it has been difficult not only to grow in a large single-crystal form or in an epitaxial layer but also to control its electrical and optical properties.[3] An important area of research, therefore, is the growth of large-area crystals of high quality and the study of the properties of these crystals.

In 1961, GaN crystals were first prepared by firing a powder of GaP in a stream of dry NH_3[4] and from Ga_2O_3 and NH_3 at temperatures from 600 to 1100°C.[5] Maruska et al.[1] reported the first achievement of growth of single crystals on sapphire substrates using the Ga–HCl–NH_3 system (hydride vapor-phase epitaxy – HVPE). Metal–organic vapor-phase epitaxy (MOVPE) of GaN has also been investigated by Manasevit et al.,[6] Duffy et al.[7] and Andrews et al.[8] The observations of stimulated emission and laser action near 3.45 eV at 2 K in single-crystal needles by Dingle et al.[2] have supported the earlier prediction[1] that GaN is a direct-gap semiconductor. In the 1970s, extensive studies were carried out on its crystal growth,[6,9,10] optical and electrical properties[2,11,12] and device

* Department of Electronics, School of Engineering, Nagoya University, Furo-cho, Chikusa-ku, Nagoya 464.
† Toyota Central R & D Labs., Inc., Nagakute, Nagakute-cho, Aichi 480-11.
‡ Matsushita Research Institute Tokyo, Inc., Higashimita, Tama-ku Kawasaki 214.
Present address: Kawasaki Lab. Opto-Electronics Development Center, Matsushita Electric Industrial Co., Ltd.

JARECT Vol. 19, *Semiconductor Technologies* (1986), *J. Nishizawa (ed.)*
OHMSHA, LTD. and North-Holland

characteristics.[3,9,11] Although the possibility of high-efficiency blue (2.6 eV) LEDs using Zn-doped GaN m–i–n (metal–insulator–n-type semiconductor) diode were reported by Pankove et al.[11] and Jacob et al.,[9,13] practical diodes had still not been realized until recently. This is probably due to mainly the lack of reproducibility for crystal growth and the difficulty in fabricating diodes. In 1980, Ohki et al.[14] succeeded in great improvement in crystal growth by HVPE and in diode structure, and realized for the first time a practical diode with high efficiency of 0.03% typically and 0.12% maximum. In HVPE, however, some serious problems remain to be solved.

This chapter consists of two parts: The first part[14] is concerned with the fabrication method and properties of the blue LED, which is made of GaN grown by the improved HVPE. The second part describes briefly more recent improvement in GaN crystal growth by MOVPE.

19.2. Crystal growth of GaN by HVPE[14]

19.2.1. *Apparatus and method*

Single-crystal GaN layers can be grown by using an open tube HVPE technique similar to that developed by Maruska et al.[1] N_2 was used as an ambient, because the surface morphology and thickness uniformity were not good when H_2 was used. The basic reactions are thought to be:[10]

$$Ga(l) + HCl(g) \rightarrow GaCl(g) + \tfrac{1}{2}H_2(g), \tag{19.1}$$

$$GaCl(g) + NH_3(g) \rightarrow GaN(s) + HCl(g) + H_2(g). \tag{19.2}$$

The structure of reactor and the temperature profile, which are very important in VPE in general, are drawn schematically in Fig. 19.1. In order to maintain the desired temperature profile, a resistance heating type furnace (#1) with six sections was used. A Ga source container was placed inside the inner tube into which HCl diluted with N_2 was passed. NH_3 was fed into the reactor through an inserted separate tube and flushed out directly at the deposition zone. Zn doping was carried out by heating a small amount of elemental Zn placed in the upper stream zone in the main carrier gas flow. This structure of reactor ensured sufficient time for reactions (19.1) and 19.2) and also prevented the mixing of Zn vapor with the Ga source. Consequently, the mixing of Zn vapor in the

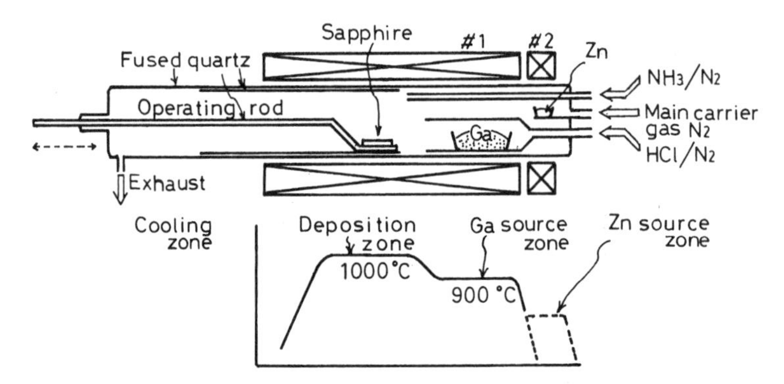

Fig. 19.1 Schematic illustration of the apparatus used for the HVPE of GaN.

carrier gas provided relatively homogeneous doping. A one-inch square (0001) sapphire (Al_2O_3) plate was used as a substrate, since preliminary experiments showed that the GaN layer grown on (0001) surface had better quality than those on ($1\bar{1}02$) surface and on such substrates as, Si, GaAs, GaP, SiC and quartz. Prior to growth, the polished substrates were cleaned with an organic solvent and rinsed in deionized water and then heat-treated as described later. The substrate was placed on a holder attached to the top of an operating rod by which it could be moved horizontally in the reactor.

The growth procedure was as follows: First, an undoped n-type GaN layer about 30 μm thick was grown. The growth rate was about 60 μm/h typically. Then the small Zn-heating furnace (#2) was turned on. Zn vapor was transported to the deposition zone to form a Zn-doped layer which became insulating (i) by compensation. After growth of the i-layer for some definite period, the wafer was moved to the cooling zone, where a single crystal could not be grown, by pulling the operating rod. This rapid cooling also prevents the outdiffusion of Zn from the i-layer. The Zn concentration, which increases with the Zn temperature, was the highest at the end of the growth, and its profile along the growth direction could be adjusted. The thickness of the i-layer thus made was of the order of 1 μm. X-ray rocking curves and back-reflection Laue patterns showed that these layers were good single crystals.

19.2.2. *Crystallographic structure of the GaN layer*

As the physical and chemical properties of GaN differ greatly from those of sapphire, the heteroepitaxial growth of GaN on sapphire has some difficulties. A major problem is the formation of macroscopic "structural defects" (hereafter called "structural defects") which can easily be observed with an optical microscope using transmission light.[15] It should be noted that these defects seriously affect the device (blue LED) properties and yield. In Fig. 19.2, typical photomicrographs of GaN layer are shown.

Fig. 19.3 is a model for the defects, which are classified into five categories:

[A] crack in sapphire substrate;
[B] crack in GaN layer;
[C] small irregular structure in GaN layer near the interface;
[D] hexagonal cone-like structure in GaN layer; and
[E] native pit on as-grown surface.

As seen in Fig. 19.3, the cracks [A] cross each other at 60° and cross the cracks [B] at 30°, while cracks [B] also cross each other at 60°. Cracks [A] are most probably caused by the large difference of thermal expansion coefficient ($\alpha_{GaN} < \alpha_{sapphire}$). In a thick epi-layer, the former cracks form a network and the wafer fractures into small pieces when it cools down to room temperature. The last three kinds of defects [C], [D] and [E] have the same cause and originate at the GaN/sapphire interface. The formation process of these defects is probably as follows: At the initial stage of growth, the small irregular structure [C] can be formed under a certain condition (not yet clearly understood). In some cases, it grows together with the epi-layer growth. If the growth of this defect ceases during the growth process of the epi-layer for some reason, it appears as a hexagonal cone-like structure [D]. On the other hand, the defect which reaches the surface of the epi-layer forms a native pit [E].

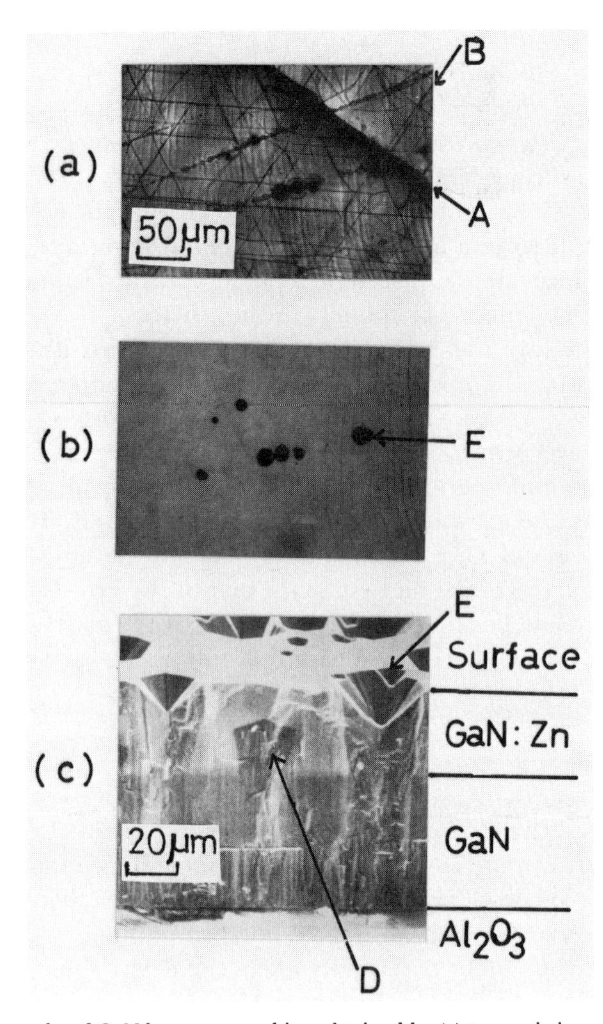

Fig. 19.2 Photomicrographs of GaN layer on sapphire, obtained by (a) transmission and (b) reflection light. (c) SEM image.

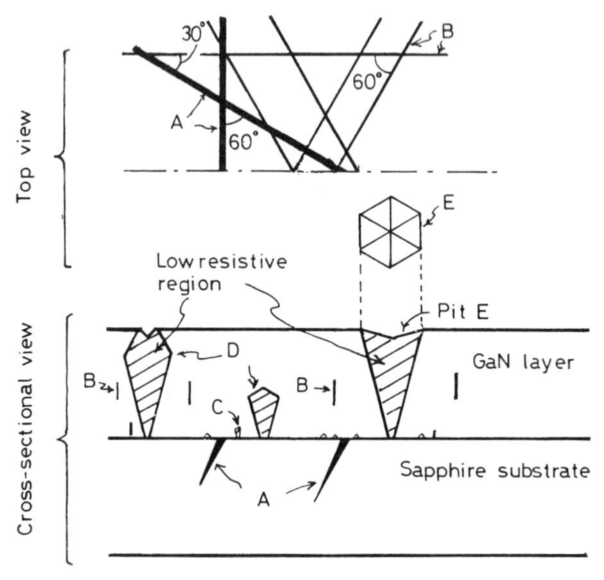

Fig. 19.3 Model for macroscopic structural defects in GaN epitaxial wafer; the columnar region with low resistivity starts at the GaN/sapphire interface.

These three kinds of defect have the same properties, for example they can easily be etched by hot alkali solution or by a mixture of H_2SO_4 and H_3PO_4, and have high electrical n-type conductivity (n^+) which cannot be reduced by compensation with a high concentration of Zn.

It is very important in fabricating LEDs to minimize such defects. The latter three kinds of defect could be markedly reduced by heat treatment of the substrate in a Ga–HCl–N_2 atmosphere at about 1000°C prior to GaN growth, although the reason for the effect of this treatment is not yet clear. In spite of various experiments, however, the number of cracks could not always be reduced.

19.2.3. *Photoluminescence and electrical properties*

It is known that the Zn atom in GaN behaves as a luminescent center[11] as well as an acceptor impurity. In this study, the temperature of the Zn source was varied up to 500°C to change the doping concentration of Zn, which was estimated by SIMS analysis, but not quantitively determined. Fig. 19.4 shows the variation of photoluminescence (PL) spectra at 77 K of GaN layers grown under different Zn partial pressure (p_{Zn}). In undoped material near-gap emission is dominant. It consists of the I_2 line originating from the recombination of an exciton trapped by a neutral donor [16] and the donor–acceptor (D–A) pair recombination line and its phonon replicas.[17] Doping with Zn reduces the intensity and forms newly broad emission bands peaking at about 430 and 510 nm in the spectra. These are called the B- and G-bands respectively in this chapter. At p_{Zn} above 0.1 Torr, the near-gap emission disappears and the B-band becomes dominant. The G band appears at $p_{Zn} \sim 0.3$ Torr. Above this value, the intensity ratio I_G/I_B increases with p_{Zn}. Fig. 19.5

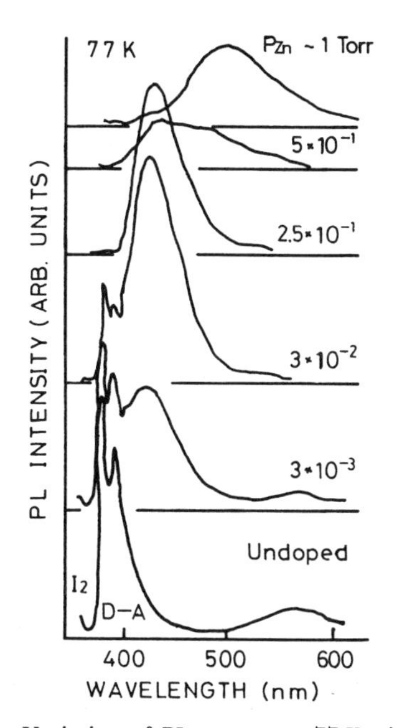

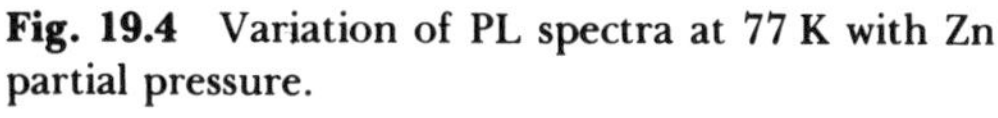
Fig. 19.4 Variation of PL spectra at 77 K with Zn partial pressure.

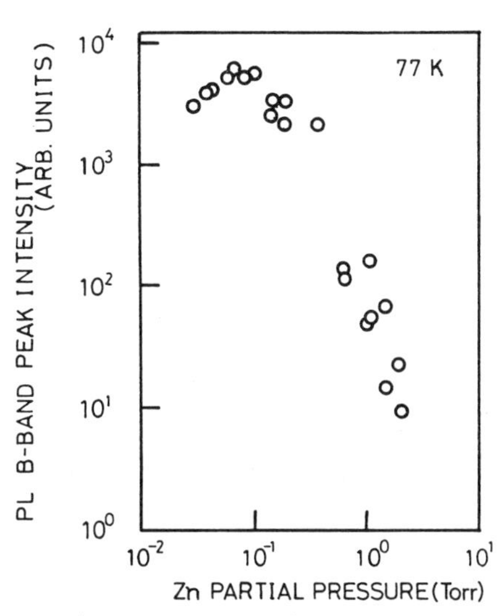

Fig. 19.5 p_{Zn} dependence of the B-band peak intensity at 77 K.

shows the p_{Zn} dependence of the peak intensity of the B-band measured at 77 K. The intensity has a maximum at $p_{Zn} \sim 0.08$ Torr. In the region of p_{Zn} below 0.08 Torr, an increase of p_{Zn} probably gives a higher Zn doping and thus increases the number of luminescent centers. At p_{Zn} above 0.08 Torr, excess Zn atoms possibly form non-radiative centers such as lattice strain or defect complexes [for example, V_N–Zn_{Ga} (V_N = nitrogen vacancy) or complexes containing interstitial Zn].

Some samples showed a yellow (Y) emission band at about 560 nm and orange (O) emission band at about 600 nm with a large half width. Some deep level containing V_N may be responsible for these bands, because crystals cooled slowly or annealed above 700°C after the growth showed these emissions. Another band peaking at 465 nm was sometimes observed in Zn-implanted samples.[18)]

Undoped GaN invariably showed n-type conductivity which is probably related to such lattice defects including V_N. The electron concentrations were estimated from the pattern of ultraviolet reflection and infrared absorption spectra,[19)] and were typically in the range (1 to 10) × 10^{17} cm^{-3}. The material grown at p_{Zn} below 0.08 Torr was still n-type, while GaN grown at p_{Zn} above 0.08 Torr became highly resistive or insulating owing to the compensation.

19.2.4. *Structure of GaN LED*

Fig. 19.6 shows a cross-section of the newly developed GaN:Zn blue LED. The m–i–n structure is essentially the same as that of Pankove,[11)] but is a more practical and novel one. The diode chip has a small area n^+-type conductive region reaching the chip surface through the n and i layers as seen in the inset. This n^+ region, which consists of defects [C], [D] and [E] described in Section 19.2.2, has been intentionally formed on the SiO_2 film[20)] which was deposited selectively on the substrate. A Ni/Al electrode was formed by evaporation and conventional photolithography. As the sapphire substrate and GaN crystal are transparent to visible light, the chip can be mounted upside-down on a conventional lead frame and coated with epoxy resin. This structure provides both n- and i-side electrodes on the top surface of the LED chip. On the other hand, all previous GaN diodes[9,11–13)] had the n-side electrode attached on the thin (several tens μm) side face of the chip, and were impractical for fabrication.

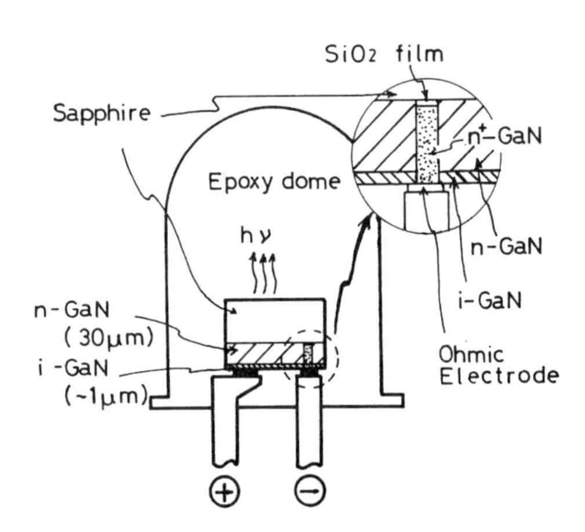

Fig. 19.6 Cross-section of a new structure of GaN:Zn m–i–n blue LED.

19.2.5. *LED Characteristics*

Typical values of the diode operating voltage V_F were between 5 and 10 V at a forward current I_F of 10 mA. This variation is mainly due to the fluctuation of the i-layer thickness, d (in μm). The diode capacitance C at zero bias and the forward voltage V_1 (in volts) at 1 mA were measured for many diodes. As the result, C varies as $C \propto V_1^{-2}$. If we approximate $C = \epsilon\epsilon_0 S/d$, where S (in μm^2) is an area of the i-side electrode and ϵ the dielectric constant of GaN, $V_1 = 1136\,(d/S)^{1/2}$ was obtained. A microscopic PL image of a cross-sectional surface of the layer showed that the Zn-doped layer (which emitted blue or green luminescence) was slightly thinner than 1 μm for low-V_F diodes. Thus controlling d is important, but it is not always easy in conventional HVPE. In our case, the use of the operating rod to move the substrate enables us to control the growth time, i.e. d.

The reverse breakdown voltage V_B varied from sample to sample between 10 and 100 V. Such a large variation of V_B was probably caused by inhomogeneities of the epi-layer, in particular the *i*-layer. The reverse breakdown occasionally caused a large leakage current and sometimes destroyed the diode.

The room-temperature electroluminescence (EL) may also consist of two emission bands near the B and G bands. The half-widths of these emissions are so large that they could not be separated from each other. Fig. 19.7 shows the change of EL spectrum with I_F. The emission peak wavelength λ_p^{EL} at low currents is nearly the same as that of PL, λ_p^{PL}. The results of Figs. 19.4 and 19.7 suggest that the emission region with low currents is the surface of the i-layer where the Zn concentration is the highest as described in Section 19.2.3, and it shifts to the inner region with somewhat lower Zn concentrations as I_F is increased.

The EL intensity L is almost proportional to the square root of I_F except for low currents and for sample D: $L \propto I_F^{1/2}$ as seen in Fig. 19.8. The current–voltage characteristics, plotted in Fig. 19.9, are expressed in the form of: (a) $\log I_F \propto V_F^{-1}$ except for very low currents in the case of low V_{F0} (V_F at 0.1 mA); and (b) $\log I_F V_F^{-1/2} \propto V_F$ for V_{F0} higher than 5 V. These facts suggest that the main conduction mechanisms of diode current with low and high V_{F0} are electron "tunneling"[14] and "tunnel-induced impact ionization",[21] respectively.

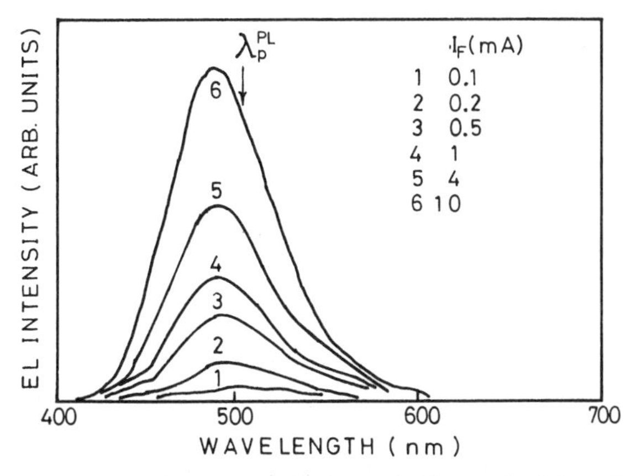

Fig. 19.7 EL spectra at various forward currents. The arrow indicates the PL peak wavelength of the crystal from which the diode was made.

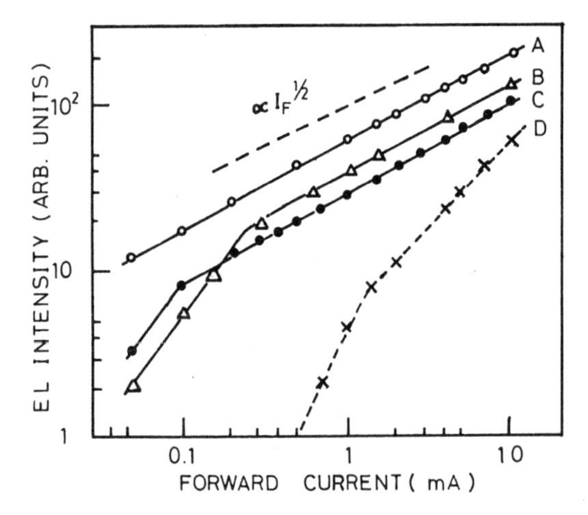

Fig. 19.8 Forward current dependence of the EL intensity in various diodes.

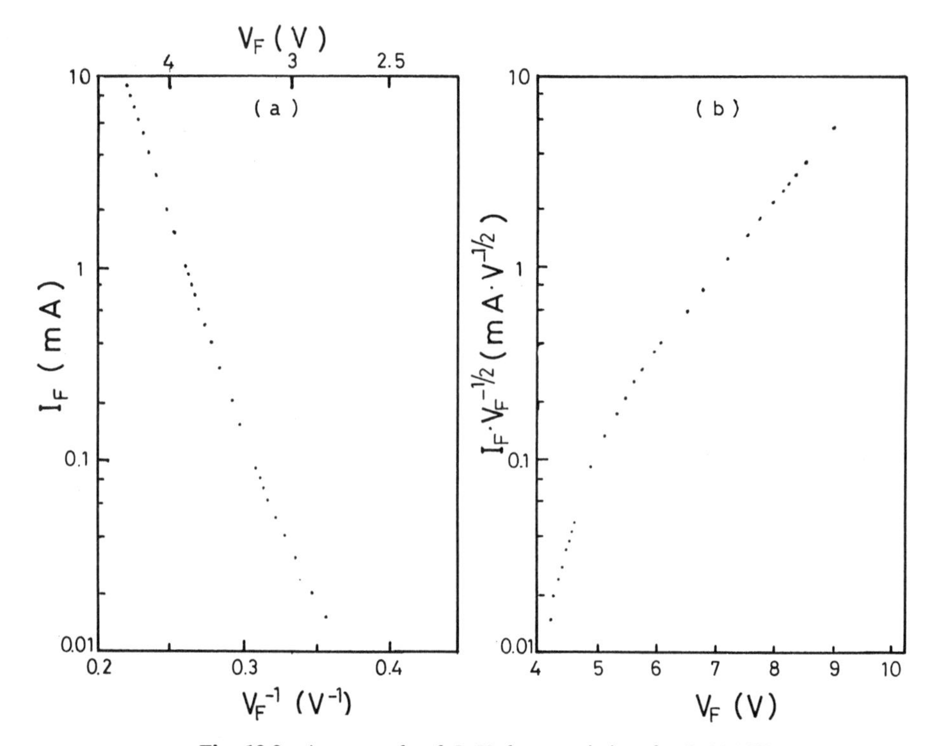

Fig. 19.9 An example of $I-V$ characteristics of a GaN LED.

In practical use, the operating current is above 1 mA and then the peak shift can be neglected. The room-temperature EL efficiency depended on I_F, and almost all the LEDs had a maximum at I_F near 1 mA. A quantum efficiency of 0.03% typically and 0.12% maximum was successfully obtained for a GaN:Zn blue LED with peak wavelengths less than 490 nm.

19.3. Crystal growth of GaN by MOVPE

19.3.1. *Aim of this study*

One of the serious problems in HVPE is the formation of macroscopic structural defects such as pits and cracks in the epi-layer. Another problem in making LEDs is that the operating voltage varies from diode to diode and is also relatively high. This is mainly due to the difficulty of controlling the i-layer thickness within 0.1 μm, because of its fairly high growth rate and its inhomogeneous growth. Reducing the growth rate causes island growth to occur, resulting in a lowering of the crystal quality and promoting inhomogeneous growth. MOVPE has been expected to be a promising technique for solving these problems. The possibility and usefulness are described in this section.

19.3.2. *Experimental*

In previous applications of MOVPE using $Ga(CH_3)_3$ (TMG) and NH_3, a large quantity of H_2(3–4 l/min)[6,7] has been used as the ambient and carrier gas for TMG vapor. In such a system, the gas flow with TMG was very fast (~500 cm/s), and turbulence arose in the reactor so the surface morphology, thickness uniformity and crystal quality were not good.[6]

We have succeeded in obtaining uniform layers with few cracks by using N_2 instead of H_2 as the ambient gas and reacting with TMG and NH_3 in a slow stream in the vertical reactor. PL data were better than those of HVPE. Zn doping has been achieved by adding $Zn(C_2H_5)_2$ (DEZ).[22] In the N_2 ambient, TMG reacts with NH_3 and trimethylgallium monamine, $Ga(CH_3)_3{\cdot}NH_3$ (mp 31–32°C) is produced[8] even at room temperature as follows:

$$Ga(CH_3)_3 + NH_3 \rightarrow Ga(CH_3)_n{\cdot}NH_3 + \tfrac{1}{2}(3-n)C_2H_6, \qquad (19.3)$$

where $n = 1, 2, 3$. Crystalline GaN can be formed by the decomposition of the adducts at temperatures above 900°C as follows:[22]

$$Ga(CH_3)_n{\cdot}NH_3 \rightarrow GaN + nCH_4 + \tfrac{1}{2}(3-\mathrm{n})H_2, \qquad (19.4)$$

where $n = 1, 2, 3$. In order to obtain single-crystal layers of GaN, therefore, reaction (19.3) at low temperatures below several hundred degrees centigrade and, in particular, in fast gas flow must be avoided, otherwise gaseous intermediate compounds run away from the hot substrate and single crystals cannot be formed. These conditions are satisfied in our system as shown in Fig. 19.10 and Table 19.1.

The ambient N_2 (1.7 l/min) and TMG vapor were passed through separate delivery tubes to reduce the velocity of the gas stream near the substrate. TMG vapor was transported by bubbling N_2 through the liquid TMG and its velocity was about 1–5 cm/s. NH_3 and DEZ vapor were also introduced separately. This growth condition differs largely from those of previous papers as shown in Table 19.1. The vertical separation between the delivery tube orifice for TMG and the substrate was about 4 cm. The orifices of the other delivery tubes for the ambient gas, NH_3 and DEZ were near the upper end of the reactor, which was not water-cooled. Substrates were also (0001) sapphire treated by the same manner as that of HVPE. The growth temperature, T_G, was varied from 930 to 1050°C by

Table 19.1 Growth conditions and crystal quality of GaN in MOVPE and HVPE

Method	Gases supplied: Flow rate of ambient (l/min)		Flow rate of NH_3 (l/min)	Velocity of TMG (cm/s)	Growth rate (μm/h)	Number of cracks	Surface morphology	Ref.
HVPE	N_2	5.4	0.4	—	60	many	smooth, pit	Y. Ohki et al.[14)]
MOVPE	H_2	4	2.5	530	3	—	irregular, uniform	M. T. Duffy et al.[7)]
MOVPE	H_2	3	—	—	6	—	irregular	H. M. Manasevit et al.[6)]
MOVPE	N_2	1.7	0.05 ~ 2	1 ~ 5	3 ~ 15	few	smooth	M. Hashimoto et al.[22]

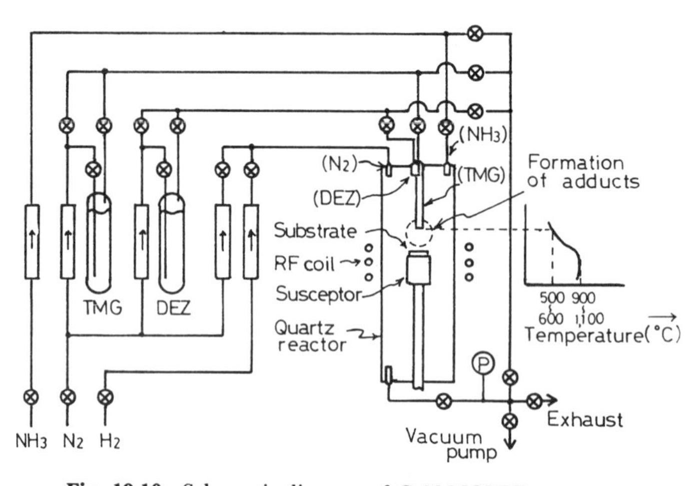

Fig. 19.10 Schematic diagram of GaN MOVPE system.

graphite susceptor heated RF inductively. Thus the formation of adducts could be avoided at the lower-temperature region far from the substrate.

19.3.3. *Crystallographic structure, surface morphology and growth rate*

Fig. 19.11 shows an example of GaN layers (1.5 μm thick) grown for 30 min at 1000°C by transporting TMG at a rate of 4×10^{-5} mol/min and NH_3 of 2×10^{-3} mol/min (0.05 l/min). It should be emphasized that macroscopic structural defects described in Section 19.2.2 are scarcely observed in the layer. The surface and GaN/sapphire interface are very flat. Such smooth, nearly crack-free and transparent GaN layers have never before been obtained in HVPE.

The growth rate increased linearly with TMG concentration, indicating that the

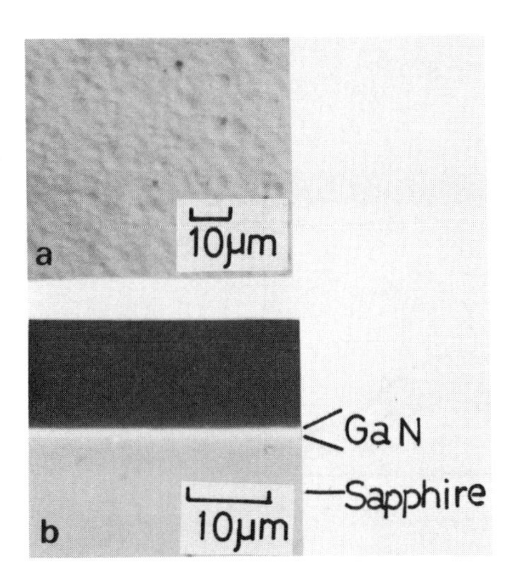

Fig. 19.11 Photomicrographs of GaN epitaxial layer grown at 1000°C by MOVPE: (a) surface, (b) cross-section. Native pits and cracks are not observed.

growth process is limited by mass transportation of the metal–organic compounds. This is similar to the cases for GaAs[23] and AlN.[24]

19.3.4. *Photoluminescence properties*

Fig. 19.12 shows the PL spectra of (a) Zn-doped and (b) undoped MOVPE GaN (a few μm thick) together with (c) undoped HVPE GaN (about 30 μm thick). The wavelengths of the Y-band of sample (b) are rather shorter than those of sample (c). The O-band, which may be due to some deep level defect and appeared sometimes in the latter as mentioned in Section 19.2.3, almost disappears in the former. Furthermore, in spite of the much thinner sample, the former intensity is comparable with that of the latter. These indicate that the MOVPE layer has higher quality than the HVPE one. In sample (a), the B-band due to Zn, which was incorporated by using DEZ, is dominant and the Y-band is weak. This suggests the possibility of Zn-doping in GaN by MOVPE.

Fig. 19.13 shows the PL spectra of undoped GaN layers grown under the same conditions as sample (b) except for the NH_3 concentration. When the NH_3 concentration is increased, the intensity of the Y-band decreases, while that of the B-band increases. This shows that MOVPE GaN can emit such a B-band without Zn-doping. This is probably due to a decrease in the number of V_N, which may be the cause of the G- and/or Y-bands, with

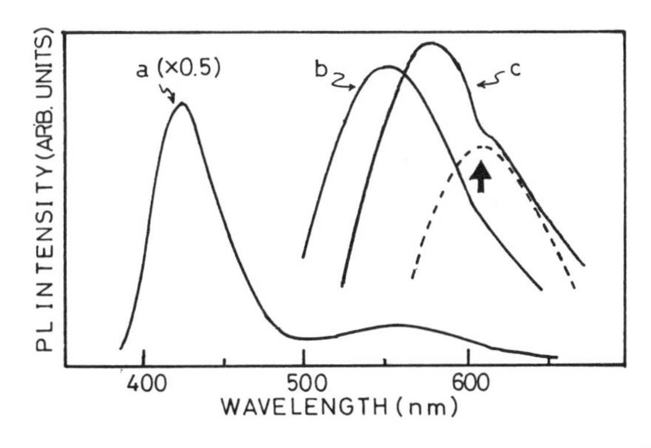

Fig. 19.12 PL spectra of: (a) Zn-doped GaN (MOVPE), (b) undoped GaN (MOVPE) and (c) undoped GaN (HVPE). The arrow indicates orange-band emission in (c).

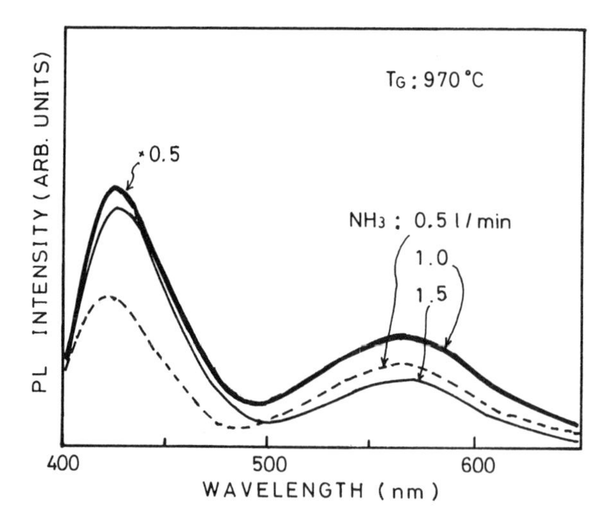

Fig. 19.13 PL spectra of MOVPE GaN grown under various NH_3 concentrations.

the increase of NH_3 concentration. The results of Figs. 19.12 and 19.13 indicate that MOVPE GaN has less deep level defects such as V_N than HVPE GaN.

19.4. Conclusions

Single crystals of GaN have been grown by an improved HVPE technique. Zn-doped insulating layers with thickness of about 1 μm were obtained by using an operating rod for the substrate holder. Structural defects characteristic of GaN were first found and were intentionally applied to the n^+ contact. These results enable a new structure of m–i–n diode to be made in which a chip can easily be mounted upside-down. A practical blue LED with an operating voltage of 7.5 V at a forward current of 10 mA with a luminous intensity of 10 mcd typically has been realized for the first time. Its $I-V$ characteristics indicated that tunneling was the main mechanism of current injection.

By using the MOVPE technique with N_2 as an ambient, major problems in HVPE have been solved. Pit-free and uniform thin GaN layers less than 0.1 μm were obtained with good reproducibility. Almost all the cracks appearing in the HVPE wafer disappeared. Deep level lattice defects can be markedly reduced, resulting in superior luminescence properties.

Acknowledgements

The authors would like to thank Messrs. Hiroyuki Kobayasi and Nobuhide Matsuda of Matsushita Electric Industrial Co., Ltd. and the staff of Kagoshima Matsushita Electronics Corp. for fabricating diodes and for help with experiments.

This work was supported in part by Hoso-Bunka Foundation.

References

1) H. P. Maruska and J. J. Tietjen: Appl. Phys. Lett., 15 (1969) 327.
2) R. Dingle, K. L. Shaklee, R. F. Leheny, and R. B. Zetterstrom: Appl. Phys. Lett., 19 (1971) 5.

3) G. Jacob, M. Boulou, M. Furtado, and D. Bois: J. Electron. Mater., 7 (1978) 499.
4) A. Addamiano: J. Electrochem. Soc., 108 (1961) 1072.
5) M. R. Lorenz and B. B. Binkowski: J. Electrochem. Soc., 109 (1962) 24.
6) H. M. Manasevit, F. M. Erodman, and W. I. Simpson: J. Electrochem. Soc., 118 (1971) 1864.
7) M. T. Duffy, C. C. Wang, G. D. O'Clock Jr., S. H. McFarlane, and P. J. Zanzucchi: J. Electron. Mater., 2 (1973) 359.
8) J. E. Andrews and M. A. Littlejohn: J. Electrochem. Soc., 122 (1975) 1273.
9) G. Jacob, M. Boulou, and M. Furtado: J. Crystal Growth, 42 (1977) 136.
10) S. S. Liu and D. A. Stevenson: J. Electrochem. Soc., 125 (1978) 1161.
11) J. I. Pankove, E. A. Miller, and J. E. Berkeyheiser: RCA Rev., 32 (1971) 383.
12) J. I. Pankove, E. A. Miller, D. Richman, and J. E. Berkeyheiser: J. Lumi., 4 (1971) 63.
13) G. Jacob and D. Bois: Appl. Phys. Lett., 30 (1977) 412.
14) Y. Ohki, Y. Toyoda, H. Kobayasi, and I. Akasaki: Gallium Arsenide and Related Compounds 1981 (Inst. Phys. Conf. Ser. 63) p. 479.
15) H. Kobayasi, Y. Toyoda, Y. Ohki, N. Matsuda, and I. Akasaki: National Tech. Rep., 28 (1982) 83. [in Japanese]
16) D. D. Sell, S. E. Stokowski, and M. Ilegems: Phys. Rev., B4 (1971) 1211.
17) M. Ilegems: Solid State Commun., 9 (1971) 175.
18) Y. Toyoda, Y. Ohki, H. Kobayasi, and I. Akasaki: Proc. 6th Symp. on Ion Implantation and Submicron Fabrication (1978) p. 79. [in Japanese]
19) M. Ilegems: Phys. Rev., B7 (1973) 743.
20) Y. Ohki, Y. Toyoda, H. Kobayasi, and I. Akasaki: US Patent 4396929.
21) J. I. Pankove and M. A. Lampert: Phys. Rev. Lett., 33 (1974) 361.
22) M. Hashimoto, H. Amano, N. Sawaki, and I. Akasaki: to be published.
23) H. M. Manasevit and W. I. Simpson: J. Electrochem. Soc., 116 (1969) 1725.
24) M. Morita, N. Uesugi, S. Isogai, K. Tsubouchi, and N. Mikoshiba: Japan. J. Appl. Phys., 20 (1981) 17.

20 ZnS BLUE LIGHT-EMITTING DIODE

Tsunemasa TAGUCHI†

Abstract

This chapter is concerned with recent developments in the study and characterization of forward-biased ZnS blue light-emitting diodes (LEDs), including appropriate interpretations for the blue electroluminescence and minority-carrier (hole) injection processes. The studies of adsorption of oxygen and chemical oxidation at the cleaved (110) faces of cubic ZnS with iodine (ZnS : I) are described and show that the formation of oxide layers is useful for the fabrication of an efficient blue LED with MπS structure. It is demonstrated that the blue LED, with an effective interfacial layer of about 300 Å, exhibits a high-efficiency injection blue luminescence at 2.76 eV at room temperature (an external quantum efficiency of about 8×10^{-4} photons/electron). The degradation characteristics of the diodes are also presented.

Keywords: ZnS, Blue LED

20.1. Introduction

Of the wide-bandgap II–VI compound semiconductors, crystalline ZnS appears to be one of the most promising for blue light-emitting diodes (LEDs). In the past decade, attempts have been made[1,2] to produce LEDs on single crystals of ZnS. However, because p–n homojunctions have not been successfully fabricated on ZnS, other technologies have had to be utilized to excite electroluminescence under forward-bias conditions. Since about 1974, in particular MIS (metal–insulator–semiconductor) structured diodes on ZnS operating under forward-bias have been investigated as blue LED.[2]

Recently, although a few publications[3,4,5] have appeared in the literature on the characterization of the blue luminescent bands and the fabrication of the blue LEDs, the technology has not matured to the point of successful commercial realization.[6] In particular, driving voltages were generally high, about 10 V, because of the high resistance of the bulk ZnS material and poor ohmic contacts. Therefore, reproducibility of neither diode preparation nor diode operating characteristics have been obtained.

Very recently, Taguchi et al.[7,8] have reported that crystals of ZnS grown from the iodine-transport method show excellent cubic structure having a lower stacking-fault density than the melt-grown crystals and allowing the production of low-resistivity LED material capable of strong blue emission. For the production of efficient blue LED using ZnS, from a practical point of view, it is desirable to find out the prerequisites for making reproducible and reliable MIS or metal–partially compensated high-resistive layer–semiconductor (MπS) structures. Since oxygen gives rise to acceptor-like levels which can result in an accumulated or a compensated semi-insulating layer due to chemical

† Department of Electrical Engineering, Faculty of Engineering, Osaka University, Suita, Osaka 565

JARECT Vol. 19 *Semiconductor Technologies* (1986), *J. Nishizawa* (*ed.*)
OHMSHA, LTD. and North-Holland

oxidation, it might be useful for making an interfacial π-layer between the metal electrode and n-ZnS to obtain efficient blue LED. The nature of the π-layer has been extensively investigated using Rutherford back-scattering (RBS)/channelling and X-ray photoelectron spectroscopy (XPS) measurements combined with an Ar^+-ion sputtering technique.[9] It has been found that the blue LED, with an effective interfacial layer of about 300 Å, exhibits a high-efficiency forward-bias injection blue luminescence around 2.76 eV at room temperature (RT) as a result of the involvement of two kinds of donor–acceptor pair recombination centres.[9,10]

This chapter deals with the fundamental properties of the forward-bias blue LEDs having the MπS structure, and gives an appropriate interpretation for the recombination and minority-carrier (hole) injection processes in the intense blue luminescence.

20.2. Crystal preparation and characterization

Cubic ZnS single crystals were grown by the iodine-transport method in a sealed quartz tube, being kept for about 1 month, in a vertical or horizontal growth furnace, at temperatures in the range from 750 to 850°C (hereafter called c-ZnS:I).[11,12] The large ZnS:I crystals were string-sawn perpendicular to the cleaved plane into long bars with an appropriate cross-section of about 2×2 mm^2, and then annealed in molten Zn at 960°C for 122 h to render them low-resistive n-type. Hall-effect measurements indicated the following room-temperature properties: (1) a resistivity of 1–10 Ωcm; (2) an electron concentration of about 10^{16} cm^{-3}; and (3) an electron Hall mobility of 120–150 cm^2/V s.

According to the PIXE (proton-induced X-ray emission)[13] measurements for as-grown high-resistivity ZnS:I crystals, the impurity content was about 1 p.p.m. for As, Fe and Ni, and less than 1 p.p.m. for Nb, Cr, Co and I. After heat-treatment using molten-Zn extraction, those residual impurities mentioned above were not detectable at p.p.m. levels, and Cu and Al were not detectable at all.

20.3. Blue LEDs

20.3.1. *Fabrication of forward-biased LED*

The crystals were cleaved into dice with a thickness of about 2 mm and heat-treated in molten Zn, then immediately separated from the Zn solution in the ampoule and subsequently quenched in liquid N_2. During this procedure, it was seen that the crystals were completely covered with a conductive Zn thin film. Zn-covered dice were then chemically etched in dilute HNO_3 for 60 s, after which chemical oxidation was performed by dipping into H_2O_2 at RT for about 20 s. To make a Schottky barrier an Au metal electrode was evaporated in vacuum on both the oxidized layers. Each die was cleaved again into two dice with a thickness of about 1 mm, and these were immediately contacted with wetted In:Hg alloy on the newly cleaved faces in purified H_2 gas at a temperature of 440°C for 40 s. The diode was finally mounted on a TO-5 holder and cast in transparent epoxy resin (NT 8020) at 120°C. The absolute RT external quantum efficiency over a range of operating current density was evaluated using a calibrated 1 inch integrating sphere connected with a photometer (photodyne model 88XLA).

20.3.2. *Cleaved* (110) *face and the effect of oxygen impurity*

The chemical oxidation layer on (110) faces produced by HNO_3 etching was characterized using the XPS measurements together with Ar^+-ion sputtering. A typical example of a compositional profile measured as a function of thickness in the layer is presented in Fig. 20.1, where both the O/Zn and S/Zn ratios are plotted. At the uppermost surface layer, the O/Zn ratio is certainly higher than 1, and it rapidly attenuates to about 1 within 15 Å, indicating that chemical oxidation takes place in this region. The S/Zn ratio, however, is about 0.5 at the uppermost layer. It is evident that the O/Zn ratio decreases with depth and approaches about 0.01 at a depth of 800 Å. We thus understand, from the reasonable decrease of the O/Zn ratio with depth, that the stoichiometric composition ratio Zn:S attains a value of 1 in the bulk region. It is therefore reasonable to consider that the apparent thickness of the π-layer is about 300 Å.

The effect of oxygen impurity at the (110) faces of ZnS crystals must be taken into account. In particular, it is noted that the n-type ZnS crystals contain a large concentration of S vacancies both at the (110) surface and in the bulk, as indicated clearly not only from the reduction in the peak due to atomic sulphur in the channelling experiment,[7] but also by an enhancement of the I_D (neutral-donor bound-exciton emission) line due to S vacancies,[14] by application of heat-treatment in molten Zn. The formation of SO_2, however, was not found in the XPS experiment. It may therefore be reasonable to point out that oxygen impurities are adsorbed at the surface via S vacancies.[15] The formation process of the oxide layer should be as follows: (1) oxygen is adsorbed on the (110) surface and then preferentially migrates via S vacancies and substitutes on the S lattice sites; (2) substitutional oxygen impurities seem to diffuse further into the bluk region via vacancies and excess oxygen impurities might react preferentially with Zn atoms to form ZnO, as inferred from the XPS results.[9]

20.3.3. *General properties of the diodes*

Fig. 20.2(a) records the forward current–voltage (I–V) characteristic, obtained at RT for two typical diodes having different thicknesses of the oxidized layer. These diodes (D_1 and D_2) exhibit a strong forward-biased blue luminescence, as will be described later. The diodes D_1 and D_2 possess intentional interfacial π- layers of thickness of about 300 and 350 Å, respectively, formed by chemical oxidation, and show that the zero-bias diffusion potential is about 1.6 V. The thickness was estimated using the method described in

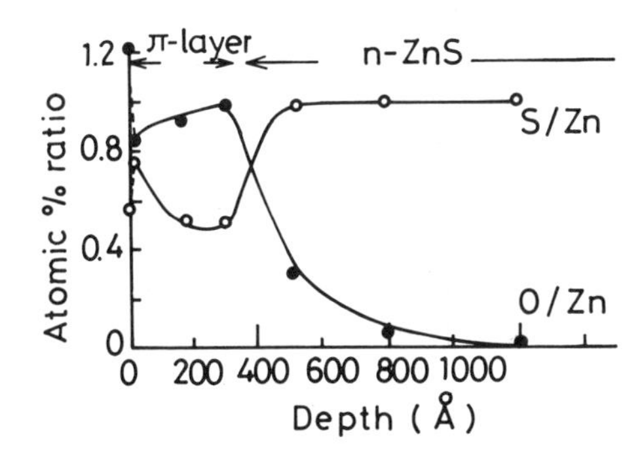

Fig. 20.1 Variations of the O/Zn and S/Zn at% ratio on the cleaved (110) face etched in HNO_3 as a function of depth.

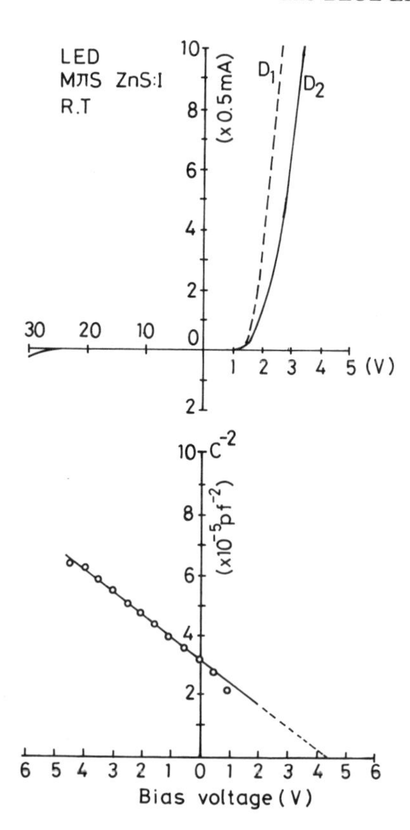

Fig. 20.2 (a) $I-V$ characteristics of LEDs at RT having different thicknesses of the oxidized layer: D_1 (300 Å) and D_2 (350 Å). (b) Typical $C^{-2}-V$ characteristics at RT of the D_1 diode.

Section 20.3.2. The $I-V$ characteristics seem to become soft with increasing thickness of the oxide layer.

Fig. 20.2(b) shows the capacitance–voltage ($C^{-2}-V$) characteristic obtained at RT for the diode D_1. From the slope of the $C^{-2}-V$ plot, the uncompensated donor density ($N_D - N_A$) is calculated to be about 4×10^{16} cm^{-3}, fairly close to the value obtained from the Hall measurement (about 2×10^{16} cm^{-3}).[19] The intercept voltages for the diode D_1 are estimated to be about 4.2 V. Using Cowley's theory,[16] the apparent interfacial layer thickness can be estimated to be about 270 Å, which is in good agreement with the value obtained for the present XPS measurement. On the other hand, for the diode without intentional interfacial layer the intercept voltage found in the $C^{-2}-V$ plot is about 2.2 V which is also the same as that obtained on a Schottky structure diode,[3,5] and as a result the interfacial layer thickness is probably much less than 100 Å.

We have reported that the n-type c-ZnS:I crystals produce a very strong blue photoluminescence (PL) band[7] around 2.65 eV at RT, which is derived from the composition of two bands separated from each other by about 18 nm as shown in the inset of Fig. 20.3. Fig. 20.3 shows the typical forward-bias injection electroluminescence (EL) spectra obtained at RT as a function of forward current across the diode. The diode did not exhibit any kind of emission bands in reverse bias. There appear clearly two peaks (designated Y_{S1} and Y_{S2}) separated from each other by about 20 nm centred at 2.6 eV and the EL band is only shifted from the PL band at relatively lower forward current. The total half-width (ΔH) is estimated to be 0.54 eV and RT, which is comparable with that (0.46 eV) for the blue PL band. It seems that the appearance of the double peaks in the

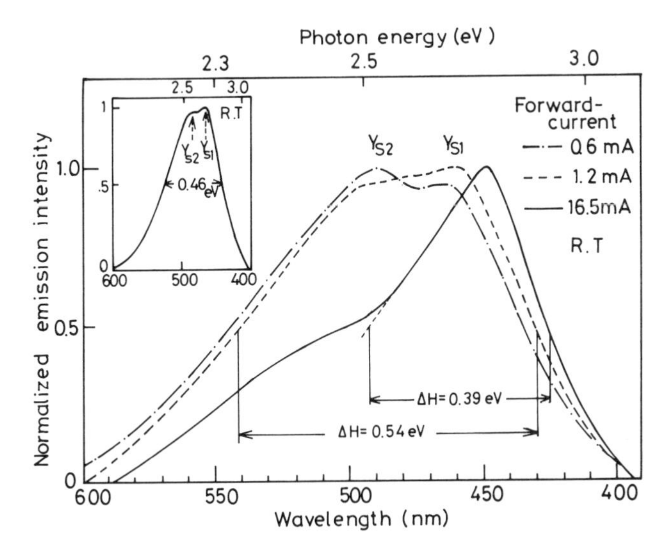

Fig. 20.3 Forward-bias injection EL spectra obtained at RT as a function of forward current in an MπS structure LED. Current (mA): (–·–·–) 0.6, (–––) 1.2, (——) 16.5. The inset shows the PL spectrum at RT under excitation by a Hg lamp (365 nm) in low-resistive n-type c-ZnS:I crystals.

injection-luminescence blue band is similar to that of the PL band. As the current increases, the broad band shifts towards higher photon energy, and as it further increases (above 16.5 mA), the band shows no further measurable shift and then becomes narrow. The typical blue band is finally located at about 2.76 eV (450 nm) with a half-width of about 65 nm (0.39 eV). The RT external quantum efficiencies were found to be (2.5–7.6) $\times 10^{-4}$ in LEDs having an effective oxidized layer thickness of about 300 Å, with low drive voltages of 5–6 V for drive current densities of 15–20 mA mm^{-2}. The blue emission was uniform over the emitting area and an extended low-energy tail of the green emission was observed.

The dependence of the forward current of the blue band at RT on emission intensities between 2 and 70 mA is presented in Fig. 20.4. The results can be expressed as

$$L = I^n$$

where L is the luminescence intensity, I is the forward current and n denotes the positive number indicating the slope of the straight lines in the figure.

The emission intensity of the blue band shows a square-law dependence ($n = 2$) with current up to 30 mA, and with further increasing current a linear relation ($n = 1$) becomes predominant. The same characteristics have already been found in the MπS structure (π:insulating ZnS film) LEDs by Lawther and Woods.[5)]

In general, it is already known that the emission intensity of the donor–acceptor pair transition on the excitation intensity depends on the production of electron and hole densities trapped at the donor and acceptor levels, respectively, in the band-gap. It therefore seems likely that our observed square-law relation is due to bimolecular kinetics. The value of forward current which indicates no measurable shift of the donor–acceptor pair is found, just like the forward current, to change from a square-law to a linear relationship, as found in Fig. 20.3. This observed linear dependence should be dominated

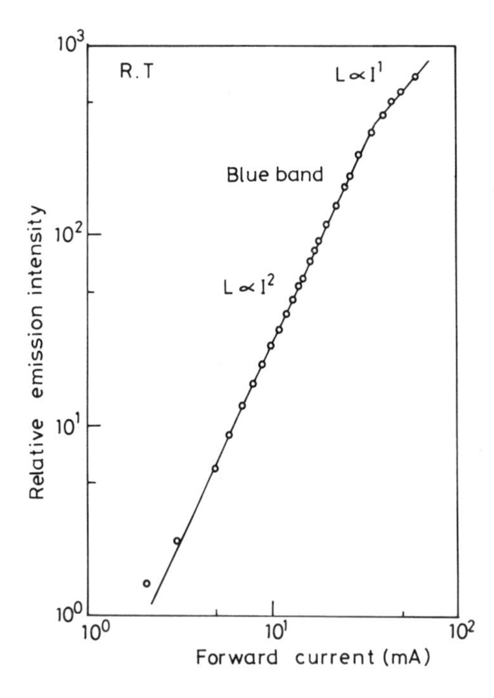

Fig. 20.4 Dependence of the forward current on the blue-emission band obtained at RT.

by a monomolecular kinetics, which can be supported by the fact that the blue band shows no measurable shift as a result of the ionization of the donor levels due to the high electric field.

20.4. Forward-bias electroluminescence characteristics

20.4.1. *Time-resolved EL*

A DC pulsed EL was excited using 0 to +10 V unipolar rectangular voltage pulses of width 10 μs to 10 ms and repetition rates of $1\text{–}10^5$ Hz generated from a Hewlett–Packard pulse generator (8013 B). Both voltage and current variations were monitored through the pulse using a 200 MHz oscilloscope. Time-resolved EL spectra were obtained at RT by a PAR 162 boxcar integrator coupled with a linear pulse amplifier using a cooled GaAs photocathode in connection with a CT-50 single-grating monochromator controlled by an NEC computer. The time resolution of the boxcar and detection system was about 1 μs and the spectral variations were obtained using a 6 μs gate.

Fig. 20.5 shows the time-resolved EL spectra obtained at RT using a 6 μs boxcar gate in an MπS ZnS LED, excited by 40 μs voltage pulses at a voltage of 2.5 V and current densities of 0.6 mA mm^{-2}. Just after an excitation pulse there appear two peaks at about 466 and 484 nm separated by about 18 nm from each other. With increasing delay time after the excitation pulse, both bands gradually shift towards higher photon energies and for delays above 200 μs no further measurable shift can be observed. This behavior is related to a donor–acceptor pair transition in nature, which has been already confirmed both in a forward-bias MπS LED under steady-state operation and photoluminescence spectra in n-type ZnS:I crystals. The peak positions of the blue EL bands are finally located at 472 and 490 nm, respectively.

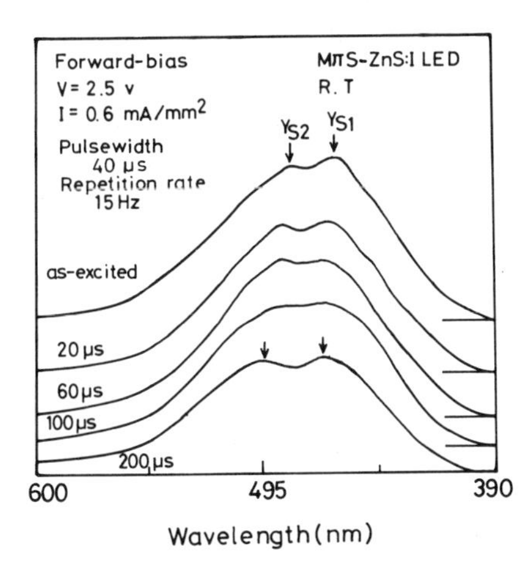

Fig. 20.5 Time-resolved EL spectra obtained at RT in the MπS diode under a forward-bias pulse voltage of 2.5 V.

It should also be noted that a weak shoulder at about 530 nm emerges during the time-resolved measurements. The present time-resolved EL characteristics give a direct proof of the donor–acceptor pair recombination both for the Y_{S1} and Y_{S2} bands, and suggest that the blue EL bands, which are predominant at relatively low current densities, consist of two types of donor–acceptor pair transitions.

With further increase in the forward current density across the diode, the EL spectra become intense and gradually shift towards higher photon energies. Fig. 20.6 shows the time-resolved EL spectra obtained at RT using 40 μs voltage pulses at a voltage of 7.5 V

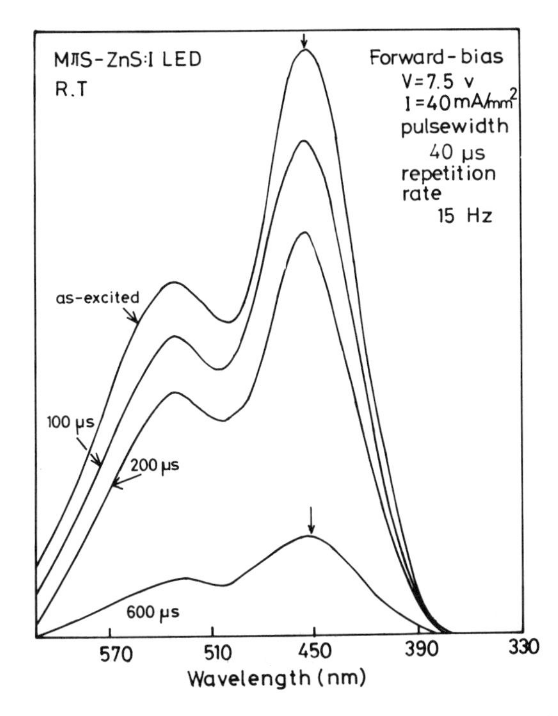

Fig. 20.6 Time-resolved EL spectra obtained at RT in the MπS diode under a forward-bias pulse voltage of 7.5 V.

and a current density of 40 mA mm^{-2}. The blue band obtained at 2.5 V in Fig. 20.5, is shifted towards higher photon energy at 7.5 V and is located at about 450 nm at RT. The low-energy extended tail, however, becomes predominant with increasing forward voltages and becomes apparent, showing a peak at 530 nm just after an excitation pulse. With increasing delay time after excitation, both blue and green bands show no measurable shift at all, but the 2.76 eV blue band shows a faster decrease in intensity than that of the 2.4 eV green band during the time-resolved measurement. It is therefore reasonable that the 2.76 eV blue band is attributed to an electronic transition between the conduction band and an acceptor level. We can estimate the acceptor level (E_a) to be about 0.85 eV above the valence band using $E_{\hbar\omega} = E_g - E_a + \frac{1}{2}kT$, where E_g is the band-gap (3.61 eV at RT), k is the Boltzmann constant and T is the sample temperature. The free-to-bound transition in the 2.76 eV band, which had previously been proposed in the steady-state EL spectra,[9)] can be confirmed by the present time-resolved studies.

From the time-resolved EL spectral measurements, it is revealed that the blue band changes its radiative recombination properties from a donor–acceptor pair to an electronic transition (free-to-acceptor) between the conduction band and the acceptor level located at about 0.85 eV. This acceptor has already been found in the PL spectra[9,13)] and its identity has been discussed.[9)]

20.4.2. *Temperature-dependent characteristics*

Fig. 20.7 shows the forward current–voltage ($I-V$) curves as a function of temperature, in which the shape of the characteristic changes remarkably with temperature.

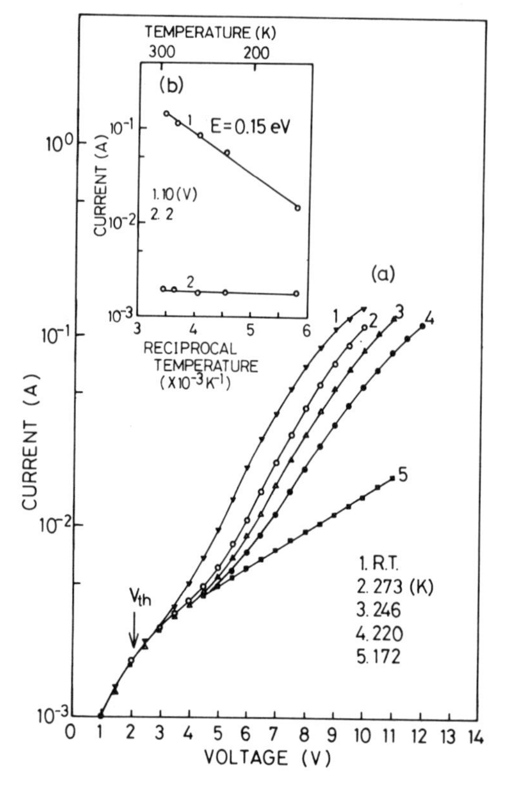

Fig. 20.7 $I-V$ characteristic at different temperatures of a forward-biased diode. The inset shows the temperature dependence of the forward current obtained at 2 and 10 V.

V_{th} in this figure indicates the threshold voltage (about 2.2 V) for the observation of the blue light emission. It is seen that the current increases approximately exponentially with applied voltage above V_{th} in forward-bias. These characteristics seem to be different from those of MIS diodes.[17] As the temperature decreases the $I-V$ characteristic becomes soft and may be divided into two exponential portions. As shown in the inset of Fig. 20.7, below V_{th} the electron current is independent of temperature, indicating that it is typical of a tunnel-recombination mechanism of the current. On the other hand, at high voltage near 10 V the activation energy for the electron current flow is estimated to be about 0.15 eV, which may correspond to a maximum potential-barrier height for electron current transport.

Fig. 20.8 illustrates the temperature-dependent blue luminescence intensity, measured from 280 to 420 K, under constant-current (1) and constant-voltage (2) conditions. As far as we are aware, it is believed that this is the first experimental report concerning the detailed temperature-dependent characteristics of blue EL in an MπS structure LED prepared from c-ZnS:I. Under a constant voltage of 6 V, the observed characteristics are divided into three regions, designated I, II and III.

In region I, as the temperature decreases the blue band drastically decreases in intensity and its activation energy for this decrease can be estimated to be 0.17 eV. In region II, on the other hand, the blue-emission intensity is independent of temperature, but with further increase of temperature the emission intensity increases monotonically; the activation energy for this increase is about 0.1 eV. It is particularly of note that the diode exhibits a maximum blue intensity at about 60°C. In region III, the blue-emission

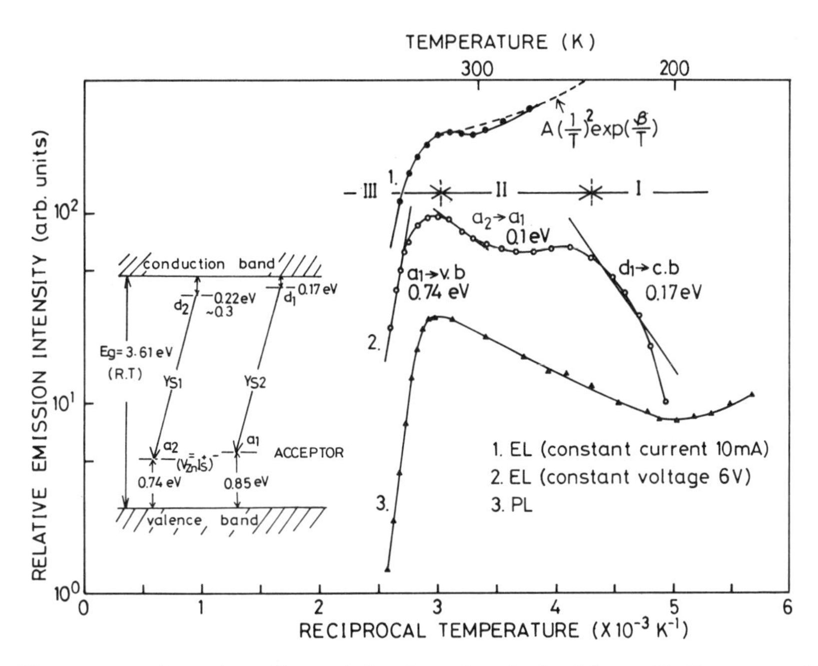

Fig. 20.8 Temperature-dependent blue-emission intensity obtained in an MπS structure blue LED under constant-current (15 mA) and constant-voltage (6 V) conditions. The PL intensity as a function of reciprocal temperature is also presented. The broken curve is calculated by Eq. (20.2). The inset represents a plausible recombination model in ZnS:I crystals encompassing two kinds of donor–acceptor pair centres.

intensity decreases remarkably due to thermal quenching; the activation energy for this decrease turns out to be 0.74 eV. It is therefore worth noting that the temperature-dependent characteristics are quite similar to those observed in the temperature-dependent blue photoluminescence.[8,13)]

We will discuss the results of the temperature-dependent blue-emission intensity observed in an MπS LED on the basis of the recombination models described in the inset of Fig. 20.8. Because of the increase in the resistivity of the bulk n-type ZnS when the temperature was reduced below 150 K, EL measurements were very difficult and none of the emission bands could be observed. The activation energy presented in region I corresponds to the ionization process of the donor levels concerned with the Y_{S1} and Y_{S2} blue band (d_1 or d_2 level$\rightarrow$ conduction band). On the other hand, the activation energy of 0.1 eV for the increase of intensity in the temperature range from 250 to 320 K is equivalent to the energy difference between two acceptor levels (at 0.76 and 0.85 eV) relating to the blue emission, implying thermal depopulation between them ($a_1 \rightarrow a_2$). Because of this effect, it is believed that the intensity of the 2.76 eV blue emission band appearing in the present MπS LED exhibits a maximum at about 60°C. The observed blue EL phenomena are similar to that of the temperature-dependent blue PL of low-resistive n-type ZnS:I. Moreover, the thermal quenching effect of the blue intensity ($a_1 \rightarrow$ valence band) occurring above 60°C yields an activation energy of about 0.74 eV, which is very close to the depth of the complex acceptor of the Y_{S1} PL band.

In general, the external radiative quantum efficiency (η_{ex}) of the diodes is given by the product

$$\eta_{ex} = \eta_r \eta_h \eta_{ie} \tag{20.1}$$

where η_r is the radiative recombination efficiency, η_h is the minority-carrier (hole) injection efficiency and η_{ie} is the ratio of internal to external efficiency.

If $\eta_r \eta_{ie}$ can be determined independently, a measurement of η_{ex} can give information about the hole injection efficiency and we can measure $\eta_r \eta_{ie}$ in photoluminescence at similar excitation densities to the EL. Curve 2 in Fig. 20.7 shows the blue light emission, which is proportional to the quantum efficiency, against temperature, and curve 1 gives information concerning the hole injection efficiency against temperature. On the other hand, curve 3 can give a measurement of $\eta_r \eta_{ie}$ as a function of reciprocal temperature. It is probable that the observed temperature variation of the quantum efficiency (η_{ex}) is largely determined both by the radiative recombination efficiency η_r and hole injection efficiency η_h. In particular, η_h has a strong temperature dependence because of the strong dependence of $\eta_r \eta_{ie}$. It seems that η_{ex} is nearly constant between 250 and 320 K as shown in this figure. In order to give the approximate constancy of η_{ex}, η_r must have an exactly inverse temperature dependence on the range 250–320 K for η_h. We can understand that the constancy of η_{ex} is deduced from Eq. (20.1) using the above experimental temperature dependences.

The minority-carrier (hole) injection efficiency in the MIS diode is expressed theoretically by[18)]

$$\eta_h \cong A(1/T)^2 \exp(\beta/T), \tag{20.2}$$

where A is a fitting parameter and β is a barrier height. The broken curve shows a typical example substituting $A = 2.5 \times 10^5$ and $\beta = 2.0$ eV in Eq. (20.2). It is therefore concluded that both the electron and hole currents are determined by mechanisms which include strong thermal activation.

20.4.3. *Electron current flow and injection processes*

The main problem for obtaining efficient LEDs is to provide an effective injection of minority carrier (holes) into semiconductors. Although a few proposals[5,19,21] have been reported so far, an unambiguous model of the injection processes responsible for the forward bias EL in the MIS or MπS structure has not yet been given.

Two models have been proposed: (1) The essential feature of the forward-bias EL model which Lawther and Woods suggested[5] is the injection of minority carrier (holes) at the insulator–semiconductor interface by the migration of hot holes from the metal via the valence band of the insulator (Fig. 20.9, ①). The measured η_{ex} of the forward-bias LEDs was shown to be a strongly varying function of the photoelectric barrier height, which in turn is a function of the electronegativity of the metal contact. Thomas et al.[22] have observed that on the basis of this concept the efficiency of the injection of the minority carriers can be increased by using a metallic material (SN_x) with an electronegativity higher than that of Au. (2) In contrast Lukyanchikova et al.[20] have claimed that the EL in MπS LEDs can be interpreted by an Auger recombination process on the contact surface of the semiconductor, not in the metal (Fig. 20.9, ②).

The blue quantum efficiency and the intensity of the 2.76 eV blue EL band were found to decrease in our diodes if the thickness of the π-layer was greater than about 400 Å. It may therefore be suggested that the effective thickness of the π-layer corresponds to the hole-diffusion distance being about 400 Å in our diodes prepared from ZnS:I crystals. Assuming that the forward-biased EL follows the injection of holes into bulk ZnS,[5] electrons entering the metal electrode with a high potential energy of about 2 eV may create deep hole states below the Fermi level in the metal and then drift to the interfacial π-layer between the Au electrode and n-type ZnS. When the π-layer thickness is greater than the hole-diffusion length, it seems that the probability of hole-injection becomes extremely low.[13] For sufficiently large forward-bias voltages, the hole-injection current is expressed as follows[18]

$$J_h = A \exp(-\chi_h^{1/2} d) V_b^2 \tag{20.3}$$

where A is a material constant, χ_h is the average barrier height of the π-layer to holes tunnelling into n-ZnS from the metal, d is the thickness of the π-layer and V_b is the forward-bias voltage.

J_h decreases with the π-layer thickness for a given constant voltage as a result of the increase in the tunnel exponent $\chi_h^{1/2} d$. We therefore believe that the forward-biased minority carrier injection luminescence is mainly due to the hole-tunnelling process and as a result an appropriate value of the quantum efficiency exists.[21,22]

The variations of the quantum efficiencies measured at a constant forward current density (0.2 A cm^{-2}) as a function of π-layer thickness are shown in Fig. 20.10. The data obtained by Lawther are also described in this figure. Both sets of data represent an increase in the external quantum efficiencies for a decrease in the π-layer thickness. However, in the case of the π-layer composed of semi-insulating ZnS, the decrease of the quantum efficiencies with increasing the π-layer thickness is not so drastic and follows a

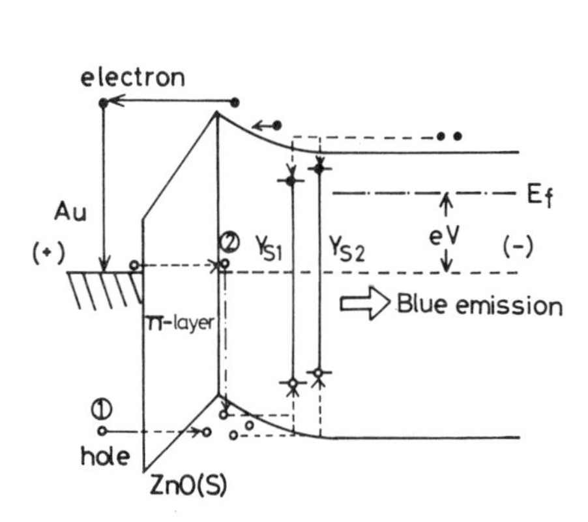

Fig. 20.9 A simplified energy band diagram in the forward-bias MπS diode: hole injection and subsequent recombination processes (① by Lawther and Woods[5]) and ② by Lukyanchikova et al.[19]).

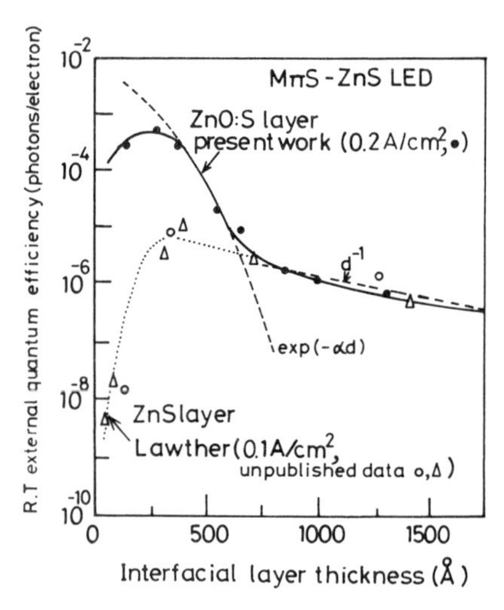

Fig. 20.10 Room-temperature external quantum efficiencies obtained in the MπS LEDs as a function of interfacial π-layer thickness. The broken curve is calculated by Eq. (20.3). The dotted curve was obtained by Lawther in the LEDs having i-ZnS π-layers (unpublished).

d^{-1} relation. On the basis of Lawther's expectation (model ① in Fig. 20.9), holes must tunnel through the π-layer to come to the semiconductor side. When i-ZnS was used as the π-layer it has been found that the hole-injection probability is not so affected by the π-layer thickness, since the tunnelling-barrier height for holes is estimated to be about 1.6 eV[23]) which is smaller than the deep hole state (about 2 eV below the metal Fermi level) created by the Auger recombination due to hot-electron–electron interactions.

On the other hand, in our case, the hole-tunnelling barrier[23]) height for the Au/ZnO:S π-layer system is estimated to be about 2.7 eV, which is larger than that of the Au/i-ZnS π-layer system. It is evident that the decrease of the quantum efficiencies becomes clearly visible when the thickness of the π-layer is larger than that of the hole-diffusion length, and qualitatively fits $\exp(-\alpha d)$ curve shown in Fig. 20.10, where α is $\chi_h^{1/2}$.

We conclude that our MπS diode composed of a ZnO:S π-layer shows a marked decrease of the quantum efficiency for the π-layer thicknesses greater than about 500 Å.

20.4.4. *Degradation of electroluminescence intensity*

An investigation of the degradation characteristics of the diodes is very important to develop useful LED production from a practical point of view. In fact, we found that after an MπS LED was operated under forward-bias conditions, the 2.76 eV blue-emission band decreased in intensity following two characteristic curves as shown in Fig. 20.11.

After passing a forward current of 20 mA for 3 h, the emission intensity decreased greatly by 60% of the initial value and then monotonically decreased to 45% of the pre-operation value. After about 10 h, it was found that the EL intensity was nearly constant, keeping one-half of its pre-operation value, and that this degradation was semi-permanent.

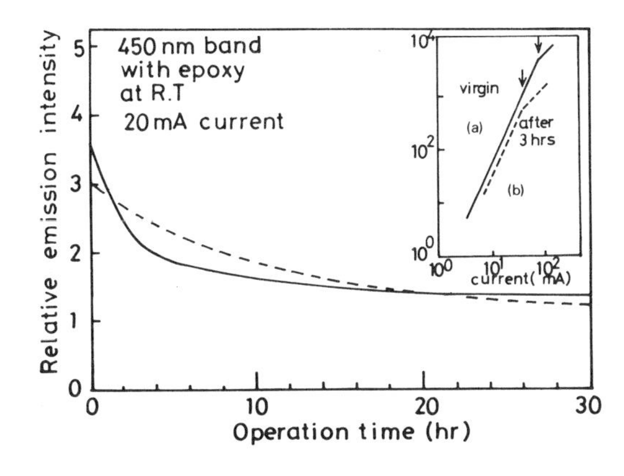

Fig. 20.11 Degradation characteristic (solid curve) of the blue-emission intensity as a function of operating time. The broken curve is given by Eq. (20.4). The inset shows the dependence of forward current on the blue-emission intensity obtained at RT before and after the LED operation.

The inset of Fig. 20.11 shows the change of the forward-current dependence of the blue-emission intensity before and after operation for 3 h. Curve (a) was obtained in the virgin LED, whilst after operation the characteristic was changed (b) and the emission intensity was found to decrease compared with the initial value at the same forward current. In particular, the point of the changeover from the square dependence to the linear dependence of the intensity on the forward current was shifted to the lower-current side (from 70 mA to about 35 mA).

The broken curve shows the degradation characteristics of the 2.76 eV emission intensity described by following the empirical relation[24] as a function of operation time:

$$B = B_0/1 + t/t_c, \tag{20.4}$$

where B_0 is a constant parameter and t_c is defined as a time that the initial intensity is decreased to its one-half value. In this case, we put $t_c = 18$ h and $B_0 = 3$ in Eq. (20.4).

We feel that t_c seems to be very short compared with that of GaP green LEDs,[25]

Table 20.1 Evaluation of the room-temperature external quantum efficiencies in various forward-biased ZnS blue LEDs. Characteristics of GaN and SiC diodes are also preserved.

Diode	Structure metal/interfacial layer/n-ZnS	Ref.	Blue-band peak (eV)	Halfwidth (nm)	Efficiency at RT (photons/electron)
ZnS MIS	Au/without a layer/melt-and vapor-grown ZnS	4)	2.7		10^{-3}
ZnS MIS	Au/NaI/melt-grown	21)	2.65	136	1.1×10^{-5}
ZnS MIS	Au/SiO_2/melt-grown	21)			1.7×10^{-7}
ZnS MIS	Au/vacuum treatment/melt-grown (Al)	3)	2.67	130	5×10^{-4}
ZnS MπS	Au/S^+-ion implantation/ZnS:I	20)	2.67	99	2×10^{-6}
ZnS MπS	SN_x/i-ZnS/CVD ZnS	22)	2.81	57	$>3\times10^{-4}$
ZnS MπS	Au/ZnO:S/ZnS:I	8)	2.76	65	8×10^{-4}
ZnS MπS	Au/O^+-ion implantation/ZnS:I	13)	2.7	100	6×10^{-5}
GaN MπS		26)	2.52	95	2×10^{-4}
SiC p-n		27)	1.61	75.5	1.3×10^{-4}

because serious degradation takes place within 3 h. After the degradation of the diode the zero-bias capacitance was much reduced and as a result the intercept voltage in the $C^{-2}-V$ plot becomes larger. Although the reason why the blue EL intensity was degraded remains unresolved, it is tentatively suggested that the decrease of the blue intensity may be related to the degradation of the π-layer as a result of which the hole injection efficiency becomes low.

Table 20.1 summarizes the characteristics of the ZnS MIS and MπS LEDs which have been published up to now and we can compare these with those of GaN and SiC. It should be noted in this regard that the efficient blue emission characteristics and the narrowest band-width presented are superior to those in GaN[26)] and SiC[27)] blue LEDs.

20.5. Conclusion

The characteristics of the ZnS blue LED based on ZnS single crystals grown by the iodine-transport method, with the MπS structure are as follows:

(1) It has been shown that the surface states on the cleaved (110) surfaces of n-ZnS crystals grown by the iodine-transport method are strongly affected by the oxidation procedures, which suggests that the chemical oxidation layer is composed of a compensated high-resistive ZnO layer with a large concentration of S atoms of about $10^{19}\,cm^{-3}$.

(2) Utilising the oxidized interfacial layer positively for the production of an efficient blue LED with external quantum efficiencies as high as 8×10^{-4} photons/electron at RT can be obtained.

(3) The efficient 450 nm (2.76 eV) blue-emission band is ascribed to an electronic transition between the conduction band and an acceptor level located at 0.85 eV above the valence band.

(4) It is suggested that the forward-bias injection electroluminescence is caused by the tunnelling-assisted minority carrier injection process, and can be interpreted by the recombination properties of the blue emission in the bulk region, as inferred from the photoluminescence characteristics of low-resistive n-type ZnS.

(5) The observed quantum efficiencies have a strong temperature dependence and show a marked decrease for the π-layer thickness greater than 500 Å. It is found that the LED is markedly degraded for a short period and results in the decrease of the blue-emission intensity during forward bias operation.

Acknowledgements

The author is most grateful to Prof. Akio Hiraki of Osaka University for the kind invitation to present this review. Particular thanks for advice, discussion and assistance are due to Prof. Masaharu Aoki of Tokyo Science University, Prof. Brian Ray of Coventry Polytechnic, Dr. Clifford Lawther, Mr. Toshiya Yokogawa of Osaka University and Mr. Mamoru Satoh of Government Industrial Research Institute Osaka. He acknowledges and expresses his gratitude to the Hattori Hokokai, Kurata, Kudo and Nippon Sheet Glass Foundations for financial support of ZnS blue LED investigations. A part of this work was also supported by Grant-in-Aid for Scientific Research for the Ministry of Education.

References

1) M. Aven and J. L. Devine: J. Luminescence, 7 (1973) 195.
2) Y. S. Park and B. K. Shin: Topics in Applied Physics, Vol. 17, Ed. J. Pankove (Springer-Verlag, New York, 1977) p. 133.
3) H. Katayama, S. Oda, and H. Kukimoto: Appl. Phys. Lett., 27 (1975) 697.
4) N. B. Lukyanchikova, G. S. Pekar, N. N. Tkachenko, Hung Mi Shin, and M. K. Sheinkman: Phys. Stat. Solidi, (a)41 (1976) 299.
5) C. Lawther and J. Woods: Phys. Stat. Solidi, (a)50 (1978) 491.
6) K. Hirahara, A. Kamata, M. Kawachi, T. Sato, and T. Beppu: Extended Abstracts of 15th Conf. on Solid State Devices and Materials (1983) p. 349.
7) T. Taguchi and D. W. Palmer: Phys. Stat. Solidi, (a)69 (1982) 55.
8) T. Taguchi, T. Yokogawa, S. Fujita, M. Satoh, and Y. Inuishi: J. Cryst. Growth, 59 (1982) 317.
9) T. Taguchi and T. Yokogawa: J. Phys., D17 (1984) 517.
10) T. Yokogawa, T. Taguchi, S. Fujita, and M. Satoh: IEEE Trans. Electron Devices, ED-30 (1983) 271.
11) S. Fujita, H. Mimoto, H. Takabe, and T. Noguchi: J. Cryst. Growth, 48 (1979) 326.
12) W. Palosz: J. Cryst. Growth, 60 (1982) 57.
13) T. Taguchi, T. Yokogawa, S. Fujita, M. Satoh, and Y. Inuishi: Physica, 116B (1983) 503.
14) T. Taguchi, T. Yokogawa, and H. Yamashita: Solid State Commu., 49 (1984) 551.
15) L. T. Brillson: Surface Sci., 69 (1977) 62.
16) A. M. Cowley: J. Appl. Phys., 37 (1966) 3024.
17) A. W. Livingstone, K. Turvey, and J. W. Allen: Solid State Electron., 16 (1972) 351.
18) W. A. Harrison: Phys. Rev., 123 (1961) 85.
19) N. B. Lukyanchikova, T. M. Pavelko, G. S. Pekar, N. W. Tkachenko, and M. K. Sheinkman: Phys. Stat. Solidi, (a)64 (1981) 697.
20) C. Lawther, S. Fujita, and T. Takagi: Japan. J. Appl. Phys., 19 (1980) 939.
21) L. G. Walker and G. W. Pratt, Jr.: J. Appl. Phys., 47 (1976) 219.
22) A. E. Thomas, J. Woods, and Z. V. Hanptoman: J. Phys., D 16 (1983) 1123.
23) S. M. Sze: Physics of Semiconductor Devices (Wiley, New York, 1969).
24) S. Roberts: J. Opt. Soc. Amer., 43 (1953) 590.
25) R. L. Hartman, B. Schwartz, and M. Kuhn: Appl. Phys. Lett., 18 (1971) 304.
26) G. Jacob, M. Boulou, and D. Bois: J. Luminescence, 17 (1978) 263.
27) K. Hoffman, G. Ziegler, D. Theis, and C. Weyrich: J. Appl. Phys., 55 (1982) 6962.

21 BLUE LIGHT-EMITTING ZnSe DIODE

Jun-ichi NISHIZAWA† and Yasuo OKUNO*

Abstract

Pure blue light emission has been obtained from ZnSe p–n junctions. ZnSe crystals are grown by the temperature difference solution growth method under controlled vapor pressure (TDM-CVP) developed by one of the authors.[1,2] p-type ZnSe crystals are grown from a Se solution by doping with a group-I element under controlled Zn pressure. The resistivity of the p-type crystal depends on the Zn pressure applied during growth and becomes a maximum at a certain Zn pressure. A p–n junction is made by the formation of an n–type layer by Ga diffusion under controlled vapor pressure into a p-type crystal. The fundamental properties of the p–n junction are as follows: the values of n in the I–V characteristics range from 1.4 to 1.8 and the diffusion potential is determined as 2.5 to 2.7 eV. The emission spectrum from a p–n junction depends on the applied vapor pressure and blue light emission without any deep level emission can be obtained under the optimum pressure treatment. The wavelength of peak is 460 nm at 77 K and 480 nm at 300 K, and the brightness is 2 mcd at 2 mA.

Keywords: TDM-CVP, II–VI Compound, p-type ZnSe, Blue LED, ZnSe p–n Junction

21.1. Introduction

II–VI compound semiconductors are one of the promising materials from which it may be possible to fabricate devices giving efficient injection luminescence in the green to blue color band. In particular, ZnSe has an energy gap of 2.67 eV at room temperature and the blue emission is expected to be from a band-to-band transition different from that in the ZnS crystal whose emission is obtained through deep levels which lowers the efficiency.

The two basic reasons why diodes of this compound have not been made industrially, are as follows:

(1) crystals of high crystallographic quality have not been obtained because no suitable growth method has been exploited;

(2) stable p–n junctions have not been obtained, hence it is difficult to prepare a p-type conduction crystal.

ZnSe crystals have usually been grown by the conventional melt growth method[3] and also solution growth methods, in which metals such as Ga,[4] In,[5] Sn,[6] Zn,[7] Bi[8] and Te[9] were used as solvent and the slow cooling of temperature segregated the ZnSe crystal have been published. However, crystals of satisfactory crystallographic quality and purity for visible light emission have not been obtained and the emission peak through deep level was

† Research Institute of Electrical Communications, Tohoku University, Sendai, 980.
* Semiconductor Research Institute, Kawauchi, Sendai, 980.

JARECT Vol. 19, Semiconductor Technologies (1986), *J. Nishizawa (ed.)*
OHMSHA, LTD. and North-Holland

dominant in almost all the diodes which have been fabricated up to now. Therefore, the guiding principle has been to seek a crystal growth technique to obtain a p–n junction overcoming the two points mentioned above.

In 1971, J. Nishizawa et al. published an experimental result[1,2] which showed the influence of the applied vapor pressure of As on top of the molten Sn on the characteristics of the segregated GaAs crystal on the substrate which settled at the lower-temperature part of the crucible under the solution. A temperature difference between the molten phase and the substrate accelerates the diffusion of dissolved GaAs towards the substrate where it segregates. This method, called the temperature difference method (TDM) under controlled vapor pressure (CVP), has been extended to the growth of GaAs and GaP using Ga solution.[10–14] A vapor pressure applied on top of the solution should not have any influence on the segregated crystal except for its growth rate. However, there exists the obvious influence on the character of the crystal, independent of the growth rate, which can be controlled by the temperature difference between the substrate and molten phase, not only by the vapor pressure.

This method was also applied for the growth of polycrystalline GaP and named as SSD by N. Watanabe et al.[15]

The authors' group conducted much experimental work on GaAs and GaP and concluded that there seems to be a control of non-stoichiometric crystal. This was described in detail in Chapter 2 of this volume.

This chapter describes the fabrication of a ZnSe p–n junction by the diffusion of n-type impurity into a p-type crystal under CVP. The dependence of the emission properties from the ZnSe p–n junction fabricated under CVP on the applied vapor pressure is investigated. The pure blue light emission from diodes made under the optimum pressure condition is reported.

21.2. Crystal growth of p-type ZnSe and formation of p–n junction

ZnSe crystals were grown from Se solvent under Zn pressure by TDM-CVP. Details of crystal growth are described in Chapter 2 of this book. The crystals grown were about 10 mm long yellowish transparent single crystals. A p-type ZnSe crystal can be grown by doping of Li in the Se solution. Its typical carrier concentration and Hall mobility measured by the van der Pauw method are $p = 6.4 \times 10^{12}\ \mathrm{cm}^{-3}$, $\mu_p = 78\ \mathrm{cm^2/V\,s}$ for one sample and $p = 3 \times 10^{15}\ \mathrm{cm}^{-3}$, $\mu_p = 20\ \mathrm{cm^2/V\,s}$ for another sample. These values depend on the doping concentration of Li and the Zn vapor pressure applied during the growth. The optimum Zn pressure applied during growth is determined as 7.2 atm at the growth temperature of 1050°C from the Zn pressure dependence of the resistivity and the photoluminescence intensity (see Chapter 2).

The p–n junction is produced by the formation of an n-type layer in the grown p-type crystal. The as-grown crystal is sliced into sections of 1 mm thickness and is mechanically polished, followed by etching in a Br–methanol solution. The slices are then placed in a Zn solution containing Ga, while Se is put in the vapor pressure chamber of a quartz tube with a two temperature zone, which is fundamentally the same system as that of TDM-CVP. The diffusion of Ga is performed at 740°C for 1 h when an n-type layer of 5–7 μm thickness can be obtained. The dependence of the electrical and optical properties of the p–n junction on the applied vapor pressure has been investigated.

21.3. Fundamental properties of p–n junctions

The properties of a p–n junction depend on its fabrication process, and are related to the CVP as follows:

(1) the growth conditions of the p-type crystal, i.e. the relation between the growth temperature and the value of applied Zn pressure and doping concentration of Li.

(2) the conditions for the formation of the n-type layer.

Usually, the resistivity of the as-grown n-type crystal grown by the conventional growth method is higher than $10^{12}\ \Omega\,\mathrm{cm}$ and the crystal is heat-treated in molten Zn to reduce the resistivity before fabrication of the devices. This technique is introduced in the formation of the n-type layer, and p–n junctions are formed by heat-treatment in molten Zn for 1 h at 1000°C and at 1100°C.

The photocapacitance spectra of these diodes and of a p–n junction fabricated by Ga diffusion in molten Zn under controlled Se pressure using a p-type crystal grown under the optimum Zn pressure is shown in Fig. 21.1. The appearance of deep levels is related to the diffusion conditions and the density of deep levels increases with the diffusion temperature. It is noted that deep levels due to the deviation from stoichiometry are produced even after annealing for 1 h. Therefore, even in the diffusion procedure, CVP is important to maintain stoichiometric composition, the same as is used for stoichiometry control during growth.

21.3.1. *Diode character*

The fundamental properties of the p–n junction have been measured. A photograph of an $I-V$ curve traced on a cathode ray tube is shown in Fig. 21.2. The relation of log I

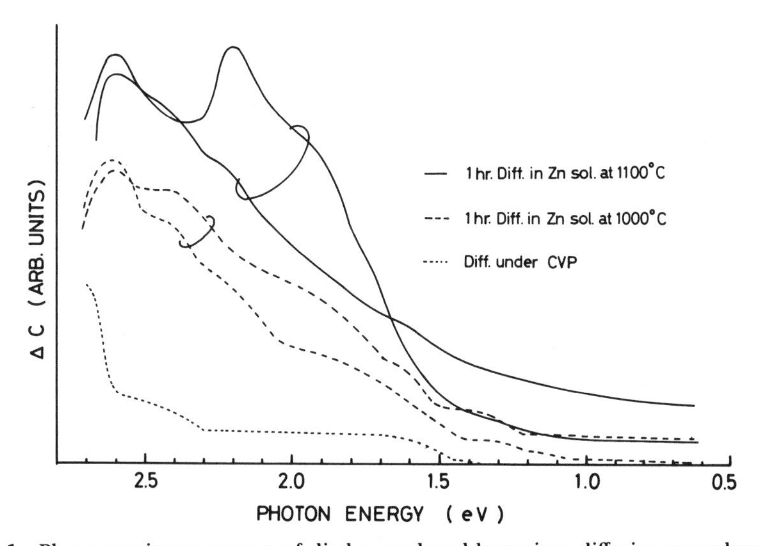

Fig. 21.1 Photocapacitance spectra of diodes produced by various diffusion procedures.

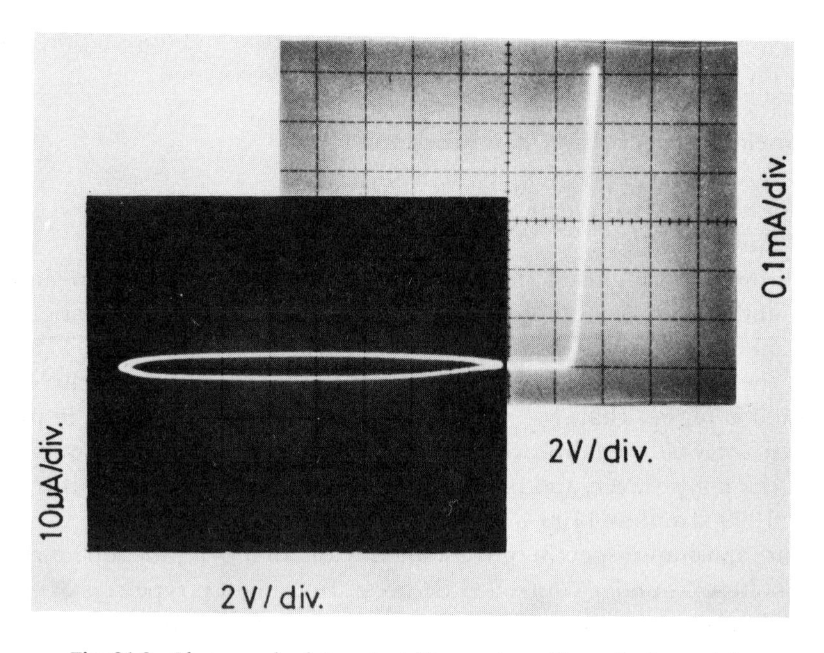

Fig. 21.2 Photograph of I against V curve traced by cathode ray tube.

against V in almost all the diodes is linear in the current range of 10^{-11} to 10^{-3} A. The value of n in the I–V characteristics expressed by

$$I = A\{\exp(qV/nkT) - 1\}$$

ranged from 1.4 to 1.8 in the 300 diodes fabricated. This result shows that the diodes obtained are different from a metal–semiconductor junction for which the value of n is 1 to 1.2.

The diffusion potential of a p–n junction, which is determined from the relation of $1/C^2$ against V, ranges from 2.4 to 2.8 eV as shown in Fig. 21.3. These values correspond to the energy gap of ZnSe at 300 K. Also, the diffusion potential determined from the measurement of the zero-volt resistance[16)] in order to provide further clarification is 2.5 to 2.7 eV. The diffusion potentials measured by the two different methods corresponds well to the energy gap of ZnSe.

The distribution of impurity concentration in the neighborhood of the junction are measured from the C–V characteristics under irradiation of monochromatic light corresponding to the photon energy of the peaks in the photocapacitance spectrum at room temperature. The density distribution in the diodes whose spectrum has deep level emission at longer wavelength is shown in Fig. 21.4, for sample GHP-80 grown under the optimum Zn pressure, and GHP-28 and -37 grown under higher Zn pressure. The density of shallow impurities ranges from 10^{15} to $10^{16}/\text{cm}^3$, independent of the Zn pressure applied during the growth. However, the appearance of deep levels depends on the Zn pressure and the deep level density of GHP-80 is less than $10^{12}/\text{cm}^3$.

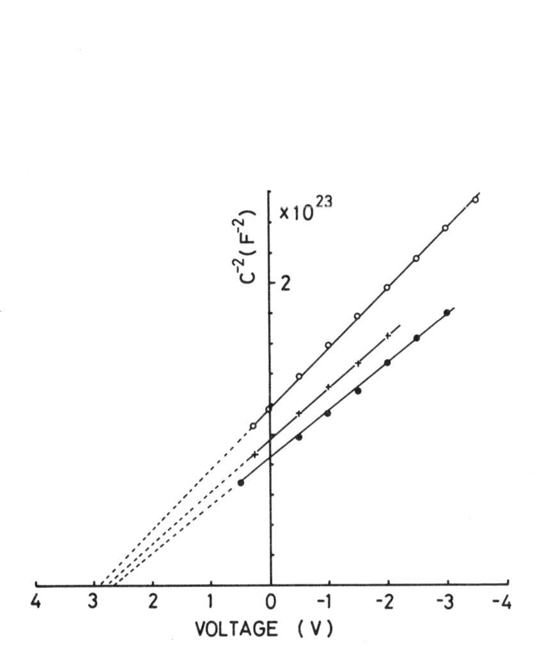

Fig. 21.3 Relation of $1/C^2$ against V. The diffusion potential can be determined as 2.4 to 2.8 eV.

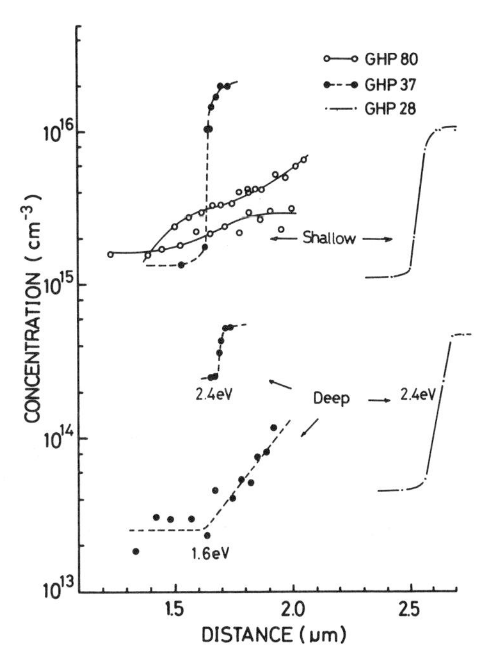

Fig. 21.4 Distributions of shallow impurities and deep levels.

21.3.2. *Emission spectra of p–n junction*

The emission characteristics of the diode under forward bias have been investigated. Emission spectra from a p–n junction have been measured at several points in the temperature range 77 to 300 K. Fig. 21.5 shows typical spectra at each temperature of diodes fabricated from crystal grown under the application of a Zn pressure deviating from the optimum pressure. Generally, the spectrum has a sharp peak in the wavelength

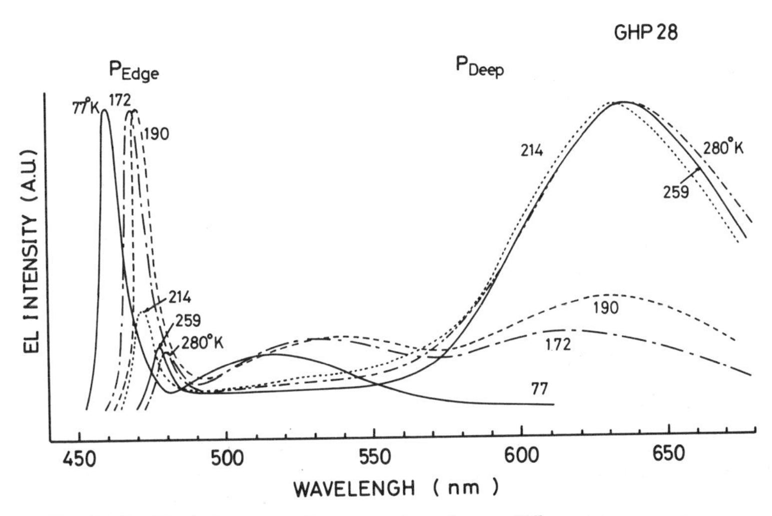

Fig. 21.5 Typical spectra from p–n junction at different temperatures.

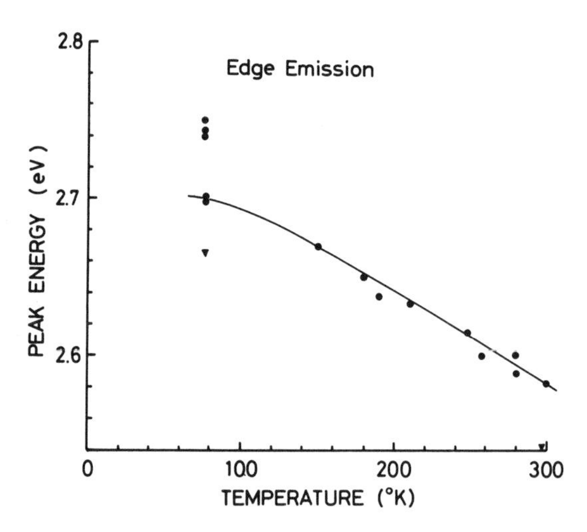

Fig. 21.6 Temperature dependence of the peak energy of near edge emission (P_{edge}).

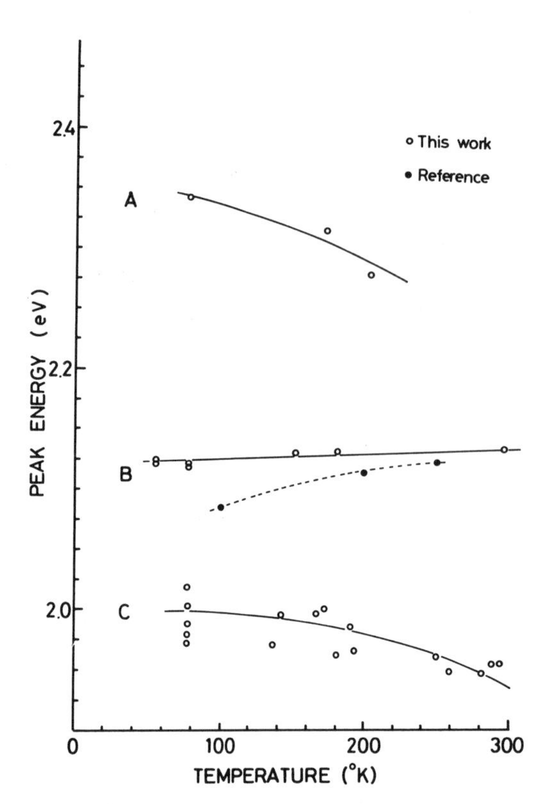

Fig. 21.7 Temperature dependence of the peak energy of deep levels (P_{deep}). ● shows the published data.[17]

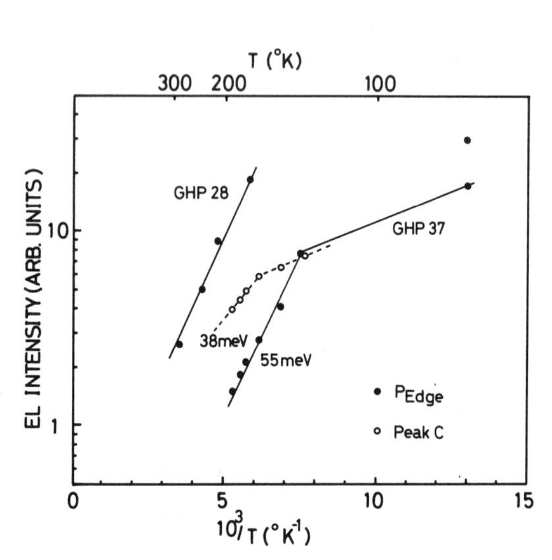

Fig. 21.8 Relation between the EL intensities and the reciprocal of temperature. The activation energies are determined as 55 and 38 meV for P_{edge} and peak C, respectively.

region corresponding to the band energy and broad peaks from deep levels in the longer-wavelength region. The two peaks are named P_{edge} and P_{deep}, respectively. The intensity of P_{deep} in most diodes is very small at temperatures lower than 172 K but it increases at higher temperatures. On the other hand, the intensity of P_{edge} decreases at higher temperatures and disappears at temperatures higher than 280 K in some diodes. Therefore, the color of emission from diodes fabricated under conditions deviating from the optimum pressure changes violet to orange as the temperature rises.

The temperature dependence of the peak energies and emission intensity of each emission peak has been examined. The temperature dependences of the peak energies of P_{edge} and P_{deep} in the spectrum are shown in Figs. 21.6 and 21.7, respectively. Fig. 21.6 shows the dependence of P_{edge}: the maximum peak shifts to lower energy as the temperature increases. A small peak at higher energy than the main peak at 77 K disappears at above 120 K and shows a shoulder in the main peak. On the other hand, deep levels appearing in the spectrum can be classified into three (peaks A, B and C) as shown in Fig. 21.7. The peak A at highest energy can be observed at lower temperatures. No diodes showed peak A at room temperature. The energy of peak B changes almost with temperature and its energy is nearly the same value as the published data[17)] whose peak seems to be self-activated emission. The peak C has a clear temperature dependence, can be observed in almost all the diodes, and shifts to lower energy by about 40 meV at room temperature.

The activation energies of P_{edge} and peak C which are observed in many diodes are determined from the relation between the peak intensities and the reciprocal of temperature. Typical data are shown in Fig. 21.8 and typical values are 55 meV for P_{edge} and 38 meV for peak C.

21.4. Vapor pressure dependence of emission properties of diodes

The relation of non-stoichiometry of crystal, i.e. the value of applied Zn pressure during crystal growth of p-type ZnSe, and the emission intensity from deep levels in the injection luminescence has been investigated.

First, the relation between the applied Zn pressure during growth and the intensity of P_{edge} and the intensity ratio of P_{deep} to P_{edge} in the emission spectrum (cf. Fig. 21.5) at 77 K has been investigated. The n-type layer of diodes used in this measurement is formed at the same temperature as without CVP. Therefore, the vapor pressure dependence of Zn applied during the crystal growth of p-type ZnSe has been examined. When several P_{deep} appear in the spectrum, the intensity of the highest one is selected.

Fig. 21.9 shows the relation between the intensity ratio of P_{deep} to P_{edge} and applied Zn pressure. The intensity ratio is closely related to the applied Zn pressure, and the emission peak from deep levels is not observed in the diode fabricated from the crystal grown under the application of the optimum Zn pressure (GHP-80). The spectrum shows a single peak at about 2.7 eV (~ 460 nm) and the emission color is blue-violet. In this figure, the growth temperature of GHP-61 and -69 deviates by 5 to 7°C from the other one. Therefore it is expected that the optimum pressures shifts downwards for GHP-61 and upwards for GHP-69. It can be said that the growth temperature range for which stoichiometry can be realized by the application of the optimum Zn pressure is very narrow. On the other hand,

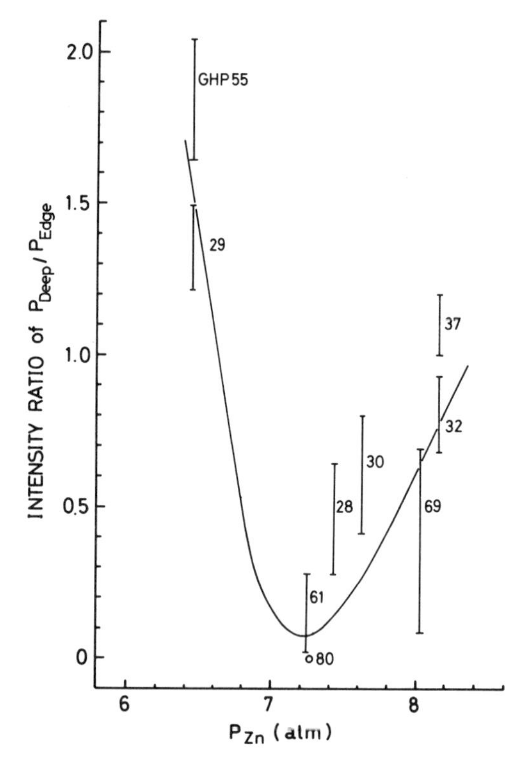

Fig. 21.9 Relation between the intensity ratio of P_{deep}/P_{edge} and applied Zn pressure. This value becomes a minimum at the optimum Zn pressure.

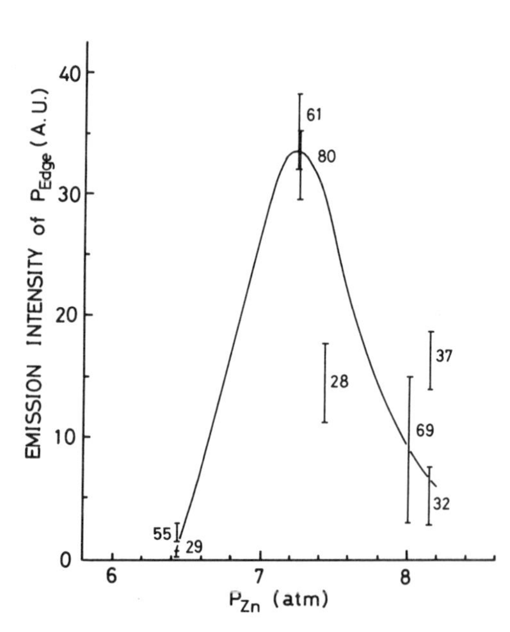

Fig. 21.10 Dependence of the emission intensity of P_{edge} on the applied Zn pressure. This value shows a maximum at the optimum Zn pressure.

the emission intensity of P_{edge} shows a maximum with the application of the optimum Zn pressure as shown in Fig. 21.10.

Therefore it becomes clear that the effect of applying Zn pressure is to suppress the emission from deep levels and to realize a blue light emitting diode. However, in this experiment, the controlled vapor pressure is not used in the formation of n-type layer, and only the optimum pressure for the growth of p-type crystal has been determined from the measurement.

The optimum Zn pressure determined from this experiment is in good agreement with the optimum pressure at which the resistivity becomes a maximum. Therefore, the diode fabricated from the crystal grown under the optimum Zn pressure shows the emission color of blue to violet at 77 K, but orange at room temperature where the emission from deep levels is dominant. Therefore, it is important in order to obtain the blue emisssion at room temperature that the n-type layer should be formed under the controlled vapor pressure.

The dependence of the wavelength of P_{edge} in the luminescence spectrum at 77 K on the applied Zn pressure is shown in Fig. 21.11. The wavelength of the peak is located at 460 ± 1 nm and is independent of the applied Zn pressure. Also, no shift of peak energy due to the difference of doping concentration of Li in the experiment is observed.

On the other hand, the behavior of deep levels is now under investigation.

The effect of controlled vapor pressure during the diffusion of Ga has been examined.

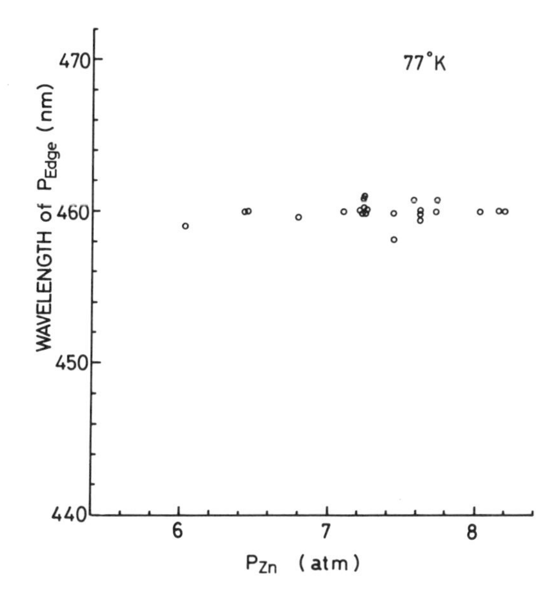

Fig. 21.11 Zn pressure dependence of the wavelength of P_{edge} in the electroluminescence spectrum at 77 K.

The diffusion of Ga is carried out under controlled Se pressure in the Zn solution. All the crystals used in this experiment were grown under the optimum Zn pressure of 7.2 atm at $T_g = 1050°C$ as discussed above.

The emission spectrum at 77 K from the p–n junction fabricated by diffusion under CVP shows a sharp single peak at the wavelength of 460 nm in all the diodes. The difference between the spectra at room temperature has been investigated. Emission peaks due to deep levels (P_{deep}) other than that due to edge emission (P_{edge}) are observed in the room-temperature spectrum as shown in Fig. 21.12.

The dependence of the edge emission (P_{edge}) intensity and the intensity ratio of P_{deep}/P_{edge} on the applied Se presssure during diffusion are shown in Figs. 21.13 and 21.14. The intensity ratio of P_{deep}/P_{edge} depends on the Se pressure applied during diffusion. It shows a minimum value at a certain Se pressure (optimum Se pressure). On the other hand the emission intensity of P_{edge} becomes a maximum when the same Se vapor pressure is

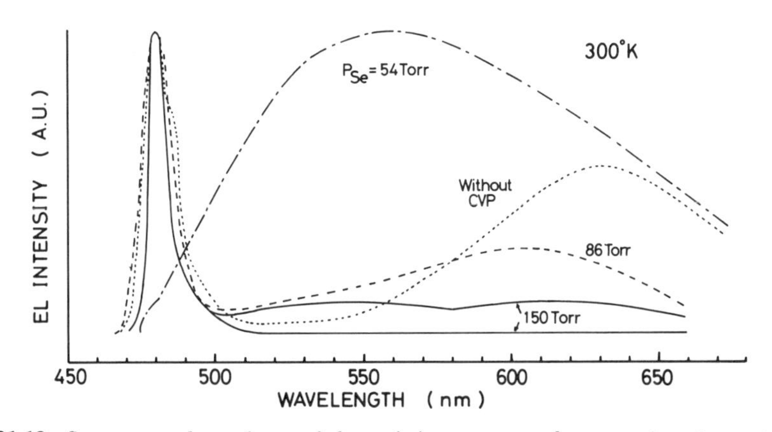

Fig. 21.12 Se pressure dependence of the emission spectrum from p–n junction at 300 K.

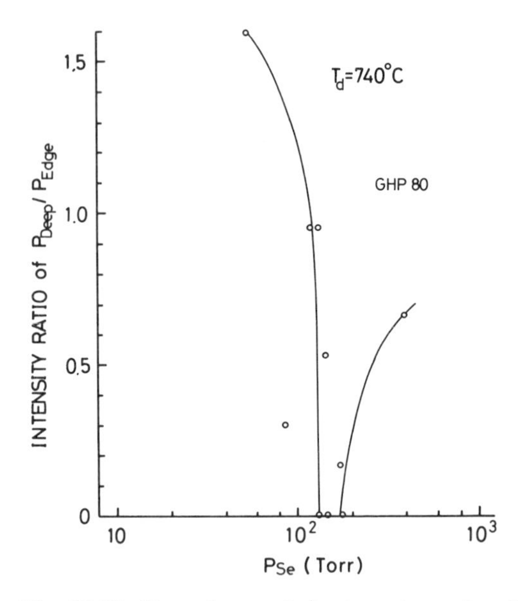

Fig. 21.13 Dependence of the intensity ratio of P_{deep}/P_{edge} in the spectrum at 300 K on the applied Se pressure. The emission intensity from deep levels disappears at the optimum Se pressure during diffusion.

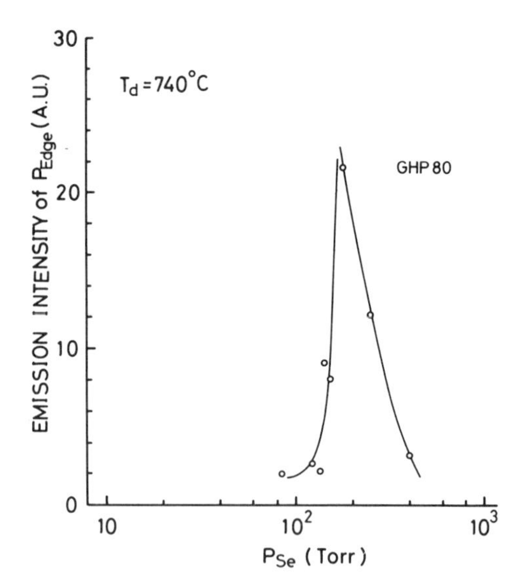

Fig. 21.14 Dependence of the emission intensity of P_{edge} on the applied Se pressure. This relation shows a maximum at the optimum Se pressure.

applied. Anyhow, it becomes clear that the emission spectrum at room temperature is changed by the controlled Se pressure. Fig. 21.15 shows the emission spectrum of a diode fabricated by impurity diffusion under the application of the optimum Se pressure into a crystal grown under optimum Zn pressure. A pure blue light emission without any emission due to deep levels can be obtained. The peak wavelength is 480 nm and halfwidth is 7 nm at 300 K.

The dependence of the output power and the brightness of this diode on the DC

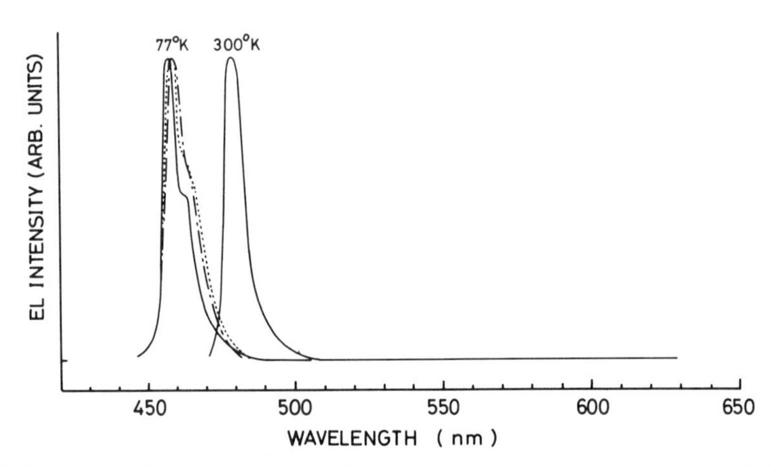

Fig. 21.15 Emission spectra from p–n junction fabricated under optimum pressure condition. The peak wavelength is 460 nm at 77 K and 480 nm at 300 K.

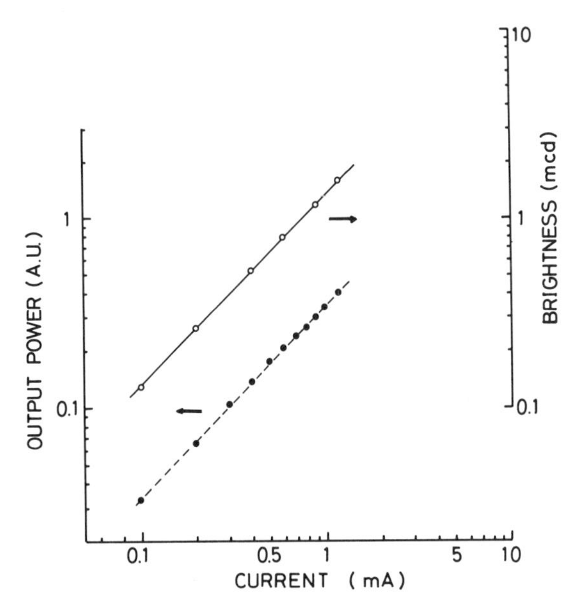

Fig. 21.16 Dependence of the output power and the brightness of the diode shown in Fig. 21.15 on the DC current.

forward current is shown in Fig. 21.16. The brightness is 2 mcd at 2 mA. These values will be further improved by using an epitaxial junction.

21.5. Discussion and conclusions

Unlike III–V compounds both constituent elements in II–VI compound have high vapor pressure, and the controlled vapor pressure becomes much more important. Therefore, the defects originating in non-stoichiometry are easily produced during all of

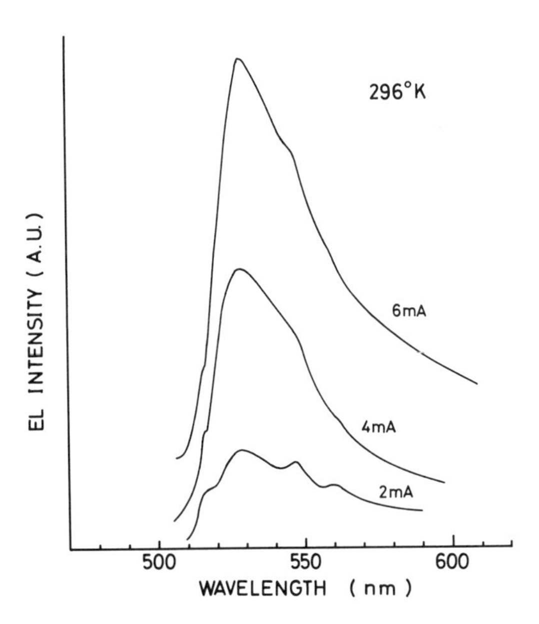

Fig. 21.17 EL spectra from ZnSe p–n junction. The green emission by the direct transition is expected.

the heat-treatment processes. It seems that p-type crystals of ZnSe cannot be obtained because of the existence of this defect. This study has shown that p-type ZnSe crystals can be grown by doping with a group-Ia element and by using the temperature difference method under controlled vapor pressure. The optimum Zn pressure to obtain stoichiometric ZnSe crystals can be determined, i.e. 7.2 atm at the growth temperature of 1050°C. The controlled vapor pressure is also important even in the diffusion process at relatively low temperatures to obtain pure blue light-emitting diode.

Of course, the effect of controlled vapor pressure can be applied not only to ZnSe but also to the crystal growth of the other II–VI compounds like as ZnTe, ZnS, CdS and CdSe. It will be possible to get crystals of all the II–VI compounds in which control of the stoichiometry and conductivity type are performed by TDM-CVP.

Consequently, it can be said that the impact on research into II–VI compounds is very great and the field of applications of these compounds will be extended. For example, the green emission from a p–n junction due to the direct transition has been obtained (Fig. 21.17) which will be published in another paper.

References

1) S. Shinozaki, K. Ishida, and J. Nishizawa: Rep. Tech. Group on SSD, IECE of Japan, No. SSD-71-10 (1971).
2) J. Nishizawa, S. Shinozaki, and K. Ishida: J. Appl. Phys., 44 (1973) 1638.
3) A. G. Fisher: Z. Naturforsch, 13a (1958) 105.
4) M. Harsy, Mater. Res. Bull. 3 (1968) 483.
5) P. Wagner and M. R. Lorenz: J. Phys. Chem. Solids, 27 (1966) 1749.
6) M. Rubenstein; J. Crystal Growth, 3/4 (1968) 309.
7) J. M. Faust, Jr., and H. F. John: J. Phys. Chem. Solids, 25 (1964) 1407.
8) M. Rubenstein: J. Electrochem. Soc., 113 (1966) 623.
9) M. Washiyama, K. Sato, H. Nakamura, and M. Aoki: Rep. Tech. Group on EFM, IEC of Japan, No. EFM-78-12 (1978).
10) J. Nishizawa and Y. Okuno: IEEE Trans. Electron Devices, ED-22 (1975) 716.
11) J. Nishizawa, Y. Okuno, and H. Tadano: J. Cryst. Growth, 31 (1975) 215.
12) J. Nishizawa and Y. Okuno: Proc. 2nd Int. School on Semiconductor Optoelectronics (Cetniewo, 1978).
13) J. Nishizawa, N. Toyama, and Y. Oyama: Proc. 3rd Int. School on Semiconductor Optoelectronics (Cetniewo, 1981).
14) J. Nishizawa and M. Koike: Int. School on Defect Complexes in Semiconductor Structures (Hungary, Sept. 1982).
15) K. Kaneko, M. Ayabe, M. Dosen, K. Morizane, S. Usui, and N. Watanabe: Proc. IEEE, 61 (1973) 884.
16) J. Nishizawa: Semiconductor Devices (Kindai-Kagakusha, Tokyo, 1961). [in Japanese]
17) M. Yamaguchi, A. Yamamoto, and M. Kondo: J. Appl. Phys., 48 (1977) 196.

AUTHORS' PROFILES

EDITOR

Jun-ichi Nishizawa was born in Sendai, Japan, on September 12, 1926. He received his B. S. degree in 1948 and his Doctor of Engineering degree in 1960 from Tohoku University, Sendai, Japan. Following a year as Research Assistant, he became Assistant Professor in 1954, Professor in 1962 and Director in 1983 at the Research Institute of Electrical Communication, Tohoku University. In 1968 he was appointed Director of the Semiconductor Research Institute in Sendai and holds the post at present.

His main works have involved the inventions of pin diodes and pnip (npin) transistors in cooperation with pin photo diode (1950), ion implantation (1950), also avalanche photo diode (1952), semiconductor injection lasers (1957), solid-state focusing optical fibers (1964) and transit-time-effect negative-resistance diodes (1954) including the avalanche injection and the tunnel injection (1958), hyperabrupt variable capacitance diodes (1959), semiconductor inductance (1957), static induction transistor (SIT) (1950, 1971), graded index optical guides (1962 et al.), etc. At present, he is carrying out work especially on development of static induction transistors to the high frequency and high power devices, the high speed thyristor, and the high speed and low power dissipation integrated circuit, and growth method of III-V compound semiconductors: temperature difference method under controlled vapor pressure (TDM-CVP) giving rise to high-efficiency LED and long-life laser diodes based on silicon perfect crystal technology by the lattice constant compensation. He also originated the electro epitaxy (1955) and photo epitaxy (1961). He discovered the avalanche effect in semiconductors and was the first in the world to explain the backward character of p-n junction by this effect (1953).

In recognition of his research and scholarship, Dr. Nishizawa received the Director's Award of the Japanese Science and Technology Agency in 1965 and 1970 for "Semiconductor devices with high resistivity thin layer" and "Invention of semiconductor junction laser and opto-electronics", respectively, the Imperial Invention Prize in 1966 for "Invention of semiconductor devices with high resistivity thin layer", the Matsunaga Memorial Award in 1969 for "Research works on semiconductor devices", the Okochi Memorial Technology Prize in 1971 for "Invention of variable capacitor diode", and 1980 for "Development of a technique for mass production of superbright LEDs by the temperature-difference liquid-phase epitaxial method", the 1974 Award of the Japan Academy for "Researches on semiconductors and transistors" the Meritorious Honor Award of the Japanese Science and Technology Agency for "Research on static induction transistors" and the Achievement Award of the Institute of Electronics and Communication Engineers of Japan for "Invention of static induction transistors" and the Purple Ribbon Medal from the Japanese Government in 1975 for "Invention of static induction transistors" in 1975 and the Inoue Harushige Award in 1982 for "Technique for continuous production of superbright LEDs" and the IEEE Jack A. Morton Award in 1983 for "Invention and development of the class of static induction transistors (SIT) and for advances in optoelectronic devices" and the honor of the 1983 Person of Cultural Merits from the Japanese Government.

Dr. Nishizawa is a member of the American Physical Society, the American Institute of Physics, the Electrochemical Society, the Physical Society of Japan, the IEE of Japan, and the Institute of Electrical and Communication Engineers of Japan. He is a fellow of the IEEE and of the Former Physical Society and the Institute of Physics.

1

Shin-ichi Akai was born on December 9, 1938 in Kyoto. He received the M.S. degree in metallurgy from Kyoto University in 1963.

After working on compound semiconductors such as InAs and Bi_2Se_3 at Kyoto University, he joined the R&D division of Sumitomo Electric Industries, Ltd. in 1964, where he worked on III-V compound semiconductors, especially GaAs single crystals. He received his doctor's degree in engineering from Kyoto University in 1975, for the thesis entitled "On the growth of low-defect-density GaAs single crystals by the 3T-HB technique". He received the Chief Award from the Agency of Industrial Science and Technology in 1975 for his distinguished service in research on dislocation-free GaAs.

He is now deputy general manager of the Semi-

JARECT Vol. 19, Semiconductor Technologies (1986), *J. Nishizawa (ed.)*
OHMSHA, LTD. and North-Holland

conductor Division of Sumitomo Electric Industries, Ltd., and responsible for general technical matters in the division.

Dr. Akai is a member of the Japanese Society of Applied Physics and Japanese Association of Crystal Growth.

Takashi Suzuki was born on April 1, 1934 in Tokyo. He received the B.S. degree from the Science University of Tokyo in 1958. He joined the Sony Corporation in 1958 and worked on Esaki diode and germanium transistors. He worked on solid state plasma at Kyushu University from 1960 to 1962. He joined the R&D division of Sumitomo Electric Industries in 1962, where he worked on III-V compound semiconductors.

He was a supervisor and general manager of R&D and production of III-V compound semiconductors from 1962 to 1983. He was general manager of the Semiconductor Department from 1978 to 1983. He received the Chief Award for his distinguished service in supervising of development of 3-in ϕ HB-GaAs in 1983 from the Agency of Industrial Science and Technology.

He is now general manager of the Semiconductor Group of the R&D Division of Sumitomo Electric Industries Ltd.

Mr. Suzuki is a member of the Crystal Technology Section of the Japanese Society of Applied Physics.

2

Jun-ichi Nishizawa: aforementioned.

Yasuo Okuno was born in Fujisawa, Japan, on September 8, 1945. He received the B.S. degree from Nihon University, Tokyo, Japan, in 1968 and his doctor's degree in electronics from Tohoku University, Sendai, Japan, in 1973.

In 1973 he became Chief Scientist at the Semiconductor Research Institute, which belongs to the Semiconductor Research Foundation, Sendai, Japan. He is working in the field of physics and technology of liquid phase epitaxy and light emitting diodes, and nonstoichiometry of III-V and II-VI compounds.

Dr. Okuno is a member of the Japan Society of Applied Physics and the Institute of Electronics and Communication Engineers of Japan.

Ken Suto was born in Urawa, Japan, on June 23, 1940. He received the B.E., M.E. and D.E. degrees from Tokyo University, in 1963, 1965 and 1968, respectively.

He has been an Assistant Professor at the Research Institute of Electrical Communication, Tohoku University. His research activities include semiconductor optoelectronics, and stoichiometry of compound semiconductors.

Dr. Suto is a member of the Institute of Electronics and Communication Engineers of Japan, the Japan Society of Applied Physics and the IEEE.

3

Hisashi Seki was born in Shizuoka on January 9, 1937. He received the M.S. degree in 1961 and Doctor of Engineering degree in 1970 from Tokyo Institute of Technology, Tokyo, Japan. In 1961, he joined the Electrical Communication Laboratory of Nippon Telegraph and Telephone Public Corporation where he worked on crystal growth of semiconductors.

He is now Professor of Industrial Chemistry at Tokyo University of Agriculture and Technology. His current research interests are concentrated on crystal growth of semiconductors. Dr. Seki is a member of the Japan Society of Applied Physics, the Chemical Society of Japan, the Electrochemical Society of Japan and the Japanese Association of Crystal Growth.

Akinori Koukitu was born in Aich, Japan, on August 28, 1949. He received the B.S. degree from Tokyo University of Agriculture and Technology, Tokyo, Japan, in 1972, and the Doctor of Engineering from Tohoku University, Sendai, Japan, in 1981.

Since 1972, he has been with Tokyo University of Agriculture and Technology, as a Research Assistant to Prof. H. Seki. He is currently engaged in research on crystal growth of III-V semiconductors.

Dr. Koukitu is a member of the Japan Society of Applied Physics, the Japanese Association of Crystal Growth and the Electrochemical Society of Japan.

4

Hiroji Kawai was born in Toyohashi, Japan, on July 17, 1946. He received the B.S. degree in Physical Chemistry from Shizuoka University, Shizuoka, Japan, in 1969. He joined the Sony Corporation Research Center, Yokohama, Japan, in 1969 where he was engaged in research on ZnS phosphors used in cathode ray tubes. Since 1980, he has been engaged in the research and development of the MOCVD for III-V compound semiconductors.

Mr. Kawai is a member of the Japan Society of Applied Physics.

5

Takafumi Yao was born in Hiroshima, Japan, on January 1, 1945. He received the B.S. and M.S. degrees in solid state physics, and the Ph.D. degree in electronic engineering from the University of Tokyo, Tokyo, Japan, in 1968, 1970, and 1985, respectively.

He joined the Electrotechnical Laboratory, Tokyo, Japan, in 1970. From 1970 to 1975, he was engaged in the experimental study of the nonstationary electron transport phenomena in semiconductors. Since 1975, he has been working on the molecular beam epitaxial growth of Zn-chalcogenides. Since 1982, he has also been engaged in research on semiconductor superlattices. He is presently the Head of the Applied Physics Section of the Electrotechnical Laboratory.

Dr. Yao is a member of the Physical Society of Japan, the Japan Society of Applied Physics, and the Institute of Electrical Engineers of Japan.

6

Toshiaki Ikoma was born in Tokyo, Japan, on March 5, 1941. He received the B.E., M.E. and Ph.D. degrees in electronic engineering from the University of Tokyo in 1963, 1965 and 1968, respectively.

In April 1968 he joined the Institute of Industrial Science, University of Tokyo as an Associate Professor and in April 1982 he was promoted to a Professor. He is also serving as a Professor at the Research Center for Function-oriented Electronics at the Institute of Industrial Science, University of Tokyo.

He is currently working on growth and characterization of compound and alloy semiconductor materials and their application to function-oriented devices, and the development of electron beam-acoustic microscopes.

He was a visiting research fellow at the Royal Institute of Technology, Stockholm, from February 1971 to August 1972, and a visiting scientist at IBM Thomas J. Watson Research Center in Yorktown Heights, New York from September 1979 to August 1980. He received the Outstanding Paper Award in 1970 from the Institute of Electronics and Communication Engineers of Japan.

Prof. Ikoma is a member of the Institute of Electronics and Communication Engineers of Japan, the Institute of Electrical Engineers of Japan, the Japan Society of Applied Physics, the IEEE and the American Physical Society.

7

Masamichi Yamanishi was born in Osaka Prefecture, Japan, on January 31, 1941. He received the B.E., M.E. and Ph.D. degrees from the University of Osaka Prefecture, Osaka, Japan, in 1964, 1966 and 1971, respectively.

He joined the Department of Electrical Engineering, University of Osaka Prefecture, as a Research Associate. In 1979, he was appointed as an Associate Professor in the Department of Physical Electronics, Hiroshima University, Hiroshima, Japan and promoted to a full Professor there in 1983. He was a Visiting Professor in the Department of Electrical Engineering, Purdue University, Indiana, U.S.A., for six months in 1984. He has been engaged in research on SAW acoustoelectric devices and acoustic DFB lasers. Currently, he is concerned with research on semiconductor lasers, LEDs and optoelectronic ICs.

Dr. Yamanishi is a member of the Physical Society of Japan, the Japan Society of Applied Physics, the Institute of Electronics and Communication Engineers of Japan, and the Institute of Electrical and Electronics Engineers.

8

Tomohiro Itoh was born in Yamaguchi, Japan, on June 13, 1951. He received the B.S. degree in electrical engineering, and the M.S. and Ph.D. degrees in electronic engineering from the University of Tokyo, Tokyo, Japan, in 1975, 1977 and 1980, respectively.

He joined the NEC Corporation, Kawasaki, Japan, in 1980 and is now Supervisor of the Ultra High-

Speed Devices Research Laboratory, Microelectronics Research Laboratories. Since joining the company, he has been engaged in research and development of InP and GaInAs MISFETs.

Dr. Itoh is a member of IEEE Electron Devices and Microwave Theory and Techniques Societies, the Institute of Electronics and Communication Engineers of Japan, and the Japan Society of Applied Physics.

Keiichi Ohata was born in Osaka, Japan on April 17, 1947. He received the B.S. and M.S. degrees in electronic engineering from Kyoto University, Kyoto, Japan, in 1970 and 1972, respectively.

He joined the NEC Corporation, Kawasaki, Japan in 1972 and is now Supervisor of the Ultra High-Speed Devices Research Laboratory, Microelectronics Research Laboratories. He has been engaged in research of ohmic contacts to GaAs, development of low-noise GaAs MESFETs, and research and development of InP MISFETs and AlGaAs/GaAs FETs.

Mr. Ohata is a member of the Japan Society of Applied Physics and the Institute of Electronics and Communication Engineers of Japan.

9

Hirofumi Namizaki was born in Toyohashi, Japan, on February 25, 1946. He received the B.E., M.E. and Dr. Eng. degrees in electronic engineering from the University of Tokyo, Tokyo, Japan, in 1968, 1970 and 1976, respectively.

In 1970 he joined the Kamakura works, Mitsubishi Electric Corporation, Kamakura, Japan, where he was engaged in development of optoelectronic instruments. Since 1971 he has been working on semiconductor lasers and integrated optics. In 1973 he joined the Central Research Laboratory, Mitsubishi Electric Corporation, Itami, Japan. From 1976 to 1977 he was a visiting scientist at the University of California, Berkeley. He is now with the LSI Research and Development Laboratory, Mitsubishi Electric Corporation, Itami, Japan.

Dr. Namizaki received the Achievement Awards in 1981 from the Institute of Electronics and Communication Engineers of Japan. He is a member of the Japan Society of Applied Physics and the Institute of Electronics and Communication Engineers of Japan.

Saburo Takamiya was born in Tokyo, Japan, on March 1, 1943. He received the B.S. degree in physics, and the D. Eng. degree from the Tokyo Institute of Technology, Tokyo, Japan, in 1965 and 1977, respectively.

In 1965, he joined the Mitsubishi Electric Corporation, where he has been working in the research and development of optoelectronic semiconductor devices. Now he is a manager of optoelectronic devices II group in LSI R&D Laboratory, Mitsubishi Electric Corporation, Itami, Japan.

Dr. Takamiya is a member of the Institute of Electronics and Communication Engineers of Japan, and the Japan Society of Applied Physics.

10

Keigo Hoshikawa received the B.S. and M.S. degrees from Shinshu University, Japan, in 1967 and 1969, respectively. He joined the Electrical Communication Laboratory, Nippon Telegraph and Telephone Public Corporation in 1969 and worked on growth and characterization of silicon crystals for LSI. Since 1982, he has been engaged in research and development on semi-insulating LEC GaAs crystal for GaAs-LSI. He is a member of the Electrochemical Society, the Japan Society of Applied Physics, the Crystallographic Society of Japan and the Institute of Electronics and Communication Engineers of Japan.

Jiro Osaka received the B.S. and M.S. degrees from Waseda University, Japan, in 1973 and 1975, respectively. He joined the Electrical Communication Laboratory, Nippon Telegraph and Telephone Public Corporation in 1975 and worked on the characterization and growth of silicon crystals. Since 1983, he has been engaged in developmental research on the crystal growth technique for semi-insulating bulk GaAs crystal by the LEC method. He is a member of the Physical Society of Japan, the Japan Society of Applied Physics and the Crystallographic Society of Japan.

11

Hiroshi Nakamura was born in Fukuoka, Japan on October 13, 1954. He received the B.S., M.S. and Ph.D. degrees in electronic engineering from the University of Tokyo, Japan, in 1977, 1979 and 1982, respectively.

He then joined OKI Electric Industry Co., Ltd., Hachioji, Tokyo, Japan, where he has been engaged in the research and development of processing technologies of GaAs MESFETs and GaAs digital ICs.

Dr. Nakamura is a member of the Japan Society of Applied Physics and the Institute of Electronics and Communication Engineers of Japan.

Masanori Tsunotani was born in Toyama, Japan on March 15, 1960. He received the B.S. degree in electrical engineering from Osaka University, Osaka, Japan, in 1982.

He joined OKI Electric Industry Co., Ltd., Hachioji, Japan, in 1982, where he has been engaged in the research and development of processing technologies of GaAs MESFETs and GaAs digital ICs.

Mr. Tsunotani is a member of the Japan Society of Applied Physics.

Yoshiaki Sano was born in Yao, Osaka, Japan, on September 1, 1949. He received the B.S.E. degree in material science from Osaka University, Japan in 1973.

In 1973, he joined the OKI Electric Industry Co., Ltd., Hachioji, Tokyo, Japan, initially working on IMPATT diodes, subsequently on bipolar power transistor. He is presently engaged in related compound semiconductor device research.

Mr. Sano is a member of the Japan Society of Applied Physics and the Institute of Electronics and Communication Engineers of Japan.

Toshio Nonaka was born in Niigata Prefecture, Japan, in 1949. He received the B.S. degree in electrical engineering from Niigata University, Niigata, Japan, in 1972.

He then joined the OKI Electric Industry Co., Ltd., Hachioji, Tokyo, Japan, where he was engaged in the development of processing technologies of III-V compound semiconductor devices and Si bipolar devices.

Since 1981 he has been engaged in the research and development of processing technologies of GaAs digital ICs and 3-dimensional devices.

Mr. Nonaka is a member of the Japan Society of Applied Physics and the Institute of Electronics and Communication Engineers of Japan.

Toshimasa Ishida was born on February 5, 1945. He received the B.S. degree in physics from Kyoto University, Kyoto, Japan, in 1968.

In 1968, he joined the OKI Electric Industry Co., Ltd., Hachioji, Tokyo, Japan, where he was engaged in the development of optoelectronic devices. Since 1981 he has been engaged in developing GaAs ICs, 3-dimensional devices, and HEMT.

Mr. Ishida is a member of the Japan Society of Applied Physics and the Institute of Electronics and Communication Engineers of Japan.

Katsuzo Kaminishi was born in Japan on January 4, 1939. He received the B.S. degree in electrical engineering from Tohoku University in 1961.

He then joined the OKI Electric Industry Co., Ltd., where he has been engaged in research and development of microwave tubes, thermal printing head and microwave diodes. He received the Ph.D. degree in electrical engineering from the Tohoku University in 1980. At present he is engaged in research and development of compound semiconductor devices.

Dr. Kaminishi is a member of the IEEE, the Institute of Electronics and Communication Engineers of Japan, the IEE of Japan, and the Japan Society of Applied Physics.

12

Naoki Yokoyama was born in Osaka, Japan, on March 28, 1949. He received the B.S. degree in physics from Osaka City University, Osaka, in 1971 and the M.S. degree in physics from Osaka University, Osaka, in 1973.

In 1973, he joined the Semiconductor Laboratory, Fujitsu Laboratories, Ltd., Japan, where he has been engaged in the research and development of GaAs field-effect transistors and GaAs digital integrated cricuits. Highlights of his work include the development of self-aligned GaAs MESFETs using refractory-metal-silicide Schottky gates. He is currently involved in developing GaAs integrated circuit fabrication processes and is responsible for the design and fabrication of these circuits.

13

Hiroshi Ishiwara was born in Yamaguchi Prefecture, Japan, on November 6, 1945. He received his B.E., M.E. and D. Eng. degrees in electronic engineering from the Tokyo Institute of Technology, Tokyo, Japan, in 1968, 1970 and 1973, respectively.

Since 1973, he has been a member of the Tokyo Institute of Technology, where he is now an Associate Professor in the Department of Applied Electronics in the graduate school. From 1978 to 1979, he was a visiting staff member at the Chalk River Nuclear Laboratories in the Atomic Energy of Canada Research Company Ltd.

He has been engaged in research on Si IMPATT diodes, ion implantation range and energy deposition distributions, ion channeling and backscattering spectroscopy, and epitaxial growth of metal-silicide, group-IIa fluoride, and Si films onto Si substrates. His current interest is in the formation of Si/silicide/Si and Si/insulator/Si structures, and the electron beam annealing of implanted and deposited Si.

Dr. Ishiwara is a member of the Electrochemical Society, the Materials Research Society, the Institute of Electronics and Communication Engineers of Japan, and the Japan Society of Applied Physics.

Tanemasa Asano was born in Ibaraki Prefecture, Japan, on September 12, 1953. He received the B.E. degree from the Ibaraki University, Ibaraki, Japan, and the M.E. degree from the Tokyo Institute of Technology, Tokyo, Japan, both in electronic engineering, in 1976 and 1979, respectively.

Since 1979, he has been with the Tokyo Institute of Technology as a Research Associate. He has been engaged in research on ion implantation in GaAs, epitaxial growth of metal-silicide films on Si, and he is presently engaged in heteroepitaxial semiconductor-on-insulator layered structures.

Mr. Asano is a member of the Japan Society of Applied Physics and the Institute of Electronics and Communication Engineers of Japan.

14

Hideaki Kohzu was born in Nagano, Japan, on April 20, 1944. He received the B.S. and M.S. degrees from Waseda University, Japan, in 1968 and 1970, respectively.

He joined the NEC Corporation, Kawasaki, Japan, in 1970 and is currently Engineering Manager of the Microwave and Optical Devices Department in the Second LSI Division. He has been engaged in the development of microwave solid-state devices such as GaAs FETs, GaAs ICs and diodes. He has also worked with fabrication process technologies.

Mr. Kohzu is a member of the Institute of Electrical and Electronics Engineers, the Japan Society of Applied Physics, and the Institute of Electronics and Communication Engineers of Japan.

Masaaki Kuzuhara was born in Osaka, Japan, on November 5, 1955. He received the B.E. and M.E. degrees in electrical engineering from Kyoto University, Kyoto, Japan, in 1979 and 1981, respectively.

He joined the NEC Corporation, Kawasaki, Japan, in 1981. Since then he has been engaged in the research and development of GaAs integrated circuit fabrication processes, including rapid thermal annealing of ion-implanted GaAs, in the Microelectronics Research Laboratories, NEC Corporation.

Mr. Kuzuhara is a member of the Japan Society of Applied Physics and the Institute of Electrical and Electronics Engineers.

Yoichiro Takayama was born in Kanagawa, Japan, on January 3, 1943. He received the B.E., M.E. and Dr. Eng. degrees from Osaka University, Japan, in 1965, 1967 and 1973, respectively.

He joined the NEC Corporation, Kawasaki, Japan, in 1967 and is currently Manager of the Ultra-High Speed Devices Research Laboratory in the Microelectronics Research Laboratories. He has been engaged in the research and development of microwave solid-state oscillators and amplifiers, using Gunn diodes, IMPATT diodes, and GaAs FETs. He also worked with IMPATT diode characterization method and loadpull characterization method for power GaAs FETs. He is currently managing the GaAs integrated circuit research group and the Josephson integrated circuit research group.

15

Hiroyuki Matsunami was born in Osaka, Japan, on June 5, 1939. He received the B.E., M.E. and Dr. Eng. degrees in electronic engineering from Kyoto University, Kyoto, Japan, in 1962, 1964 and 1970, respectively.

He joined the Department of Electrical Engineering at Kyoto University as a Research Associate in 1964 and became an Associate Professor in 1971. He was a visiting Associate Professor in North Carolina State University, U.S.A. from September 1976 to July 1977. He is now a Professor at Kyoto University responsible for semiconductor materials and energy conversion devices. His current research work involves preparation and characterization of semiconductor materials and application for photoelectronic devices.

Dr. Matsunami is a member of the Japan Society of Applied Physics, the Institute of Electronics and Communication of Japan, the Institute of Electrical Engineers of Japan and the Institute of Electrical and Electronics Engineers.

16

Hitoshi Matsumoto was born in Osaka, Japan on March 1, 1942. He received the B.Eng., M.Eng. and D.Eng. degrees from Osaka University, Osaka, Japan in 1964, 1966 and 1984, respectively. His thesis research was in ceramic type CdS/Cu_{2-x} and CdS/CdTe solar cells.

He joined the Wireless Research Laboratory, Matsushita Electric Industrial Co., Ltd. in 1966 and is now a Senior Engineer in the Laboratory. From 1966 to 1975, he worked on ceramic type $CdS/Cu_{2-x}S$ solar cells, and since 1976 has been working on screen printed CdS/CdTe solar cells.

Dr. Matsumoto is a member of the Chemical Society of Japan.

Seiji Ikegami was born in Kagoshima, Japan on March 13, 1929. He received the B.Eng. and D.Eng. degrees from Kyushu University, Fukuoka in 1952 and from Osaka University, Osaka, Japan in 1965, respectively.

Since 1952 he has been with the Wireless Research Laboratory, Matsushita Electric Industrial Co., Ltd. From 1952 to 1973, he worked on ferroelectric material, and since 1974 has been working on II-VI compound solar cells, especially screen printed CdS/CdTe solar cells. He is head of the solar cell research group in the Laboratory.

Dr. Ikegami is a member of the Institute of Electronics and Communication Engineers of Japan, the Institute of Electrical Engineers of Japan, the Japan Society of Applied Physics, and the Physical Society of Japan.

17

Masanori Okuyama was born in Osaka, Japan, on March 4, 1946. He received his B.S., M.S. and Ph.D. degrees in electrical engineering from Osaka University, Osaka, Japan, in 1968, 1970 and 1973, respectively.

In 1973 he became a postdoctoral fellow of the Japan Society for the Promotion of Science, and in 1974 he became a Research Associate at Osaka University. From 1977 to 1978 he worked at the Carnegie-Mellon University, Pittsburgh, Pennsylvania, U.S.A. He has been engaged in the field of analysis of modulation spectroscopy, switching and memory device, photo-thermal selective absorber, preparation and application of ferroelectric thin films and deep impurity in semiconductor. His current interest is Si-monolithic pyroelectric infrared sensor and ultrasonic sensor, $Si\text{-}SiO_2$, interface and deep impurities in semiconductor.

Dr. Okuyama is a member of the Japan Society of Applied Physics, the Physical Society of Japan and the Institute of Electrical and Electronic Engineers.

Yoshihiro Hamakawa was born in Kyoto, Japan, on July 12, 1932. Professor Hamakawa has been engaged with Osaka University since his completion of M.S. degree in 1958, and received a Ph.D. degree in 1964 from Osaka University.

He is now Professor of Electrical Engineering Science at Osaka University. He held visiting posts at the University of Illinois (1965–1967) engaged as a VRA Professor both at Department of Electrical Engineering and the Material Research Laboratory, Urbana, Illinois, USA.

He has performed research in the field of semiconductor physics, optoelectronics and solar photovoltaic conversion, particularly, on optical properties and band structure of solids, opto-electronic devices, solar cells, amorphous semiconductors and devices.

He has authored or co-authored eight books and more than 120 papers, including a text book on "Semiconductor Physics and Devices, Volume 1 & 2" and the academic review volume; "Recent Advances in Modulation Spectroscopy" in "Optical Properties of Solids – New Developments" (North-Holland Pub. Co., 1976).

Prof. Hamakawa is a member of Board of Japan Society for the promotion of Science No. 125, Optoelectronic Energy Conversion Committee (1970–79), the chairman of the Solar Photovoltaic Committee, Sunshine Project (1978), a neutral committee member (1974–79) and the chairman of Workshop for Amorphous Silicon Solar Cells (1979) in Sunshine Project, AIST, MITI, the chairman of the Amorphous Material R&D Survey Committee in Japan Society of Electronic Industry and a member of Board of "Japan Society of Applied Physics" (1979).

He is a member of the Institute of Electrical and Electronics Engineers, the American Physical Society, ISES, SID Japan Society of Applied Physics, the Institute of Electrical Engineers of Japan, Physical Society of Japan. He was awarded the 1970 RCA Fundamental Research Grant for his work on "Electro-optical effects and band structure in mixed compound semiconductors" (1970) and also the 1977 Yamada Science Foundation Grant for his research on "Valency controls in amorphous semiconductors" (1977).

18

Yoshihiro Hamakawa: aforementioned.

Yoshiro Oishi was born in Kyoto on December 22, 1958. He received the B.S. degree in electronic engineering in 1982 from the Fukui University, Fukui, Japan, and the M.S. degree in electrical engineering in 1984 from Osaka University, Osaka, Japan.

His current interest is electroluminescence of ZnS.

Mr. Oishi is a member of the Japan Society of Applied Physics.

Takatoshi Kato was born in Aichi on March 14, 1961. He received the B.S. degree in electrical engineering in 1983 from Osaka University, Osaka, Japan, and now is a student on the Master's Course.

His current interest is electroluminescence of ZnS.

Mr. Kato is a member of the Japan Society of Applied Physics.

19

Isamu Akasaki received the B.S. degree in 1952 from Kyoto University, Kyoto, Japan, and the Dr. Eng. degree in 1964 from Nagoya University, Nagoya, Japan, respectively.

In 1959, he joined the Department of Electronics, Nagoya University, Nagoya, Japan as a Research Associate. In 1964, he became an Assistant Professor and an Associate Professor, both in the same department. From 1964 to 1981, he worked at Matsushita Research Institute, Tokyo, Inc., Kawasaki, Japan. He was a Head of the Semiconductor Department.

He is currently a professor in the Department of Electronics, Nagoya University, Nagoya, Japan, and is engaged in research in crystal growth, characterization and device physics of III-V and II-VI semiconductors and their alloy semiconductors.

Dr. Akasaki is a member of the Japan Society of Applied Physics, the Institute of Electronics and Communication Engineers of Japan and the Japanese Association of Crystal Growth.

Hiroshi Amano was born in Shizuoka, Japan on September 11, 1960. He received the B.E. degree in electronics engineering from Nagoya University, Nagoya, Japan, in 1983. He is now a M.E. student in the Department of Electronics, Nagoya University, Nagoya, Japan, working in the area of the crystal growth of III-V compound semiconductors.

Mr. Amano is a member of the Institute of Electronics and Communication Engineers of Japan and the Japan Society of Applied Physics.

Nobuhiko Sawaki was born in Gifu, Japan, on December 18, 1945. He received the B.E., M.E. and Dr. Eng. degrees in electronics from Nagoya University, Nagoya, Japan, in 1968, 1970 and 1973, respectively. From 1973 to 1977, he was a research assistant, and from 1977 to 1982, an Assistant Professor in the Department of Electronics, Nagoya University. Since 1982, he has been an Associate Professor in the Faculty of Engineering, Nagoya University and has been engaged in research on the physics and technology of semiconductor devices and materials. His current responsibilities include high field transport and optical properties of semiconductors and superlattices.

Dr. Sawaki is a member of the Physical Society of Japan, the Institute of Electrical Engineers of Japan, and the Institute of Electrical and Electronics Engineers Inc.

Masafumi Hashimoto was born in Nagoya, Japan, on December 6, 1939. He received the B.S. and M.S. degrees in electronic engineering from Nagoya University, Nagoya, Japan, in 1963 and 1965, respectively. From 1965 to 1982 he worked in the Matsushita Research Institute, Tokyo Inc., Kawasaki, Japan, on development of LED and crystal growth of III-V compound semiconductors.

Since 1982 he has been employed at Toyota Central R&D Laboratories Inc., Aichi, Japan and has been engaged in the development of light-emitting devices. He is a leader of III-V compound semiconductor device group.

Mr. Hashimoto is a member of the Japan Society of Applied Physics.

Yoshimasa Ohki was born in Yamanashi, Japan, on January 3, 1946. He received the B.S. and M.S. degrees in applied physics from Tohoku University, Sendai, Japan, in 1968 and 1970, respectively.

In 1970 he joined the Matsushita Research Institute Tokyo, Inc., Kawasaki, Japan. Since then, he has been engaged in research on vapor phase epitaxy of III-V semiconductors and on light-emitting diodes. Since 1981, he has been at the Kawasaki Laboratory Opto-electronics Development Center, Matsushita Electric Industrial Co., Ltd., Kawasaki, Japan, where he is concerned with research on MOCVD growth of III-V semiconductors.

Mr. Ohki is a member of the Japan Society of Applied Physics.

Yukio Toyoda was born in Shimosuwa, Nagano, Japan, on August 22, 1943. He received the B.S. and M.S. degrees in Physics from Niigata University, Niigata, Japan, in 1966 and 1968, respectively and the Ph.D. degree in Physics from Nagoya University, Nagoya, Japan.

In 1974 he joined Matsushita Research Institute Tokyo, Inc., Kawasaki, Japan, where he worked on ion implantation in III-V compounds and characterization of III-V compounds, and engaged in the development of light-emitting diodes. Since 1981, he has been at the Kawasaki Laboratory Opto-electronics Development Center, Matsushita Electric Industrial Co., Ltd., Kawasaki, Japan, where he has been engaged in the research on 1.3 μm semiconductor lasers.

Dr. Toyoda is a member of the Physical Society of Japan and the Japan Society of Applied Physics.

20

Tsunemasa Taguchi was born in Akita, Japan on March 2, 1947. He received the M.E. degree in Radiation Physics in 1972 from Nagoya Institute of Technology and a Ph.D. degree in Solid State Physics in 1976 from Osaka University. Since 1976, he has been a Research Assistant Professor in the Faculty of Engineering at Osaka University and has been actively engaged in the study of II-VI compounds. From 1979 to 1981, he joined the atomic-collision group in the Physics Department of the University of Sussex in the U.K. as a Visiting Research Scientist to investigate the principal mechanisms of point-defect production in II-VI compounds. His research interests center on the studies of ZnS blue light-emitting diode and of point defect-creation in MBE-grown ZnSe and CdHgTe. He has currently received a Nishina Memorial Foundation Scholarship Award for the study of point defects in II-VI compounds.

Dr. Taguchi is a member of the Physical Society of Japan, the Japan Society of Applied Physics, the Japanese Association of Crystal Growth and the British Association for Crystal Growth.

21

Jun-ichi Nishizawa: aforementioned.

Yasuo Okuno: aforementioned.